RECENT FORAMINIFERA

RECENT FORAMINIFERA

by
Esteban Boltovskoy & Ramil Wright

Dr. W. JUNK b.v. – PUBLISHERS – THE HAGUE 1976

ISBN 978-90-481-8510-8

Cover design Charlotte van Zadelhoff

Contents

CHAPTER CONTENTS

INTRODUCTION TO THE SECOND EDITION

The first edition of this book (*Los Foraminíferos Recientes;* biologia, métodos de estudio, aplicacion oceanografica: EUDEBA, Buenos Aires, 510 pp, 114 text figs.) was published in the Spanish language in 1965. The work was received warmly by colleagues and reviewed favourably in specialized scientific journals in 9 different countries. One consistent commentary both in reviews and in numerous letters from colleagues was the desire to see a second edition of this book published in the English language. This desire is now realized thanks to the kind suggestion of Junk Publishers.

However, because of the 13-year delay between preparation of the first and second editions, it became necessary to update the book as well as translate it. The bibliographical explosion in the sciences in the last decade made the renovation of this work a time-consuming task. To aid in this labour the original author invited his colleague, R. Wright, to participate in the revision. The work of bringing the text up to date consisted of a revision of all chapters by evaluating and summarizing those data which have appeared in more than a thousand articles published since 1963–1964. Several new sub-chapters have been added to describe new concepts and techniques developed since the first edition appeared. Many illustrations have been revised and new ones added. This work was done jointly for which reason the book appears under our dual authorship. The translation of data from the first edition was the responsibility of R. Wright.

The revision of the text was completed in January, 1975, and information coming to our attention after that date is not included in the book.

The techniques described in this book for collecting samples, separating the fauna, analysing the fauna, etc., are not always the most sophisticated available. We recognize that the development of new equipment has led to the development of more refined and efficient techniques, but we wish to address the book to all students of foraminifera at all levels, from the beginning student to the academic or industrial researcher. We also address the book to workers in all kinds of institutions, from the industrial corporations, large universities and government laboratories with large research budgets to those individuals working in organizations operating on modest funds. Consequently, we have tried to describe techniques and methodology appropriate at all levels.

We have abandoned the practice followed in the first edition of saving space by excluding pagination of references in the bibliography. In the preparation of this edition we found it necessary to review references cited in the first edition. The abbreviated entries were often less than helpful in locating a reference. Consequently, where possible, we have expanded the amount of information included in the bibliography.

Following the practice of the first edition we prefer to cite the generic and trivial name as written by the original authors. Fewer exceptions are made than in the first edition. This is a rather important point and requires some additional explanation.

There are two common ways in the current literature of handling this nomenclatural problem. One is to cite the original name and to place after it, in parenthesis, the changed name which according to the writer is the correct one. A second way is to simply cite only the changed name, omitting the original designation. Neither of these methods is desirable. They both reflect the taxonomic taste of the writer but without the necessary ingredient of examination of the original specimens whose names are being changed. The first method also has the disadvantage of enlarging the text.

To better understand why we disagree with both of these methods we cite the following examples. In the recent past several hundred species ascribed to (or which would have been ascribed to) *Bolivina* were divided among *Bolivina*, *Bolivinita*, *Bolivinitella*, *Brizalina*, *Coryphostoma*, *Sigmavirgulina* and *Suggrunda* and there is no uniformity or agreement in the placement of these previous species of *Bolivina* in the other genera. We can not place the species which used to be (or would have been) in *Bolivina* into the genera now in use unless we can examine the specimens ourselves. In the case of trivial names the situation is yet worse as can be seen in the following example. *Sagrina virgula*, which was established by Brady (1884), is divided into three species, *Rectobolivina virgula* (Brady), *Stilostomella bradyi* (Cushman) and *S. antillea* (Cushman), according to Barker (1960). If someone cites *Sagrina virgula* (Brady) how can we decide to what species these specimens should be assigned in the modern classification without first examining the specimens?

Therefore, we are strongly convinced that, as a general rule, the original names should be maintained. Consultation of the species index, which contains the original author's name, should clarify any confusion.

However, as we said above, in this book some exceptions were made to this practice. They were made in the case of very old names not in use for several decades, and only in such cases when the change could not create even the most minimal of misunderstanding. Such a case is mentioned in the footnote on page 140. They are also encountered in the discussion of data from Ehrenberg, Rhumbler and some other authors, mainly from the last Century. There are only a few of these exceptions.

A number of persons helped make this book possible and we wish to acknowledge their contributions. To Haydeé Lena we owe our greatest debt of gratitude for her immeasurable help in the preparation of the sections on cytology (Chapter II,1 and XI,3). We wish to acknowledge A. Bé, F. Haake, C. Hemleben, H. Lena, G. Lutze, R. Röttger, P. Thompson and K. Towe who graciously provided illustrative material for the book.

For her help in drafting we thank Isabel Riobó de Magaldi and in typing, Ana María Leverone, Estela Orma and Luis María Mendez. We also wish to thank Cynthia Wright for long hours of editorial assistance.

Buenos Aires, February, 1976

E. Boltovskoy & R. Wright

INTRODUCTION TO THE FIRST EDITION

The great development of study of the oceans resulted in the investigation of marine organisms. The importance placed on these investigations is reflected in the creation of numerous marine biological stations and the organization of various oceanographic expeditions by nations of both hemispheres.

The Protozoa, particularly the Foraminiferida, play a significant role in the organic world of the oceans. These typically marine organisms are numerous in both the benthos and the plankton.

Unfortunately, the bibliography in Spanish dedicated to these organisms is relatively small. There are works which analyse certain regions, faunas or specific problems, but there is no manual or text of a general character dedicated principally to the Recent forms. The work of Colom (1946) and the recently published work by Charlton de Rivero Palacio & Bermúdez (1963) are excellent analyses of fossil foraminifera and well serve their geologic purpose. The summary of Douglass (1960) has the same goal. However, this last work is only available in mimeographed form as it was prepared for internal use in the University of Chile. Copies, therefore, are difficult to obtain. These circumstances stimulated me to write the present work.

Some words about the content and purpose of this book: 1. It is designed particularly for those interested in biological oceanography in general and in Recent foraminifera in particular. At the same time, it should prove of some value to micropaleontologists, since conclusions based on study of Recent species are commonly of great interest to those who study fossil specimens. Moreover, there are various methods of investigation that are similar for both groups of foraminifera. Consequently, some of the pages of this book are devoted to methods of treating fossil foraminifera.

2. This book does not include a systematic section dealing with the description of the families and genera that comprise the Order Foraminiferida. Such descriptions are easily found in several manuals published in various foreign languages. I do not think that comprehension of systematics presents any great language difficulties since taxonomic descriptions use a limited number of often repeated terms.[1]

[1] In 1956 a special dictionary was published to help those who work with the foraminiferal bibliography in foreign languages: E. Boltovskoy, Diccionario foraminiferologico plurilingue (in 5 languages): English, Spanish, German, French and Russian (with alphabetic indicies): Ministerio de Marina, Servicio de Hidrografia Naval, Argentina, Publ. Misc. No. 1001.

3. The text contains many generic and specific names obtained from the works of many authors. In several cases more than one name has been employed for the same taxon. In those cases where the generic names are undoubtedly synonyms; e.g. *Ammonia – Streblus – Rotalia,* I have not considered it convenient to follow the authors choice since this might produce difficulties in understanding the text; I chose the most common name. When there were doubts about synonomy of generic names, I cited the taxon as the original author did. The specific names have been conserved in every case as originally cited.
4. A great deal of attention has been given to the practical side of foraminiferal study. It has become clear that recent university graduates possess a good understanding of biological and protozoological theory, but are often at a loss to handle the more common tasks arising from the treatment of foraminiferal material. Because of this lack, this book contains detailed counsel on different procedures and techniques used by the foraminiferologist. Moreover, some advice is tendered which is not directly related to foraminiferal study but which may prove of great value to those working with these organisms; e.g. the construction of simple plankton nets and elementary devices for the extraction of bottom sediments, instructions on the cultivation of algae for the maintenance of foraminiferal cultures, etc.
5. The bibliography dedicated to the description of various methods used in the different stages of foraminiferal study is rather ample. There are numerous articles dealing with the procedures of sample disaggregation, test removal, etc. It is practically impossible to describe all the techniques available. I have therefore limited myself to describing only the most common techniques and ones which I have used in my laboratory.
6. When treating various themes, especially those poorly understood (ecology, distribution, etc.), I have described the various observations and opinions from the literature before presenting conclusions. In each case I have indicated the origin (author, date) of this information so that the reader may examine the original source to obtain a deeper understanding.
7. For reasons of space and because of the purpose of this book, some types of investigations, indirectly associated with foraminifera, have been omitted. Among these omissions are: the exposition of various zoological nomenclatural problems (e.g. types), advice on photomicrography, etc. Nevertheless, bibliographic sources of information on these omitted topics are always given.
8. When it became necessary to employ illustrative examples to best explain certain themes, I preferred to cite examples from South America. There are two reasons for this preference. Above all, this region is not well known from a foraminiferal perspective. In addition, this work is designed principally for those marine biologists who speak Spanish, the majority of whom reside in Central and South America.
9. All species names are cited in the text without reference to their authors. However, a taxonomic index is provided at the end of the book which contains this data.
10. The bibliolography requires two clarifications: (a) Unfortunately, there is no

universally accepted system for the transliteration of names written in Cyrillic script. There is, therefore, confusion over the transcriptions of the Slavic authors. Some of these authors may appear to be different persons because of different transcriptions of their names. To lend some uniformity to this situation I have adopted the system that was published in the Bulletin of Zoological Nomenclature, London.[2] The only exceptions are made for those Slavic authors whose names have appeared frequently in Latin characters based on a different transliteration system. (b) In order to save space, the bibliographic entries are abbreviated, without pagination and illustration data.

This work has been in preparation for several years but the lack of some information has postponed its publication. In the year 1961, thanks to a subsidy granted by the Consejo Nacional de Investigaciones Cientificas y Technicas (Argentina) I was able to visit various institutions in Europe and finish the manuscript. I consider it a very agreeable obligation to express my sincere appreciation to the Consejo. In addition, I most cordially thank Dr. R. Ruhstaller, Srtas. Haydeé Lena and Anna-Maria Coronado and Sr. F. Theyer for their help in the preparation of the manuscript. I also wish to thank Sr. J. C. Riobó for the drafting of the majority of the illustrations.

Buenos Aires, April, 1963

E. Boltovskoy

[2] Almasov, A. & Boltovskoy, E. 1965. On the treatment of words written with cyrillic characters, for the purposes of zoological nomenclature, bibliography, references, indicies, etc.: Bull. Zool. Nomencl., v. 11. no. 1.

Rerum natura, nusquam magis quam in minimis, tota est.
Pliny the Elder

I. THE SYSTEMATIC POSITION AND IMPORTANCE OF THE FORAMINIFERA

The foraminifera are members of the Order Foraminiferida, Subclass Rhizopoda, Class Sarcodina, Phylum Protozoa. They constitute about 2.5% of all animals known from the Cambrian to the Recent.

The closest relatives of the foraminifera are the amoebas and the thecamoebians (Testacea). The characteristic traits of these three groups are as follows:

Order Amoebida: the organism does not possess a fixed cell wall and the protoplasmic body therefore has no definite form. The pseudopodia are, for the most part, short and fat, having a lobose form (lobopodia).
Order Thecamoebida: the organism has a sac-like single chambered test with an aperture through which the pseudopodia are emitted. The fundamental part of the test is a chitin-like substance which is secreted by the protoplasm. This test is able to agglutinate pieces of foreign material to its exterior. The organism possesses elongate filamentous pseudopodia (filopodia) which seldom anastomose. The pseudopodia have a right angled shape and exhibit rapid movements. This may be an artificial, polyphyletic group.
Order Foraminiferida: these organisms are generally unattached although some are sedentary. They possess a simple or multichambered test. The test walls are composed of different materials and have different architectures. There is at least one and occasionally more than one test aperture. The pseudopodia are slender and ramified and exhibit a filamentous streaming motion. Most of the species are characterized by an alternation of reproductive modes, sexual and asexual, although the former may be absent.

There are not well-defined limits between these three groups. Some genera which had been considered as foraminifera have only recently been reclassified as thecamoebians. There are rhizopods whose systematic position is still doubtful. These problems are not unique to the rhizopods as the absence of fixed boundaries between taxonomic groups of organisms is a well known phenomenon in zoology.

The generally accepted distinction between the foraminifera and the thecamoebians is the presence of an alternation of reproductive generations among the former group. But to make this distinction it is necessary to study living material or, at the very least, to examine a very large number of empty tests. Neither of these is always possible. Moreover, not all foraminifera possess both reproductive modes. And to complicate matters even more, there are some investigators who state that some thecamoebians can reproduce both sexually and asexually. As is so often the case in taxonomic discrimination, a simple, single trait does not serve particularly well to distinguish between the two groups.

The granular structure of the pseudopodia may constitute a much more useful taxonomic trait for distinguishing major groups among the rhizopods. Most, but unfortunately not all, foraminifera possess this trait. Cytoplasmic and pseudopodial examinations of foraminifera have not been very numerous to date.

We should not worry excessively over the inexactitudes of the distinctions between the foraminifera and thecamoebians. Taxonomy is a man-created system which imposes a convenient order on nature. There are even investigators for whom it is more convenient not to separate these two orders; they consider the thecamoebians as fresh water foraminifera.

The Foraminiferida are very numerous taxonomically. An estimate based on a study of the Zoological Record (1865–1958) indicated that about one-half of the named protozoa are foraminifera and that an estimated 4791 living species had been described by 1958 (Levine, 1962). We estimate that about 38 000 fossil and Recent foraminiferal species have been named of which about 10 000–12 000 are found living in the Recent seas and oceans. This number is much greater than the actual number of species. Many of the named species have no zoological value and are representatives of previously described species (synonyms). The actual quantity of valid living species is probably somewhat more than 1000.

The study of foraminifera is of importance in various branches of natural science, especially in biology, geology and oceanography.

1. Biological importance

Foraminifera possess several attributes which make them appropriate objects of biological study: (a) they are numerous and lend themselves to statistical analysis, (b) there is more than one morphological state per species, (c) their highly visible external morphology is sculptured into a complex architecture which yields taxonomic and ecological information, (d) most species are relatively easy to cultivate, and (e) their small size permits cultures to be maintained in a small laboratory.

2. Geological importance

Because of their extensive variability, abundance and rapid evolution, foraminifera are excellent biostratigraphic indicators. Because of the sensitivity of some species to their environment, foraminifera are useful in the reconstruction of paleoenvironments and in paleoecologic studies. Their small size is helpful because many specimens can be obtained from small samples, e.g. well cuttings and cores. A malacologist will encounter, on the average, some dozen specimens in a kilogram of sediment whereas the same sample might yield more than 200 000 foraminifera. This abundance can prove quite useful in the field where a small sample can be examined with a 30 power microscope and a sufficient number of specimens can be found to draw preliminary conclusions about a stratigraphic zone or a paleoenvironment. The methods of removing the foraminiferal tests from the sediment are relatively simple and can be accomplished with a minimum of equipment.

During the last 40 years the study of foraminifera has become an important branch of geological study, particularly in the area of petroleum exploration. All large petroleum companies have at least one and often several specialists in foraminifera, and special laboratories for the preparation and examination of samples. Other geological areas in which foraminifera play a significant role are the preparation of geologic and hydrologic maps, exploration for various mineral resources (in addition to petroleum), etc.

3. Oceanographic importance

Foraminifera are almost exclusively marine organisms and have a widespread geographic (horizontal) and bathymetric (vertical) distribution in the oceans. This widespread occurrence coupled with their biological importance explains why the foraminifera are so useful in oceanographic studies. Foraminifera can be used to investigate (a) the movements of water masses, (b) ecologic and zoogeographic problems, (c) paleooceanographic problems, e.g. the opening and closing of straits, age of different land bridges, etc., (d) paleoclimatologic problems, and (e) problems of marine geology. This is but a partial listing of the oceanographic problems for which the foraminifera prove of help and value.

The importance given to foraminifera in oceanographic investigation is evidenced in the activity of the great oceanographic voyages from the original *Challenger* to its modern-day counterpart, the *Glomar Challenger.* Collection and examination of foraminifera was an integral part of all these expeditions. A number of voluminous works detailing the foraminiferal collections of these expeditions have been prepared. Most modern oceanographic institutions have special laboratories where teams of specialists work on foraminifera.

The number of persons occupied in the study of foraminifera, including those who help in the preparation of material is about 3000. Several hundred publications appear each year dealing with foraminifera or mentioning them.

II. THE LIVING ORGANISM

1. General and cytological description

The soft parts of foraminifera are composed of protoplasm. This material takes on different forms in the various chambers of the multichambered forms. For example, the food particles as well as food residue occupy the last chambers forming a 'vegetative segment'; the earlier chambers contain the nuclei and other inclusions and constitute the 'chromatic segment'. The endoplasm of the single chambered forms is more or less homogenous with uniformly distributed inclusions. Prior to reproduction, specifically sexual reproduction, the entire endoplasm often undergoes a process of 'purification' in which all the waste material and pigments are expelled beyond the test and the endoplasm becomes white and translucent.

The protoplasm of foraminifera contains all the organelles and inclusions typical of other animal cells as well as some which are unique to foraminifera. The protoplasm is bounded by a cellular membrane, the plasmalemma. This membrane, like most intercellular membranes, is three-fold, consisting of two dense protein-rich bands and a central low density layer of lipids. The plasmalemma also surrounds the pseudopodia.

The description of foraminiferal cytoplasm depends on the method used to examine it. In optical microscope studies it is described as a granular mass in which various zones can be differentiated. In *Allogromia crystallifera* the cytoplasm adheres to the internal part of theca forming a vacuolated layer which is very delicate but easily recognizable (Dahlgren, 1962). Lena (1972) also differentiated an external zone in *Allogromia flexibilis* with a system of canaliculate sinuses 1.3–1.7 μ in diameter. These canals communicate with the space between the protoplasm and the theca. The function of this zone of canals and the vacuolated layer is unknown; they may be preparation artifacts, although Angell (1971) using electron microscopic techniques observed a vesicular and vacuolated cytoplasmic layer just beneath the test (theca) of *Myxotheca* sp. which is actively involved in the test forming process.

When studied by electron microscopy, much greater detail is revealed in the cytoplasm. The structure is permeated, as are all animal cells, by a complex system of tubules, vesicles and flattened sacs which are formed by membranes and are frequently in communication with each other. This intracellular membranous organization is called the cytoplasmic vacuolar system by some investigators and the endoplasmic reticulum by others.

Dahlgren (1967a), in a very detailed study of *Ovammina opaca*, distinguished four

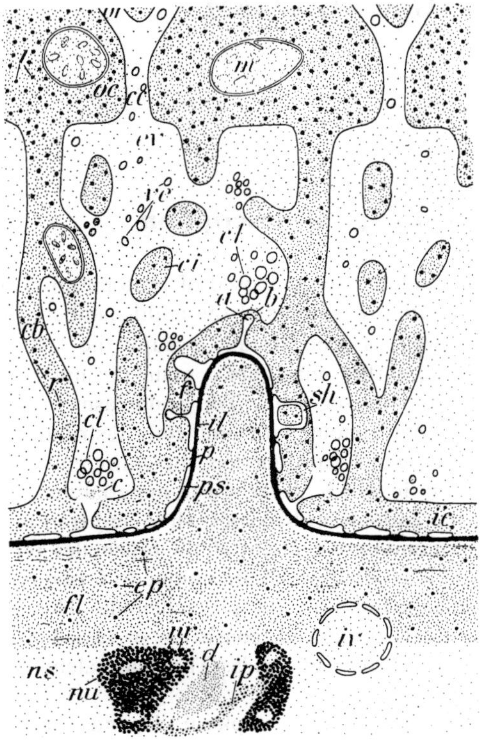

Fig. 1. Radial saggital section through edge of nucleus and adjacent cytoplasmic structures, *Ovammina opaca* (Dahlgren, 1967a); a. ampulla; b. bulge caused by ampulla; c. crater from inner agranular reticulum; cb. cytoplasmic bridge; cc. cónnecting canal from exonuclear vacuole to outer agranular reticulum; ci. cytoplasmic 'island'; cl. cluster of vesicles; d. DNA-axis of nucleosomic aggregate; ep. electron-opaque particle in fibrous lamina; ev. exonuclear vacuole with substance; f. fissure or canal of inner agranular reticulum; fl. fibrous lamina; ic. inner membrane layer; ip. intermediate part; iv. intranuclear vesicle; m. mitrochondrion; nu. nucleolonema with nuclear ribosomes; nr. nuclear ribosome; ns. nuclear sap; oc. outer cytoplasm; or. outer agranular reticulum; p. pore in nuclear cover; ps. perinuclear space; r. ribosome; sh. 'shunt canal' in inner agranular reticulum; ve. vesicles in exonuclear vacuole.

aspects of the cytoplasmic vacuolar system: (a) a reticular system extended throughout the cytoplasmic mass, (b) a vacuole surrounding the nucleus (exonuclear vacuole), (c) rounded empty vacuoles and (d) vacuoles with trophic particles in various states of digestion (Fig. 1). All the membranes of the cytoplasmic vacuolar system are agranular in the adult; granular membranes were observed only in the asexually reproducing form (schizont). These features are not common to all foraminifera and may even be unique to *O. opaca.*

The 2–5 μ exonuclear vacuole is distributed centrifically about the nucleus, is in

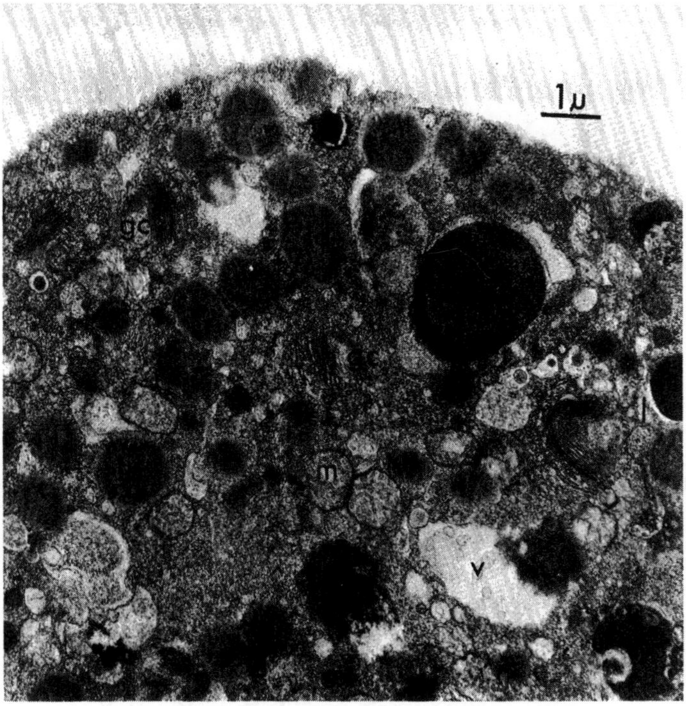

Fig. 2. Cytoplasm of *Globigerina bulloides* (Lee *et al.*, 1965); gc. Golgi complex; l. lysosome; m. mitochondrion; v. vacuole.

direct contact with the agranular reticulum and consists of the following elements: (a) islands and bridges composed of cytoplasmic materials, (b) small vesicles of different sizes and (c) electron opaque materials. The cytoplasmic bridges are arranged radially about the nucleus and often exhibit an anastomosing pattern. The vesicles have an average size of about 575 Å and tend to form in small clusters. These vesicles are circular or elliptical in section and their wall is composed of a single membrane within which is an electron opaque substance.

The mitochrondria of the foraminifera are tubular with a distinct outer and inner membrane (Figs. 2 and 4); no apparent differences have yet been observed between them in different species. They occur scattered throughout the protoplasm including the pseudopodia. They range in size from 0.5–1 μ.

The Golgi complex (Fig. 2) in foraminifera appears similar to the classic Golgi structure as observed in most metazoans. Two morphological components can be differentiated: (a) flattened sacs looking like dense parallel membranes in the section, (b) accumulations of vesicles intimately associated with these sacs.

The lysosomes (Figs. 2 and 3) are enzymatic packets surrounded by a membrane.

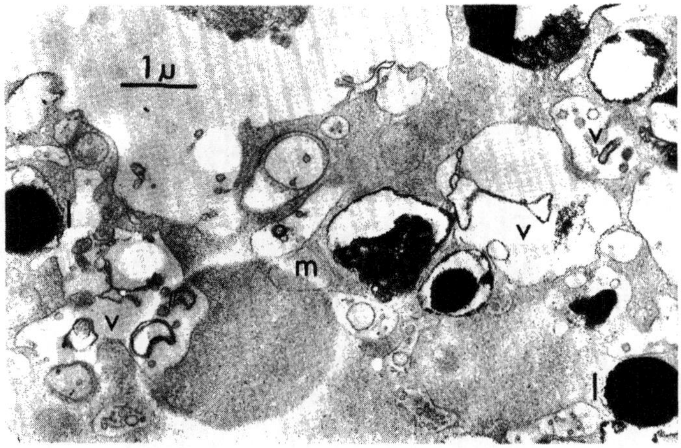

Fig. 3. Cytoplasm of *Allogromia laticollaris* (Lengsfeld, 1969b); l. lysosome; m. mitochondrion; v. vacuole.

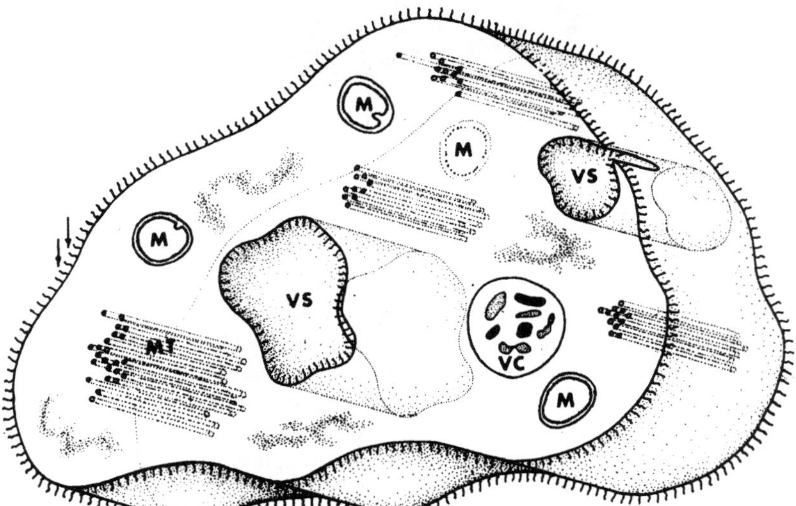

Fig. 4. Pseudopodial cross-section of *Iridia diaphana* (Marszalek, 1969); m. mitochondrion; mt. microtubules (composed of 30 Å diameter subunits); vc. food vacuole; vs. vesicle (originating as extension of plasmalemma); Arrows indicate filamentous covering observed on outer surface of plasmalemma and inner surface of vesicles.

They occur in great abundance in *Allogromia laticollaris* and may have been described as osmophilic granules in *Myxotheca arenilega* (Schwab, 1969) and as opaque particles in *A. laticollaris* (Lengsfeld, 1969b).

Marszalek (1969) introduced the term 'vesicles' to describe some membranous objects that occur in the pseudopodia of *Iridia diaphana* (Fig. 4). The structures form from the invagination of the plasmatic membrane in a manner similar to the process which forms the vesicles and canals that originate by pinocytosis in *Amoeba*. They probably function as devices to facilitate the extension of the pseudopodia and increase the ability of the organism to take up sea water and its dissolved nutrients.

The cytoplasmic vacuoles often contain diatoms, multilocular algae, crustacean muscle fibers and non-identifiable structures which may be the partially decomposed remains of food.

Electron microscopy of *Iridia diaphana* (Marszalek, 1969) and *Shepheardella taeniformis* (Hedley, Parry & Wakefield, 1967) revealed the existence of similar looking microtubules in the pseudopodia (Fig. 4). Each microtubule in *I. diaphana* has a diameter of about 250 Å and consists of various filamentous subunits (30 Å) arranged in circular form. From 9–13 subunits comprise each microtubule. These microtubules are arranged parallel to the long axis of the pseudopod and appear to give rigidity to the pseudopod.

A set of poorly understood structures was described as cryptosomes by Lee *et al.* (1965) in *Globigerina bulloides* and *Globigerinoides ruber*. This name was applied provisionally because it was not known if the structures were homologous in the two species and because their origin and function was unknown. In *G. ruber* the cryptosomes are undifferentiated bodies of large size which can be stained by Naphthol yellow S. They are surrounded by a PAS-positive reticulum and by small particles which can be stained by azure A-Schiff. The cryptosomes of *G. bulloides* vary greatly in size (15–30 μ) and morphology. They may occupy as much as $\frac{2}{3}$ of the total protoplasm in a chamber and more than one may occur in a single specimen. When they are fully developed the cryptosomes are Feulgen negative. The cryptosomes may develop from the hypertrophic somatic nucleus.

A vesicular system was also described in *Globigerinoides ruber* and *Globigerina bulloides* (Lee *et al.*, 1965). In *G. ruber* this system is intensely stained by hematoxylin, iron haematin, aceto-carmine or azure A-Schiff suggesting that this structure is composed of proteins. It is formed of pairs of 3–4 μ tubules which are twisted around themselves. There is no connection between the pairs of tubules although they commonly cross each other. Thin sections of the walls of these tubules reveal the presence of a helicoidal protein with a periodicity of 1–2 μ.

The vesicles of *Globigerina bulloides*, as those of *Globigerinoides ruber*, are oriented radially. They are stained by azure A-Schiff, aceto-orcein, hematoxylin and aniline blue. The walls of the vesicles are thicker than those of *G. ruber*. Lying along the outside of the vesicles is a spiral band of material similar in appearance to the vesicle walls. The vesicles are not paired and each coils gently around its own axis.

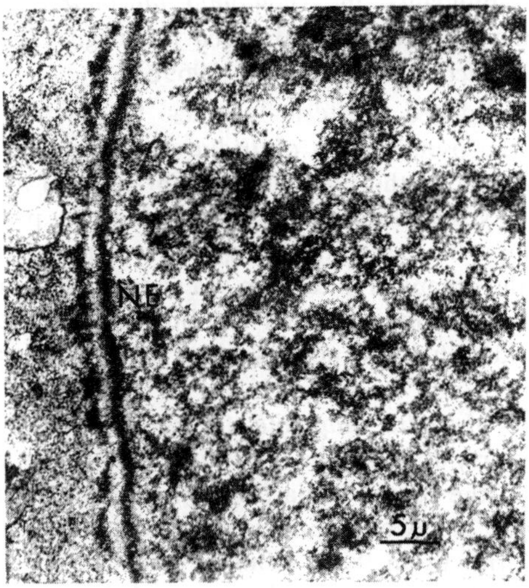

Fig. 5. Nuclear envelope, *Globigerina bulloides* (Lee *et al.*, 1965).

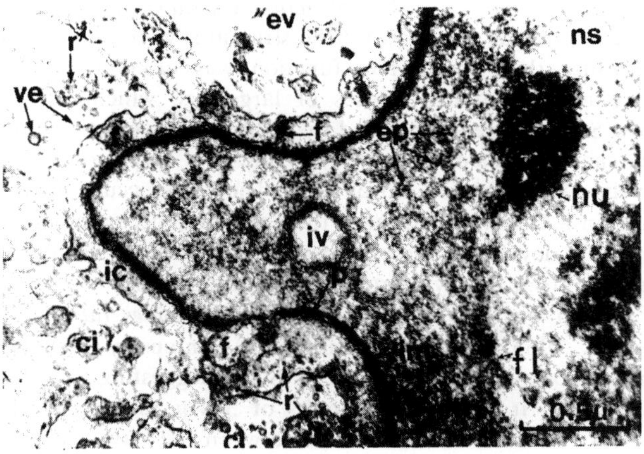

Fig. 6. Nuclear villus with adjacent structures, *Ovammina opaca* (Dahlgren, 1967a). The inner cytoplasm, *ic.* is irregularly interrupted by components of the inner agranular reticulum consisting of fissures or canals; *f.* Ribosomes; *r.* are sparsely distributed in the inner cytoplasm as well as in the cytoplasmic 'island'; *ci.* Vesicles; *ve.* in the exonuclear vacuole; *ev.* occur in quantity, forming at least one typical cluster; *cl.* Pores; *p.* are not conspicuous, they are partly concealed by the inner membrane layer; *im.* which is superseded inwards by the fibrous lamina; fl. In the fibrous lamina, electron-dense particles; *ep.* very similar to the nuclear ribosomes are seen, visible in a small part of a nucleolenema; *nu.* of the nucleosomic aggregate; *iv.* intranuclear vesicle; *ns.* nuclear sap; *d.* DNA-axis of nucleosomic aggregate.

In both species the vesicular system is best developed in the last chamber. Its function is unknown.

The number of nuclei in the foraminifera is generally variable and depends upon the reproductive generation. Those specimens that reproduce asexually possess a large number of nuclei. Those reproducing sexually commonly have but one nucleus.

In general, the vegetative nuclei of the sexual (gamont) and asexual (schizont) foraminiferal stages are indistinguishable. Within the nucleus are (a) the nuclear envelope; (b) nucleoli and dispersed chromatin; and in some species, (c) differentiated zones found adjacent to the internal membrane of the nuclear envelope (inner membrane layer, fibrous lamina and spongy layer).

The nuclear envelope is made up of two membraneous layers (Fig. 5) and is punctured throughout by pores which provide contact between the cytoplasm and caryoplasm. The surface of the nuclear envelope is not always smooth. Dahlgren (1967a) observed a series of evaginations or nuclear villi on the surface of the nuclear envelope in *Ovammina opaca*. These villi are 1–15 μ long and about 1 μ in diameter (Fig. 6). Nuclear villi are found in gamontic stages but have not been observed in schizont stages. These villi increase the surface area of the nucleus by as much as 200% over that of smooth surfaces.

The nucleus of *Allogromia laticollaris* (Fig. 7) exhibits no differentiated zone beneath the nuclear envelope, whereas the nucleus of *Ovammina opaca* contains an inner membrane layer and a fibrous lamina. The membrane is a somewhat irregular, electron opaque layer about 100–300 Å thick. Its inner boundary is often vague and

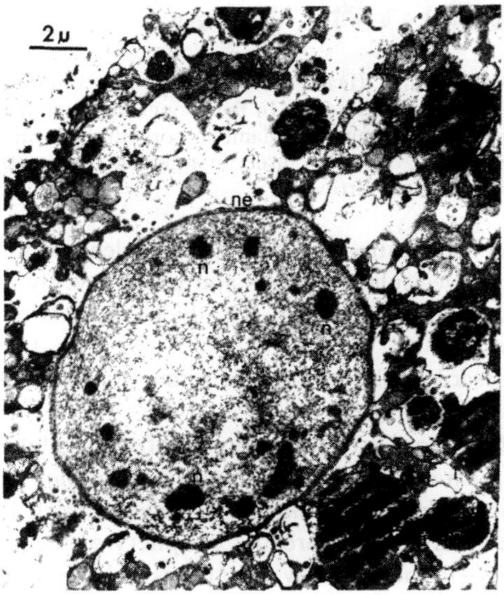

Fig. 7. Nucleus of *Allogromia laticollaris* (courtesy H. Lena); n. nucleoli; ne. nuclear envelope.

is usually found adhering to the nuclear side of the envelope. In immediate contact with this layer is the fibrous lamina, about 0.5 μ thick which contains occasional filamentous structures.

The nucleus of *Hippocrepinella alba* is more elaborate, exhibiting a fibrous lamina and a spongy layer occurring from the nuclear envelope inward toward the center of the nucleus (Fig. 8). The fibrous lamina is in the shape of a honeycomb layer and is

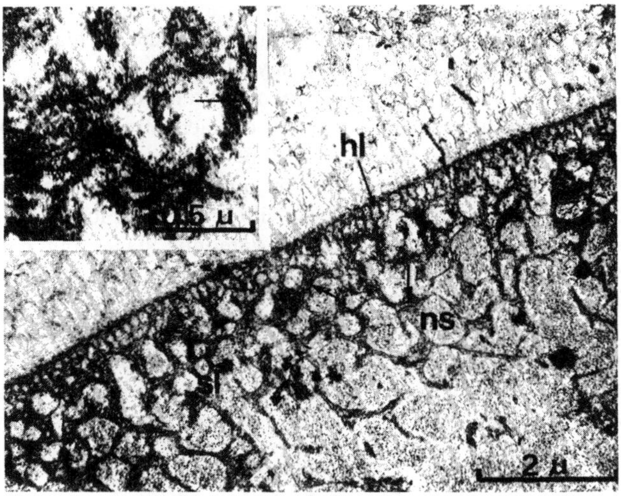

Fig. 8. Nucleus *Hippocrepinella alba,* (Dahlgren, 1967b); Honeycomb layer (fibrous lamina); *hl.* is shown as a fairly well delineated structure. In the spongy layer; *sl.* the filamentous character of the mesh-bars is visible (arrows), and the filaments sometimes seem to frame the compartments containing nuclear 'sap'; *ns.* The inset enlargement shows the coarse, undulating filaments or filamentous aggregates (arrows).

connected with, or an elaboration of, the inner nuclear membrane. This layer is composed of very small, vesicular elements and grades evenly into a more irregular spongy layer. This inner spongy layer (3–5 μ thick) consists of coarse filaments or aggregates of filaments.

Rhumbler (1911) described the nucleoli in terms of the kinds of nuclei in which they occurred:

(a) suspensoid nucleus – found in the majority of the agglutinated monothalamic species (*Bathysiphon, Rhabdammina, Technitella, Pelosina, Saccammina, Hippocrepinella, Rhynchosaccus, Iridia, Myxotheca, Webbinella,* etc.). The nucleoli are small spheres whose number increases toward the periphery of the nucleus.

(b) emulsoid nucleus – found in many calcareous polythalamic species (*Eponides, Nubecularia, Peneroplis, Cibicides, Discorbis, Planorbulina, Elphidium*). The nucleoli are much larger in this type of nucleus.

(c) alveolar nucleus – found in planktonic foraminifera (*Orbulina universa*, *Globigerina sp.*, *Hastigerina pelagica*). The nucleoli are characterized by a mixture of nucleosomatic and caryosomatic material.

Dahlgren (1967a) described a new nuclear structure in *Ovammina opaca* which he called the nucleosomic aggregate (Fig. 9). This structure is tubular with a RNA sheath surrounding a DNA axis.

As is apparent from the above discussion, the number of taxa whose cytoplasm has been examined is quite small. Many more species need to be studied before any broadly based cytological generalizations can be drawn.

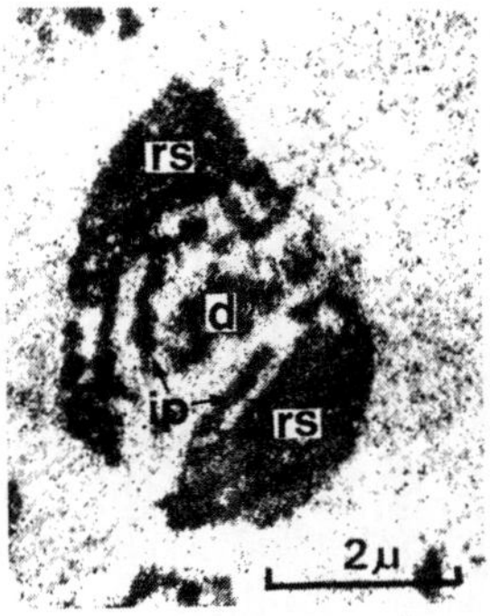

Fig. 9. Nucleusomic aggregate *Ovammina opaca*, (Dahlgren, 1967a); d. DNA-axis; ip. intermediate part; rs. RNA-sheath.

Foraminiferal protoplasm can be divided into internal protoplasm (endoplasm) occurring within the test, and external protoplasm (ectoplasm) located outside the test in contact with the environment. The ectoplasm, because of its microscopic texture, is somewhat more homogeneous than the endoplasm. Nevertheless the division between the two is somewhat conditional as it is not possible to describe a permanent or sharp demarcation between them. Immediately prior to reproduction in most species all the protoplasm withdraws into the test and the ectoplasm appears to unify with the endoplasm.

The ectoplasm is intimately involved in the secretion of the test and across it most of the interchange between the organism and the environment occurs. One of the remarkable characteristics of this endoplasm is its apparent ability to simultaneously dissolve calcareous material on its external surface and secrete calcareous test material on its inner layer (Banner, 1971). The ectoplasm forms elongate pseudopodia which give the organism the ability to move and attach itself.

Loeblich & Taplan (1964) describe the pseudopodia of foraminifera as being granuloreticulate and forming elongate linear extensions of protoplasm which can bifurcate and anastomose. However, reticulation by means of bifurcation and anastomosis is not always observed. The most frequent pseudopodial

Fig. 10. Pseudopodia of *Heterostegina depressa*, Kaneohe Bay, Oahu, Hawaii (courtesy R. Röttger), diameter 3.05 mm; age $6\frac{1}{2}$ months.

arrangement in *Elphidium incertum*, for example, consists of 'separate, straight strands converging to meet as a point away from the test' (Sheehan & Banner, 1972).

The idea that the foraminifera are sarcodins with granuloreticulopodia is derived from the fact that the majority of the genera whose pseudopodia have been studied are monothalamic forms with a simple skeleton and structure but these genera are not necessarily representative of all foraminifera. Of the approximately 1200 genera listed by Loeblich & Tappan (1964) only about 3 dozen species have been the subject of pseudopodial studies. More than half of these are monothalamic and therefore are not representative of the majority of foraminifera.

The cytoplasm that comprises the pseudopodia contains, besides granular bodies, the same structures recognizable in the endoplasm: mitochondria, membraneous systems, vacuoles, etc. The function of the granules is not known.

The length attained by pseudopodia is variable; e.g. those of *Elphidium crispum* and *Rotalia tepida* are only two or three times as long as the test diameter whereas those of *Iridia lucida* are up to 20 times the diameter. If a foraminifer is disturbed the pseudopodia retract but are re-emitted from the test after a relatively short time.

Figure 10 illustrates a living foraminifer and its pseudopodial net.

The imperforate foraminifera project their pseudopodia from the aperture. Ectoplasm usually surrounds the entire test and this envelope of ectoplasm also forms small delicate pseudopodia. In the perforate foraminifera the aperture is the principal pseudopodial orifice but may not be the only opening through which pseudopodia project as the pores may also serve this function. Sheehan & Banner (1972) observed in *Elphidium incertum* that pseudopodia may be emitted from four different test openings; the aperture, the pores, and the umbilical and retral fossettes. They also noted an 'external layer' of cytoplasm from which pseudopodial extensions are emitted. To this list can be added the relict sutural apertures of the last whorl in *Ammonia spp.* (Banner & Williams, 1973). There is some question as to whether or not all of these 'openings' are in fact open during the entire life-time of the individual (see Chapter III,5).

The pseudopodia perform various functions including tactile, food gathering, locomotion, respiration and test construction functions.

(a) Sense of touch. The pseudopodia act as sensory devices warning the organism of nearby objects and changes in the chemical environment.

(b) Capturing food. The pseudopodia surround particles of food and incorporate them into the ectoplasm. Within the pseudopod there are two currents simultaneously moving toward and away from the test (Jahn & Rinaldi, 1959; Wohlfarth–Bottermann, 1961, 1964). Great variations occur in the velocity of these currents. Sheehan & Banner (1972) reported that the pseudopodial granules in *Elphidium incertum* move at velocities of 5–10 μ/sec and Arnold (1964) calculated velocities of 1.7–5 μ/sec in *Spiroloculina hyalina* and 12–20 μ/sec in *Allogromia laticollaris*. Sheehan & Banner suggested that both phagocytosis and pinocytosis occur within the pseudopodia. The pseudopodial currents carry the nutritive material to the endoplasmic layer and then into the test proper either through the aperture or other test perforations. The nutrient particles are probably undergoing digestion even as they travel toward final assimilation in the interthalamous cytoplasm.

Based on the study of numerous well preserved specimens of the agglutinated foraminifer *Trochammina ochracea*, Frankel (1974) suggested a mode of feeding for this species which consists of test attachment to depressions in detrital grains and the production of an organic layer or '*Puffermasse*' (Rhumbler, 1938) which partially isolates this depression from the external environment. The foraminifer then presumably feeds off the extracellular products of diatoms and other micro-organisms which inhabit this space.

Foraminifera are omnivores. Many different kinds of organic materials serve as nutrition including bacteria, coccolithophores, diatoms, dinoflagellates, algae, spores, algal gametes, infusoria, other sarcodina, radiolaria and parts of other plants and animals such as copepods, mollusca, nematodes and echinoderms (Winter, 1907; Sandon, 1932; Myers, 1935a, 1937, 1943c; Jepps, 1942; Arnold, 1954c; Grell, 1954, 1973; Bradshaw, 1955, 1957; Nyholm, 1955b; Hedley, 1958; Lee *et al.*, 1961; Murray, 1963; Lee *et al.*, 1966; Lee & Pierce, 1963; Boltovskoy, 1969; Lengsfeld, 1969b; Muller & Lee, 1969; Lipps & Valentine, 1970; Christiansen, 1971, 1974 Ms; Röttger, 1972; Ross, 1972). Some species may utilize other foraminifera for food. Christiansen (1971) described *Spiculosiphon radiata* extending its pseudopodia into the test of its prey and then incorporating pieces of the victim's test into its agglutinated wall. *Bulimina marginata* is also a typical carnivor. Christiansen observed (1974 Ms) how this foraminifer settled on a specimen of *Technitella* or *Hippocrepinella* (which were considerably larger than *B. marginata*), made an orifice in the wall of the prey and in a short while removed the protoplasm. It is interesting that the orifice is made not by means of boring but by pushing the grains apart. Other species may utilize organic materials in the substrate in the same manner as some browsing metazoans.

Some cultured species exhibit a high selectivity toward the kinds of food they will eat. Certain foraminifera species will eat only certain species of diatoms; other will eat different species. Some species will consume only living food whereas others prefer dead food. One primitive genus, *Allogromia*, has been grown for more than a year in monobacterial cultures.

Rhumbler (1911) reported that *Globorotalia menardii* consumes diatoms and small radiolaria whereas *Globigerina spp.* generally eat crustacean fragments. *Globigerina*, *Orbulina* and *Hastigerina* contain long filamentous muscles of copepods in various degrees of decomposition but never remains of their tests. Either the digestive mechanism of the foraminifera dissolves the test or the organism is able to open the copepod test and ingest only the soft parts.

In experiments conducted on *Quinqueloculina sp.*, Christiansen (1971) noted that different dead metazoans were accepted as food but when the specimens had the opportunity to select they preferred crustaceans, especially copepods. 'In one experiment 10 foraminifera were placed at random in a glass vessel. Next morning all of them were found crowded around one dead copepod buried 6 mm in the mud' (p. 471).

The feeding character of the different species depends in part on the type, size and position of the aperture and test perforations. If the aperture is large, as in the miliolidae, food bits may be ingested intact through it. If the aperture is very small, the foraminifer has two options: to be satisfied with minute particles, e.g. bacteria, or to conduct digestion outside the test. In the latter case the specimen forms a food cyst outside the test wall in which digestion occurs. Jepps (1942) observed this type of digestion in *Elphidium crispum* when the diatom *Navicula mutica* was the food source. Various cysts were formed daily by the foraminifer (Fig. 11).

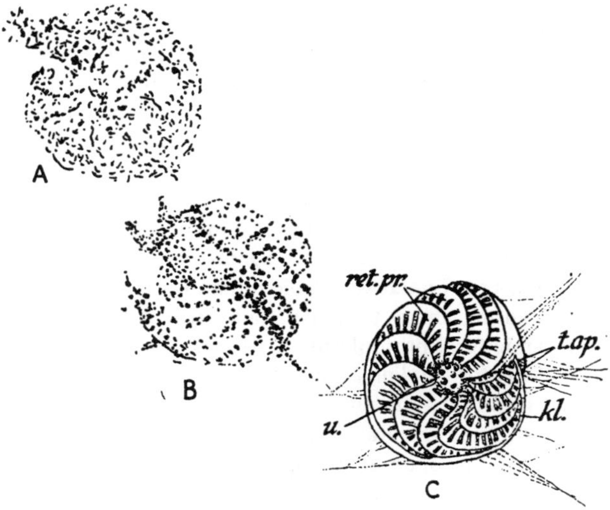

Fig. 11. *Elphidium* food cysts (Jepps, 1956); A. Abandoned food cyst composed primarily of empty diatom frustules and some xanthosomes; B. A more recent cyst; C. *Elphidium* moved to new site; t.ap. terminal aperture; ret.pr. retral processes; kl. keel; u. umbo.

Jepps (1942) also noted that specimens of *Elphidium* denied food for a period began to loose colour and became almost completely white. If food was re-administrated, the specimens returned to their original pale brown colour. Bradshaw (1961) observed that the protoplasmatic mass retracted to earlier formed chambers under conditions of low nutrition.

Although the foraminifera are not commonly considered as predatory and aggressive in their food gathering habits, as early as the middle part of the last century various observations were made of feeding foraminifera which indicated the intense food gathering activity of some species. Stuart (1866) noted that the pseudopodia of *Orbulina universa*[3] were so strong and sticky that they could easily capture and hold copepods that swam by. Schultze (1854) mentioned that the pseudopodia of *Gromia*

[3] Stuart erroneously called this species, *Coscinosphaera ciliata*, a radiolarian.

and *Elphidium* were capable of paralysing infusoria. Winter (1907) observed that a species of *Peneroplis* could kill copepods larger than itself in a period of 60–90 seconds. Within two hours only the empty transparent test of the copepod remained outside the *Peneroplis* test. The soft parts had been ingested by the foraminifer.

Some large specimens of certain foraminifera such as *Astrorhiza limicola* are capable of holding relatively large living crustaceans in their pseudopodial net. Despite its arduous resistance the victim was retained in the net until it died. *Nemogullmia* is thought to hold nematodes with its 'glutinous pseudopodia' until they die (Nyholm, 1956).

There are also observations which suggest that foraminifera can kill their prey with a special paralysing secretion. Similar kinds of secretions have been observed in a heliozoa (*Actinophryx*) and some dinoflagellates. The observations of Schultze were mentioned above. Nevertheless, not all infusoria seem to be equally affected by this substance. Winter (1907) and Rhumbler (1911) observed infusoria which seemed to thrive in the midst of pseudopodial nets. Jepps (1942) observed the ability of *Elphidium* to kill at a distance, but added that crustaceans were never used as food in her experiments. Winter noted that any venomous substance used by *Peneroplis* against copepods was completely ineffective against infusoria. Lee *et al.* (1961) noted that cultured specimens of *Rotalia* were able to paralyse specimens of the biflagellate *Dunaliella parva* causing them to cease their irregular movements and remain adhered to the foraminiferal protoplasm. A special case of predation was described by Christiansen (1971) for *Pilulina argentea*. Some specimens of this species burrow quickly into a mud substrate and create hollows just beneath the mud surface. These hollows are lined with small mineral particles cemented together to form small living cavities. These 'houses' are open at the upper end but covered over with fine mud. The pseudopodia extend from the test and spread out in the mud layer over the cavity. When copepods land on the mud over a *Piluina* 'house' they are firmly held and unable to free themselves from the sticky pseudopodia. They are slowly drawn into the cavity where they are consumed by the foraminifer (Fig. 12). All these observations indicate that the foraminifera have a certain capacity for active predation.

The food particles that are not digested, such as diatom frustrules, radiolarian capsules, etc., are often carried by cytoplasmic currents and expelled from the test. The genera *Peneroplis* and *Gromia* form fecal pellets from this residue. Similar clumps were observed by Rhumbler and Schaudinn and later by other investigators in various benthonic foraminifera on muddy bottoms. Schaudinn called these clumps stercome. They are spherical and ovoidal grayish-brown particles within the cytoplasm and are composed of quartz shards, diatom frustrules and other organic and inorganic colloidal fragments common to muddy substrates. They are commonly 5–50 μ in size but exceptional cases are known up to 350 μ. They may be very numerous in an individual. The cement which binds this material is probably secreted by the foraminifer and was analysed in *Gromia* (Hedley, 1962a) as a mucopolysaccharide, the same material which composes the inner organic lining of many

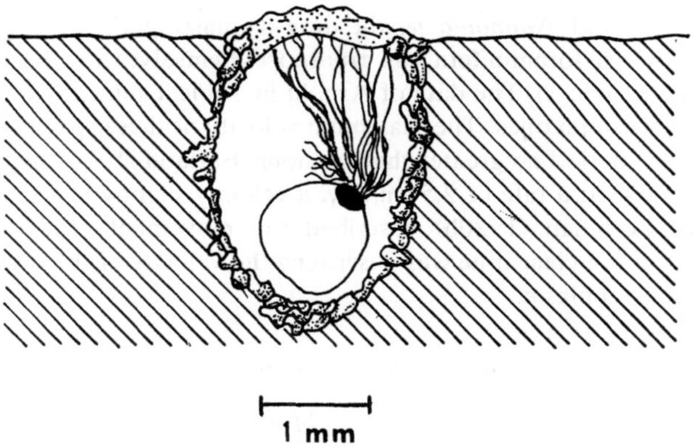

Fig. 12. Pilulina argentea living in a chamber built of mineral particles buried in a mud substrate (after Christiansen, 1971).

foraminifera. The biological function of the stercomata is not well understood. Rhumbler suggested that they serve to consolidate the protoplasm. But they may also serve as devices to augment the density of benthonic foraminifers, keeping them more firmly on the bottom. Those of the last formed chamber are always expelled prior to reproduction, forming an agglomerated pile along with digestive residues. These piles have been called fecal pellets although their principal constituents are stercomata. Banner, Sheehan & Williams (1973) suggested that the stercomata are indigestible waste material which has accumulated in the early formed chambers and can not be easily expelled because they are being formed at a rate greater than the rate at which they can flow through the small septal openings.

Foraminifera may possess the ability to feed osmotically by directly absorbing nutritive organic substances dissolved in sea water, substances that are the product of the vital processes of various algae (Cyanophyceae and Chlorophyceae). But if osmotic nutrition plays any role, it is probably a minor one as the following observations indicate. At the end of spring and the beginning of summer a considerable quantity of dissolved nutritive salts is present in sea water as is a rich phytoplankton population. The foraminiferal population increases and individuals grow rapidly at this time. However, in the winter, when the nutritive salts are again present, the phytoplankton population is low because of low illumination, and the foraminiferal population undergoes little growth activity. Banner & Williams (1973) suggest that osmotic exchange may occur in *Ammonia* but would probably be restricted to the last formed chamber by the thick multilaminar chitinoid membrane lining all earlier formed chambers. The passage of large organic molecules across this membrane is unlikely.

(c) Locomotion. Foraminiferal movement is effected by the pseudopodia. The pseudopodia lengthen, attach themselves to some point, and pull the remainder of the organism along. The strength of pseudopodia is illustrated by Banner & Williams

(1973) who observed *Ammonia tepida* drag a quartz grain as large as itself. The velocity with which foraminifera can transport themselves has been calculated for various species (Table 1). The data of Arnold in Table 1 refer to vertical movements as well as to horizontal ones. The maximum velocity of these specimens was 1 cm/hr, maintained for a period no greater than an hour. Earland (1934) suggested that most foraminifera move at a rate of 2.5 cm/day, a velocity slightly less than that observed in the table. Cushman (1920b) described the movement of *Iridia diaphana* as 1 mm/min, but it not clear if he was describing the velocity of the organism or that of the streaming protoplasm.

Table 1. Velocity of foraminifera

Author	Species	Velocity, cm/hr
Banner & Williams, 1973	*Ammonia tepida*	0.18
Narchi, 1963	*Elphidium articulatum*	0.2
Arnold, 1953a	*Allogromia laticollaris*	0.25–0.29
Sliter, 1965	*Rosalina globularis*	0.3–0.4
Jepps. 1942, 1956	*Elphidium crispum*	0.4–0.5
Dujardin, 1841	*E. sp.*	0.4–0.8
Narchi, 1963	*Bolivina punctata*	0.6
Myers, 1943c	*E. crispum*	1.0
Buchanan & Hedley, 1960	*Astrorhiza limicola*	1.0
Schultz, 1915	*A. limicola*	1.0
Lutze, Grabert & Seibold, 1971	*Heterostegina depressa*	1.0
Stephenson & Rees, 1965	*Discobotellina biperforata*	2.0
Narchi, 1963	*Massilina secans*	2.0
Winter, 1907	*Peneroplis pertusus*	3.0
Narchi, 1963	*Poroeponides lateralis*	12.0

We can conclude that the velocity of foraminiferal movement is distinct in different species but the variation is not extreme.

(d) Respiration. The interchange of gasses between the organism and the external environment can probably take place across the pseudopodial membrane, but the mechanics of exchange have not yet been studied. Oxygen consumption data are given in Chapter IX.

(e) Test construction. The role played by the pseudopodia in test construction is examined in detail in Chapter III.

The foraminiferal literature contains occasional passing references to the ability of some adult foraminifera, e.g. *Iridia diaphana*, to voluntarily abandon their test. In his cultures of *I. diaphana*, Marszalek (1969) noted a nude specimen which was similar in appearance to an ameboidal rhizopod but there was no indication of whether it had a theca or not. Although there are no confirmed observations of a foraminifera leaving its test, it is clear that the loss of the test is not necessarily fatal for some species. Buchanan & Hedley (1960) artificially removed the test from the agglutinated form *Astrorhiza limicola* and placed the specimens in a container without test building material. The specimens secreted a substance similar to the agglutinating cement around themselves. When placed in a container with mineral particles, the nude specimens constructed their typical tests. Angell (1967c) and McEnery & Lee (1970) decalcified living specimens of *Rosalina floridiana* and found that the specimens not only remained alive but were capable to recalcifying their test. The same observations were made on *Ammonia tepida* (Bradshaw, 1961). However, Bradshaw found that specimens of *Spirillina* did not survive exposure to low pH conditions. *Rosalina* and *Ammonia* both possess organic linings between the calcareous test and the plasmalemma whereas *Spirillina* does not. This suggests that the membrane may be an effective environmental buffer.

A similar experiment involving the removal of a test from a living foraminifer was carried out by Christiansen (1974 Ms). He broke a test of *Saccammina sphaerica* and took the broken test away. After two weeks Christiansen put small particles of red and green glass in the culture dish and the organism used these particles to construct a new test.

A good review of the pre-1964 literature on foraminiferal protoplasm, pseudopodia and physiology was prepared by Hedley (1964).

2. Life cycle

D'Archiac, as early as 1853, called attention to the observations that there are differences between the tests of foraminifera belonging to the same species. These differences, while usually not very great, are almost always present. This observation was confirmed in 1861 by Parker & Jones and in 1879 by de la Harpe and Hantken. Munier-Chalmas labelled this phenomenon dimorphism in 1880. Schlumberger, after studying some Miliolidae, proposed that the dimorphic forms be called the megalospheric and microspheric forms. The early history of the development of this concept concluded in 1894 when Lister and Schaudinn, simultaneously and independently, explained the cause of the dimorphism as the alternation of generations.

In some cases the microspheric and magalospheric generations of the same species are so distinct from each other that they have been considered as different species; e.g. *Discorbis patelliformis* and *D. erectus* (Fig. 13), or different genera; e.g. *Discorbis ornatissima* and *Pulvinulina columbiensis* (Steinker, 1973). There are cases even more extreme in which members of different generations have been interpreted as different genera from different families; e.g. numerous species of *Articulina* (family Miliolidae) from the Neogene of the Caucasus Mts. have megalospheric forms which were classified as *Nubecularia* (family Opthalmidiidae). There are undoubtedly many species, now considered as valid, which are only different forms of another species.

A. Conventional Reproduction

1. Aspects of the microspheric form

This form is the asexual one and is sometimes called form B, the agamont, or the schizont form. It generally has a relatively large test with an initial chamber

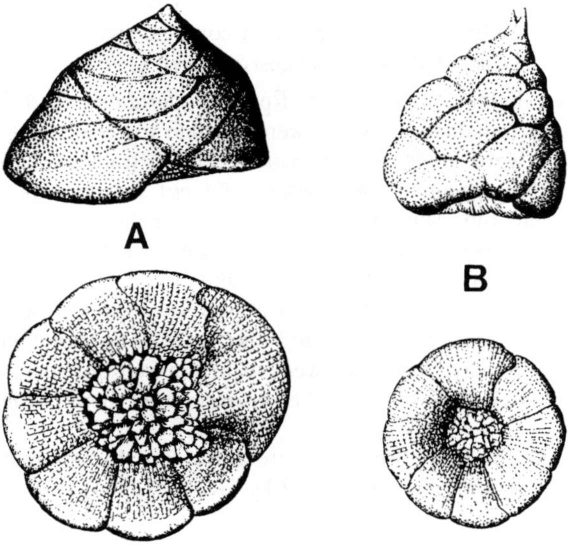

Fig. 13. Schizont (A) and gamont (B) form of *Discorbis patelliformis.* This latter form has been described by Sidebottom with the name *D. erectus* (after Le Calvez, 1938b).

(proloculus) which is relatively small, often no larger than 0.02 mm. It is characterized by several nuclei, generally of different sizes. When the organism matures, the protoplasm collects in the last chamber or beyond the test. Each nucleus gathers some protoplasm around it and forms the initial chamber of a new individual. These initial chambers abandon the microspheric test and form new individuals, usually megalospheric (Fig. 14A). The reproduction is thus asexual (schizogony). In some species; e.g. members of *Discorbis*, *Heterostegina*, *Tinoporus* and *Cibicides*, the juvenile forms have developed two or more chambers prior to leaving the maternal microspheric test (Fig. 14B). The number of offspring varies in different species and may be a function of the size of the parent species. Small forms such as *Spirillina*, *Patellina*, and *Cornuspira* usually emit from 8–12 young, whereas *Rotalia beccarii tepida*, a larger species, has 28–32 offspring. Yet larger forms including *Discorbis*, *Amphistegina*, *Calcarina* and *Elphidium* yield up to 225 young and *Heterostegina depressa* has produced 10–600 offspring in culture.

2. Aspects of the megalospheric form

This form is the sexual one and is sometimes called form A, the gamont, or sporozont form. It is commonly characterized by a relatively small test and a relatively large proloculus (0.2–0.5 mm). There is usually but one nucleus and in the mature form this nucleus divides into many small nuclei which carry a small quantity of cytoplasm with them when they leave the test as flagellated zoospores (gametes).

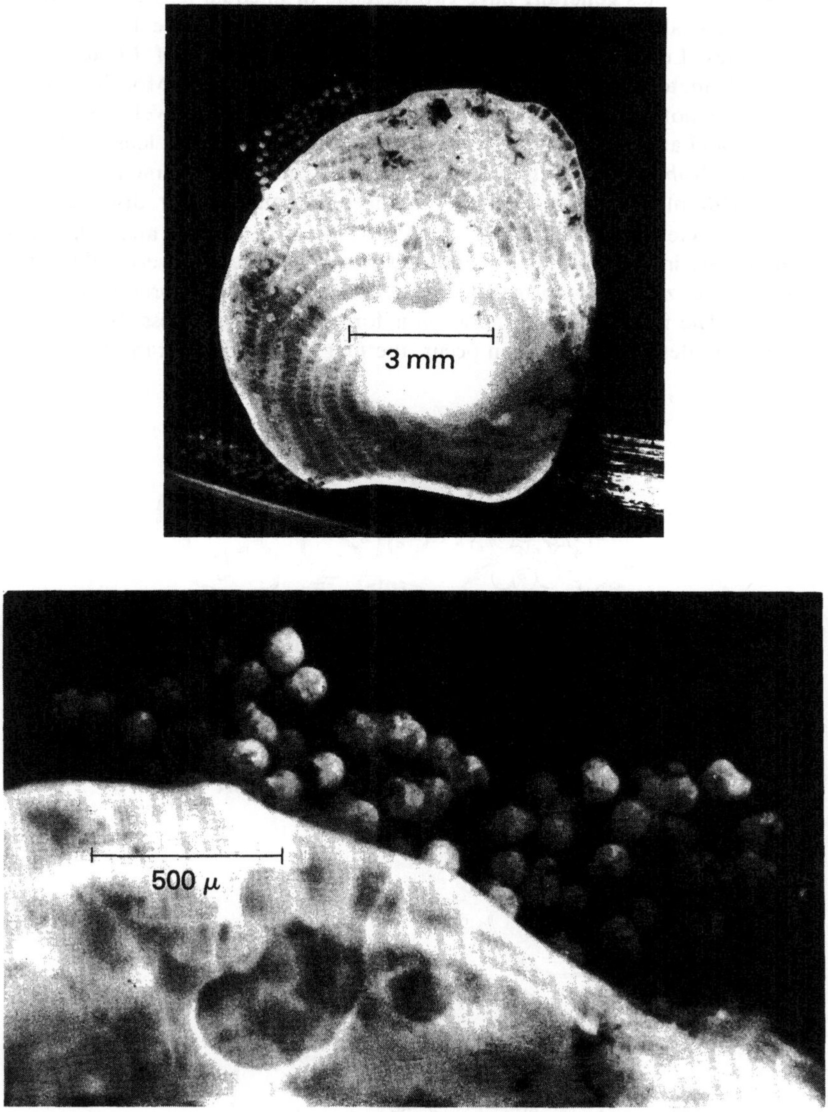

Fig. 14. A sexual reproduction *Heterostegina depressa*, (Lutze, Grabert & Seibold, 1971); A. Adult specimen with offspring leaving test; B. Enlargement showing multichambered offspring.

These small gametes generally have two flagellae of unequal length protruding from the same side although triflagellate and ameboid gametes have been observed in some species. Le Calvez (1938b) calculated that a specimen of *Iridia lucida* only 1 mm in diameter could produce about 70 000 000 gametes. Morphologically the gametes can not be differentiated sexually. They move relatively quickly (up to 150 μ/sec) and are able to survive outside the parental test for at least 24 hours. The gametes of *Nubecularia lucifuga* are sometimes enclosed in membraneous cysts (Arnold, 1967b) which are capable of longer life and greater dispersal. Hedley, Parry & Wakefield (1968) and Angell (1971) described and photographed gametogenesis in *Boderia turneri* and *Myxotheca* sp., respectively. When two gametes join, a zygote is formed. This process of sexual reproduction is called gamogony. The microspheric generation is formed by this process. If an alternation of the two modes of reproduction occur, the first and third generations will be much

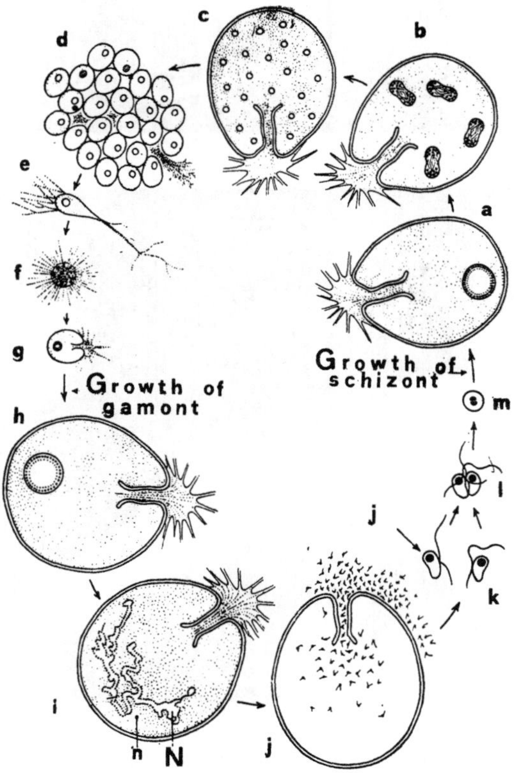

Fig. 15. Reproductive cycle of *Iridia lucida*, a monogametic species whose gametes are biflagellate (after Le Calvez, 1953).

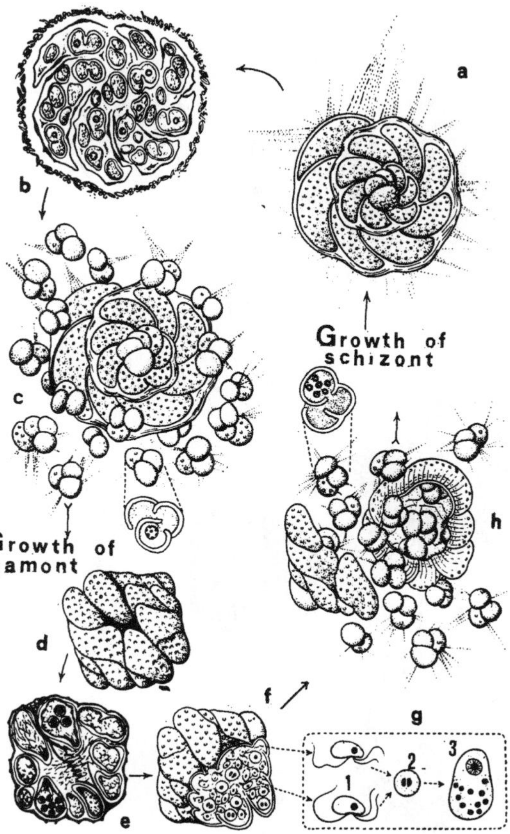

Fig. 16. Reproductive cycle of *Discorbis patelliformis*, a plastogametic species whose gametes are triflagellate (after Le Calvez, 1953).

more similar to each other than to the second generation. Once the gametes have vacated the parental test, the parent dies and the test remains empty.

Alternation of reproductive mode is known in primitive species; e.g. *Hyperammina*, but has been studied in more detail in the more advanced species (Figs. 15–17).

B. Variations in Reproduction

Some species exhibit peculiar reproductive traits differing from those described above. Several examples of these variations from the general scheme are discussed below.

Berthold (1971) described gametogenesis in the gamont stage of *Patellina corrugata* in which the number of developing gametes do not correspond to the number

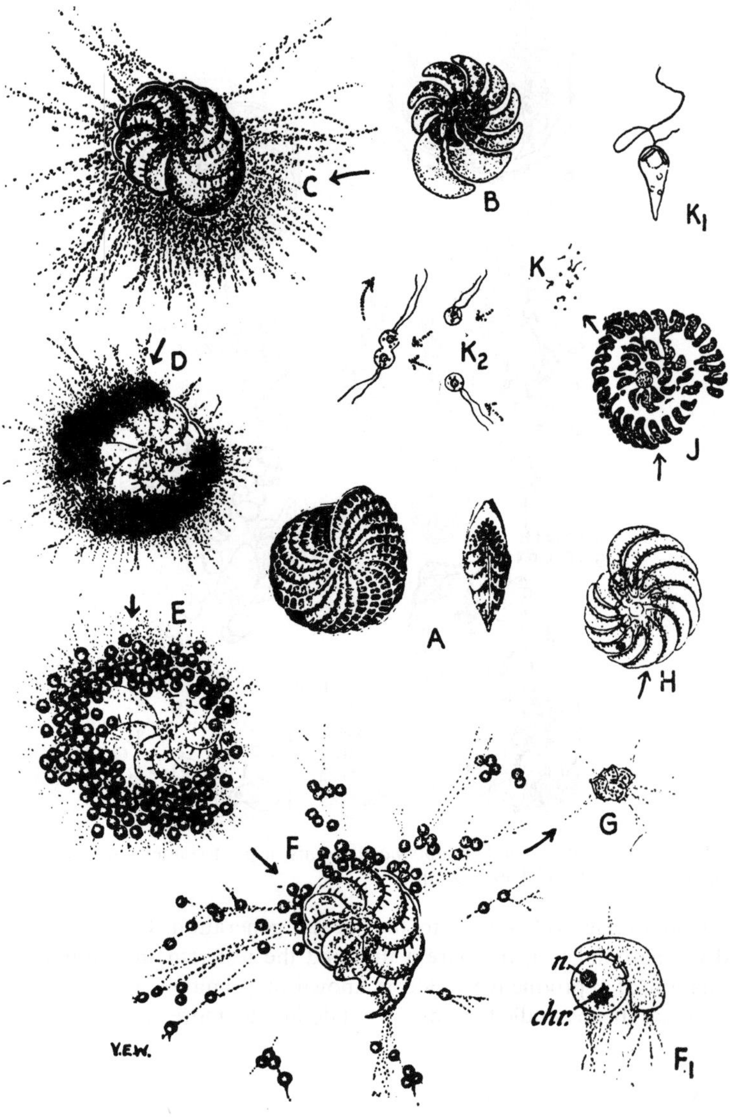

Fig. 17. Reproductive cycle of *Elphidium crispum* (after Jepps, 1956); A. Empty test; B. Decalcified fixed microspheric specimen with several nuclei; C. Living specimen with several pseudopodia; D. Dark colored protoplasm leaving the test; E. Formation of megalospheric proloculi; F. The new individuals leave the empty, partially destroyed maternal test. Time elapsed since stage 'C', 12 hours; F_1. Decalcified, fixed juvenile (1–2 hours to form the second

of parental nuclei after division; some gametes receive two nuclei and others appear to contain none.

The megalospheric form is not always smaller than the microspheric form. There are species, e.g. *Glabratella sulcata*, in which this relationship is reversed and the megalospheric form is the larger.

The same is true of the size of the proloculus. Schaudinn (1895a) was the first to note that the smallest proloculus of the megalospheric form of *Elphidium* was the same size as the largest proloculus of the microspheric form of the same species. Subsequent investigations have revealed that not only does this observation hold for some other species but that there are examples of some species in which the megalospheric proloculus is smaller than the microspheric one. Grell (1958c) suggested that the marked difference in size of proloculus between sexually and asexually produced populations holds only for species which produce gametes capable of free swimming (*Elphidium*, *Peneroplis*, etc.) The terminology 'megalospheric' and 'microspheric' may not be as precise as the words indicate, and perhaps the terms should not be used.

Another long held conviction, excellently described by Le Calvez (1953), was that the departure of gametes from the parent necessitates the death of the parent. However, only two years later Arnold (1955b) described a small cultured miliolid species which was capable of reproducing at least twice and perhaps more often prior to death. Bradshaw (1957) described parent specimens which retained small quantities of protoplasm with pseudopodial activity even after sexual reproduction occurred. Sliter (1970b) found that *Bolivina doniezi* specimens could asexually reproduce a number of times in culture. Similar conclusions were reached by Okropidze (1957) who said that fossil forms of *Nummulites* appeared as if they had reached sexual maturity early in life and had reproduced several times.

The gametes of foraminifera from two different species do not unite with each other; fusion is possible only between gametes belonging to the same species. However, this observation raises the interesting question; is it possible for gametes arising from any two megalospheric specimens of the same species to unite? The experiments conducted in this area are few but the evidence seems to indicate that the answer is no. There appears to be sexual differentiation among some species. For example, when pairs of megalospheric *Discorbis mediterranensis* were isolated and left alone, some pairs began to crawl about until united, whereas other pairs remained indifferent to each other. Le Calvez (1950) noted that of 17 specimens of the same species and generation only 7 were able to unite with the remaining 10. Grell (1957a) and Berthold (1971) observed the same phenomena in a rich sample of cultured *Patellina corrugata*.

---

chamber); n. nucleus; chr. chromidial shreds; G. Two-day old juvenile (4 chambers); H. Decalcified and fixed megalospheric individual (after several months); J. Section of megalospheric form while gametes are being released; K. Flagellate gametes leave the adult test; K_1. Flagellate gamete immediately after leaving test; K_2. Gamete after short period of time. Two gametes unite to form a zygote which grows to an adult microspheric form.

At the same time, there are species characterized by autogamy. The gametes do not abandon the maternal test and the copulation occurs within this test between gametes of the same parent. This phenomenon has been observed in *Allogromia laticollaris, Rotaliella heterocaryotica* and *R. roscoffensis* (Arnold, 1955a; Grell, 1954c, 1957b; Lee & McEnery, 1970). Something even more curious occurs in the copulation of other foraminifera. The parental specimens unite so that only a very narrow space occurs between them. It is within this space that the gametes unite to form the zygote (Fig. 16). This type of union has been called *Patellina*-type union by Le Calvez (1950), and plastogamy by Schaudinn (1895b), and has been observed in *Patellina, Spirillina, Discorbis, Gromia, Glabratella, Buliminoides* and *Globigerina* (Myers, 1935b, 1936; Grell, 1958a,b; Arnold, 1966a; Boltovskoy & Theyer, 1970; Seiglie, 1970a; Berthold, 1971; Steinker, 1973). In some species more than two (up to 10) megalospheric forms join together. Le Calvez thought that this mode of union was an adaption to life in the littoral zone where environmental changes are rapid and severe. The reproductively mature adults must position themselves closely together to ensure union of the gametes even though the gametes are flagellated. A typical example of plastogamy was described by Lipps & Erskian (1969) in *Glabratella ornatissima.* They noted that two or more adults are joined along their apertural margins by a monolamellar organic membrane less than 1 μ thick. The apertural faces and internal septa are dissolved, providing a large open area in which exchange and union of gametes occurs. When the zygotes have grown to the 2–3 chamber stage they secrete an enzyme which degrades the organic membrane and thereby liberate themselves.

The classic alternation of generation is also more complicated than originally thought. As a general rule the number of microspheric specimens is considerably less than the number of megalospheric forms. As an illustration, the examination of 1339 tests of living *Planorbulina mediterranensis* collected throughout the year revealed only 39 microspheric specimens. Specimens of *Patellina* and *Calcarina* exhibited a 1:5 relationship between microspheric and magalospheric forms and *Elphidium* exhibited a 1:30 ratio. This preponderance of magalospheric forms is explained by the observation that in some species asexual reproduction occurs repeatedly (Arnold, 1964, 1967b) with the result that megalospheric young are produced in several successive generations. Moreover, when a microspheric form reproduces it often undergoes considerable dissolution and becomes unrecognizable.

The number of megalospheric young produced by the microspheric parent, the microspheric:megalospheric ratio, along with a knowledge of the time involved, can give us an idea of the reproductive rate. Myers (1942a) calculated that in cultures of *Elphidium crispum* in which the population increased some 40 times annually, about 1000 specimens per square foot would be produced annually.

Although the above examples indicate that megalospheric specimens often outnumber the microspheric ones, the opposite situation sometimes prevails. Microspheric forms are more numerous than megalospheric ones when the specimens live under unfavorable conditions. In some cases megalospheric forms are not present at all. These situations seem to be confined to winter, in the littoral zone, or in mid-latitude sublittoral zones. Hofker (1930b) claimed that only the microspheric specimens of *Rotalia beccarii flevensis* in the Zuidersee were able to survive the harsh winter conditions or hot mid-summer conditions when the temperature exceeded 20 °C. Megalospheric forms were present only during the spring and autumn. These observations concur with those of Myers (1943b) who calculated the following microspheric: megalospheric ratios for *Tretomphalus:*

autumn	1:140
winter	1:24
spring	1:340

Moreover, it appears that the greatest quantity of microspheric specimens occur in species with plastogamic reproduction; i.e. the absence of free swimming gametes.

Paleontologic studies give some indication that the differences between megalospheric and microspheric generation increases during the evolutionary history of a phyletic group. Phylogenetically young species tend to be characterized by apogamic reproduction. In this cycle of reproduction the microspheric, sexually reproduced forms disappear completely. The Fusilinidae, Alveolinidae and some other families, however, exhibit the greatest difference between these two generations in their most evolutionarily advanced representatives. The absence of a sexual generation in the less evolved foraminifera is a phenomenon observed in another class of protozoa, the Mastigophora.

The absence of a sexual reproductive phase is not a generic trait but rather a specific one. There are cases in which species of the same genus; e.g. *Discorbis*, can be differentiated by the absence of a sexual process. Unfortunately, the reproductive cycle of the overwhelming majority of the foraminiferal species is unknown and the distribution of apogamic reproduction among the species is therefore unknown although it appears to be more widely distributed than is commonly believed.

Laboratory studies on various protozoa other than foraminifera, have shown that if hundreds of asexually reproduced generations are created uninterrupted by a sexual generation, the brood begins to show signs of decay, the period between reproductions increases, the resistance of the offspring diminishes and the organisms perish. These observations have not held true for those foraminifera which repeatedly reproduce without a sexual generation. In one case where the successive offspring were consistently smaller than the parent (Röttger, 1972a), the size reduction was attributed to the artificiality of the conditions and possible lack of sufficient food for growth in the crowded culture. It may be that the foraminifera possess some mechanism, as yet unknown, which maintains their vital processes in equilibrium despite the recycling of the same genetic material.

The duration of the reproductive cycle depends on the species and the environmental conditions. Small species living in shallow mid-latitude waters may complete their reproductive cycle in several days, whereas some larger species may live two years. The studies of Myers (1938, 1941, 1943a) on *Elphidium crispum* are among the most extensive in this regard. In temperate zones the reproductive cycle requires two years to be completed. In tropical zones where food is less abundant but more consistently supplied, the species grows at a slower rate than during the temperate spring and completes its reproductive cycle in one year. Table 2 shows the duration of the reproductive cycle for various benthonic species.

The reproductive cycle data for planktonic species are much scarcer. Berger & Soutar (1967) concluded that life spans of planktonic foraminifera are on the order of one month. Berger (1971a) later revised this figure downward to 2 weeks. These

Table 2. Duration of reproductive cycle, various benthonic species

Species	Author	Length of reproductive cycle
Spiroloculina hyalina	Arnold, 1964	2 weeks
Nubecularia lucifuga	Arnold, 1967b	few weeks
Rosalina sp.	Hedley & Wakefield, 1967b	3–5 weeks
Streblus beccarii tepida	Bradshaw, 1957	4 weeks
Patellina corrugata	Myers, 1935b	41 days
P. corrugata	Berthold, 1971	40 days
Bolivina doniezi	Sliter, 1970b	9–13 weeks
Heterostegina depressa	Röttger, 1972a	8 months
Epistominella exigua	Boltovskoy & Lena, 1969b	<1 year
Elphidium gunteri	Boltovskoy & Lena, 1969b	<1 year
E. articulatum	Boltovskoy & Lena, 1969b	<1 year
Quinqueloculina seminulum	Boltovskoy, 1964b	1 year
E. macellum	Boltovskoy, 1964b	1 year
Buccella frigida	Boltovskoy & Lena, 1969b	1 year
Marginipora vertebralis	Ross, 1972	>1 year

calculations are based on studies of test size distribution at various depths and resulting calculations of species turnover rates. However, Bé (in press) is of the opinion that 'life spans of planktonic foraminifera are probably in the order of months rather than weeks or days'.

The question must surely arise: of what importance is the study of the 'alternation' of reproductive generations in the foraminifera? First, it has become evident that since the microspheric generation generally goes through more stages of development it is more useful in studies of phylogenetic linkages than the megalospheric forms. Secondly, there is an important practical consideration to the varied reproductive behavior of foraminifera. As has been described, there is a dimorphism that accompanies the alternation of generations. An understanding of the presence and characteristics of these different forms of the same species is essential for taxonomic studies.

Nevertheless, even the concept of dimorphism is not well understood and may be a simplification of the reality. Lister (1903) and Schaudinn (1895a) noted that it is often possible to isolate separate and distinct megalospheric forms from the typical micro- and megalospheric generations. Hofker (1925) described this phenomenon in detail as trimorphism. According to his point of view, each species has 3 forms, A_1, A_2 and B, the first two of which are megalospheric forms differentiated by the size of their proloculus, and the last of which is the microspheric form with a proloculus yet smaller than that of the A_2 form.

There have been numerous studies of trimorphism in both fossil and Recent forms. The following 3 'species' from the Miocene of the Vienna Basin may all belong to the same species: A_1: *Nodosaria aculeata* (4 chambers); A_2: *Dentalina floscula* (6 chambers); B: *Marginulina hirsuta* (9 chambers). Hofker (1922) described trimorphic forms from the Recent sediments of the Zuidersee, a zone characterized by very abrupt changes in temperature and salinity. Hofker suggested that the variety of environmental conditions there was able to provoke the morphological differences noted. Studies of the same species in the Malayan Archipelago by Keijzer (1936) revealed no trimorphism at all. Ecological conditions in this area are very stable and constant.

Hofker persisted in his views regarding trimorphism. He maintained (1951b, 1956) that those species which are geologically old or more primitive demonstrate trimorphism more often than the more advanced and recently evolved species. These 'newer' species exhibit dimorphism, 'paratrimorphism' or are apogamic. He illustrated these conclusions with several examples. As he used only fossil forms, his conclusions were based on various mathematical considerations of proloculus diameter, test size and number of chambers.

The concept of trimorphism has its adherents (Hofker, *op. cit.*; Sigal, 1959) and its adversaries. The latter appear to have a stronger case based on statistical studies (Tan Sin Hok, 1932) and culture studies (Myers, 1938; Le Calvez, 1938b). Myers claimed that only two or more asexual generations of some *Sorites* could be considered as trimorphic. Le Calvez said that the idea of trimorphism ought to be modified. One should not speak of 3 different forms but of the possibility of encountering 3 biologically different individuals. He maintained that the alternation of sexual and asexual reproductive modes in some species did not even produce dimorphism (all proloculus sizes were the same) much less trimorphism. Adams (1957), after studying various representatives of the Lagenidae from the Lias, concluded that no trimorphism had been observed although dimorphism was evident. Among the species studied by Adams were several recently evolved species (of Triassic origin) and these are the very forms which Hofker claimed should best illustrate the phenomenon of trimorphism.

In summary, we see that there is not agreement about the concept of trimorphism. The affirmation of Hofker that the recently evolved species are apogamic does not correspond with the observed data. Whether or not some species do reproduce in such a way to form 3 distinctly different morphological populations is not wholly clear. Ghose (1974) described a trimorphic population of *Pellatispira glabra* in which the megalospheric forms were dimorphic. He concluded that this megalospheric dimorphism was due to a biannual reproductive cycle during which environmental conditions changed and the two asexually produced forms developed at different rates.

In concluding this section on the reproductive cycle it is necessary to note that foraminiferal reproduction does not occur only in the ways described above. There are species of Astrorhizidae which can reproduce by regeneration from their own fragments if the fragments contain a nucleus. This mode of reproduction is probably accidental. Pokorny (1958) describes this type of reproduction in *Nodosaria scalaris.*

Christiansen (1974 Ms) observed some very interesting behaviour in an agglutinated tubular foraminifer. The organism begins to grow as an elongate tube sealed at one extremity. At a particular stage of development, it breaks in half leaving two tubes, one of which is open on both ends, contains protoplasm and continues to live, and the other of which is closed at one end and is void of protoplasm. These two forms, under the existing artificial criteria of taxonomic classification, would certainly be placed in different genera.

The ideas of Buchner (1940) about the reproduction of *Oolina* are also interesting. He noted that a new individual grows near the oral extremity of the parent, adhering to the parent by its aboral extremity. Later the new specimen frees itself from the parent and leaves a small scar on the parental test. Occasionally the new individual has a connection with the parent via a small short tube. These tubes are frequently found on the aboral portion of *Oolina* tests. Joined specimens of *Oolina* are even found occasionally. As all these conclusions are based on studies of dead tests they are speculative.

C. Death

There are four principal modes of death for foraminifera. The first of these is a result of reproduction. Although some species may yield more than one generation of offspring the most common situation is parental death immediately after and caused by the reproductive process. To form a new generation the parent expends not only its protoplasm but may utilize some of its test. This is especially true in asexual reproduction where the young leave the parental test with their first few chambers already formed. The parent must sacrifice part of its test material to the young (Arnold, 1967b). Species of *Patellina, Discorbis* and *Cymbalopora* generally dissolve the septa and ventral walls of the larger chambers to provide this material. The walls of *Bulimina, Cornuspira* and the Miliolidae become thinner during sexual reproduction. *Elphidium* dissolves part of the apertural face with a consequent enlargement of the aperture.

The second mode of death is capture by some predator. Remains of foraminiferal tests are commonly encountered in the alimentary tracts of annelids, holothurians, crustaceans, echinoderms and fishes. Upon passing into the digestive tracts of these animals the foraminifer is usually killed. Nevertheless, Walker (1971) observed examples of *Rosalina floridana* which were consumed by the gastropod *Littorina littorea* but were capable of resuming normal activity upon being removed from the gastropod gut or fecal pellets. Similar observations were made by Mateu (1969a) on foraminifera found in an echinoderm. These specimens were apparently able to seal their apertures with detrital material, thus protecting themselves from the abrupt chemical and physical changes caused by ingestion.

Third, foraminifera die by being buried in sediments agitated by waves and currents. The foraminiferal organism is capable of vertical migration through loose sediment but in many cases dies before reaching the surface or is crushed by the shifting sediments. Myers (1943c) described situations in Plymouth, England in which up to 80% of the benthonic foraminifera are buried during large storms. These specimens die within a few days or weeks and their protoplasm is decomposed by bacteria. Occasionally nodules of pyrite begin to form within the tests as a result of decay. The anaerobic bacteria reduce sulphate to obtain sulphur and produce hydrogen sulphide which in turn generates pyrite at the expense of ferric salts present in the environment. Frankel (1964) offered a supplementary explanation for pyrite formation. He encountered pyrite filled specimens near Durban, South Africa and concluded that the pyrite was formed in anaerobic reducing conditions characterized by low salinity waters and a kaolinite-illite rich substrate containing finely disseminated carbon particles.

However, the discovery of pyrite in a foraminiferal test is not certain evidence of death by burial in anaerobic conditions; Seiglie (1973) described living specimens of *Ammonia tepida*, *Fursenkoina pontoni*,[4] *Florilus grateloupi*, *Cribroelphidium poeyanum*, *Discorbis*(?) *bulbosa* and *Quinqueloculina candeiana* which also contained pyrite in their tests. Seiglie noted that the pyritization is due to anaerobic bacterial activity occurring within the living organism. Le Calvez (1951) first mentioned the presence of iron sulphides in the protoplasm of foraminifera but attributed the material to unusual metabolic activity rather than to the presence of bacteria within the test.

Finally, it is possible that foraminifera can die from disease. Seiglie (1973) interpreted the pyritization by bacterial activity inside foraminiferal tests described above as a disease. We can suppose that other kind of diseases can also take place in foraminiferal specimens.

3. Seasonal occurrence

Both quantitative and qualitative changes in the foraminiferal fauna occur during different seasons of the year. These are usually explained by seasonally controlled

[4] As an example of the confusion that can result from the transliteration of Cyrillic characters as mentioned in the introduction to the first edition, we cite the case of the genus *Fursenkoina* which was introduced by Loeblich & Tappan in 1961. It was named after A. V. Furssenko but the transliteration system used by these authors was one which resulted in a single ‘s’ in the generic name. However, when Furssenko writes his name in the Roman alphabet he used a double ‘s’. The original designation of the genus must stand, however, although it does not correspond to the exact name of the person honoured.

physio-chemical factors and/or relationships with other organisms which exhibit seasonal control.

Fluctuation in planktonic populations is especially illustrative of seasonal changes. Numerous studies in mid-latitude waters have shown that phytoplankton exhibit two abundance peaks; in the spring and in the autumn, and that the zooplankton peaks lag behind the phytoplankton ones.

With respect to the benthonic foraminifera, Myers (1935a) concluded that reproduction in the sea is limited almost exclusively to the warm months of summer. The same author later noted that *Elphidium crispum* only reproduces during a few weeks in March.

Walton (1955) determined that the maximum development of foraminifera in a Baja California, Mexico bay occurred in June and August. The June peak was related to the phytoplankton bloom and the August peak was controlled by the occurrence of optimum water temperature.

The study of a different California gulf by Reiter (1959) indicated that the maximum foraminiferal abundance occurred in September, October and November with a gradual decrease in abundance from December–March.

A study of the marshes of Poponesset Bay, Massachusetts by Parker & Athearn (1959) revealed that the living population was greatest in June and least in December. The seasonal fluctuations of *Quinqueloculina poyeana* in some Texas bays was clearly described by Phleger & Lankford (1957). Lutze (1968a) observed the seasonal changes of abundance of *Cribrononion articulatum* and *Miliammina fusca* off the German coast. Haake (1967) found that *Cribrononion excavatum* reached its greatest abundance during the spring and summer months in Kiel Bay, Germany. Murray (1968a) noted that *Quinqueloculina dimidiata* and *Reophax moniliformis* are much more abundant during the summer and fall in Christchurch Harbour, England than in winter when these species are in 'dormancy'. Wright (1968) showed that the living foraminiferal population in the Río Quequén Grande, Argentina reached its greatest abundance in summer. Studies of *Elphidium excavatum* in North Wales (Haman, 1969) revealed three abundance peaks, May–June, October and January. This latter high, in full winter, is the least predominant and was considered the result of sexual reproduction. Ingmanson & Ross (1969) noted peak densities in Seahorse Key, Florida in mid-winter and speculated that mid-day summer water temperatures of 32 °C may have been near lethal for some of the foraminifera. Ramanathan (1969, 1970) found peak abundances in April in the Vellar Estuary, India. Daniels (1970a,b) found that the agglutinated:calcareous specimen ratio in the Limski Canal, Yugoslavia increased in the winter months, perhaps due to lower temperatures and less available calcium carbonate. Ross (1972) noted that *Marginopora vertebralis* reaches its peak reproductive activity in the late spring.

The observations cited above were generally based on monthly examination of an area for one year. Weekly collection from Puerto Deseado, Argentina for a period of several years (Boltovskoy, 1964b, 1965d; Boltovskoy & Lena, 1969b) provided a vast wealth of detailed data about seasonal phenomena. The greatest abundance of

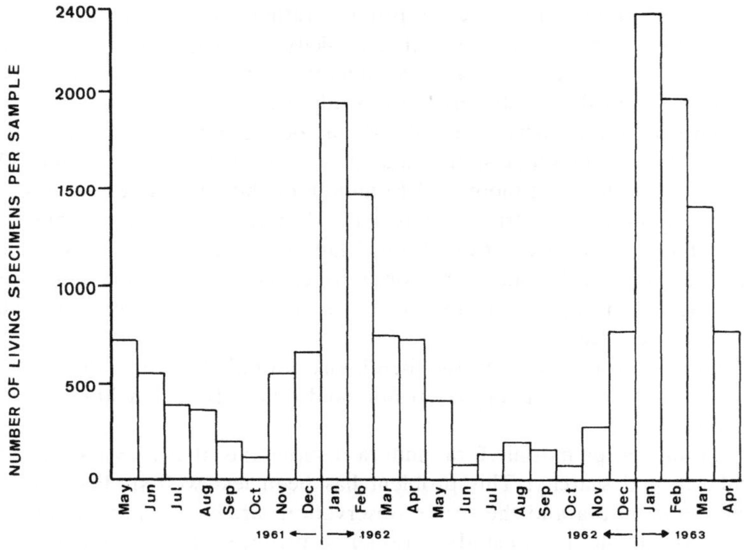

Fig. 18. Seasonal variation in the living benthonic foraminiferal population, Puerto Deseado, Argentina (after Boltovskoy, 1965d).

the total living population occurred in mid-summer (Fig. 18). When individual species were examined 4 different groups were found: (a) those with mid-summer abundance peaks, (b) those with various peaks throughout the year, (c) those which did not exhibit any seasonal fluctuation, and (d) those which unexpectedly demonstrated their main bloom in winter. To this latter group belong *Buliminella elegantissima, Elphidium articulatum, Epistominella exigua* and *Rotalia beccarii.* These 4 species behaved this way only in 1964–1966 and it is not known if this is typical or atypical behaviour.

Forti & Roettger (1967) and Closs & Madeira (1968) studied the seasonal foraminiferal abundances in the brackish waters of the Patos Lagoon, Brazil. They had unusually large and frequent collections and noted that reproduction and maximum total populations occurred at different times for different species. When the entire fauna was considered, the greatest reproduction occurred in the fall and winter, thus verifying the observations of Boltovskoy & Lena that some species have their main bloom in the coldest season.

Unfortunately, the overwhelming majority of species encountered in the investigations of seasonal phenomena occur in small quantities and it is therefore difficult to draw reliable conclusions about the presence or absence or the character of their periodicity.

Christiansen (1958) claimed to have observed no seasonal phenomena at all among the benthonic foraminifera of the Oslo Fjord. If one were dealing with species which reproduced very frequently the absence of peaks would be understandable.

However, as benthonic foraminifera reproduce rather 'slowly' (Table 2) they should yield evidence of some sort of abundance cyclicity. The explanation for the absence of such peaks in Christiansen's data is difficult to explain but may lie in a better understanding of all the conditions in the study area.

The great variation in these observations may be explained by the strong influence that the local environmental setting and even microhabitat has on reproduction. Waldron (1963) noted that blooms of foraminifera shift from one locale to another following fluctuations in nutrient availability. Detailed long-term examination of a Long Island salt marsh (Lee *et al.*, 1969; Matera & Lee, 1972) showed that many foraminifera probably bloom in response to changing environmental conditions. Different species will bloom at different times and the same species may bloom at different times in different years.

The seasonal occurrences of the benthonic foraminifera are still very poorly understood and represent a very interesting and relatively easy field of investigation for researchers.

Investigations on planktonic foraminifera demonstrate that they too have periods of abundance and scarcity. This problem has been less well studied than for the benthonic foraminifera and the scarce observations are sometimes contradictory.

Bé (1960a) was the first to study seasonal occurrences of planktonic foraminiferal assemblages. He found that March and October were the months of maximum development and December the minimum in the Sargasso Sea.

Tsi-chung and Sau-yee (1964) conducted observations in the northern South China Sea. They noted that in some areas (Kwang-tung coast) the peak abundances occur in the winter and spring whereas the reverse is true in other areas (eastern and southern Hainan coast) where the maximum population occurs in the summer and autumn.

The observations of succeeding authors were concerned mainly with seasonal occurrences of particular species (see Table 3).

Table 3. Seasonal occurrences of some planktonic foraminifera

Species	Area	Peak Abundance	Author
Globigerina bulloides	Antarctic	spring and summer	Bé, 1969a; Bé & Tolderlund, 1971
	Subtropics	late winter	Tolderlund & Bé, 1971
G. pachyderma	Antarctic	spring, summer and early fall	Bé, 1969a
	Beaufort Sea	summer	Vilks, 1973 Ms; Bé (in press)
G. pachyderma incompta	North Atlantic	summer	Cifelli, 1965
	North Atlantic	summer and fall	Cifelli & Smith, 1969
G. quinqueloba egelida	North Atlantic	winter	Cifelli, 1965; Cifelli & Smith, 1969

Globoquadrina dutertrei	North Atlantic	fall	Cifelli, 1965
Globigerinella aequilateralis	North Atlantic	spring	Cifelli, 1965; Cifelli & Smith, 1969
Globigerinita glutinata	North Atlantic	winter	Cifelli, 1965
G. uvula	South Atlantic, South Pacific	fall fall	Boltovskoy, 1966d
Globigerinoides conglobatus	North Atlantic	fall	Bé & Hamlin, 1967
G. ruber	North Atlantic	summer	Cifelli, 1965; Cifelli & Smith, 1969
G. trilobus trilobus	North Atlantic	fall	Cifelli, 1965
Globorotalia hirsuta	North Atlantic	winter	Cifelli, 1965; Cifelli & Smith, 1969
G. menardii	North Atlantic	summer	Cifelli, 1965
G. inflata	Antarctic North Atlantic North Atlantic	early August–early January fall summer	Bé, 1969a Cifelli, 1965 Bé & Hamlin, 1967
G. punctulata	North Atlantic	summer	Cifelli, 1965
G. truncatulinoides	Antarctic North Atlantic Equatorial and North Pacific	early May–late October winter winter (cooler season)	Bé, 1969a Cifelli & Smith, 1969 Banerji, Schafer & Vine, 1971
G. scitula	Equatorial and North Pacific	winter (cooler season)	Banerji, Schafer & Vine, 1971
Hastigerina pelagica	North Atlantic	summer	Cifelli, 1965
Orbulina universa	North Atlantic	summer	Cifelli, 1965
Pulleniatina obliquiloculata	North Atlantic	winter	Cifelli, 1965; Bé & Tolderlund, 1971

Bé (in press) emphasized that seasonal fluctuations in abundance of planktonic foraminifera follow the phytoplankton development. The phytoplankton bloom takes place in winter months in the tropical seas, and in spring and fall in subtropical and temperate zones. The spring maximum is progressively delayed towards the higher latitudes until in the Arctic Ocean only one phytoplankton bloom takes place in August. Consequently, in the temperate zone the maximum abundance of planktonic foraminifera takes place between December and March and in the subarctic latitudes the maximum concentration is reached between May and October. Bé also stated that the seasonal contrast in abundance increases from low to high latitudes.

As we can see, the existing observations are still very incomplete and contradictory. Nevertheless, even this preliminary and fragmentary data prove the existence of seasonal variation in the whole planktonic foraminiferal fauna and in individual species. It is important to emphasize that different species may have their greatest occurrence in different seasons and that the same species may have its peak development at different seasons in different areas.

In summary, we conclude that the foraminifera, like the majority of marine organisms, are subject to seasonal fluctuations. One of the tasks awaiting future investigators is the detailed study of this phenomenon.

4. Biological cycles

Although cyclical phenomena in the biosphere are well known, their causes and satisfactory descriptions of their occurrences are not. Cycles of different duration have been described. Ottestad (1942) described periodicities of 11, 17, 23 and 57 years in the organic world. Various authors have agreed with these cycles and others have suggested cycles of 33 and 90 years. There are investigators who deny the existence of any cyclicity. The debate over this phenomenon is long standing and has produced a large volume of literature. An interesting general synthesis on the topic was produced by Klinckowstroem (1952).

The phenomenon of cyclicity has been little studied by those interested in foraminifera. This is unfortunate as the great interest in foraminifera and the extensive data on them indicate that foraminiferal studies could help solve this problem to some extent. There are only two foraminiferal studies which treat this phenomenon.

Riveroll & Jones (1954) investigating argillaceous rocks of the Puente Formation (Upper Miocene), California discovered a varved sequence containing foraminiferal tests. They determined that each varve represented one year's deposition and that the foraminifera in the layers were subjected to a 10 year cycle of abundance. As the purpose of this investigation was to examine the stratigraphic occurrence of the foraminifera, the observation of their periodicity was only made in passing.

Using data from the literature on the occurrence of *Buccella frigida* from the Argentine continental shelf as well as his own data, Boltovskoy (1956c) suggested the existence of a rhythmic cycle of occurrence of this species. He proposed that the phenomenon of periodicity may play some role for all foraminifera and noted that the existence of a rhythmic cycle may help explain many of the contradictory data published on the distribution and abundance of different species at different locations.

5. Parasites, epibionts and predators

The first data on the presence of other organisms in foraminifera were presented by Carpenter (1876) who described the invading organisms as gregarinids. Somewhat later Rhumbler (1893) and Schaudinn (1896) found different organisms in *Saccammina*. Rhumbler (1911) made detailed comments on these findings and also decided that some of the foreign bodies were gregarinids. He suggested that the arrangement between Foraminifera and gregarinids was a case intermediate between true parasitism and the simple penetration of foraminiferal tests that were dying or already dead.

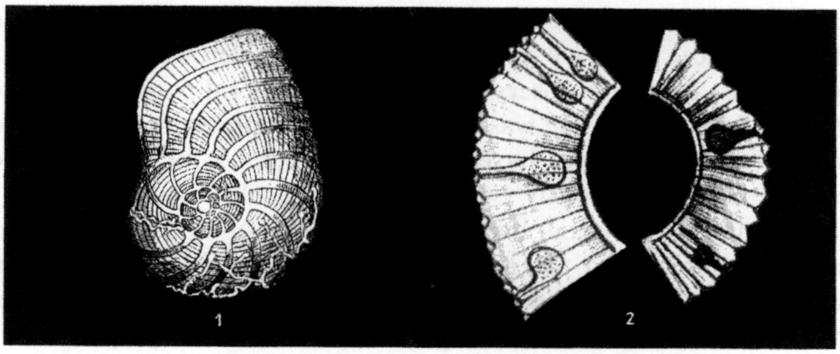

Fig. 19. Foraminiferal parasites (after Rhumbler, 1911); 1. *Peneroplis* with canals excavated by *Thalamophaga;* 2. Wall of *Globigerina pachyderma* with *Marsupophaga* as a parasite.

He noted that this 'partial parasitism' is not limited to unilocular forms (*Saccammina*, etc.) but is also observed in the multilocular tests (*Rotalia*, etc.). The penetrating organisms belonged to different groups, e.g. nematodes, metazoan eggs and other foraminifera.

Apart from this, Rhumbler described other phenomena which he attributed to parasitic activity. The parasitic organisms penetrate the calcarous test walls of foraminifera and there construct a series of fine canals utilizing the organic material found there. One of these parasites, *Thalamophaga ramosa*, is frequently found in empty tests of *Globorotalia* and *Globigerina* but never in living tests. Rhumbler also noted that another parasite, which he named *T. incerta*, prospered in the tests of *Peneroplis* (Fig. 19).

Moreover, Rhumbler described structures which he interpreted as the work of parasites. The structures were rounded holes which perforated the walls of *Globigerina* and *Orbulina* as well as the lateral keel of *Globorotalia menardii*. Occasionally these perforations would not reach the inner test wall and would terminate in a saclike pouch. This phenomenon was observed in *Cristellaria* and *Globigerina pachyderma* and was attributed to a parasite called *Marsupophaga* by Rhumbler (Fig. 19). The construction and character of *Thalamophaga*, *Marsupophaga* and other 'imaginary genera' described by Rhumbler are not described and their taxonomic position is completely unknown.

Later investigators have ascribed various anomalies in fossil foraminiferal tests to the action of parasites or predators. Some authors said that the sponges *Cliona* and *Vioa* were responsible for such kinds of activities.

Derville (1935) described the results of the activity of algal parasites which perforated tests of *Nubecularia*. He suggested that the chemical destruction of the tests was for the purpose of oxygen liberation.

Livan (1937) showed some interesting illustrations of fossil foraminifera with wall perforations presumably made by parasites. In some cases the distribution of these perforations on the tests was random but in some tests the holes were in geometric patterns. On one *Lenticulina* the holes were only on one side and were concentrated along the sutural margin between two chambers. Livan suggested that the holes were produced by chemical corrosive activity of young carniverous gastropods.

Among the most extensive observations on foraminiferal parasites are those of Le Calvez (1939, 1940, 1947a). He was able to study the phenomenon of parasitism on living foraminifera and did not have to deduce the cause of various structures on empty tests as did the previously cited investigators. He noted that spirally wound nematodes, apparently parasitic, live in the cytoplasm of *Rotorbinella turbinata* and *Iridia lucida*. He also noted the amoeba *Vahlkamphia discorbini* as a parasite in *Discorbis mediterranensis*. In addition, he observed various other protozoans of uncertain taxonomic status occupying positions in the cytoplasm of various foraminifera.

Le Calvez was also able to demonstrate experimentally the phenomenon of parasitism between two foraminifera. He observed *Fissurina marginata* behaving as an ectoparasite on the host *Discorbis vilardeboanus*. The parasitic species would rake its pseudopodia over the surface of the extoplasm of the host, trapping the nutritive particles circulating there. It would then divert these particles to its own aperture.

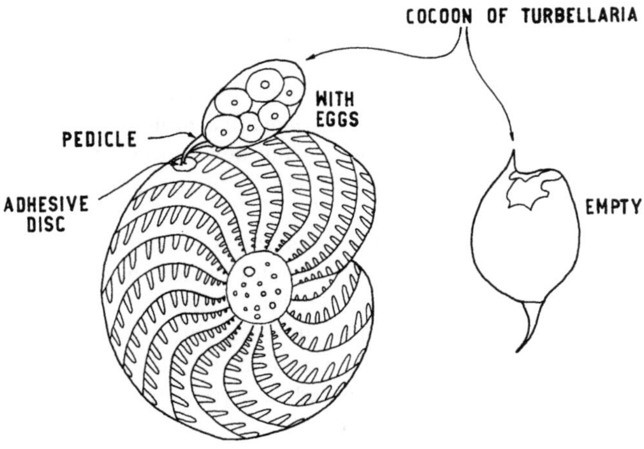

Fig. 20. Turbelarian cocoon growing on *Elphidium* (after Jepps, 1942).

During reproduction the parasite would abandon its host and give birth, asexually, to 2–6 young, each of which would attach itself to the host. It was evident that the parasitic species was adapted very intimately to this mode of life and would die without the presence of the host species in the culture. Le Calvez suggested that this mode of life explained why so few *Fissurina marginata* specimens were found in the fossil record.

Myers (1943c) observed cases of parasitism in foraminifera. He found nematodes as parasites in several different species. He also noted that between 3–5% of the sublittoral *Elphidium crispum* population was infested by a coccid parasite, *Trophosphaera planorbulinae.*

Todd (1965) described *Rosalina carnivora*, a foraminifer which lives on the pelecypod *Lima angolensis*. She suggested that the foraminifer bores into the pelecypod valve to obtain calcium carbonate for its test construction. As only the foraminifer benefits from this relationship but neither of the participants suffer, Todd refers to *R. carnivora* as a commensal parasite.

Adshead (1967) described a fungal parasite on the pseudopodia of planktonic foraminifera. She suggested that the fungus must be a well adapted parasite because of the slow rate at which it depleted the host.

Banner (1971) described an assemblage of *Alveolinella quoii* which was attacked by a foraminiferal parasite, *Planorbulinopsis parasita*. He emphasized that this was the first observation of endoparasitism by a polythalamous rotaline foraminifer. The parasite bores into its host along the intercameral sutures or at the suture at the base of the apertural face. Banner said that the parasite may remove calcium carbonate, nutritive material, symbionts, protoplasm, biochemical components and trace elements from the host. Because Banner studied only empty tests, Le Calvez (1972) criticized the conclusions he drew. She suggested that *P. parasita* should be considered as a 'renter' or 'squatter' rather than an endoparasite.

There have been no studies of the seasonal nature of parasitism but one might suppose that its incidence may be greater during winter months when less food is available in the environment.

Beyond the realm of parasitism but nevertheless a case of interelationship between organisms is the relationship between turbelarian worm cocoons and the foraminifera to which they are often found attached by means of a pedicle. This phenomenon was observed for the first time by Ehrenberg (1839) on the surface of *Elphidium* and *Nonion*. He considered the cocoons to be foraminiferal eggs and described them as a product of foraminiferal reproduction. Schultze (1854) described a similar structure on *Elphidium* as the cocoon of the ciliate *Cothurnia*. Jepps (1942) not only was able to collect numerous specimens of *Elphidium crispum* in Plymouth Harbour, England with attached cocoons but was able to cultivate them, thereby proving that the cocoons were inhabited by turbelarians (Fig. 20). She claimed that the turbelarian cocoons are not very intimately related to the foraminiferal tests because they are able to survive and develop even if separated from the test.

Boltovskoy (1963b) found numerous examples of *Quinqueloculina seminulum* in

Puerto Deseado, Argentina with the same attached cocoons. Some of these had a large pedicle whereas others were apparently attached directly to the test surface. In this latter case the test was etched, sometimes to a great extent, at the site of attachment (Fig. 21). A detailed study of this material with thin sections made it clear that the turbelarian cocoons were not parasites but were epibionts. The turbelaria use the test walls as a substrate for the attachment of the cocoon containing the young. If the test is an empty one, the cocoons remain until knocked

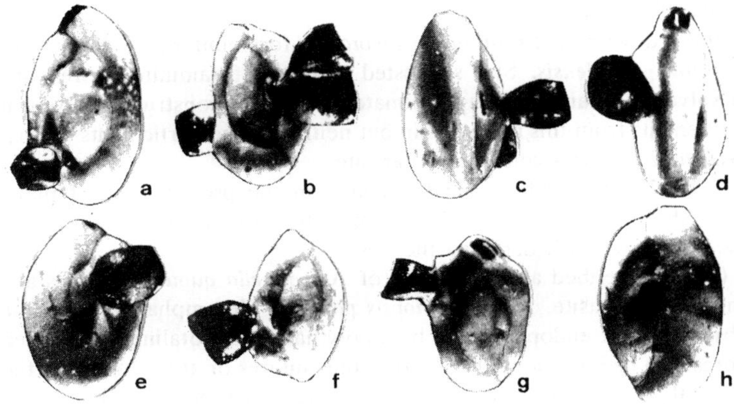

Fig. 21. Quinqueloculina seminulum from Puerto Deseado, Argentina with *Turbelaria* cocoons; a. pedicle and circular support visible; b. 3 cocoons on one test, on the right side the cocoon support is partially covered by material of the test wall; c. pedicle completely covered by test material; only the edge of the circular support is visible; d.,e. pedicle and support completely covered; f.,g. the foraminifera has secreted excess material around the cocoon; h. scar on the wall marking a previous point of cocoon attachment.

off by the movements of the test as it is swept across the substrate by currents and wave action. If the test is alive, it will try to envelop the cocoon when a new chamber is added. This is usually impossible to accomplish with only one new chamber, although the new chamber may envelop the pedicle and the small circular adhering plate (Fig. 20). This partial envelopment gives the cocoon the appearance of lacking a pedicle. The new chamber is slightly deformed by the pedicle. When the living organism tries to completely cover the cocoon with a new chamber, the deformation is usually severe. The calcareous material aquires a very irregular topography in the vicinity of the cocoon. If the cocoon falls off the test, a great depressional scar remains in the side of the test. The cocoons usually fall off only if the pedicle is subjected to drying. Under water, in its natural habitat, the pedicle is quite tough.

In the case of cocoons attached to *Elphidium* and *Nonion*, no observations have been made on the effect of chamber addition. This is because for these species an entire whorl of new chambers must be added prior to noting any effect of the cocoon whereas in *Quinqueloculina* the next one or two chambers will amost certainly react with the cocoon.

Although Boltovskoy notes that the presence of these cocoons on foraminiferal tests is an example of epibiosis, not parasitism, their presence may, nevertheless, cause some inconvenience to the foraminifer. Since more than one cocoon may attach itself to a test, there may be a restriction of test mobility. Their presence may also cause a greater expenditure of energy and material during test secretion when the foraminifer is forced to try to cover the cocoon or cocoons.

Le Campion (1970) encountered ovoidal clear chitinous turbelarian cocoons on living foraminifera in the Arcachon estuary, France. He interprets them as true parasites rather than epibionts.

Algal epiphytes belonging to the chlorophytes have also been observed as epibionts on *Elphidium* by Jepps (1942) and one of us (EB). In both cases, the foraminifera, although alive, were inactive.

The predators of foraminifera are very numerous. The remains of foraminifera are found in the guts of many benthonic metazoans but it is difficult to tell if they are selectively chosen or randomly ingested. They are found in almost every group of invertebrates as well as in fish.

We previously mentioned the holes often found excavated in foraminiferal tests. These holes may be the work of predators as well as of parasites. In order to consume the protoplasm of a foraminifer, a predator must pass through the aperture or create a new orifice. As most species have either a very small aperture or have some structure (tooth, cribbing, tegillae, flap) which impedes entry (see Chapter III,8), it may be easier for the predator to bore a new opening.

It is difficult to specify all those organisms which feed on foraminifera. Undoubtedly many gastropods do, especially in shallow tropical waters. Sponges and nematode worms are also responsible for a good many observed borings.

Livan (1937), Pozaryska (1957), Vyalov & Kantolinskaja (1968) and Collen (1973) described the effects of predation on fossil foraminifera.

Table 4 gives a list of some of the observations on animals which feed on Recent benthonic foraminifera.

The quantity of natural predators of foraminifera in a given locale will have an effect on the foraminiferal population in that area. Said (1951b) established a negative correlation between molluscan fecal pellets and foraminifera in Narragansett Bay, Massachusetts. This may be due to the reduction of the foraminiferal population by the ingestion of bottom sediments by the browsing molluscan population.

Christiansen (1958) suggests that the presence of many organisms (molluscs, polychaetes, crustaceans) which consume foraminifera in the Oslo Fjord is one of the causes for the low number of foraminifera found there.

The observations of Saidova (1958) with respect to predation of benthonic foraminifera at great depth are of special interest. She found that the alimentary tract

Table 4. Predators of benthonic foraminifera

Author	Foraminiferal species	Predator
Myers, 1940	not listed	polychaetes
Myers, 1943c	not listed	crustaceans, annelida
Perkins, 1958	not listed	*Nereis diversicolour* (polychaete)
Reyment, 1966	*Spiroplectammina* sp., *Quinqueloculina* sp.	gastropods
Green, 1968	not listed	*Gobius* (teleost)
Mateu, 1968a	many	holothurians
Mateu, 1968b	many	*Pinna noblis* (pelecypod)
Mateu, 1969b	many	*Spatangus purpureus* (echinoderm)
Saidova, 1968	many	echinoderms
Lutze, Grabert & Seibold, 1971	*Heterostegina depressa*	echinoida
Sliter, 1971	*Rosalina globularis*, *Bolivina doniezi*	nematodes
Walker, 1971	*Rosalina floridana*	*Littorina littorea* (gastropod)
Ross, 1972	*Marginopora vertebralis*	holothurians

of many echinoids and holothurians collected at depth between 3000 and 9300 m contained numerous specimens of benthonic and planktonic foraminifera. They were in an excellent state of preservation, although in the bottom sediments of the same sites no planktonic forms were found. The most interesting observation was that many of tests contained protoplasm. However, if the foraminifer was in the last section of the intestine, it contained no protoplasm. According to Saidova, this signifies that gastric juices of benthonic predators do not dissolve calcareous tests but do allow the digestion of protoplasm. Saidova concluded that benthonic foraminifera should represent an important component of food for many deep-water dwellers.

Unfortunately, there are not more specific data about the role of foraminifera in the nutritive patterns of other organisms. Therefore this is a ripe field for investigation.

6. Symbiosis and commensualism

There are groups of unicellular aquatic algae which have been known for some time as symbionts in various protozoans. They are generally spherical with a diameter of 3–10 μ. Based on the colour of their chromatophores they are called zoochlorellae (green) or zooxanthellae (yellow or brown). The former are usually found in fresh water protozoans and are represented primarily by *Chlorella*. The zooxanthellae for the most part are dinoflagellates, although some may be diatoms (Dietz-Elbrächer, 1971).

For some time the presence of these algal symbionts has been known in the radiolaria of the euphotic zone where photosynthesis occurs. Margalef (1961) estimated that they may comprise up to 10% of the biomass in the acantharid radiolaria.

Their presence in foraminifera (*Rotalia*) was first observed by Schultze (1856) who erroneously interpreted them as protoplasmic inclusions. Later a number of investigators, among them Stuart, Bütschli, Murray and Schaudinn, observed and correctly identified them as symbionts.

The first detailed study of these organisms in foraminifera was conducted by Winter (1907). He observed cultures of *Peneroplis* that were infested by zooxanthellae and found that the large tests contained up to 100 000 symbionts.

Rhumbler (1911) noted the presence of these organisms in planktonic foraminifera. He noted that the zooxanthellae found in *Globigerina* and *Orbulina* were 8.9–19.4 μ long and possessed a fusiform corpuscular shape. He described the presence of various other types of symbionts in planktonic foraminifera.

Myers (1943a) observed that symbiotic algae are confined to foraminifera with transparent walls with large apertures (*Peneroplis, Sorites, Marginopora*). *Quinqueloculina* does not possess them because the thickness of its walls does not permit transmission of sufficient light. If this species ingests symbiotic algae it will expell them with other non-digested particles.

Boltovskoy (1963a) described symbiotic algae in the following littoral foraminifera from the Patagonian coast: *Buccella frigida, Elphidium macellum, E. gunteri, E. articulatum, Buliminella elegantissima* and *Nonion depressulus*. All these species possess transparent or translucent walls and have a small aperture. None of the numerous examples of *Quinqueloculina* has symbionts. This observation confirmed the deduction of Myers that the symbiotic algae lived only in tests with hyaline walls. However, his restriction that only large apertured species possess symbionts was not consistent with Boltovskoy's observations. Boltovskoy said that the apertural size should not play any significant role because most algae are small enough to easily pass through most foraminiferal apertures, especially in their autospore stage.

Haynes (1965) suggested that the transparent walls served as a greenhouse for the algae, thereby enhancing their growth.

Symbionts have been observed in planktonic foraminifera as well as benthonic and may even occupy a significant fraction of each chamber (Lee *et al.*, 1965). So far, zooxanthellae have been observed in 18 benthonic and 11 planktonic species.

The coexistence of the foraminifer and its symbiont is advantageous to each. The symbiont supplies oxygen to the foraminifer and consumes the carbon dioxide expelled by the foraminifer. Other metabolic relationships have also been observed; e.g. the zooxanthellae may remove excess nitrogenous wastes produced by the foraminifer.

Lee & Zucker (1969) found that the maximum uptake of food and of calcium in

Archaias angulatus occurred when the symbiont zooxanthellae were present and receiving high light intensity. These investigators also confirmed the observations of Myers and Boltovskoy that the thick, opaque test of *Quinqueloculina* does not support zooxanthellae. Ross (1972) noted that the dinoflagellate symbiont of *Marginopora vertebralis* aids in removing waste material from this large species.

Alldredge & Jones (1973) described 3 dinoflagellate species which attach themselves to the surface of the planktonic foraminifer *Hastigerina pelagica*. The authors suggested that the foraminifer may provide the dinoflagellates with (a) mobility during the vegetative stage, (b) increased buoyancy, (c) a nutrient source from its waste, (d) protection and (e) a means of dispersal. The potential advantages to the foraminifera were less clear but may consist of utilization of waste products, including oxygen.

The importance of symbiotic algae was demonstrated by Pringsheim (1946) who showed that the infusoria *Paramaecium* could prosper and multiply in an algal solution without any other nutrient. Observations on other invertebrates support the idea that the symbiont can provide the entire food source to the animal. Numerous observations indicate that zooxanthellae can supply the entire oxygen needs to some shallow water corals even under low light intensities (Drew, 1973). Radiolaria have been known to survive by consuming their symbiotic algae under times of nutritive stress. Two observations have been made of this sort on foraminifera. Dietz-Elbrächter (1971) and Röttger (1972b) both observed that *Heterostegina depressa* could grow without the uptake of nutritive particles simply by digesting the zooxanthellae. Both investigators noted the dependence of the specimens on a light source which provides radiant energy for the algal photosynthesis. Röttger observed that the algal symbionts are transmitted to the offspring during foraminiferal reproduction.

Bé (in press) observed movement of the symbionts into and out of the test of *Globigerinoides conglobatus* as well as within the bubble like capsule of *Hastigerina pelagica*. In the latter species they moved freely along the spines and the bubbles obviously did not hinder them. Another observation of Bé is that those planktonic species which prefer the uppermost water layer usually contain symbionts, but species which occur at greater depths (below the euphotic zone) are without them.

The importance of the symbionts should not be overemphasized. Although their presence is an advantage to the foraminifer, it is not an absolute requirement. In any foraminiferal population there are specimens of the same species with and without symbionts. However, those with the symbionts are often in a better state of health than those without.

Because of their need for light energy, symbiotic algae show a preference for foraminifera which have transparent walls and live in shallow water environments. Other than this selectivity, it is not clear if there is any preference for some foraminiferal species rather than others. Because the taxonomy of the symbiotic algae is confused, it is also not clear if particular algae species and particular foraminiferal species have affinities for each other. It seems reasonable that such preferences should exist.

It should be also mentioned that some researchers do not share the point of view that unicellular algae found in planktonic foraminifera are symbionts. Hansen (in press), for example, found different ingested algae in all stages of decomposition and the degree of decomposition was greater in the chambers than in the umbilical cytoplasm where it in turn was greater than in the surface cytoplasm. On the basis of these observations Hansen deduced that a symbiotic relationship between foraminifera and algae is not likely.

Commensalism between foraminifera and other animals has been studied only slightly. Aside from the case described by Todd (1965) of 'commensual parasitism' cited in section 5 of this chapter, the following cases may be considered as typical examples of commensalism. Bock & Moore (1968) found that many specimens of the small pelecypod *Notocorbula operculata* had a foraminifer, *Hanzawaia strattoni*, attached to it. The attachment point on the foraminifer was always on the ventral surface but was randomly distributed on this side. The attachment is accomplished by a weak cement. The advantage gained by the foraminifer in this case is that the mollusc's continual search for an optimum position on the substrate, free of silting and offering a vantage point for food ingestion, benefits the foraminifer as well. Dobson & Haynes (1973) described foraminifera living on hydroids beneath the photic zone. The hydroids provide both shelter and an attachment point.

7. Foraminifera and algae

In many foraminiferal investigations observations are made on the relationships between foraminifera and algae. The algae are commonly referred to as seaweed or seagrass (Cushman, 1921, 1930, 1941b; Schmidt, 1953; Christiansen, 1958; Mateu, 1965). The calcareous alga, *Corallina officinalis*, and the 'seagrass', *Thalassia testudinum*, are often cited as providing an attachment point for foraminiferal tests. Some investigators list not only the specific names of plants but give rather detailed lists of the foraminiferal species found on them (Hedley, Hurdle & Burdett, 1967; Bock, 1968, 1969; Dommasnes, 1969). There are also studies where comparison is made between the foraminiferal standing crop of different biotopes (Behm and Grekulinski, 1958; Murray, 1970a, Boltovskoy, 1971d; Wright & Hay, 1971). These studies usually note that the richest populations occur on the algae. There are a few studies in which a specific foraminiferal assemblage is ascribed to a specific type of algae (Blanc-Vernet, 1969; Lee *et al.*, 1969; Furssenko & Furssenko, 1970). There are however, discrepancies in the results of these studies. Matera & Lee (1972) demonstrated that the foraminiferal fauna that lives on the algae is not continuous with that living on the sediment below.

According to Atkinson (1969, 1971) and Boltovskoy, Lena & Asensi (Ms) it is difficult to correlate particular foraminiferal species to particular algal species, but it is quite clear that some algae support quantitatively richer foraminiferal assemblages than others. *Corallina* appears to exhibit the richest foraminiferal fauna because the morphology of this alga offers the most suitable protective environment for foraminiferal tests.

At the present time it is impossible to describe a more specific relationship between individual foraminifera and individual algal species.

8. Changes produced by human activity

It is well known that human activities can have significant effects on flora and fauna, both aquatic and terrestrial. The foraminifera have not escaped these effects. As yet the noticable effects on foraminifera have not been in the open ocean but have been confined to closed and semiclosed bays, gulfs, lagoons and other shallow water coastal zones. The changes which occur are of different origins and character. We will describe three different types of effects.

A. Introduction of a New Species

This situation is well known in marine biology with respect to other plant and animal groups. These kinds of introductions have sometimes been provoked intentionally to enrich the population of a species useful in the human economy. But the phenomenon occurs occasionally independent of human design. Examples of such transport are numerous but the details are not always well known. The 'chinese crab' was unknown in Europe until 1912 when several were carried, accidently, on merchant ships to the mouth of the Elbe River. In a short period of time this crab multiplied and disseminated not only along the North Sea and Baltic coast but also upstream into many of the rivers emptying into these water bodies.

The intentional transport of foraminifera for purposes of cultivating a new locality has never been recorded. However, the involuntary introduction of a new species has probably occurred several times, but such introductions are difficult to document. Harrington (1956) suggested that *Ammobaculites* were transported from Japan to California along with commercial oysters in the latter part of 1930. Arnal (1961) interpreted the entire foraminiferal fauna of the Salton Sea, California as a result of an accidental introduction by man. The first specimens were carried to this area when fish cultivation began to flourish or by hydroplanes during World War II. Resig (1974) suggested that the foraminiferal population of a landlocked saline lake in the Hawaiian Islands may have been accidentally introduced by the transport of coastal mud by man or birds.

B. Changes in the Fauna as a Result of Hydrologic Change

As a result of large scale man-made constructions the hydrologic regimen of some bodies of water may change drastically and cause a consequent change in the biological populations. Dihdkov'skij (1958) conducted two studies of the Azov Sea, the first in 1950 and the second in 1957. The second study revealed a number of foraminiferal species unknown in the area at the time of the earlier study. He ascribed this influx of new species to the construction in 1952 of a canal which united the Don and Volga Rivers. A large part of the Don was diverted to the Caspian Sea and there was a consequent increase in the salinity of the Azov Sea from 10.56‰ to 12.31‰. Whether or not this change in salinity was sufficient to provoke the faunal change noted is debatable. Since Dihdkov'skij was studying only dead tests his data may not have reflected the living faunas. As the coast of the Azov Sea is composed of Neogene sediments, it is possible that some of the species he encountered were reworked fossil forms.

C. Changes in the Fauna as a Result of Pollution

In areas that are industrialized and/or have high population densities, the resulting waste products may change the hydrologic conditions of nearby water bodies and consequently change the composition of the fauna and flora. The problem of polluted water and its influence on the biota has become a significant problem of world-wide proportions. The bibliography with respect to this problem is voluminous. Nevertheless, because of the complexity of the problem, our understanding is still in a state of infancy.

The study of foraminifera in contaminated water is important not only from a theoretical and academic point of view but also because of its practical significance. Knowledge of the influence of various types and degrees of pollution on various foraminiferal species can be utilized to establish certain species as pollution indicators. Species present in a certain abundance in a certain environment may be indicative of particular types of aquatic contamination.

Unfortunately, the studies of polluted waters and their effects on foraminifera are not numerous. Beginning with the work of Bandy and his students on the sewage outfalls along the Southern California coast, a number of studies have been conducted to determine the effects of pollution on foraminifera in that area (Zalesny, 1959; Resig, 1960; Watkins, 1961; Bandy, Ingle & Resig, 1964a,b, 1965a,b; Lidz, 1966b). Additional investigations have been conducted along the northeast American estuaries (Schafer & Sen Gupta, 1969; Schafer, 1970, 1973; Schafer & Cole, 1974; Buckley *et al.*, 1974). These investigations indicated that the introduction of domestic and industrial waste into shallow water environments can have marked effects on the distribution, productivity and morphology of both benthonic and planktonic foraminifera species. Immediately adjacent to sewage outfalls, especially those with high discharge rates (10^9 l./day) the fauna becomes impoverished; probably due to low oxygen and pH values. However, an aureole of high abundance and diversity generally surrounds the outfall area in which both benthonic and planktonic specimens flourish. High phosphate concentrations (up to 10 times the norm) encourage high phytoplankton production and thus a larger population of planktonic foraminifera. Bandy, Ingle & Resig (1965b) and Schafer (1970) suggest that a bit more organic matter is helpful to the foraminiferal population but a great deal more is very harmful. Calcareous benthonic forms seem to be more favourably affected by the effluent than agglutinated forms even though the low pH of the pollutants tends to remove the calcareous tests from the dead population in the substrate. The relict population thus does not reflect the living population.

There seems to be a higher proportion of deformed specimens near polluted areas. The phenomenon was noted in the California studies cited above and by Lidz (1965), Boltovskoy & Boltovskoy (1968), Wright (1968) and Seiglie (1971b).

The diversity of the benthonic population is reduced near pollution sources (Rouvillois, 1970, 1972). Certain species appear to react favourably to pollution and become dominant near the contaminated areas. *Buliminella elegantissima* is so affected in the California areas and *Quinqueloculina rhodiensis, Ammonia beccarii* and *Fursenkoina pontoni* increase in abundance in polluted Puerto Rican bays (Seiglie, 1971b). Seiglie (1968b) noted that Caribbean assemblages of *Buliminella, Bulimina, Florilus, Fursenkoina, Nonionella* and perhaps *Uvigerina* seem to be associated with organic carbon content and may be indicative of marine pollution. McCrone & Schafer (1966) cited the pollution tolerance of *Ammonia beccarii* in the Hudson estuary, New York and Schafer (1970, 1973) listed *Buccella frigida* and several species of *Elphidium* as being tolerant of contaminated waters in the Restigouche estuary, Canada. Other species appear to be intolerant of pollution: *Nonionella spp.* in California, *Fursenkoina spinicostata* in Puerto Rico and *Ammotium cassis, Ammomarginulina fluvialis, Pseudopolymorphina novangliae, Reophax fusiformis, R. arcticus, R. scotti* and *R. nodulosus* in Nova Scotia.

An increase in the *Elphidium excavatum – Cribrononion incertum* ratio seems to be related to increasing pollution in the Byfjorden, Sweden (Olsson, in press).

Matoba (1970) noted that sewage with a high sulphur content in a Japanese bay precludes the presence of living foraminifera. Clark (1971) was not able to detect any significant effect of effluent from a fish farm in Nova Scotia on the foraminiferal population.

Despite this interesting work, knowledge of the reactions of foraminifera to specific pollutants is in a very preliminary stage and deserves a great deal more attention than it has received to date. There is a need to initiate culture studies of pollution sensitive species to understand better the particular responses of these species to various pollutants.

III. THE TEST

The foraminifera differ from many other exoskeletal invertebrates in that the soft parts of the organism sometimes extend beyond the exoskeleton. For many years fossil foraminifera were the main subject of foraminiferal study, and consequently the only part of the foraminifer that was studied was the test. For a long time the study of living forms was also concentrated only on the test, even though protoplasmic material was available for study. Because of this history our knowledge of the test is much greater than that of the protoplasm.

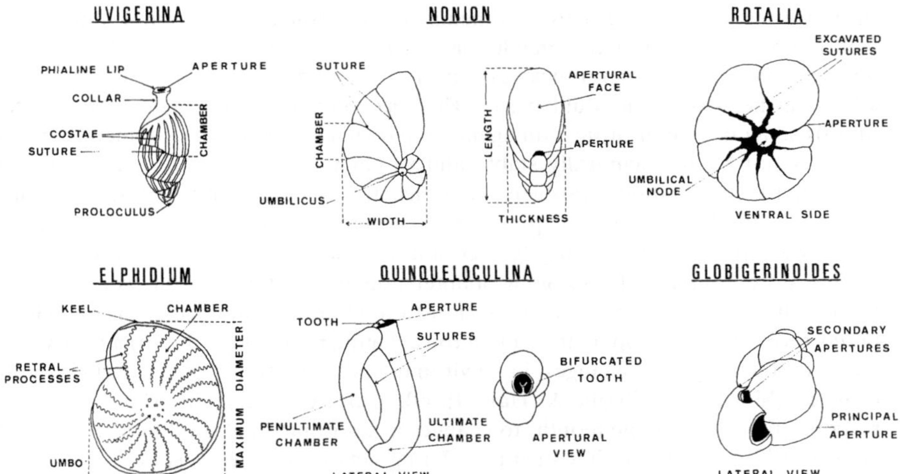

Fig. 22. Test morphology.

The test is composed of one or more chambers. When more than one chamber is present, the initial one is generally the smallest and is known as the proloculus. The chambers are separated by septa whose intersection with the test wall produces a line of contact or suture. The primary orifice through which the protoplasm enters and leaves the test is called the aperture. The basic morphological features of the test are shown in Fig. 22. The morphological characters of the test are used by specialists to distinguish one species from another. The remaining sections of this chapter will treat each morphological trait separately.

1. Form and size

The oldest and most primitive form of test was probably an oval, sac-like shape with a simple rounded aperture; the form corresponding to a simple drop of protoplasm. The protoplasm is able to secrete an organic membrane to which various bits of particulate matter can adhere. It is possible that this was the manner in which the primitive test evolved; perhaps as a means of counteracting the natural bouyancy of well-fed protoplasm. It is reasonable that the later development was such that newly acquired particulate material near the aperture gave the test an elongate form along which further extension could occur. The formation of an elongate tube-like test affords some degree of protection against rapid changes in the environment but such a test is not hydrodynamically stable and would tend to rotate in a current. One solution to this rotational problem, in which the same amount of test material is utilized, is the formation of a spiral tube. The most primitive of these spiral tubed forms were probably irregularly formed and only later became symmetrical.

The more primitive forms grew uninterruptedly. If this growth were, for any reason, interrupted, the test would be transformed into a multilocular or multichambered form. The development of numerous chambers provides the organism with a natural buffer against changes in environmental conditions, particularly osmotic changes (Marszalek, Wright & Hay, 1969). Cultured specimens show very slow response times (up to one month) to changes in environmental conditions (Röttger & Berger, 1972) indicating that such a buffering action may be at work.

The adaption of the organism to different kinds of environments and habitats led to a magnificent diversity of forms, some of which are displayed in Fig. 23.

The range of sizes of foraminiferal tests is enormous. There are Tertiary forms of great size, among them *Nummulites gizehensis* (110 mm) and *Lepidocyclina gigas* (120 mm) as well as the Cretaceous species *Loftusia elongata* (118 mm). The largest Recent species are *Neusina agassizi* (190 mm), taken from a depth of 300 m in the Pacific Ocean, and *Protobotellina cylindrica* (120 mm) from the Malvin Islands area. Only empty tests of these two Recent species have been found and there is no proof that they are living at present. One of the largest specimens found with protoplasm was *Syringammina tasmanica* from New Zealand (Lewis, 1966) whose size is 44 mm.

It is difficult to determine which are the smallest foraminifera because of the

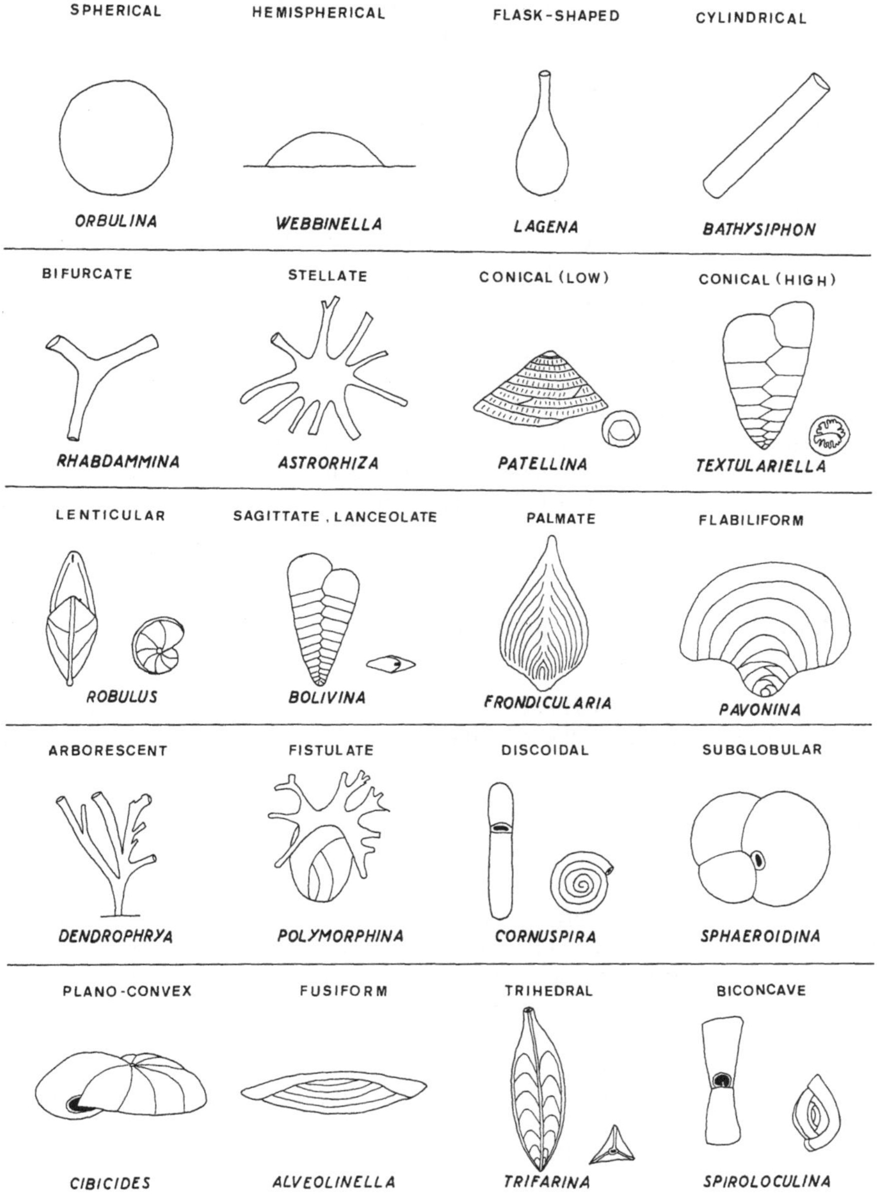

Fig. 23. Principal test shapes.

difficulty in distinguishing juvenile from adult forms. Two species of the genus *Silicotextulina*, *S. diatomitarum* (0.042 mm) from the Tertiary of California and *S. deflandrei* (0.078 mm) from the Recent of Argentina, are examples of small adult forms but smaller yet are the 'microforaminifera' of Wilson & Hoffmeister (1952) whose minimum adult size is about 0.020 mm. There are diverse opinions about the taxonomic positions of these small foraminifera and this subject is treated in more detail in Chapter VIII,4.

According to Bergmann's Law (1847) the largest examples of any group of cold-blooded invertebrates or terrestrial plants occur in tropical or sub-tropical climates. The law appears to hold true for foraminifera, as the largest specimens of both benthonic and planktonic forms are found in tropical and sub-tropical waters.

The size of a given test is, of course, dependent to a large extent on environmental conditions. It is not, therefore, unusual to find specimens occurring in one habitat to be several times larger or smaller than specimens of the same species living under different conditions. Hofker (1932) described a foraminiferal fauna from Naples among which there were representatives of the Astrorhizidae whose size was twenty times that of specimens of the same species found in Arctic waters. Boltovskoy & Gualancañay (in press) described fully mature specimens of *Cassidulina subglobosa* that varied in size from 200–800 μ.

2. Chambers and their position

The extreme variability of chamber shape is shown by the highly irregular, often fantastic shapes and positions taken by the chambers in some sedentary species. This irregularity in shape is due to the nature of the substrate on which the chambers rest. An illustration of this condition is seen in the multifaceted forms of *Cibicides variabilis*, *Erichsenella*, *Rosalina leei* and several *Parrina* (Hedley & Wakefield, 1967b).

Olsson (1971a) described in great detail the logarithmic nature of chamber addition in some trochospiral planktonic species, among them *Globigerina pachyderma*. The form of the last few chambers of this species is at variance with that of the earlier formed chambers. These irregularly positioned and shaped last few chambers are common in several species. This modification of the chambers, which is often accompanied by a reduction in wall thickness, is discussed more fully in the section on pathological specimens (section 11).

Changes in chamber shape within a species have been described by Kennett (1968a) and Takayanagi, Niitsuma & Sakai (1968) for the planktonic species *Globorotalia truncatulinoides*. The changes in the shape of the chamber result in a change in the overall shape of the trochospiral test and seem to be a function of water temperature. The benthonic species *Nonionellina flemingi* has been found to vary from trochospiral to planispiral and to have a variable number of chambers

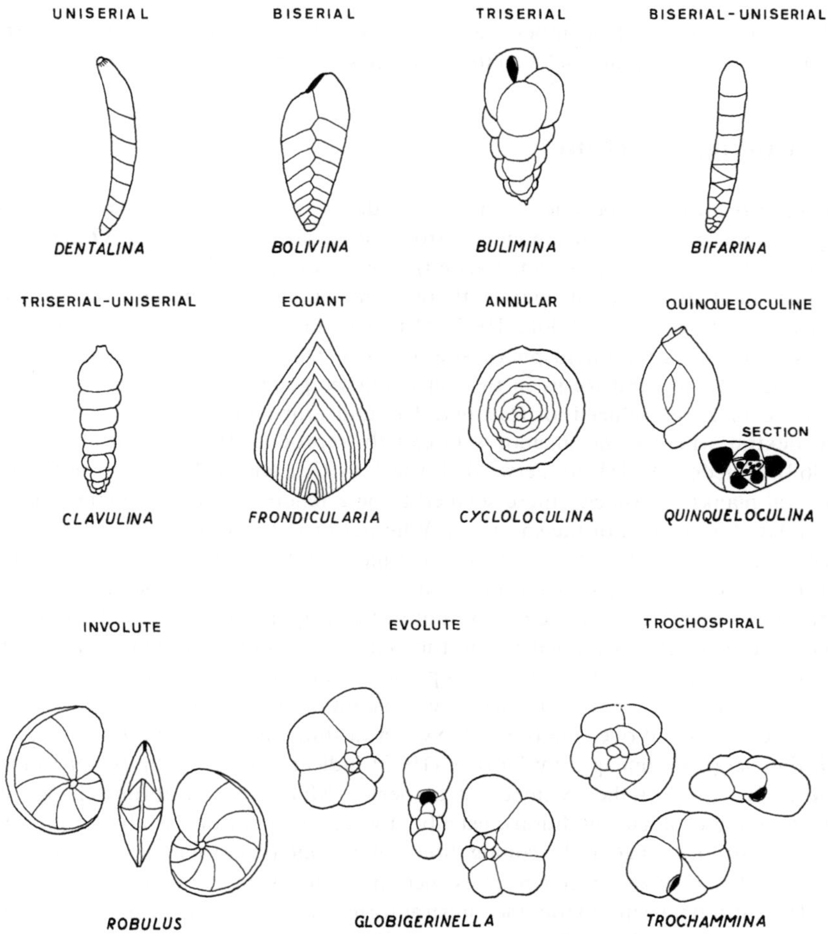

Fig. 24. Principal modes of chamber disposition.

depending on the water temperature (Lewis & Jenkins, 1969). For some foraminifera, especially those belonging to the Orbitoididae, the shape of the chambers is of considerable taxonomic importance.

The position of the chambers with respect to one another can vary greatly and was for many years considered as a major, if not primary, feature in the erection of classification schemes. At the present time, however, wall structure and microstructure are considered of greater taxonomic value. Many spatial arrangements are utilized by the foraminifera and several of these are shown in Fig. 24. Many genera change their form of chamber addition as they mature, although the microspheric forms commonly exhibit all variations whereas the megalospheric forms may not.

3. Coiling direction

Some trochospiral specimens have a tendency to coil in a clockwise (dextral; right-handed) direction when viewed from the spiral side of the test. Others coil in a counterclockwise (sinistral; left-handed) direction (Fig. 117).

For many trochospiral species there is no apparent preference for dextral or sinistral coiling. Føyn (1936b, 1937) observed the same proportion of dextral and sinistral forms in the microspheric specimens of *Discorbis vilardeboanus* but noted a dominance of dextral forms in the macrospheric generation. The gamontic generation (sexually reproduced) of *Rosalina floridana* grown in laboratory cultures tend to be dextral whereas the agamonts (asexually reproduced) are sinistral (Lee *et al.*, 1963). Thiede (1971c), Steuerwald & Clark (1972) and Vella (1974) noted that in several planktonic species there seemed to be a tendency for the coiling direction to be related to the size of the specimen. Whether or not this phenomenon is connected with the reproductive cycle as in *R. floridana* is not known. Longinelli & Tongiorgi (1964) suggested that sinistral and dextral forms of benthonic species may differ in that they secrete their calcite at different water temperatures, i.e. at different times of the year. How this is related to the fundamental nature of coiling is unexplained.

Bolli (1950, 1951) proposed that a phylogenetically young species coils randomly, but that during the evolutionary development of the taxon one form of coiling becomes dominant over the other. However, a study of the *Globigerina cretacea* and *G. daubjergensis* lineages by Hofker (1972) indicated that coiling is not random in the early stages of the existence of a species. Moreover, an abrupt change in the surrounding ecologic conditions, usually in water temperature, is able to produce an equally abrupt change in the coiling direction in some planktonic species. Bolli (1971) suggested that those species near extinction often revert to random coiling.

The ratio between dextral and sinistral forms has been used by many investigators as a clue to paleoclimatology and stratigraphic correlations. This subject is treated at greater length in Chapter XIV,4.

Dextral and sinistral coiling is not unique to foraminifera but is a condition that often occurs in nature. Coiling in opposite directions is found in gastropods, certain bacterial colonies (*Bacillus mycoides*), the spiral vascular conductors of some plants, the spiral distribution of leaves, and the mineral quartz.

Alpatov (1957) proposed that the coiling characteristics of a population be given by designating the percentage of dextral forms in a sample. If, for example, 35 of 75 tests of the same species are dextrally coiled, the 'dextrality' would be 47%. Alpatov, after studying the molluscan literature, decided that a 50% dextrality was rare enough to warrant special notice and he utilized the organic chemistry term racemic to describe this state.

Alpatov also offered an interesting explanation of dextral:sinistral coiling phenomena. He suggested that organic molecules comprising the protoplasm are disymmetrical (dextral or sinistral), giving the protoplasm an optical activity. This optical activity might be advantageous to the organism. Alpatov noted that this problem requires a great deal of study by biologists, chemists, physicists and crystallographers.

4. Chamber wall

The composition and construction of the chamber wall are of great systematic importance for the foraminifera. Many classification schemes have been based on a simple four-fold distinction: (a) chitinous, (b) agglutinated, (c) calcareous, (d) siliceous.

Recent detailed studies of tests have revealed new information about both the composition and the construction of test walls with the result that there have been changes in the classifications.

A. Chitinous Walls

'Chitinous' walls appear to be the most primitive. Fossil examples are not well known, although their rarity is probably due to the ease with which erosive and transportative agents destroy this delicate material rather than to the lack of such forms in the past. The term chitin is not exactly correct. Awerinzew (1903) first noted that the material is not the true chitin found in the exoskeletons of crustaceans but rather is a tectinous substance with different chemical characteristics. He considered tectin to be a combination of proteins and various hydrates of carbon. The material reacts chemically as a carbohydrate and is probably a mucopolysaccharide (Hedley, 1958). The organic lining of the less primitive foraminifera has a similar composition (Angell, 1967b) as do the fossil forms of the 'microforaminifera' (Wetzel, 1957). Other authors have called this substance pseudochitin, chitinoid and keratin. Nevertheless, following the customs of the past, we continue to call these tests chitinous. Structural and chemical studies of *Allogromia laticollaris* indicated that this test consists of fine lamellae of uniform composition including a nucleic acid (Pierce *et al.*, 1968).

Only the Allogromiidae have tests composed primarily of chitin, but some authors place this family with the Thecamoebians rather than in the Foraminiferida. C. Ogden (personal communication) noted that *Cryptodifflugia*, a thecamoebian, has a calcareous lining inside the outer organic covering. The Astrorhizidae also have chitinous tests but the test is partially or completely covered by detrital material. Some groups whose normal form is calcareous occasionally exhibit chitinous specimens. This phenomenon occurs among some estuarine Miliolidae found in brackish water with low calcium carbonate concentration, and in marsh, bay and lagoonal environments where the pH is low (Echols & Wegweiser, 1967; Kane, 1967). Calcareous species that live attached to the substrate often display an adhering wall composed only of chitin.

Chitinous walls are usually transparent and lack pores.

B. Agglutinated Walls

Agglutinated walls are commonly considered as being composed of a layer of chitinous material covered by a cap of cemented detrital material (Fig. 25). In reality,

Fig. 25. Agglutinated surface of *Haplophragmoides canariensis* (Towe, 1967).

not all species contain the chitinous layer. This was demonstrated by Buchanan & Hedley (1960) who observed that *Astrorhiza limicola* exhibits an interior chitinous sheath only when the test is breached. Under these conditions the organism secretes a chitinous membrane over the breach. This membrane is then covered by agglutinated particles taken from the substrate. The same authors noted that after death the desiccated protoplasm changes from a cream colour to a dark reddish grey and adheres as a fine film to the interior walls of the test. This phenomenon may have given the false impression of the existence of an internal chitinous lining in this species.

Both the detrital material and the cementing agent of agglutinated foraminifera are highly variable: The grains are usually composed of locally available materials and often are a good reflection of the character of the bottom sediment. Quartz, mica plates, coal, volcanic glass, rutile, ilmenite, feldspar, tourmaline, zircon, amphibole, sponge spicules, diatoms, molluscan fragments, coccoliths and even foraminiferal fragments are among those materials utilized. These grains usually form a tightly interlocking mosaic with the spaces between them filled with finer grains and cement. The most common cement is an impure calcareous, ferruginous material.

The cementing agents of *Haplophragmoides canariensis* are a stable acid mucopolysaccharide and a very fine grained, amorphous ferric oxide material that

Table 5. Chemical composition of agglutinated foraminifera (after Vingradov, 1953; Hedley, 1963)

Species	SiO_2 (%)	$CaCO_3$ (%)	CaO (%)	Fe_2O_3 (%)	Al_2O_3 (%)
Astrorhiza crassatina	96.2	2.5	†	1.3	†
Rhabdammina abyssorum	94.7	2.9	†	2.4	†
Rhabdammina abyssorum	88.3	4.0	†	7.4	†
Rhabdammina abyssorum	95.7	1.9	†	2.4	†
Hyperammina friabilis	93.6	4.0	†	2.0	†
* *Protobotellina cylindrica*	81.0	†	3.3	1.0	8.3
* *Saccammina sphaerica*	78.0	†	4.9	0.7	5.1
* *Pilulina jeffreysii*	87.0	†	4.2	0.6	4.6
* *Cyclammina cancellata*	—	†	—	2.0	2.0
* *Cyclammina cancellata*	84.8	5.5	†	9.4	†
* *Cyclammina cancellata*	84.7	6.8	†	8.1	†
Haplophragmium latidorsatum	76.1	7.3	†	16.3	†

* analyses by Hedley, 1963
† no test performed for this oxide

contains no iron or calcium carbonate (Towe, 1967). The observations of Hedley (1963) and Towe (1967) suggest that the iron is not actively secreted metabolically by the organism but is chemically removed from the detrital grains used by the organism for nutrient and test material. The incorporation of detrital grains to remove adsorbed organic compounds from their surface was suggested for the non-agglutinated species *Edentostomina durrandii* which selectively ingests zircon grains (Mattiat & Zobel, 1968).

The colour of the test is not a reliable indicator of the amount of iron in the cement. Pure white and reddish brown tests may be equally ferruginous. The chemical composition of agglutinated tests was summarized by Vinogradov (1953) and Hedley (1963) whose data are shown in Table 5. The high percentage of SiO_2 is due to the fact that quartz is the mineral most often available in the substrate.

The selectivity of various species with respect to the agglutinated material is interesting. More primitive genera such as *Astrorhiza* exhibit no mineralogical selectivity and cover their tests with whatever material of appropriate size is available. The thecamoebians also show no selectivity. But *Technitella* choses sponge spicules, *Rhabdammina* and *Psammosphaera fusca* use quartz grains, while *P. parva* will use sponge spicules as well as quartz; *P. testacea* uses bits of other foraminifera; *P. bowmanni* uses mica flakes; and *P. rustica* sponge spicules. A curious example of selectivity is found in *Reophax bilocularis* which uses bits of *Globigerina* tests. *Spiculosiphon radiata* is capable of consuming the protoplasm of other foraminifera and incorporating pieces of the test of the prey into its own test wall (Christiansen, 1964).

In addition to selectivity in the kinds of material chosen some species are selective in the arrangement of the particles. *Marsipella spiralis* uses sponge spicules only between 60–100 μ in length and arranges them always at a 30–45° angle to the test axis. *Discobotellina biperforata* appears to be capable of shifting the surface position of grains as the specimens grow (Stephenson & Rees, 1965a). A detailed statistical study of the agglutinated material on *Reophax curtus* revealed that the only significant preferences shown by the organism was the selection of larger grains for its later (larger) chambers (Smith & Kaesler, 1970). Similar observations were made by Sliter (1968) on cultured specimens of *Trochammina pacifica* which chose larger grains for its later chambers. This species also showed no selectivity with respect to mineral composition and did not require the presence of mineral grains in the culture in order to grow and add new chambers.

These examples of selectivity should not be considered as having great taxonomic value. Slama (1954) and Buchanan & Hedley (1960) found that when agglutinated forms were cultured without the usual test-building material, substitute materials were readily utilized by the organism. The causes of selectivity are not at all clear but appear to be related to the physical chemistry of the protoplasm.

The internal structure of the agglutinated wall has not been subject to extensive study but those examinations which have been made reveal a variety of structures. Murray (1973a) examined 19 Recent agglutinated species and found that although the majority possessed a simple arrangement of detrital grains, some species showed various internal complexities. *Cyclammina cancellata* has a complex alveolar structure visible only from the inside of the chamber wall. This structure is discussed in more detail in section 6. *Clavulina pacifica, Textularia sagittula, Gaudryina rudis* and *Siphotextularia flintii* contain small pores penetrating the inner chamber wall but not the outer. These pores are lined and closed off by an organic membrane and appear to branch as they approach the outer wall. These four species all have a calcareous cement binding the grains.

As far as we know all agglutinated foraminifera are benthonic; however, Saidova (1970b) described a form which she interpreted as an agglutinated planktonic genus, *Orbulinoides agglutinatum*. From plankton samples in the Kurilo–Kamchatka trench at depths between 1055 and 4610 m she described 44 specimens of this species which are spherical and agglutinated (chitinous cement) and possess numerous circular apertures.

C. Calcareous Walls

Both the hexagonal (calcite) and orthorhombic (aragonite) crystal forms of calcium carbonate are used in the construction of calcareous walls. Todd & Blackmon (1956), after studying the walls of several species, concluded that the phylogenetic origin of the species, rather than the ecology, determined the calcium carbonate crystal system. They also noted that only a few Recent foraminifera (Robertinidae and Ceratobuliminidae) have aragonite tests.

Table 6. Chemical composition of calcareous foraminifera (after Vinogradov, 1953).

Benthonic species	$CaCO_3$ (%)	$MgCO_3$ (%)	$FeCO_3$ (%)	SiO_2 (%)
Quinqueloculina auberiana	90.1	9.3	0.3	0.3
Pyrgo sp.	92.1	—	—	—
Orbiculina adunca	89.8	10.0	0.1	0.1
Orbitolites complanata lacinata	86.5	12.5	0.7	0.6
Orbitolites complanata lacinata	88.2	8.8	—	0.3
Orbitolites complanata lacinata	88.7	9.6	—	0.1
Orbitolites complanata lacinata	87.9	10.5	—	0.1
Orbitolites marginata	89.0	10.6	0.1	0.3
Operculina complanata	93.6	4.8	0.1	0.9
Amphestegina lessonii	92.9	4.9	trace	0.3
Miniacina miniacea	88.8	11.2	trace	trace
Tinoporus baculatus	88.7	11.1	0.2	trace
Planktonic species				
Globorotalia menardii	77.0	3.7	4.0	15.3
Globigerina bulloides	93.1	0.6	1.7	1.6
Globigerina bulloides	91.3	0.3	2.7	1.8
Globigerina bulloides	92.5	0.9	1.3	1.4
Sphaeroidinella dehiscens	84.3	1.8	4.9	8.9

In addition to calcium carbonate there are numerous other chemical components in test walls. Vinogradov (1953) prepared a table of chemical analyses of test walls drawing from the work of several authors (Table 6). These data indicate that $CaCO_3$ comprises about 90% of the test wall. Numerous qualitative and some quantitative analyses have been conducted on a large number of species (Wood, 1949; Said, 1951a; Switzer & Boucot, 1955; Emiliani, 1955a; Boltovskoy, 1956b; Krinsley, 1960; Knauff, 1963; Hooper, 1964; Barchuk *et al.*, 1970; Belyaeva, 1973; Savin & Douglas, 1973) and a great number of elements have been encountered (B, Na, Mg, Al, Si, P, K, Ca, Ti, V, Cr, Mn, Fe, Ni, Cu, Sr, Ag, Sn, I, Ba, Bi). It is likely that many of these elements are contaminants and that the number of biogenic trace elements is much smaller. The use of an electron microprobe (Lipps & Ribbe, 1967) to solve the contaminant problem, revealed only the presence of Mg, Fe and Sr in the tests of 9 planktonic species.

Studies of the microstructure of the calcareous tests utilizing polarized light microscopy were initiated with the studies of Sorby (1879), Ebner (1887), Awerinzew (1901), Emerson (1916), Sollas (1921) and culminated with the detailed investigations of Wood (1949, 1963), Smout (1954) and Reiss (1958–1963). The widespread use of the transmission electron microscope for ultrastructure studies beginning with Jahn (1953) and Hay, Towe & Wright (1963) and the introduction of the scanning electron microscope for similar investigations by Malapris & Jannin (1967), Honjo & Berggren (1967) and Hay & Sandberg (1967) has provided a wealth of new information about wall structure.

There are several different types of calcareous wall structure, each of which will be treated below.

1. Microgranular walls

These are walls composed of equidimensional, tightly packed, small grains of crystalline calcite; they lack pores and in some cases have agglutinated material as a covering. The Endothyracea, Fusilinacea and Parathuramminacea exhibit this type of wall structure, but there are no Recent species in this category.

2. Porcellaneous walls

Foraminifera exhibiting this type of wall structure, primarily the Miliolidae, are characterized by a milky white porcelain-like surface. These cryptocrystalline walls are white in reflected light and yellowish-brown in transmitted light. The walls are imperforate. Serova (1961b) described a three-layered wall with a total thickness of 2–5 μ, the middle layer of which is composed of needle shaped calcite crystals 2–15 μ long whose optic axes are randomly arranged. The outer and inner layers are composed of calcite crystals whose long axes are parallel to the surface of the test. The general structure described by Serova utilizing the optical microscope has held up well under the scrutiny given these walls with the electron microscope (Hay, Towe & Wright, 1963; Lynts & Pfister, 1967; Towe & Cifelli, 1967; Haake, 1967).

The thin outer layer consists of a 'tile roof', 'brick' or 'parquet' pattern of calcite rhombohedra (Fig. 26A) and/or a random two-dimensional array of calcite needles (Fig. 26B). This outer covering is not always visible on empty or fossil tests as it is subject to removal by the gastric juices of predators (Walker, 1971) and alteration by post-mortem acidic environments (Murray & Wright, 1970). The internal layer is made up of a 3-dimensional random array of calcite needles (Fig. 26C). It is this random array of crystals which gives both the peculiar heterogeneous pattern typical of this wall when seen in polarized light and the milky white opacity when seen in reflected light. The inner wall appears to consist of lath-like calcite needles in a two-dimensional random arrangement similar to that of the outer wall.

The lack of organization of the internal layer suggests the absence of a well-organized organic matrix on or in which the calcite crystals can nucleate. Calcite

Fig. 26. Wall structure of porcellaneous foraminifera; A. Outer wall, *Triloculina tricarinata* (Haake, 1971); B. Outer wall, *Quinqueloculina seminulum*, ×16 500 (Towe & Cifelli, 1967); C. Internal layer, *Q. subrotunda*, ×8500 (Towe & Cifelli, 1967).

precipitation in this structurally incoherent and colloidal matrix was described as spontaneous and random by Towe & Cifelli (1967) although the discovery that crystallization sites in the walls of some porcellaneous species are not randomly distributed may alter this view (Oliver, Very & Vuagnat, 1973).

Diatom frustules have been found in the organic layer of some Miliolidae (Very, 1970) and have been interpreted as the waste products from ingestion of living diatoms.

3. Radial hyaline walls

The calcite crystals in these walls are oriented with their c-axes perpendicular to the test surface. This orientation gives the wall a glassy transparent reflective appearance. The walls contain pores. Most of the Nodosariacea, Orbitoidacea, Buliminicea, Rotaliacea and Globigerinacea exhibit radial hyaline walls. According to Krasheninnikov (1956), there are 3 subdivisions of this wall structure based on the size of the crystals:

finely radial: 0.5–1.0 μ crystals with pores of 1 μ diameter
coarsely radial: 2–3 μ crystals with pores of 3–4 μ diameter
indistinctly radial: 2–3 μ crystals placed vertically and irregularly; no one crystal is flush with both the inside and outside wall; the wall has a fibrous, somewhat disordered structure.

4. Finely granular hyaline walls

These walls are constructed of multi-sized calcite crystals which are randomly oriented optically. This arrangement disperses incident light rather than reflecting it as do the crystals in the radial hyaline wall. However, some tests of this type are transparent because of their thin walls.

The last two types of calcareous wall structure have received the greatest amount of investigation and have recently been reexamined using more sophisticated equipment, particularly the electron microscope which allows direct observation of the crystals rather than indirect observation based on their optical properties in polarized light.

Since calcite is in the hexagonal crystal system, a radial wall is made of crystals whose basal pinacoid (0001) is coincident with the plane of growth, i.e. the wall (Hansen, 1968b). One of the characteristics of granular walls is that no optic figures are visible when they are viewed in polarized light. Towe & Cifelli (1967) stated that this does not necessarily imply a random arrangement of crystals. They suggested that the crystals could be oriented in such a way that the optic axes are neither perpendicular nor parallel to the test wall but nevertheless exhibit a preferred orientation. One of the rhombohedral faces ($ho\bar{h}l$) so characteristic of calcite would provide a likely plane of growth. The observations of Hansen (1968b) that the cleavage rhombohedron ($10\bar{1}4$) was parallel to the wall supported this suggested mode of crystal orientation.

However, one of the serious problems of wall description among radial and granular calcareous forms is the confusion between the optical orientation and the morphological orientation of crystal grains. The elaborate discussions over the wall structure of *Rotalia beccarii* is a good example of this confusion (Wood, 1949; Krasheninnikov, 1956; Cifelli, 1962b; Towe & Cifelli, 1967; Hansen & Reiss, 1971; Banner & Williams, 1973). The walls of this species have been described as radial by some and granular by others but appear to be morphologically granular (granules of Banner & Williams; stacked plates of Hansen & Reiss) but optically radial; the optic axis of the granules/plates is arranged normal to the test wall.

The observations of Wood (1949) and Hansen (1972a) that members of the same genus can have both radial and granular walls casts doubt on the usage of wall

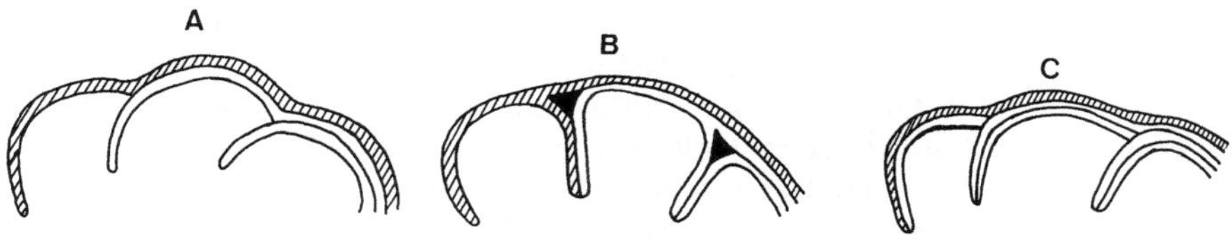

Fig. 27. Wall construction of the lamellar foraminifera (after Loeblich & Tappan, 1964), (cross hatched areas represents the most recent addition); A. Monolamellar; B. Rotalid; C. Bilamellar.

microstructure as a major taxonomic tool. These discoveries lend further credence to the postulation of Towe & Cifelli (1967) that the differences in wall structure lie primarily in variations within the stereochemistry of the organic compounds involved in the calcification process.

The tests of these calcareous forms also differ from those of other foraminifera in that they are lamellar. A new lamella, or calcareous layer, is added to the entire specimen each time a new chamber is added (Smout, 1954). There is no general agreement regarding the nature of these lamellae. Studies of lamellae utilizing the electron microscope have resulted in a great number of articles and raised several questions. The intensive recent investigations were initiated by Reiss (1957), who proposed three types of lamellar walls; a monolamellar wall in which a new chamber wall extends to cover the entire test (Fig. 27A); a 'rotalid' wall in which the exposed part of the penultimate chamber is covered with an extension or septal flap giving all but the most recently formed chamber a secondarily doubled wall (Fig. 27B); and a bilamellar wall (Fig. 27C) which consists of an outer and an inner wall separated by a dark line which appears to be a spongy organic layer (Hansen, Reiss & Schneidermann, 1969). Some forms exhibit a modification of the bilamellar wall in which the inner lining, which usually stops at the wall of the earlier formed septum, also coats the previous septum giving the wall a trilamellar structure (Reiss, 1957, 1963a; McGowran, 1966; Reiss & Schneidermann, 1969).

The lamellar character of wall structure has been considered of primary taxonomic importance in more recent classification schemes (Reiss, 1963b; Loeblich & Tappan,

1964; McGowran, 1968). Recent investigations indicate that there may be many more lamellae in the inner and outer units of the 'bilamellar' wall than previously thought (Bé & Hemleben, 1970), although some species appear to be non-lamellar (Towe, 1971). Hansen & Reiss (1972) suggested that all perforate, lamellar foraminifera are bilamellar. One of the difficulties that has led to such extensive disagreement is the recognition of lamellae. There may be a problem in the distinction between test walls with layers and test walls with lamellae separated by an organic membrane. This membrane is often difficult to detect.

The discovery that some planktonic species acquire a calcite crust or cortex when they inhabit deeper waters coupled with the concept of a bilamellar test wall also resulting in shell thickening has led to some controversy about the fundamental structure of planktonic tests. Hofker (1962, 1967, 1971) and Blow (1969) rejected the idea of bilamellar tests, suggesting that all thickening is secondary. Recent photographs published by Premoli-Silva (1966), Takayanagi, Niitsuma & Sakai (1968) and Reiss & Luz (1970) seem to provide good evidence that the lamellar layers do exist and are structurally different from one another.

If lamellae are to be used as taxonomic traits they should be present in some foraminifera and absent in others. However, as the references cited above indicate, there is a decided lack of consensus over their presence or absence in groups of foraminifera. Consequently, the use of lamaellar wall structure as a tool in classification is undesirable at the present time.

5. Monocrystalline walls

A few species, members of the Spirillinacea, have walls composed of a single crystal or several large crystals of high-magnesium calcite.

6. Spicular walls

One genus, *Carterina*, has a test composed of elongate calcareous spicules, each a single crystal usually aligned parallel to the periphery of the test.

D. Siliceous Walls

Foraminifera exhibiting siliceous walls are rare and are usually described from among deep water species. It is not clear whether the silica has been secreted by the organism, is adhering to the test as agglutinated material, or in the case of fossil species, is the result of diagenetic activity. Lipps (1973) suggested that the traditional test for siliceous material in the test, the absence of a reaction to strong acids, was not a conclusive test. He pointed out that the organic material acting as a cement or binder in the tests of the 'siliceous' Rzehakinidae would behave in the same way. He suggested that there is no evidence that silica is directly secreted by the organism.

A curious opaline silica test of a living '*Miliolinella*' was described from the Scotia Sea, Antarctica by Echols (1971).

The attention that we have given to composition and construction of test walls is based on the belief that these are fundamental factors in taxonomic discrimination.

Both Krasheninnikov (1956) and Blackmon & Todd (1959) concluded that the microstructure and mineralogy of foraminiferal tests are invariable features within genera. The latter authors also said that among hyaline species the wall microstructure is independent of the environment and method of fossilization. They also attribute the overall appearance of the test to the microstructure.

Le Calvez (1953) suggested that the ability to secrete calcareous or siliceous tests developed prior to the ability to agglutinate. He argued that chitinous tests in a primitive form gradually became enriched in calcium carbonate and calcareous tests were produced. A somewhat contrary view was proposed by Miklukho-Maclay (1956) based on a study of Paleozoic foraminifera. This author claimed that the capacity to agglutinate is quite variable and may depend on the environment, but that observations in various families indicated that the ability to agglutinate diminishes from the more primitive forms to the more advanced. Consequently, Miklukho-Maclay concluded that the agglutinated form is more primitive than the secreted one because a secretionary process is more complicated than simple adhesion of detrital particles to an organic membrane.

5. Pores

The presence of pores or perforations in the test wall of the non-miliolid calcareous foraminifera has long been recognized. Their existence in agglutinated forms has been repeatedly 'discovered' and denied. They were well described in the last century (Carpenter, Parker & Jones, 1862) but many subsequent investigators have classified the agglutinated species as wholly imperforate. Examination of agglutinated forms with the electron microscope (Murray, 1973a) clarified this problem to some extent. Although most of the species he examined lacked pores, some possessed pore-like structures which opened on the inner test wall but failed to penetrate the outer wall. Murray pointed out that literally hundreds of agglutinated species have not been examined in detail and it is difficult to draw any general conclusions about the porosity of agglutinated forms.

Calcareous foraminifera can be divided into two categories on the basis of their pore diameters: from 1–6 μ (Nodosariidae, Buliminidae, Polymorphinidae and Heterohelicidae) and from 6–15 μ (Rotaliidae, Acervulinidae and Nonionidae). These categories are only general and there are many examples of specimens with pore sizes outside these ranges (*Nonion, Bolivina*). It has been suggested (Hofker, 1951c, 1968) that more advanced forms have larger pores and that within a given species at a given time the pore density is constant. However, Bé (1968) demonstrated that pore density may be environmentally controlled in some Recent planktonic species (Chapters IX,1 and XIV,11).

Further relationships between environmental conditions and pores were described by Murray (1973a). He noted that porous agglutinated species occupy stenohaline or hypersaline environments and are apparently in osmotic equilibrium with their surroundings. Species without pores are able to occupy a wider range of salinity habitats perhaps because they can isolate themselves from environmental changes by plugging the only opening to the test, the aperture, and gradually adjusting to the new conditions (see Marszalek, Wright & Hay, 1969).

The pores were considered for many years as supplementary apertures allowing (a) free passage of the cytoplasm to the exterior, (b) excretion of waste products and (c) gaseous diffusion between cytoplasm and sea water. They were even considered as passageways for gametes. Protoplasmic filaments have been observed apparently

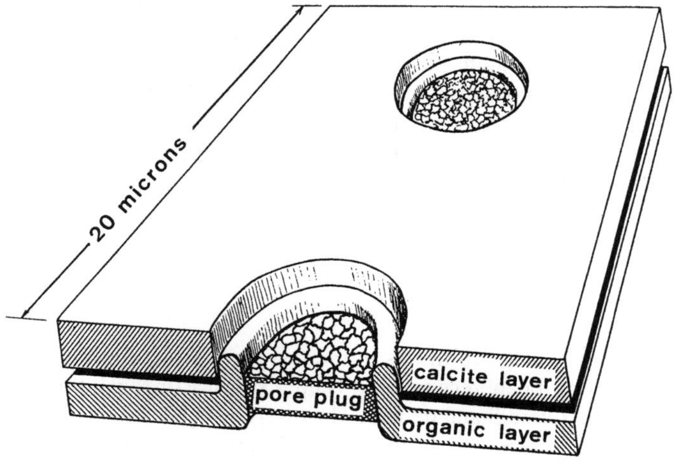

Fig. 28. Schematic section of the wall of *Discorinopsis aguayoi* (after Arnold, 1954d).

extending through the pores although it is often difficult to directly examine pseudopodia extruding from particular openings in the test (Sheehan & Banner, 1972). The observations of Jepps (1942) and many others on *Elphidium* indicate that it is exceedingly difficult to chemically fix or stain the protoplasm in the interior of the test. This would seem to indicate that the numerous pores covering the test surface are not connected to the interior and perhaps that the only means of cytoplasmic communication with the exterior in this genus is through a complex canal system (Marszalek, Wright & Hay, 1969).

More powerful microscopes have revealed a more complicated relationship between pores, wall structure and cytoplasm than was previously thought. Many calcareous forms were discovered to have two chemically distinct wall layers; an outer one composed of calcite and containing pores and an inner organic membrane lacking pores. This observation created a great deal of discussion about the function of pores. Arnold (1954d) in his studies of *Discorinopsis aguayoi* observed a two-layered wall (Fig. 28) in which the pores penetrating the outer layer range in size from 1.5–7.0 μ and are rather uniformly distributed. The organic layer has a series of truncated cones that partially fill the pores in the calcareous layer. These truncated cones contain an organic filling, called a 'pore plug' by Arnold, and occupy a position slightly distal to the base of the organic layer. This plug is pierced by a series of irregularly placed micropores of 0.5 μ diameter. Arnold thought that these micropores were large enough not only to permit osmotic metabolism but also to serve as passageways for pseudopodia. He speculated that these openings could serve as valves which regulate the flow of pseudopodial movement. These 'pore plugs' were first noticed by Le Calvez (1947b) who described them as dark discs in pore openings.

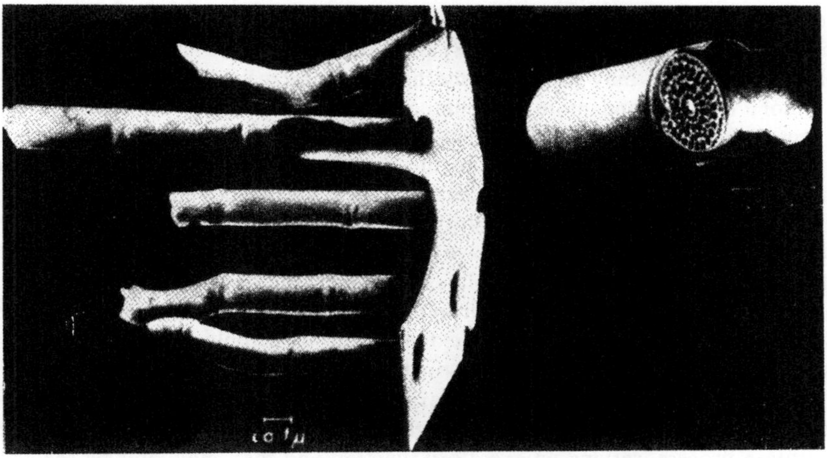

Fig. 29. Reconstruction of pore canals and *Siebplatte* (after Jahn, 1953).

Jahn (1953) observed similar features in the Nonionidae and Nummulitidae. Her examination of these foraminifera with an electron microscope revealed discoidal plates ('*Siebplatte*') 1 μ in diameter (Fig. 29). The micropores within the sieve plate are rounded or triangular with a diameter of 0.1 μ or less.

The observations of Arnold and Jahn were confirmed by Wood & Haynes (1957) in a study of *Cibicides*. They discovered numerous irregularly shaped pores which had a tendency to anastomose. Among these pores there were many circular micropores in plates which correspond to the pore plugs of Arnold. The presence of microporous sieve plates has been noted by Hay, Towe & Wright (1963) in the fossil form *Robulus midwayensis*. Pore plugs were observed in the Recent species *Cymbaloporetta squammosa* (Lynts & Pfister, 1967), *Rosalina floridana* (Angell, 1967b), *Amphistegina madagascariensis* (Hansen, 1972b) and in planktonic species (Towe, 1971). The significance of these complicated structures is not yet clear. Perhaps the sieve plates act as filters for numerous cytoplasmic inclusions known to exist within the test but which have never been observed in the pseudopodia.

Detailed studies of *Ammonia* (Banner & Williams, 1973) revealed that the organic membrane of *Ammonia* spp. protrudes into the base of the pore as in *Discorinopsis aguayoi* described above. However, there is no evidence for any perforation of this lining. The bolivinid species examined by Sliter (1974), like *Ammonia*, exhibited multiple pore diaphragms extending across the pores. These lamellae are two-layered and appear to be continuous with the organic membrane of the wall.

Various Miliolidae and Soritidae exhibit pseudopores that do not reach the inner surface of the wall. These structures appear to be primary in origin (Lynts & Pfister, 1967; Bignot, Depeche & Neumann, 1971).

The function of pores is not well understood. They may provide passageways for cytoplasm or gases. They do seem to be involved in the calcification process. They unite the calcite, organic membranes and linings into a single structure by giving organic continuity to the test. Bé (1968) and Frerichs *et al.* (1972) suggest that pores in planktonic species may serve as a mechanism to increase buoyancy.

As more detailed and more frequent examination of wall structure is made it becomes increasingly apparent that most pores are not simple openings in the inorganic wall but are closed off from the chamber by the primary organic membrane. How much restriction this places on nutrient and pseudopodial material is not well known. Hansen (1972b) illustrated minute pseudopodia in *Amphistegina madagascariensis* which appear to have passed through the pores and to have been separated into fine strands by sieve plates. However, the observations of Sheehan & Banner (1972) on *Elphidium incertum* indicated that the pseudopodia which radiate outward from all parts of the test are extensions of an ectoplasmic layer and do not pass through the pores. There is no reason, of course, to believe that each species has a similar type of pore system. After examination of ultrathin sections of living planktonic species Towe (1971) proposed that the pores may remain open during the calcification process, serving an excretory function and then may later be sealed off by the organic lining. Based on the hypothesis that pores serve a metabolic exchange function in hyaline calcareous forms, Cherif & Flick (1974) suggested that porcellaneous forms are imperforate because the unoriented calcite needles that comprise the wall are so permeable that pores are unnecessary.

Some species may have pores which serve specialized functions. *Orbulina universa* contains two different sizes of pores, the larger of which may serve as apertures. Le Calvez (1936a) observed gametes leaving the test by way of these larger pores. The smaller pores appear to be analogous to pores in other planktonic foraminifera. Myers (1943b) described the larger pores on the hemiplanktonic genus *Tretomphalus* as exit sites for gametes.

Greiner (1974) proposed that the pores originated in hyaline calcareous walls as a consequence of the separation of organic material and calcite during wall crystallization. The calcite grows in an orderly fashion upward from an organic template (see section 10) and the organic material which is excluded from the calcite crystal lattice collects in equally spaced organic plugs (the pores). As this is a purely physical process, Greiner ascribed no adaptive significance to the pores and suggested that the pores therefore serve no purpose.

The study of pores and sieve plates utilizing optical equipment requires very high magnification, transmitted light and the use of some clarificant such as glycerine or castor oil. The presence of a raised border about the pore as in *Globigerina* or a conical shaped pore opening as in several planktonic species can give erroneous impressions about the nature of the pore. The use of X-ray microscopy (Bé, Jongebloed & McIntyre, 1969) and electron microscopy greatly enhances the ability of the investigator to study these microstructures.

6. Internal skeleton

There are various kinds of internal skeletal features in agglutinated and calcareous foraminifera. These features usually have little effect on the gross form and position of the chambers. Occasionally they influence the ornamentation of the external wall.

The alveolar structures of *Cyclammina cancellata* (Fig. 30) and several members of the Spirocyclinidae as described by Brönnimann (1951) and Banner (1970) are an example of one kind of internal skeletal feature. The primary alveoli are connected with the secondary and tertiary alveoli, thus giving the internal test wall a very complicated labyrinthic structure and resulting in a wall which is thick with respect to the size of the test.

Various calcareous species exhibit secondary layers, stolons, pillars and complicated umbilical structures all of which have been designated in the literature as 'supplementary skeletons', 'endoskeletons' or 'internal skeletons'. These structures are found most often in the 'larger' foraminifera which are now extinct, although some Recent species, e.g. the Rotaliidae, exhibit these features on a greatly reduced scale.

Other complex internal features consist of various systems of radial and spiral canals (Fig. 30B,C,D). These are found in the families Rotaliidae, Elphidiidae and Nummulitidae. The spacial arrangement of these canals can be studied by filling the empty tests with some dark coloured material (e.g. Canada Balsam) and later dissolving the test to obtain a mould of the internal canal system. Examination of broken tests utilizing the scanning electron microscope is another way to view the internal canal system.

The first studies of radial and spiral canals were conducted by Carpenter, Parker & Jones (1862) who referred to the canals as sinuses because they thought they were areas where two contiguous chamber walls were not completely united. Hofker (1927, 1930) described the canal system of *Rotalia polystomelloides* (Fig. 30C) as consisting of a spiral canal and various interseptal canals. The spiral canal has the form of a conical spiral tube whose diameter gradually increases. It begins blindly near the initial chambers and ends in a small orifice at the surface near the aperture on the ventral (umbilical) side of the last chamber. The spiral canal is connected with the ventral edge of each chamber and is joined between each chamber by interseptal canals which in turn divide into small canals that reach the surface near the sutures.

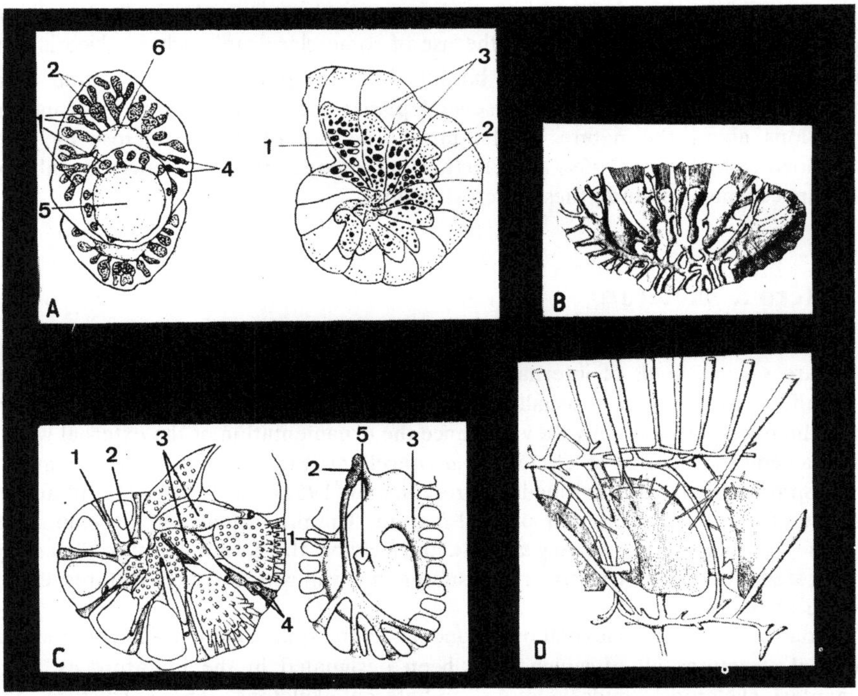

Fig. 30. Internal features of foraminifera; A. Alveolar wall in *Cyclammina cancellata* (after Brönnimann, 1951); 1. Primary alveoli; 2. Secondary alveoli; 3. Tertiary alveoli; 4. Orifices of the tertiary alveoli; 5. Proloculus; 6. Chamber interior; B. Canal system in *Rotalia schroeteriana* (after Hofker, 1951a); Internal mould of Canada Balsam; C. Canal system in *Rotalia polystomelloides* (after Hofker, 1951a); Internal mould of Canada Balsam; Left: ventral view; right: schematic section of chamber; 1. Interseptal canal; 2. Spiral canal; 3. Ventral lobe; 4. Ventral pores; 5. Foramen (opening in chamber face); D. Canal system in *Pellatispira orbitoidea* (after Umbgrove, 1928); Internal mould of Canada Balsam.

The function of internal structures such as pillars and stolons is not well understood. They may serve as structural supports. Hofker (1927) observed that the canal system of *Siderolites* is directly related to the pore system. Jepps (1942) proposed an excretory function for the canal system in *Elphidium*, whereas Marszalek, Wright & Hay (1969) interpreted the canal system of *E. macellum* as a buffer system which cushions the effects of rapid physio-chemical changes in the environment. They and Ujiié (1956) noted that the apertural openings in this species are blind and that the only access to the protoplasm is through the tortuous spiral canal system.

The organic endoskeleton or lining which is found in most foraminifera is also an internal skeletal feature. Its relationship with the wall, the pores and growth are discussed elsewhere (sections 4, 5 and 10).

The studies of Angell (1967b), Arnold (1967a) and an excellent review by Banner, Sheehan & Williams (1973) allow some generalizations and speculations to be made about the organic membrane. Since these generalizations are based on the study of very few species, they should be taken with reservation.

In those species which possess an organic lining it appears to be composed of an acid-resistant mixture of mucopolysaccharides, proteins and indeterminate substances. It seems to be structurally independent of other organic components, the wall and the protoplasm. It lines the walls of all the chambers (except perhaps the last in *Ammonia*), and is thicker in the earlier formed chambers which suggests growth as the organism grows. Sheehan (in Banner, Sheehan & Williams, 1973) suggested that this thickening is due to accretion of waste material on the periphery of the cytoplasm in each chamber where it is cemented together by mucopolysaccharides.

Apart from this proposed excretory function, the organic membrane has been linked with chamber formation (see section 10), protection from the physical environment (Hedley, 1958), protection during reproduction (Nyholm, 1961), protection of the protoplasm from harmful ultraviolet radiation (Haynes, 1965) and protection from the chemical environment (Marszalek, Wright & Hay, 1969).

7. Ornamentation

Calcareous tests exhibit a varied superficial sculpture, examples of which are shown in Fig. 22. Agglutinated forms rarely exhibit ornamentation and when present, it is usually poorly developed.

The presumed functions of ornamentation are quite diverse. The long spines of various planktonic genera (Fig. 31A) may serve as devices to increase buoyancy; i.e. to maximize the surface area with minimal mass. These spines are separate structures, are not continuous with the wall, and apparently do not grow concurrently with the wall. The spines in several planktonic species may serve as supports for new chamber walls (Fig. 31B) as was shown in a beautifully illustrated study by Hemleben (1969b). The spines of *Hastigerina pelagica* exhibit barbs along their length (Fig. 31C) which may act as holdfasts or support the frothy ectoplasm noted in this species (Chapter VII,1).

It is interesting to note that the spinose planktonic foraminifera live in the upper part of the euphotic zone whereas the non-spinose species commonly live below 50 m. When the upper-layer species descend they tend to lose their spines.

Spines on planktonic species are quite flexible during their early growth stages probably because they have a large proportion of organic material mixed with the calcite (Rhumbler, 1911; Hemleben, 1969c). The presence of this organic component also explains why spines are more resistant to solution than the main mass of the test.

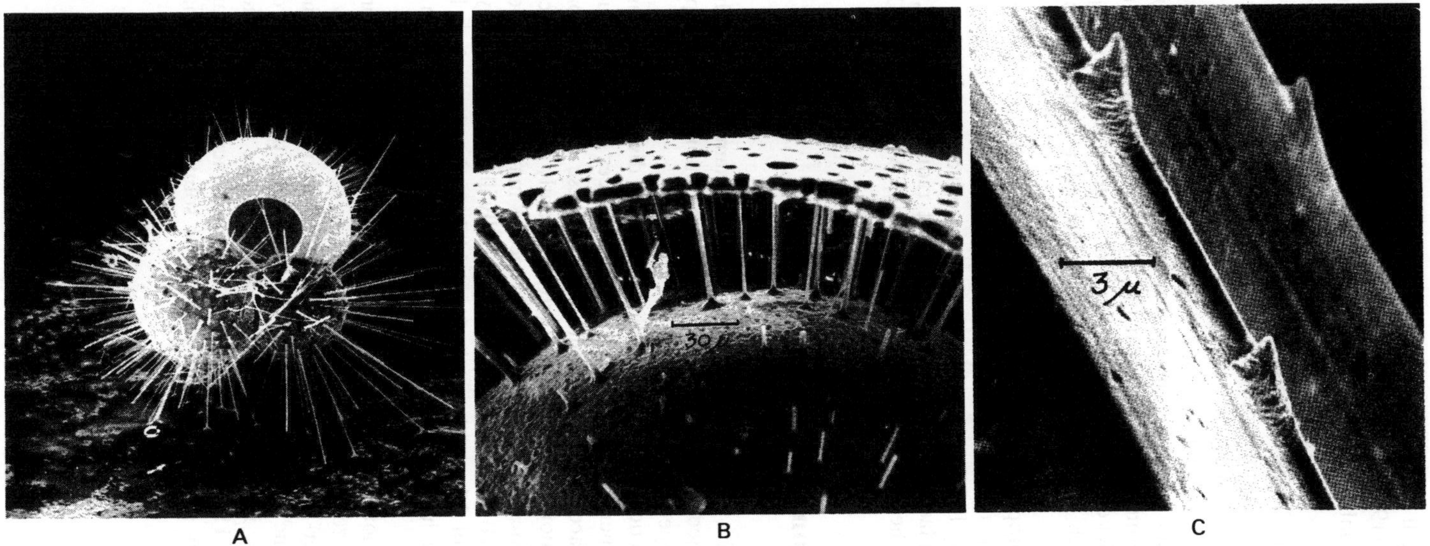

Fig. 31. Spines on planktonic foraminifera (after Hemleben, 1969b); A. *Globigerinoides ruber*; B. *Orbulina universa*, last and penultimate chamber; C. *Hastigerina pelagica*, spines and barbs.

The shape of spines on planktonic species varies among species and within the same specimen. Individual spines may change shape from their base to their tip. The change is different for different species but always constant in direction: circular to triangular to triradiate (Bé, Harrison & Lott, 1973; Walker & Vilks, 1973). From a structural point of view, this change is not surprising since this progression of cross-sectional shapes provides increasing strength (moment of inertia) relative to the amount of material; i.e. energy used.

The spines and spicules of some benthonic species (*Meandroloculina aculeata*, *Baculogyspina sphaerulata*) may serve as protective devices or as mechanisms to maintain stability on soft bottoms. Ornamentation may also act as a means of reinforcing the test with a minimal expenditure of material and energy. These proposed functions are speculative. There are numerous species which live on soft bottoms that lack spines and there are types of ornamentation that do not provide structural support to the test; in fact, some types appear to weaken it. The ornamentation within a single species may vary considerably under different environmental conditions. Specimens of *Quinqueloculina milletti*, s.l. living under different salinity and pollution conditions exhibited a wide variation in their surface ornamentation (Wright, 1968).

In some species ornamentation does not appear until the specimen reaches an adult state. Certain planktonic species in late ontogeny develop a thick layer of euhedral calcite (cortex) which is secreted over the entire test. It has been suggested that this secondary thickening may be necessary to maintain a constant density as the foraminifera grows. A newly added chamber whose wall thickness is the same as that of earlier chambers would have a greater proportion of its volume as low density protoplasm and would therefore be less dense. For this reason, secondary thickening of the test would be necessary to maintain a constant density or to increase the density in order to promote sinking. Secondary shell thickening has even been reported in a benthonic species, *Rosalina leei*, in laboratory cultures (McEnery & Lee, 1970). Loose (1970) noted shell thickening on benthonic specimens living in turbulent water and suggested this was a survival adaption.

Poorly developed ornamentation may be due to poor nutrition (Boltovskoy, 1954c), low salinity (Kurc, 1961), or other anomalous conditions. Theyer (1966) reported the development of a keel on *Buccella frigida* in deep cold (more hostile?) waters.

A rather unique type of ornamentation occurs in some 'larger' foraminifera and Rotaliidae. This ornamentation consists of superficial granulation and conical pillars that penetrate several whorls of the test (Cifelli, 1962b; Ujiié, 1965). These plugs and cones are probably related to the internal structures discussed in the previous section.

8. Aperture

The aperture is the principal orifice by which the protoplasm gains access to the exterior. Almost all species possess at least one aperture. The variety of form,

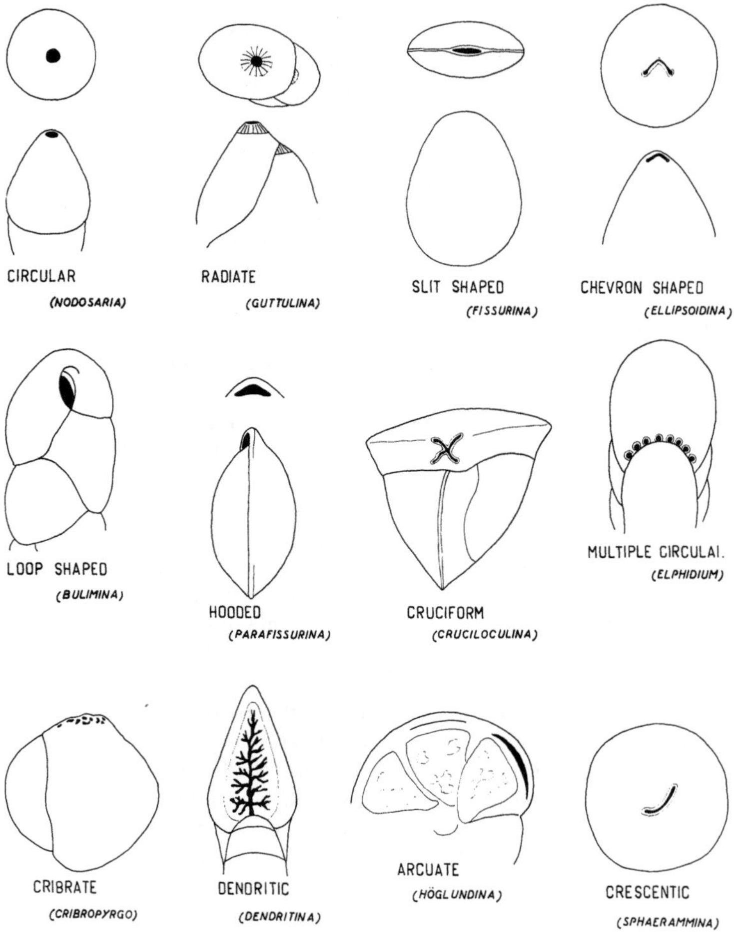

Fig. 32. Shape and form of aperture.

position and number of apertures is shown in Figs. 32 and 33. In the cases of multiple apertures, all may be the same size or one may be dominant (primary). In the latter case, the remaining ones (secondary or supplementary) are found in a peripheral, basal, dorsal, ventral or terminal position.

In some foraminifera the apertures of the first and last chambers are connected by plates (Buliminidae, Caucasinidae and Bolivinidae), siphons (Pleurostomellidae) or other similar features.

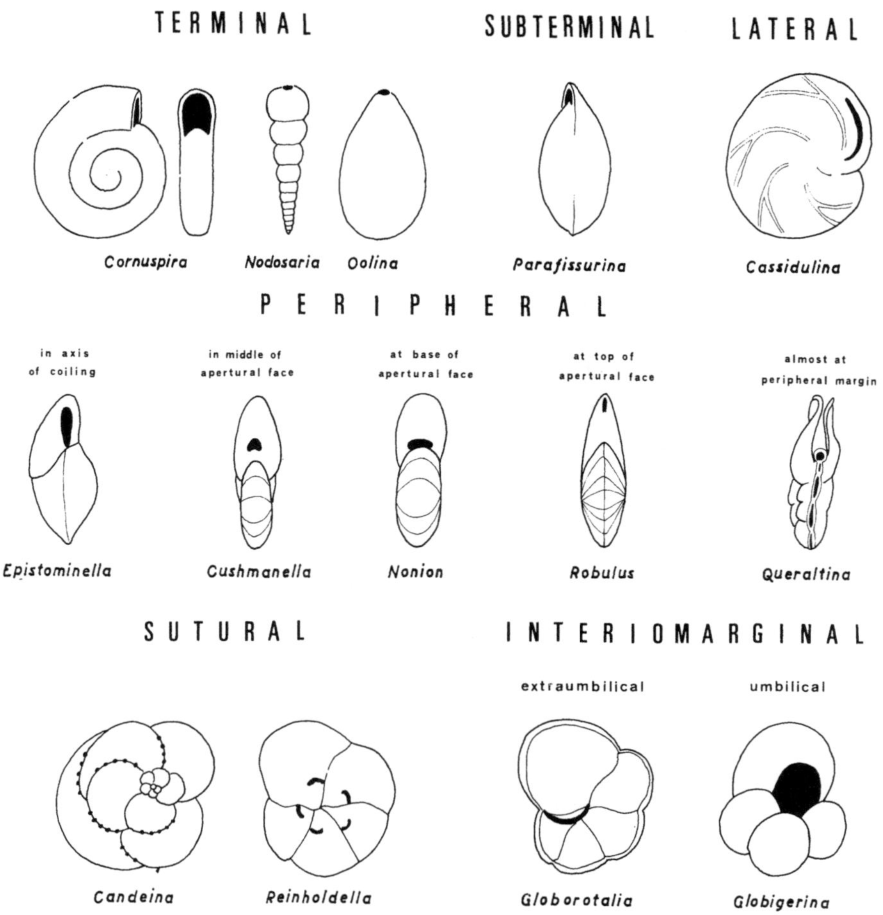

Fig. 33. Position of aperture.

Many apertures contain distinct supplementary structures such as teeth, tooth plates, tubular collars, rims, chamberlets, tegilla and bullae (Fig. 34). The functional role of these features is poorly known. It is possible that some serve as devices to channel or retain the protoplasm or to prevent the entry of parasites or predators. They may also serve as devices to mechanically strengthen the apertural margin, or, in the case of planktonic species, to increase the specific gravity of the individual in later life so that it sinks.

The aperture may be important from a phylogenetic point of view as various changes are sometimes noted in ontogeny. The genus *Pyrgo* undergoes quinqueloculine, triloculine and biloculine stages in its ontogenetic development with

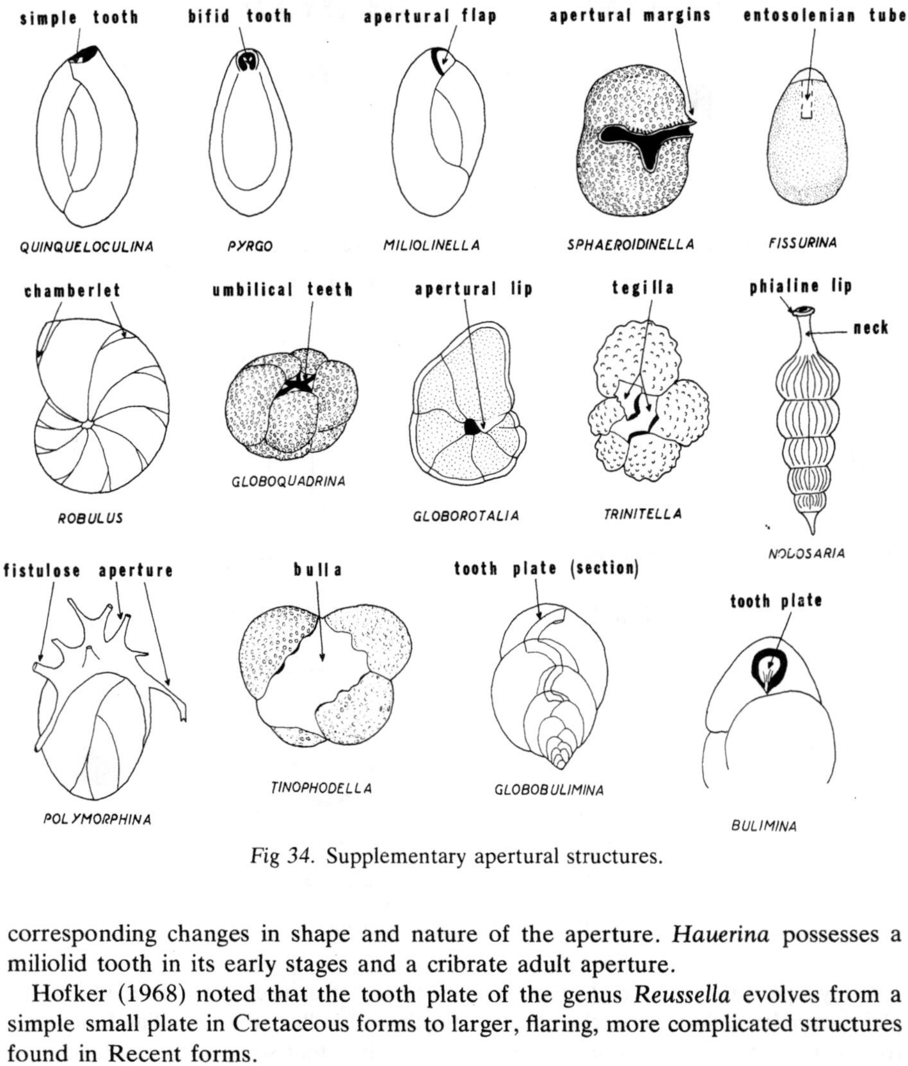

Fig 34. Supplementary apertural structures.

corresponding changes in shape and nature of the aperture. *Hauerina* possesses a miliolid tooth in its early stages and a cribrate adult aperture.

Hofker (1968) noted that the tooth plate of the genus *Reussella* evolves from a simple small plate in Cretaceous forms to larger, flaring, more complicated structures found in Recent forms.

The aperture has been cited as a stable morphological trait having taxonomic value. However, it should be pointed out that certain genera exhibit different apertural arrangements under different environmental conditions. Wright (1968) described such a situation for *Quinqueloculina milletti.* Even within culture populations maintained under uniform conditions Arnold (1964) noted that apertural configurations may vary considerably. He suggested that the toothlike structure in the miliolid aperture may serve: (a) as a check valve to restrict the movement of cytoplasm or nuclei from chamber to chamber or to prevent the entry of large predators; (b) as a device to orient particles that are ingested; (c) to facilitate segmentation of multinucleate protoplasmic masses during schizogony; (d) to spread or divide the extruding protoplasm into pseudopodia; (e) to subdivide the incoming mass of protoplasm during feeding; (f) as a digestive aid by restricting the velocity of incoming food-bearing protoplasm and complicating the circulation pattern; (g) as a plow or wedge for the organism as it moves along the substrate; (h) as a holdfast mechanism to the substrate; (i) as an anchor to the food cyst that often forms around the aperture.

9. Colour

Empty tests have a different appearance than those containing protoplasm and we will treat the two cases separately.

A. Empty Tests

Agglutinated tests are, for the most part, dark yellow, although some specimens are grey. The yellow colour has often been explained as being due to the presence of a ferruginous cement.

Calcareous tests are usually white or gray-white, although specimens of some taxa are dark pink (*Homotrema rubra*), pink (*Globigerinoides ruber*) or brown (*Discorbis*). The basic colour is apparently a function of the chemical composition or microstructure of the test. The pink colour of *G. ruber* is attributed to a pigment (pheophytin) in the organic wall lamellae (Bé & Hamlin, 1967). This pigment is believed to be related in some way to the presence of zooxanthellae in the chambers (Lee *et al.*, 1965).

There appears to be little or no taxonomic significance to the colour since agglutinated species which are brown in one geographic area may be gray in another and both pink and white specimens of calcareous planktonic species *Globigerina rubescens* and *Globigerinoides ruber* are found coexisting in the same population.

The opacity and hence the apparent colour of the test can be altered by the substrate chemistry. Murray (1967b) noted that at a pH of 7, transparent calcareous tests can be altered to a white opaque condition in only 15 days. A pH of 7 or less is not uncommon in some organic rich substrates. Black foraminiferal tests (Murray, 1966a, 1966b) and pyritized tests (Seiglie, 1971b) have been noted but these appear to be due to some form of post-mortem chemical alteration.

The Miliolidae present an interesting colour problem. In reflected light the whole test has a white or pale yellow porcellaneous appearance but in thin section the colour is a yellow-brown. Krasheninnikov (1956) suggested that the miliolid wall is composed of small (0.5 μ) unoriented crystals which disperse and reflect light to a high degree, giving the wall its opaque and porcelain-like luster. He attributed the yellow-brown colour in thin section to the same phenomenon saying that a slightly larger grain size would not produce this colour. The results of more recent studies indicate that the thin surface veneer of lath-like calcite 'tiles' probably accounts for the characteristic luster (see section 4).

B. Tests with Protoplasm

If the wall is transparent, the most frequent colours encountered in most planktonic and in many benthonic foraminifera are green and brown and all intermediate shades. These colours are due not only to the actual colour of the protoplasm but also to the colour of the symbiotic algae living in the test or algae captured for food. There are also dark blue (*Alveolina*), brown (*Discorbis*) and pale blue (*Rupertia*, *Carpenteria*) forms.

The following colours have been observed in planktonic foraminifera in the South Atlantic:

Globorotalia inflata *Globorotalia truncatulinoides* *Hastigerina pelagica*	orange, or even bright orange
Globigerinoides ruber	green-yellow
Globigerinella aequilateralis	bright green
Globigerina quinqueloba	brown
Globigerinoides trilobus	green

These variations are probably due to variations in nutrition and symbionts.

Bé (1960a) observed that living planktonic foraminifera, with the exception of *Hastigerina pelagica*, exhibit different colours at different times of day. During the daylight hours they are an olive green whereas at night they are a light green-yellow. Bé reported that this phenomenon may be due to changes in the carotenoid: chlorophyll ratio; during the night, chlorophyll pigments break down, the ratio increases and the specimens become light green-yellow. Carotene wax pigments have been observed in *Allogromia laticollaris* (Schwab & Schlobach, 1973). These kinds of pigments are widely distributed in plants but generally are unknown in animals. They may be synthesized by the foraminifer from particular kinds of food.

The colour of some foraminifera may be due to a defensive response (mimicry). Some forms exhibit a brownish-gray colour which is quite similar to that of the substrate. Rhumbler (1911) mentioned the dark violet and black colours of deep dwelling specimens of *Reophax nodulosus*. He also noted the red colour of *Miniacina miniacea* and the blue of *Rupertia* and *Carpenteria* in brightly coloured polychromatic reef zones.

10. Growth

The mechanism by which unicameral foraminifera grow is not well understood. Christiansen (1974 Ms) conducted experiments on living specimens of *Saccammina sphaerica*, an agglutinated form. He applied methyl-eosin stain to the specimens and noted that a band of material about the circumference of the test stained red, indicating living protoplasm. He concluded that new detrital grains were being added to the test in this region. When stain was applied to the same test at a later time, a similar but differently oriented band of material about the circumference stained. These observations would seem to indicate that this unicameral, spherical, agglutinated test grows by adding new material about its circumference, reorienting these bands to maintain a spherical shape.

No observations have been made regarding the growth of unicameral calcareous tests. How these tests attain adult size is unknown. Do they grow by accretion, by molting, or some other mechanism? This is an area that needs to be studied.

Many investigators have observed and speculated about the addition of chambers to multicameral calcareous tests. Most of these data pertain to hyaline tests.

Le Calvez (1938a) was among the first to describe in detail the addition of a new chamber to a multilocular hyaline calcareous test (*Discorbis bertheloti*). The pseudopodia are originally situated in the form of a fan; later they withdraw slightly leaving behind a halo of detritus. This newer, smaller fan corresponds to the size and shape of the new chamber. Around this new pseudopodial net the organism constructs a delicate organic film which is then covered with calcareous material. This new chamber is at first vacant but is quickly filled with protoplasm. Similar observations were made by Sliter (1970b) on *Bolivina doniezi*. Le Calvez noted that the volume of the new chamber is equal to the volume of protoplasm that existed in the previous chambers.

Angell's (1967a) observations on *Rosalina floridana*, another rotaliform benthonic form, indicated that it constructs an algal growth cyst between the test and the substrate. The pseudopodia which built the cyst coalesce to form a frothy framework (analge) on which an organic matrix is secreted. The cytoplasm from the test flows into the new chamber forcing the analge out the new aperture where it covers the new chamber surface. Calcite is then secreted on the organic membrane. The frothy analge breaks up to form new pseudopodia.

Chamber formation in *Heterostegina depressa* as observed by Spindler & Röttger (1973) is similar in many ways to that described above. The site of the new chamber is covered by an ectoplasmic sheath ('*Ektoplasmahülle*' of Röttger, 1973). Cytoplasm flows out of the test, fills the sheath and secretes an organic membrane in the shape of a new chamber between the sheath and the edge of the test. Calcification occurs on this membrane for 15–20 hours to give the new chamber its firm support.

Towe & Cifelli (1967) proposed that calcification processes in foraminifera occur on an organic membrane consisting of a 'passive' carbohydrate (the sulfated acid mucopolysaccharide described earlier) which gives form and shape to the wall and an 'active' protein which acts as a template for the nucleation of calcite crystals. Angell (1967b) and Reiss & Schneidermann (1969) suggested that calcification proceeds from this organic lining to produce a chamber wall. On either side of this calcareous wall there are organic membranes which are united by cylindrical organic pore

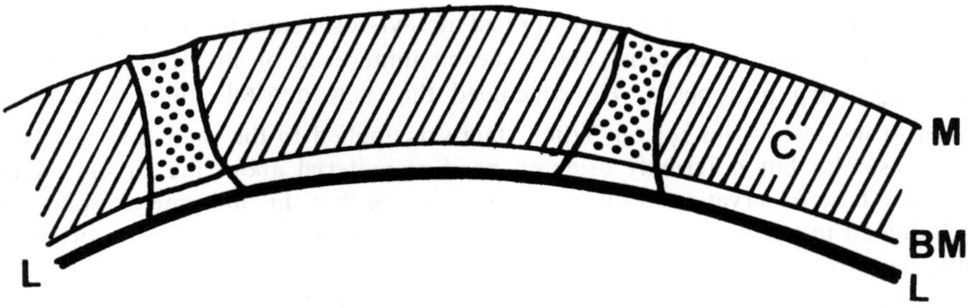

Fig. 35. Diagram of the test wall, final chamber, *Rosalina floridana*, showing organic membranes (after Angell, 1967b); L. Organic lining; BM. Basal membrane; C. Calcite layer; M. membrane covering exterior surface of calcite layer (stippled areas represent pore processes).

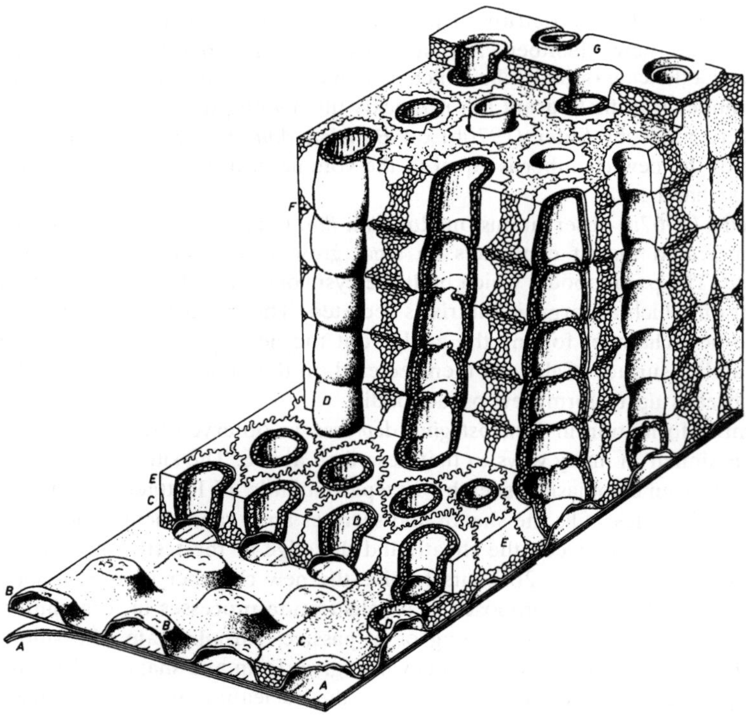

Fig. 36. Construction of wall in *Ammonia* (after Banner & Williams, 1973), (see text for explanation).

linings (Fig. 35). Other investigators (Arnold, 1964; Lynts & Pfister, 1967; Hansen, 1970; Banner, Sheehan & Williams, 1973; Stapleton, 1973) suggested that the organic layer may represent a three-dimensional matrix *in* which calcification occurs rather than a template *on* which crystals grow.

Based on extensive studies of electron micrographs and cultured specimens of *Ammonia* spp., Banner & Williams (1973) proposed a complex mode of chamber addition (Fig. 36). The following letters refer to this figure. The outer zone of cytoplasm (plasmalemma) secretes a chitinoid wall (B) in the shape of the new chamber. This organic membrane contains small evaginations (pustulae) analogous to the truncated cones of Arnold (see section 5). Cytoplasm streams out of this new framework, covers the pustulated membrane and secretes fine granular calcite as a lamellar sheet covering all but the raised pustalae on the initial membrane (C). The ectoplasm secretes cylindrical 'tectinuous' sheets (D) around each pustula (the organic linings of pores). The ectoplasm secretes calcite granules which become fused into blocky grains near the organic lining (E) but retain their granularity away from this zone (F). The wall is completed when the ectoplasm secretes a veneer of calcite granules over the distal part of the calcareous zone (G). Pauses occurring during the secretion of the calcareous portion of the wall could result in 'tectinuous' membranes being laid down over the pores.

It is interesting to note that according to this mechanism of wall construction, the pores are initiated by the pustulae on the organic membrane, a mechanism which contrasts with that of Greiner (1974) (section 5) in which the pores are no more than a non-adaptive consequence of separation during calcification of inorganic calcite from organic materials.

Hemleben (1969b) and Bé & Hemleben (1970) proposed that the bilamellar nature of the planktonic wall is due to crystal growth on both sides of the organic layer.

The absence of oriented crystals in porcellaneous walls was explained by Greiner (1974) as being due to the absence of an organic template (Fig. 37). He explained the presence of oriented calcite rhombs on the surface of porcellaneous specimens as the result of surface tension at the seawater-cytoplasm interface.

The rate of addition of new chambers is variable, changing not only from species to species, but also within a given species according to age and environmental conditions.

Myers (1943b) observed that very young specimens of *Tretomphalus* could add two new chambers in a 24-hour period, but when the specimens had grown to 5–6 chambers the velocity of chamber addition decreased to one chamber every 18–24 hours. To reach the adult size of 13–17 chambers required a period of 42 days. The same author noted that *Elphidium*, *Rotalia* and *Discorbis* are able to add a new chamber in a 24-hour period but that 5–7 days of active metabolism are necessary prior to the chamber addition. Jepps (1956) observed an even more rapid rate of addition in *Elphidium*. She noted that chamber formation began a few hours after the young leave the parent and proceeded at a rate of two chambers per day, a rate

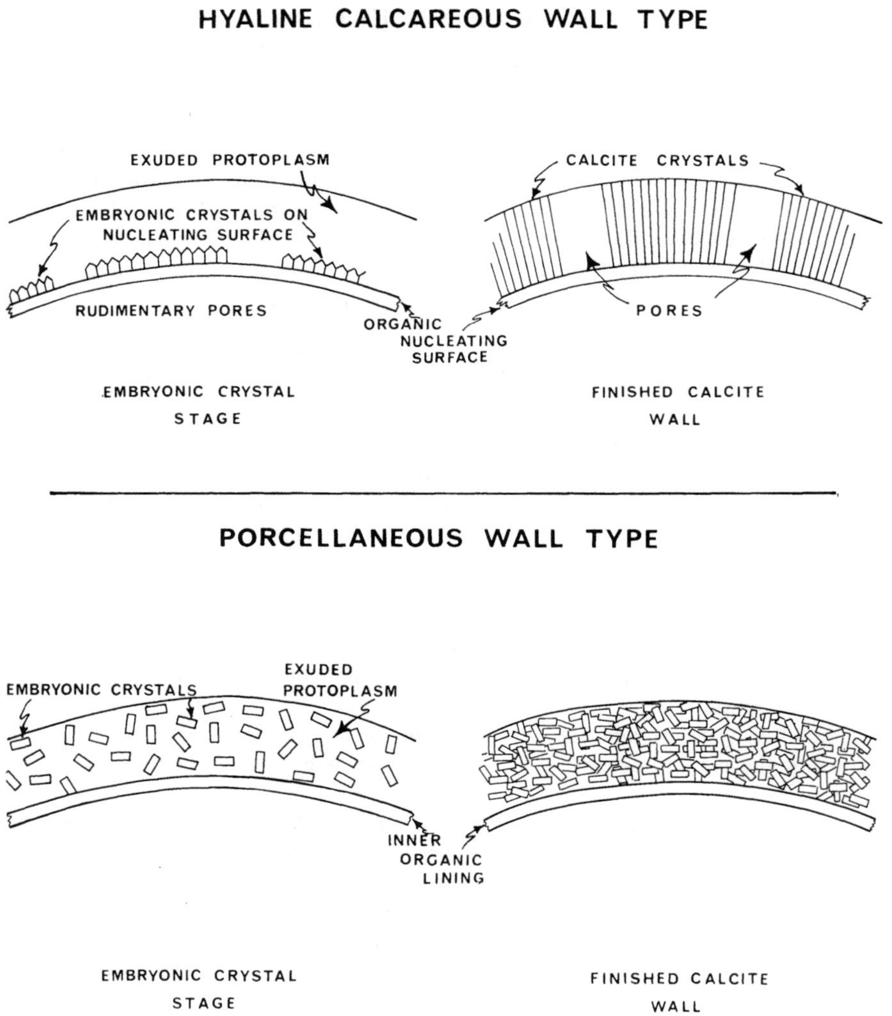

Fig. 37. Differences between wall construction in porcellaneous and hyaline calcareous foraminifera (after Greiner, 1974).

which gradually diminished with time. Ross (1972) explained the decrease in rate with time by suggesting that whereas protoplasm was formed at a steady rate, additional chambers became increasingly larger. Consequently, an increasingly longer waiting period is required before each new chamber is added.

Cultured specimens of *Bolivina doniezi* required about 11 hours for the development of a new chamber. A new chamber was added every five to ten days (Sliter,

1970b). Bradshaw (1957) observed the growth rate of cultured *Streblus beccarii tepida*. Under optimum conditions a new chamber could be added in 24 hours, but with an increase or decrease in either temperature or salinity the rate would diminish and ultimately reach a point at which the organism would live but not add chambers. *Heterostegina depressa* added a new chamber every two to three days when grown under optimum conditions (Röttger, 1972c), but required a longer period when either the temperature or illumination were reduced. A decrease in temperature will decrease the metabolic rate and consequently reduce the digestive and growth rates.

Several interesting morphological studies have shown that foraminifera follow regular growth patterns which can be numerically (Růžička, Hajkr & Grmela, 1966; Olsson, 1971a) and geometrically (Hofker, 1968; Berger, 1969a; Chang, Kaesler & Merrill, 1974) defined.

11. Pathology and regeneration

In most foraminiferal populations there are several specimens which lie beyond the normal range of morphological variation. Some of these abnormalities develop during the life span of the individual and others are the result of post-mortem changes. Examples of the second case are various agglutinated forms such as *Trochammina* and *Jadammina* which are often found in a flattened and collapsed state after death.

The abnormalities of the first case are more interesting and diverse from the point of view of their origin and environmental significance. These unusual forms may be due to either (a) mechanical or (b) ecological causes.

A. Mechanical Causes

The test may suffer many kinds of mechanical disfigurements which are healed or patched by the organism. These impairments commonly occur in the last few chambers. At the place of damage the protoplasm immediately swells to fill the break and secretes material which tends to heal the injury. The newly constructed portion of the wall is initially very delicate, but thickens and even regains some of the original ornamentation in time. Nevertheless, it is always possible to distinguish the reconstructed part of the wall because it is thinner and exhibits scars and abnormal contours. Pozaryska (1957) noted that the regenerated areas always demonstrate a marked tendency toward simplification. Laboratory cultures of *Heterostegina depressa* have shown that healthier and more rapidly growing specimens have thinner walls in the last few chambers and are therefore more susceptible to breakage (Röttger & Berger, 1972).

In some cases the broken portion of the last chamber is not healed and a second aperture is created (Amicis, 1895; Cushman & Jarvis, 1930; Ellison, 1953; Livental[1], 1953; Boltovskoy, 1954b, 1966b; Dhillon, 1969, 1970).

A summary of observations on abnormalities was published by Majzon (1936). Some of his cases and more recent ones are cited below.

Apertural breakage in specimens of *Articulina* and *Meandroloculina* seems to induce new chamber growth at angles different than the usual (Bogdanowicz, 1950b, 1973; Roth, 1962). Chamber damage to *Elphidium* may result in the growth of a digitated patch at the breakage site. Schultze (1854) was the first to partially destroy tests of a living foraminifer (*Elphidium*) in order to observe the reaction of the organism. Similar observations made by Verworn (1889) indicated that only those portions of the broken test containing a nucleus were able to regenerate new test material. Pazdrova (1958) found that *Opthalmidium* has a great capacity for the regeneration of mechanically broken tests. The first observations of regenerative abilities in planktonic forms were made by Rhumbler (1911) who also gave a brief classification of abnormalities.

In studying the fauna of the estuary of the Rio de la Plata, Boltovskoy (1957a) found that the percentage of regenerated forms varied in different parts of the estuary. Where the fauna was depauperate and under environmental stress, the percentage of regenerated forms was low, presumably because of a reduced ability to regenerate. Boltovskoy used the percentage of regenerated forms ('regeneration index') as a measure of ecological conditions in the estuary.

B. Ecological Causes

In addition to mechanical breakage the test may exhibit morphological deformations due to ecological factors. Many of these irregularities are capricious ones reflecting the influence of the substrate on attached tests. An illustrated example is *Cibicides lobatulus* as shown in Fig. 76.

A rapid change in the physio-chemical environment (salinity, nutrient supply, temperature, trace element concentration, etc.) may provide a change in the normal rate and nature of test growth. Myers (1943c) observed this on a small scale noting that chambers added in the winter are often smaller than those formed during the summer. This gives the test a somewhat irregular form. Changes in the smoothness of the periphery of *Elphidium discoidale* in the estuary of the Rio de la Plata have been explained by Boltovskoy (1957b) as responses to changes in salinity. The same kinds of observations were made in the brackish waters along the North Sea coast by Bartenstein (1938), on specimens of *Elphidium gunteri* by Closs & Madiera (1968), and on cultured forms of *E. crispum* by Murray (1963). Some changes are of a more dramatic nature; deformation of the aperture (Bartenstein & Brand, 1938; Boltovskoy, 1958a), or deformation of the entire test (Millett, 1898–1904; Bartenstein & Heinemann, 1954; Dzhanelidze, 1957; Arnal, 1957; 1958, Kurc, 1961; Closs, 1962; Glaçon, 1964; Forti & Roettger, 1967; Tufescu, 1968). Studies of populations around sewage outfall areas (Watkins, 1961; Lidz, 1965) have revealed large numbers of monstrous specimens. A large number of teratological or pathological specimens of *Ammonia beccarii* were reported in the hypersaline lagoons of the eastern Mexican coast (Seiglie, 1964; Ayala-Castañares & Segura, 1968).

Bé (1965) suggested that the diminutive final chamber in some planktonic species may be considered as an aberrant or pathological growth feature. Berger (1969a) modified the German term *Kummerfauna* (depauperate of starved fauna) and introduced it to the foraminiferological literature. He applied the term kummerform to those foraminiferal tests in which the last chamber is equal to or smaller than the previous one. In a later study (1970b) he noted that this definition does not apply to *Orbulina universa*. Kummerforms of *Orbulina* are those in which the penultimate chamber is not completely enveloped by the last. According to Berger and other investigators (Hecht & Savin, 1970; Bandy & Theyer, 1971; Bandy, 1972a) kummerforms are the result of environmental stress. For *Globigerinoides ruber*, a shallow water tropical species, the stress condition appears to be related to a decrease in water temperature, whereas for *Globoquadrina dutertrei* it may be an increase in water temperature (Hecht & Savin, 1971, 1972). However, we should note that the latter species is not only warm-temperate but is also well known as a warm-water species.

The term kummerform has received ample usage in the planktonic foraminiferal literature. Nevertheless, not all investigators hold the same opinion regarding its interpretation. Banerji, Schafer & Vine (1971), for example suggested that kummerform phenomena are controlled by ontogeny rather than by the environment.

After a mathematical study of the growth patterns of *Globigerina pachyderma*, a species well known for its kummerform development, Olsson (1973) showed that a reduction in chamber size occurs at the same stage of ontogeny in all specimens. He concluded that these observations do not support the idea that kummerforms are caused by environmental stress. But the observations do support the hypothesis that kummerforms develop as a consequence of old age.

Two other interesting types of pathological forms have been described in the planktonic foraminifera. The first is the phenomenon of 'twinned' specimens; i.e. two specimens firmly attached to each other (Fig. 38). Specimens exhibiting this feature

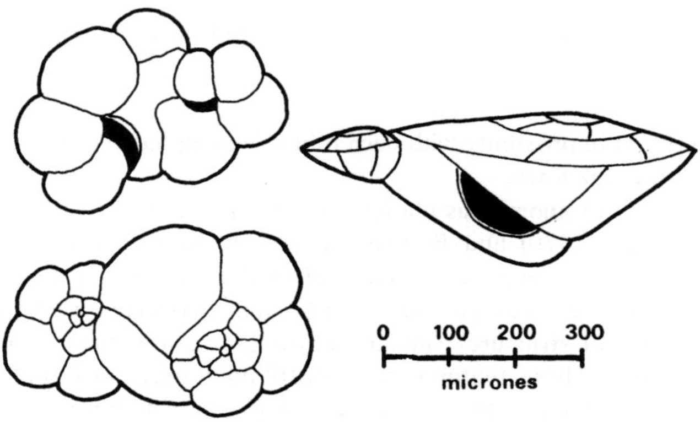

Fig. 38. 'Twinned' specimens, Southwestern Atlantic Ocean.

have three things in common: (a) both specimens always belong to the same species, (b) the tests are always of different size and (c) as far as we know, the phenomenon has been observed only in planktonic foraminifera typical of cold and cold-temperate waters. This feature was observed by Earland (1933, 1934, 1936), Boltovskoy (1966b) and Boltovskoy & Boltovskoy (1970). Earland explained its origin as a result of plastogamy, but Boltovskoy considered this explanation unsatisfactory because of the size differences between the two participants and because the

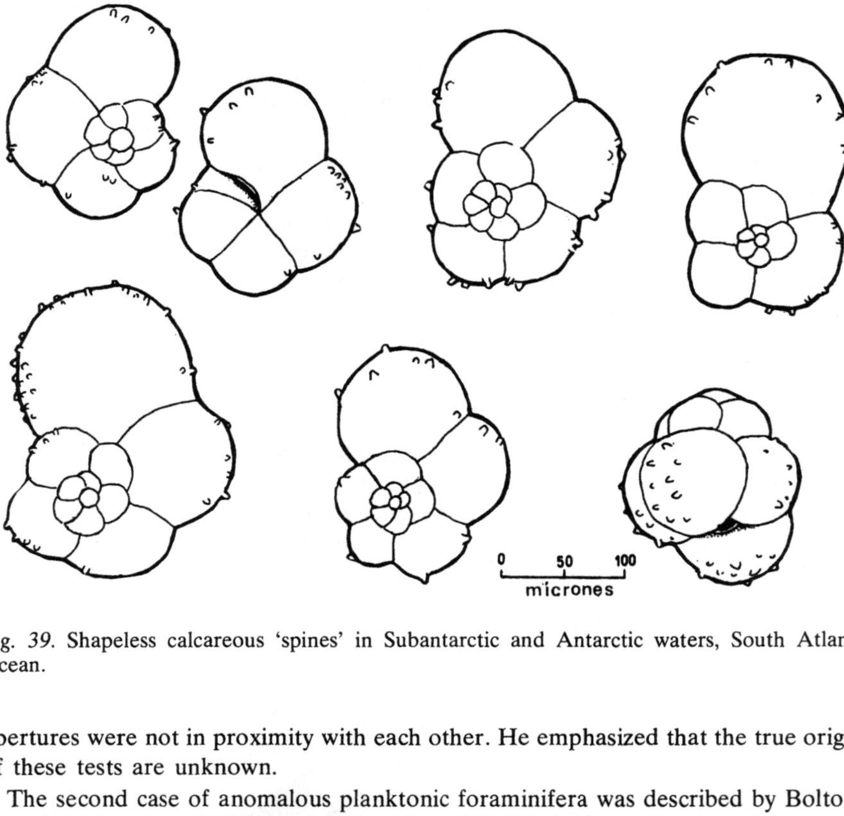

Fig. 39. Shapeless calcareous 'spines' in Subantarctic and Antarctic waters, South Atlantic Ocean.

apertures were not in proximity with each other. He emphasized that the true origins of these tests are unknown.

The second case of anomalous planktonic foraminifera was described by Boltovskoy & Boltovskoy (1970) and Boltovskoy (1971b) in Subantarctic and Antarctic waters. It consists of numerous specimens of *Globigerina quinqueloba*, *G. bulloides*, *G. pachyderma*, *Globigerinita glutinata* and *G. uvula*, collected in plankton samples, which exhibited wall structures similar to strong and sometimes almost shapeless calcareous 'spines'. These features were distributed irregularly and sparsely, but mainly near and on the periphery of the test (Fig. 39). Another striking feature was their orientation, as many spines were not perpendicular but oblique to the surface.

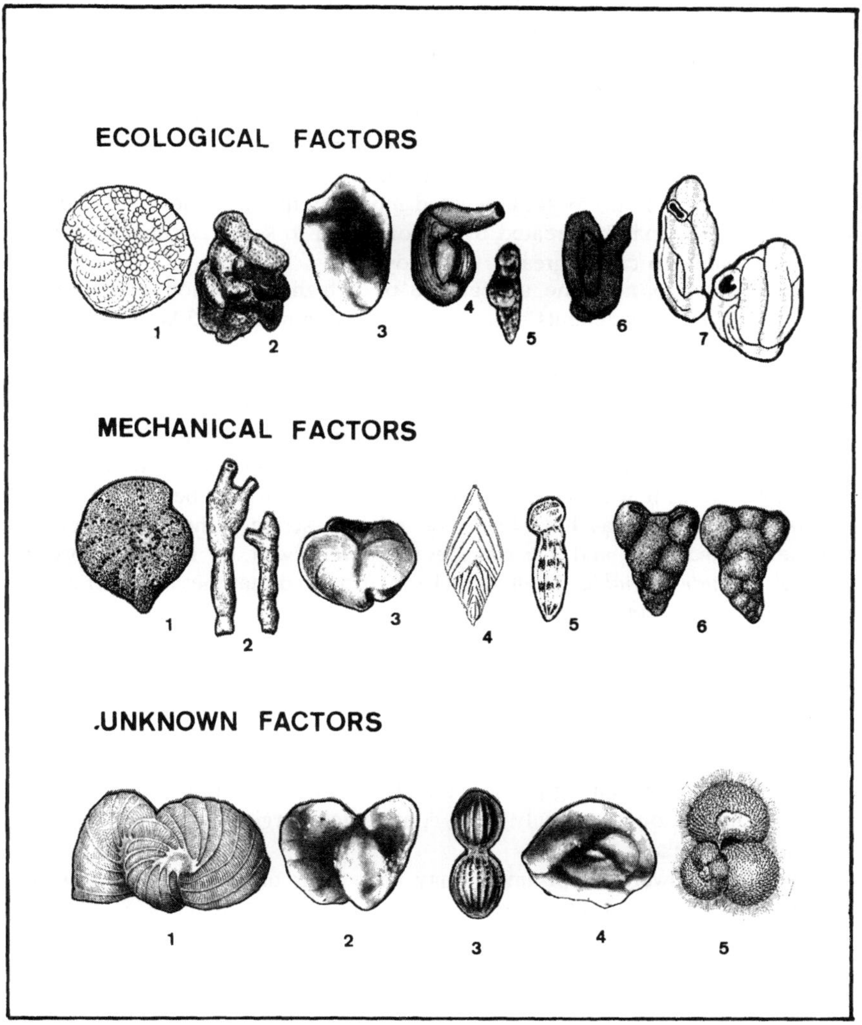

Fig. 40. Pathological specimens; Ecological factors: 1. *Elphidium* (Jepps, 1942); 2. *Cibicides* (Earland, 1934); 3. *Quinqueloculina* (original); 4. *Quinqueloculina* (Arnal, 1958); 5. *Bolivina* (Bartenstein & Heinemann, 1954); 6. *Miliammina* (Closs, 1962); 7. *Quinqueloculina* (Bogdanowicz, 1952); Mechanical factors: 1. *Elphidium* (Boltovskoy, 1957b); 2. *Articulina* (Bogdanowicz, 1952); 3. *Lenticulina* (Heron-Allen & Earland, 1932); 4. *Frondicularia* (Pozaryska, 1957); 5. *Nodosaria* (Pozaryska, 1957); 6. *Bulimina* (Boltovskoy, 1954b); Unknown factors: 1. *Peneroplis* (Rhumbler, 1911); 2. *Rotalia* (original); 3. *Oolina* (Earland, 1934); 4. *Spiroloculina* (original); 5. *Globigerinoides* (Rhumbler, 1911).

These structures are quite unlike the long delicate 'traditional' spines typical of many planktonic foraminifera. There has been very little note of these structures in the literature. Heron-Allen & Earland (1929) figured *Globigerina cristata* and Parker (1962) illustrated several specimens of *Globigerinita humilis* which exhibited approximately the same (but less developed) structures. They did not mention these 'spines' in the text. In addition, Earland (1934) figured a specimen of *G. bulloides* (misidentified as *G. triloba*) covered with the same type of structure. He called it 'spinose variety'. Boltovskoy & Boltovskoy devoted more attention to these structures and suggested that they may be created by the organism to serve as supports for cytoplasmic expansions to create greater test buoyancy.

It is interesting to note the appearance of a high proportion of pathological specimens in Tertiary sediments described by Bogdanowicz (1950a) and Dmitrieva (1952). The sediments in which these specimens were found represent a time of abrupt hydrological change in the basin of deposition; a change which appears to be correlated with the morphological changes in the foraminifera.

Figure 40 shows some of the variations exhibited by pathological specimens.

It seems certain that many abnormal specimens of species already described have been 'discovered' as new species. The degree of distortion may be so great in some cases that new genera may have been created to describe pathological specimens. The genera *Cibicedella* and *Dyocibicides* may very well be nothing more than variants of *Cibicides variablis*. Thalmann (1952) proposed that these sorts of misidentifications be called *nomina caduca*.

12. Dwarfism

Foraminiferal specimens which are considerably smaller than normal are well known in both fossil forms and among living populations. Bandy (1960a) demonstrated that such specimens are found not only along continental margins but on the continental slope and abyssal plains as well.

Populations of 'dwarfed' specimens may have their origins in various selective sorting processes that occur after the death of the organism. Tasch (1955), summarizing the work of various investigators on this subject, suggested that this sorting can be due to the work of various natural sieves including current and wave action, vegetation traps and organisms such as sponges and algae.

Of more interest for the study of living foraminifera are the ecologic conditions that could cause dwarfism. In addition to being smaller than usual such specimens often have asymmetrical tests and exhibit poorly developed ornamentation. There appear to be quite diverse opinions regarding the ecological causes of dwarfism. Do unfavourable environmental surroundings create a stress which precludes complete growth and development or do unusually good conditions result in earlier sexual maturity and thus produce smaller adults? It is around these two points of view and their derivitives that most arguments have been presented.

A. Dwarfism Due to Unfavourable Conditions (Environmental Stress)

Myers (1936, 1943c) accredited the reduction in size to a lack of nutrition. He observed that specimens of *Elphidium crispum* found in the sublittoral zone, where nutritional conditions are very good, tend to be much larger and have more chambers than specimens found in the littoral zone, where less nutritional material is available. Boltovskoy (1954–1959) cited numerous examples of unusually small specimens on the Argentine continental shelf. He attributed this phenomenon to a lack of nutrition and a disequilibrium in the quantity of trace elements, Lalicker (1948), in a discussion of dwarfism in protozoa, suggested that reduced size is due to insufficient nutrition, low temperature and changes in pH. He cited experimental data on *Paramaecium bursaria* which showed a reduction in size correlated to pH reduction. Mjatliuk (1974) also suggested that insufficient nutrition creates undersized foraminifera which lack ornamentation and exhibit various teratological features.

A lack of dissolved oxygen was cited by Said (1953), Pokorny (1958) and Bettenstaedt (1962) as a factor which would result in a reduced metabolic rate and a consequent reduction in the rate of growth. In support of these conclusions Miller (1953) found a dwarfed fauna at one collecting site where anaerobic conditions prevailed.

Reduced salinity has been cited by Morishima (1955) as the factor contributing to the reduced size of the foraminiferal fauna in Tokyo Bay. The same factor was cited by Kurc (1961) in a study of the small Miliolidae and Buliminidae in the Thau Lagoon. The effects of reduced salinity on foraminiferal size were also discussed by Le Calvez & Le Calvez (1951), Myers & Cole (1957), Furssenko (1959) and Tufescu (1973). They, as well as Moore (1957), suggested that reduced size is one of the consequences of salinity reduced below normal marine levels. Specimens of the same species living under more normal salinity conditions are larger. Specimens of *Elphidium crispum* cultured by Murray (1963) developed smaller chambers when the salinity was reduced below normal. *Miliammina fusca*, a species adapted to low salinity waters, was observed to have smaller specimens when found in areas of higher salinity (Wright, 1968).

Troithkaja (1970) suggested that the presence of abundant volcanic glass in various lagoons along the Sea of Japan is the environmental factor that accounts for the small size of the foraminiferal fauna.

Kraskeninnikov (1958a) mentioned that a very sparse and undersized foraminiferal fauna (primarily *Cibicides lobatulus*) was found on calcareous sediments in the White Sea, Russia in areas where there were large quantities of H_2SO_4 released by decaying organic material. On the same sediment in an oxygen rich environment the fauna is more diverse and larger in size.

Todd (1957) suggested that benthonic specimens living near their ecological tolerance limits will be smaller than normal. This suggestion was supported by the observations of Lena (1966) on the fauna of Tierra del Fuego, Argentina. In a study of dwarfed specimens from a stagnant pond in Massachusetts, Todd & Low (1961) indicated that the unfavourable environment was responsible for the size reduction. One of us (EB) has observed that planktonic specimens in the southwest part of the Atlantic Ocean become much smaller near the limits of the water masses which they characterize. This size reduction can even be used to mark the zone of convergence between two water masses.

It appears, therefore, that there is a sizeable body of opinion supporting the theory that environmental conditions which are restrictive to one degree or another are the primary cause of dwarfism among foraminifera. But, at the same time, there is ample support for the idea that favourable living conditions tend to produce specimens that are smaller than normal.

B. Dwarfism Due to Optimal Reproductive Conditions

This idea was originally proposed by Nicol (1944) for foraminifera and was amplified by Bradshaw (1957) based on the study of *Streblus beccarii tepida* in laboratory cultures. He noted that those specimens kept at 15–20 °C (less than normal for this species) were considerably larger than those maintained at normal temperatures of 25–30 °C. *Bolivina doniezi*, whose laboratory reproductive temperature range is 18–22 °C, grows to a larger size when cultured at temperatures of 22–26 °C (Sliter, 1970b). These larger specimens grow at a slower rate than normal but have a postponed sexual maturity. Bradshaw later (1961) reaffirmed his earlier observations and concluded that foraminifera living near the limits of their tolerances grow to a larger than normal size. He also cited the observations of Parker that planktonic specimens larger than normal are often found in high latitude bottom sediments outside their normal distribution areas. Lewis & Jenkins (1969), in a study of *Nonionellina flemingi* in New Zealand waters, found that specimens collected in the cooler higher latitude waters south of New Zealand are larger than those collected near the islands. They attributed the size difference to the fact that the specimens found in the warmer waters reached reproductive maturity earlier than the others. Berger (1969b) gave evidence that planktonic specimens are smaller in the cold, fertile, phosphate rich regions of the oceans and larger in the warmer, less fertile areas.

Lankford (1959) and Phleger (1960b) suggested that very small specimens develop under optimal conditions. Similar conclusions were drawn by Gunter (1957) regarding the small sizes of various groups of organisms. He suggested an inverse relationship between optimal temperature conditions and size. The relationships between salinity and size were less clear, but he demonstrated that small specimens are usually found in areas with optimal salinity.

In a summary of data from several environments, Lankford (1967) noted that often thin shelled, relatively unornamented, small specimens occur in areas where the maximum living benthonic population is found. Presumably, these are areas of optimum environmental conditions. The specimens described above are in contrast to those thick shelled, highly ornamented, large specimens found near the boundaries of the distribution of a species.

The preceding observations indicate that there is no general agreement regarding the nature of the effects that various environmental factors have on specimen size. It is possible that large specimens are a reflection of unfavourable conditions in which the specimens reach reproductive maturity later than usual; and that 'normal' or small specimens are produced under optimal conditions which result in early reproduction; and that truly dwarfed specimens are the product of very inhospitable surroundings in which the metabolism rate is reduced to such an extent that individuals grow very slowly. These explanations are supported by the observations of Bé, Harrison & Lott (1973) on living specimens of *Orbulina universa* in the Indian Ocean. They found that under optimum reproductive conditions (greatest relative abundance of the species) specimens ranged in size from 450–650 μ. In areas of increased water temperature, the abundance decreased and size increased. In areas of decreased water temperature, the abundance decreased and size decreased.

IV. BENTHONIC FORAMINIFERA OF NORMAL MARINE WATERS

1. General characteristics

Benthonic foraminifera are those that live on the sea bottom. Taxonomically they are much more diverse than the planktonic foraminifera but the number of individuals per species is less among the benthonic forms.

To better understand the conditions of life under which the benthonic foraminifera live and to become familiar with the terminology applied to the oceanic zones, the reader is referred to Fig. 41.

Benthonic foraminifera live at all depths of the oceans including those areas subject to periodic inundation and desiccation; e.g. the intertidal zones. Stschedrina (1957) noted that one species of *Trochammina* prospers under such conditions. Voorthuysen (1960) found several species living in abundance in areas which are devoid of water for up to 4 hours a day. One of us (EB) has observed that some species can survive in the moist sand of some small tidal pools along the Patagonian coast near Puerto Deseado, Argentina. Normal high tide does not reach these pools and only during exceptionally high tides or heavy surf do these pools receive marine water. The genus *Pyrgo* dominated the living foraminiferal fauna in these pools.

Fischer (1966) cited various foraminifera, notably *Peneroplis* in southeast Asia, and *Calcarina* and *Baculogypsina* in the Great Barrier Reef, which could survive the absence of water for periods of 2–3 hours. Experiments with these specimens revealed that they were able to survive up to 20 hours without water while sitting on the surface of sandy sediment in a Petri dish. Both Fischer and Smith (1968) observed that *Marginopora vertebralis* is a hardy shallow water species. Smith noted that this species seems to thrive when exposed to direct sunlight above low tide levels. Exposure to fresh water (rainfall) had no adverse effect. Experiments conducted by Dietz-Elbrächter (1969) on *Ammonia beccarii* showed that 45% of the specimens placed in a completely dry Petri dish survived one hour and 10% survived 3 hours. When a piece of wet filter paper was used in the dish, 50% survived for 26 hours and 10% for 56 hours.

It seems clear, therefore, that some benthonic species are well adapted to live in the shallowest of environments. On the other hand, they are also found in the ocean deeps; e.g. up to 10 415–10 687 m in the Tonga Trench (Beljaev, Vinogradova & Filatova, 1960).

Benthonic foraminifera can be divided into two groups, those that move freely over the substrate, and those that attach themselves to sediment or vegetation. Species belonging to the first category are the most numerous. Among the latter

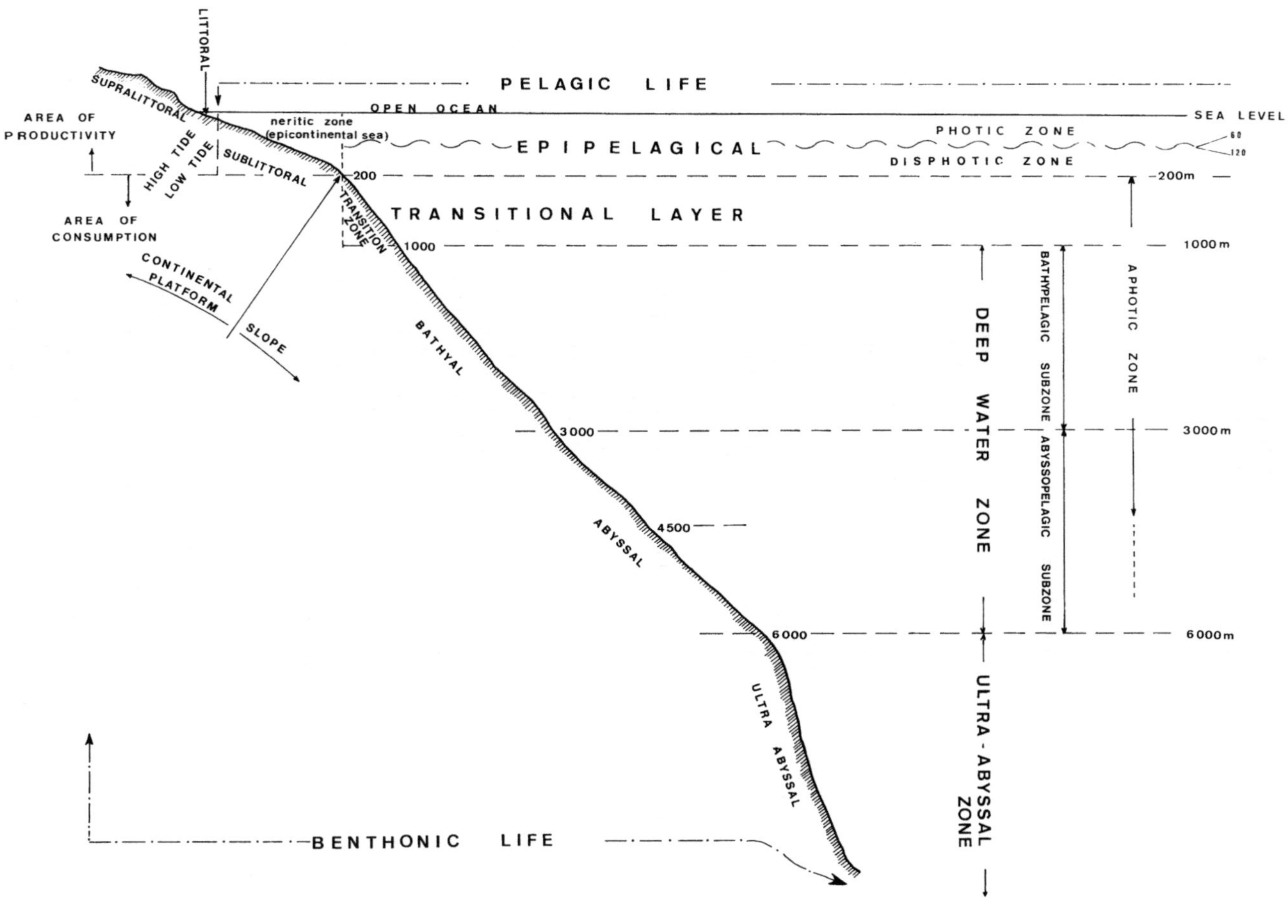

Fig. 41. Vertical biological zones of the oceans (after Belyaev, Vinogradova & Filatova, 1960).

group are those that are permanently attached by means of their test (*Carpenteria*, *Rupertia*) and those attached temporarily by their pseudopodia (*Cibicides*, *Discorbis*). There are distinct adaptive advantages in being able to move the test; e.g. the specimen can move to areas of better nutrition, stability, etc., or away from unfavourable conditions. The main advantage offered to specimens which remain attached is stability in turbulent water. Free, unattached tests run the risk of being transported and even destroyed by waves and currents.

The variability of conditions in the benthos produces great morphological variability among the benthonic foraminifera. The first foraminifera appearing in the fossil record are agglutinated benthonic forms in Lower Cambrian rocks. The first planktonic forms did not appear until the Jurassic.

The optimum conditions for benthonic foraminiferal life occur at depths of 150–300 m where there are thousands or even tens of thousands of specimens per square metre.

Benthonic foraminifera generally live on or above the sediment-water interface but can penetrate the bottom sediment and prosper there. Myers (1943c) suggested that foraminifera which were buried to depths no more than 5–7 times their diameter could work their way to the surface, but if buried deeper than that would not survive for more than 2–3 months. More recent investigations have revealed that various species can penetrate and survive at much greater depths. Richter (1961) found active specimens of *Nonion depressulus* at depths of 4 cm in sediments along the North Sea coast. Boltovskoy (1966c) found a few specimens of *Elphidium* and one of *Epistominella exigua* at depths of 16 cm in Puerto Deseado, Argentina. Schafer (1971) and Ellison (1972) found living specimens as deep as 9 cm. The most spectacular discoveries were those of Akers (1971) who found living specimens in intertidal marsh samples 40–50 cm below the surface.

Data pertaining to the rate of movement of foraminifera are presented in Chapter II and discussion of the relationships between various environmental factors and foraminifera are given in Chapter IX. The remainder of this chapter deals primarily with the distributional characteristics of benthonic foraminifera in normal marine waters.

2. Qualitative horizontal distribution

When describing the area of distribution of a species it is necessary to distinguish between the zones in which the species normally lives and reproduces. The species may be transported by currents, etc. to a zone beyond its normal area. The species may even survive and grow in that zone but not be able to reproduce.

There are some species which are found distributed in all the world's oceans. These cosmopolitan species, relatively few in number, generally prefer to live in certain depth zones. Their distributional area is not necessarily contiguous. Among these cosmopolitan species are *Quinqueloculina seminulum*, *Bulimina marginata*, *Buliminella elegantissima*, *Nonion affine*, *Elphidium crispum* and *Cibicides lobatulus*.

Some species live only in one specific part of the ocean, a particular bay or coastal zone. These endemic species are unknown in other areas. Examples of such species are *Elphidium lessonii* on the southern continental shelf of Argentina, *Jullienella foetida* in the Guinean coastal zone and *Bifarina hancocki* on the continental shelf of Ecuador.

The term biogeography is applied to studies of the areal distribution of organisms. If the organisms are animals living in the sea, the term marine zoogeography is used. Biogeography is concerned with the particularities of geographic distribution of various organisms, the relationship between biological factors and these distributions, and the history of these distributions. Consequently, it is not surprising that the results of such studies are usually shown on biogeographic maps based on (a) qualitative, (b) quantitative and (c) historical criteria.

(a) Qualitative criteria. Qualitative studies result in the division of an area into different zones characterized by different faunas. The principal ecological factor that controls this separation is temperature. There are, of course, other influential factors, but they generally play a subordinate role. The preparation of a map based on these qualitative distributions is generally the first step in any biogeographic study.

(b) Quantitative criteria. The availability of nutritive material is probably the most significant factor affecting the numerical distribution of individuals, although other factors may exert considerable influence in particular areas. Maps prepared on a quantitative base are a logical second step in a biogeographic study as they show not only the distribution of taxa but the frequency of these taxa in different areas.

(c) Historical development criteria. Maps prepared using this perspective show the routes by which the Recent fauna migrated into a particular area and how the fauna has changed with time. These maps are particularly complex as they require not only a knowledge of the qualitative and quantitative aspects of the fauna but also an understanding of the geological history of the area and its adjoining regions.

Perhaps one of the best known, extensive and detailed biogeographic investigations is that of Ekman (1953). Using the data of many investigators as well as his own observations, he produced an excellent synthesis of zoogeographic ideas.

The data collected and used by Ekman were utilized by Hedgpeth (1957) to prepare a zoogeographic map of the littoral zones of the world. He established provinces using only coastal data. The fauna of the deeper parts of the oceans is much more monotonous and data is less available from these areas. Four major types of provinces are shown on Hedgpeth's map: (a) arctic–antarctic (cold-water fauna), (b) boreal–antiboreal (cold-temperate fauna), (c) warm-temperate and (d) tropical.

Foraminifera lend themselves very well to zoogeographic investigations since they tend to be representative of particular geographic areas. They are rather sensitive to changes in the environment in terms of species diversity and abundance.

One of the first maps of shallow water foraminiferal distribution was prepared by Cushman (1928, 1948). The map bears the date 1921 so probably contains data collected up to that time. Two maps were prepared, one showing the distribution of typical cold water foraminifera, and the other dealing with the warm water fauna.

This latter map contained 4 provinces: (a) West Indian, (b) East African, (c) Mediterranean and (d) Indo-Pacific.

Thanks to the zoogeographic studies of Cushman and later investigators,[5] it has been possible for us to update and refine this first map. We realize that our map (Fig. 42) too is of a preliminary nature since many areas have not been well studied from a zoogeographic perspective. We also lament that we are unable to describe more than 3 major foraminiferal regions (compared to the 4 of Cushman). The state of the littoral fauna precludes any finer major division. There is no uniformity in the literature regarding the names of the faunal subdivisions.

The divisions we find most convenient and logical are

A. AREAS DOMINATED BY WARM WATER SPECIES

1. West Indian Province; 2. Panamanian Province; 3. Lusitanian Province; 4. Mediterranean Province; 5. West African Province; 6. East African Province; 7. Indo-Pacific Province.

B. AREAS DOMINATED BY COLD WATER SPECIES

1. Arctic Province; 2. Antarctic Province

C. AREAS DOMINATED BY TEMPERATE WATER SPECIES
(warm temperate and cold temperate)

1. Pacific North American Province; 2. Atlantic North American Province; 3.

[5] A list of all these investigators would be too immense since all those who have examined Recent littoral and shelf foraminifera have contributed to our better understanding of faunal provinces. We cite only those investigations which were of particular help and/or which specifically dealt with zoogeographic problems: Albani (1965, 1968a,b, 1970), Ayala-Castañares (1963), Bandy (1963b, 1964), Barbieri & Medioli (1969), Bhatia (1956a), Berthois & Le Calvez (1959), Bhalla (1968, 1970), Boltovskoy (1954–1970), Braga (1960), Braga & Galhano (1965a,b), Brenner (1962), Cebulski (1969), Chatterjee & Gururaja (1968), Chekhovskaja (1973), Chiji & López (1968), Closs (1962), Closs & Barberena (1960a,b, 1962a,b), Collins (1958), Davis (1964), Drooger & Kaasschieter (1958), Dupeuble (1963, 1964a,b), Dupeuble *et al.* (1965), Earland (1933, 1934), Ferreira (1958), Gabel (1971), Galhano (1963), Guilcher *et al.* (1965), Haga (1963), Haman (1971a), Haynes (1973), Hedley, Hurdle & Burdett (1965), Heron-Allen & Earland (1922, 1932), Huang (1971a,b), Hulme (1964), Iaccarino (1967–1969), Kameswara (1971), Kim, Kim & Kim (1970), Lankford & Phleger (1973), Le Calvez (1958, 1963), Le Campion (1968), Lidz & Lidz (1966), Mangin, Desbrosses & Lecolle (1969), Margarel (1966), Mateu (1970), Matoba (1970), Montcharmont-Zei (1956–1968), Moura (1965), Murray (1965–1970), Narchi (1956), Parker (1948, 1958), Phleger (1947–1964), Rao (1969–1971), Rocha & Mateu (1971), Rocha & Ubaldo (1964a,b), Rouvillois (1970), Said (1950–1953), Said & Kamel (1957), Saidova (1964, 1966b, 1970a,c), Segura (1963), Seiglie (1965, 1966), Sellier de Civrieux (1970b), Sen Gupta (1972), Sen Gupta & Schafer (1973), Sozeri (1966), Stockton (1973), Stschedrina (1947–1962), Takayanagi (1953), Thiede (1971), Todd (1957–1958), Todd & Low (1971), Tufescu (1974), Ubaldo (1967) and Uchio (1960, 1967).

Chilean–Peruvian Province; 4. Argentine Province; 5. South African Province; 6. Atlantic European Province; 7. Sinian Province; 8. Tasmanian Province.

The map prepared from this analysis is rather similar to that of Hedgpeth. This is not surprising since the geographic distribution of all these organisms is controlled by similar factors. There are, however, some differences, particularly along the western coasts of the Americas, the eastern coast of South America, and in European waters. These differences are partially due to the major four-fold division of Hedgpeth and our three-fold division. Some differences may be due to a lack of data in particular areas or to a deviation from the regional pattern because of the local geological history. An understanding of regional geological history can not be overemphasized for its role in clarifying zoogeography. Discrepancies between zoogeographic systems based on different animal groups may be explained by local geological events (see the discussion of foraminiferal zoogeography versus malacological zoogeography of South America p. 105).

The majority of the zoogeographic provinces can be divided into subprovinces. For foraminifera, this subdivision is possible only in some regions since insufficient data prevents subdivision in all parts of the world. As an example of subprovincial categorization, we will describe the South American continent.

Cushman considered the coastal areas of Columbia, Venezuela, the Guianas and Brazil as parts of the West Indian Province whereas the foraminifera south of this area were typical cold water forms. He did not make any finer division and only an approximate boundary line was drawn on his small scale map. Later studies in this area permitted a much more detailed provincial division of South America: Cushman & Wickenden (1929), Cushman & Kellett (1929), Cushman &Parker (1931), Heron-Allen & Earland (1932), Earland (1933, 1934), Carvalho & Chermont (1952), Boltovskoy (1954–1964), Tinoço (1955, 1967a,b, 1971), Narchi (1956, 1962a,b, 1963a,b), Closs & Barberena (1960a,b, 1962a,b), Boltovskoy & Theyer (1965, 1970), Seiglie & Bermúdez (1963), Bermúdez & Seiglie (1963), Miró (1965), Closs & Medeiros (1967), Lena (1966), Rodrigues (1968, 1972), Fossatti Dutra Pereira (1969), Madeira (1969), Roettger (1970), Saidova (1971), Seiglie (1971a), Khusid (1971) and Boltovskoy & Gualancañay (in press).

Even the considerable amount of data from these sources does not result in a map free of errors and gaps. We consider this division as a working one subject to constant revision and refinement. We must also note that the boundaries between subprovinces are not sharp even though the lines on the map may be. There is always a mixed faunal zone occurring near these boundaries. These mixed or transitional zones may be wide or narrow depending on local conditions. We have not placed these mixed zones on the map in order to avoid confusion. Therefore, the boundaries on the map between subprovinces are indicative of the position of the transitional zones between two different faunas.

A detailed examination of the provinces and subprovinces of South America (Fig. 43) will serve as an example of the kind of conclusions that have been drawn for all littoral and shelf areas of the world.

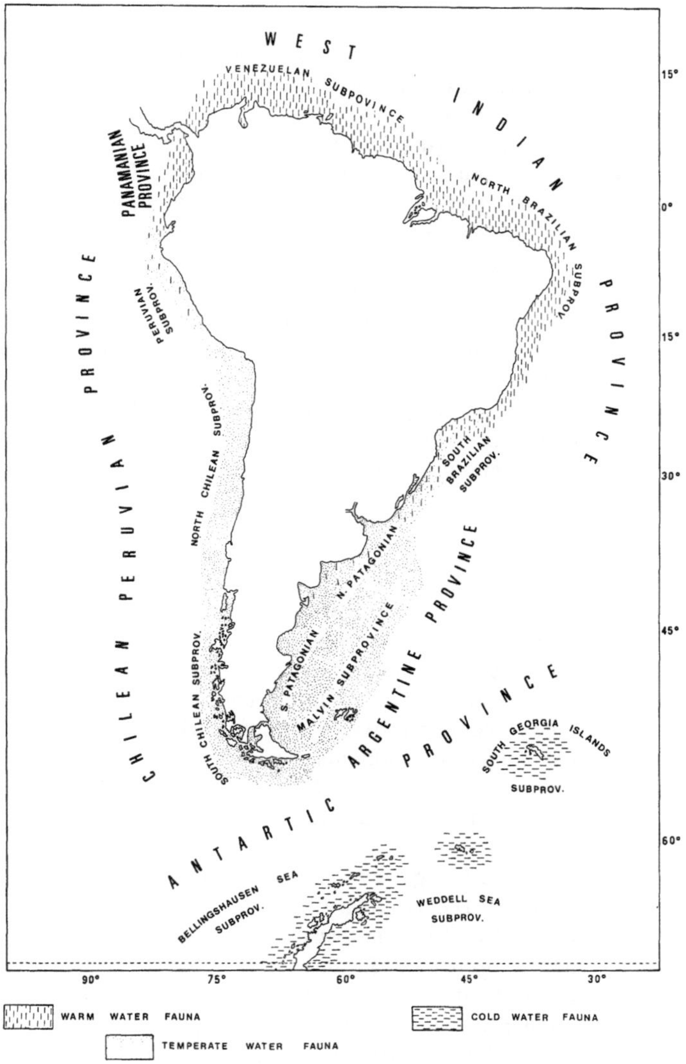

Fig. 43. Foraminiferal zoogeography of South America.

A. West Indian Province

This province is also known as the Caribbean or Antillean. It is geographically extensive, occupying the marine areas of the eastern coast of Central America, the Caribbean Sea, the Columbian, Venezuelan, Guianan and Brazilian coasts as far south as 32–33°S. The South American parts of this province can be divided into 3 sub-provinces: (a) Venezuelan, (b) North Brazilian and (c) South Brazilian. Not only does the fauna of the coastal zones belong to the West Indian province but so does the fauna of the coastal islands; the Lesser Antilles, Fernando da Noronha Island, Trinidad and others in the area.

The foraminiferal fauna consists of typical warm water species and is best characterized by *Amphistegina lessonii, Archaias angulatus, Bolivina subaenariensis, Borelis pulchra, Chrysalidinella dimorpha, Eponides antillarum, Spiroloculina* spp., etc. The limits of these 3 subprovinces are: from Panama to about 4°N (Venezuelan); from 4°N to Cabo Frio (North Brazilian); from Cabo Frio to 32–33°S (South Brazilian). The faunal differences between the Venezuelan and North Brazilian subprovinces are not great and the 4°N latitude boundary is only approximate. The differences between the faunal associations of the North and South Brazilian subprovinces are much clearer. South of Cabo Frio some temperate water species (*Bulimina patagonica, Cibicides dispars, Nonionella auris, Oolina borealis, O. costata*, etc.) typical of the Argentine Province are found mixed in with the abundant typical West Indian fauna. To a certain extent the South Brazilian subprovince is a zone of transition.

B. Argentine Province

South of the West Indian Province hydrologic conditions are somewhat more complicated and the biogeographic divisions reflect this complexity. The Argentine Province is an area of temperate water characterized by the dominance of *Buccella peruviana*, s.l.. This species also occurs along the Pacific coast of South America but its true 'realm' is in the Argentine Province. Other species also characteristic of this province are *Bolivina compacta, Bulimina patagonica, Buliminella elegantissima, Elphidium* spp. *Epistominella exigua, Miliolinella subrotunda, Quinqueloculina seminulum, Pyrgo ringens* and others, but the dominant species is *Buccella peruviana*, s.l.

Three subprovinces can be recognized in this area, (a) North Patagonian, (b) South Patagonian and (c) Malvin. The boundary between the first two is drawn at 41°S (Valdés Peninsula) and is based on the distribution of two species of *Elphidium*. In the North Patagonian subprovince *E. discoidale* is very abundant whereas *E. macellum* is absent. South of the Valdés Peninsula the situation is reversed; *E. macellum* becomes numerous while *E. discoidale* disappears rapidly.

Another characteristic of the North Patagonian subprovince is the presence of some warm water species (*Globulina caribaea, Massilina secans, Poroeponides lateralis, Spiroloculina planulata*) which do not penetrate south of the Valdés Peninsula.

Figure 44 shows the distribution and abundance of the most important non-cosmopolitan species in the Argentine coastal area.

The northern limit of the Malvin subprovince occurs at approximately 52°S in the coastal areas but is extended north of the Malvin Islands on the continental shelf in

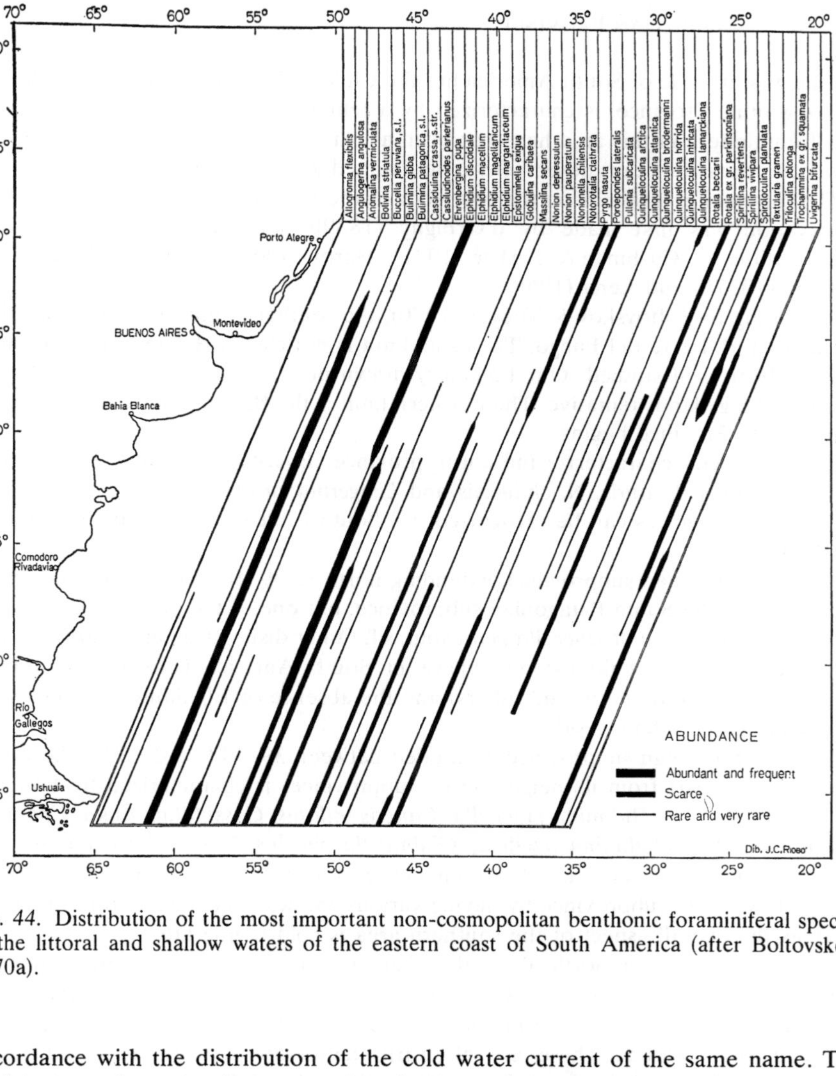

Fig. 44. Distribution of the most important non-cosmopolitan benthonic foraminiferal species in the littoral and shallow waters of the eastern coast of South America (after Boltovskoy, 1970a).

accordance with the distribution of the cold water current of the same name. The Malvin fauna has a 'cooler' aspect than do those of the subprovinces to the north. It can also be differentiated from the other subprovinces by the abundance and large size of its constituents. Typical representatives of the Malvin fauna are: *Buccella peruviana*, s.l. (typical of all the Argentine subprovinces), *Cassidulina crassa*, *Cassidulinoides parkerianus*, *Ehrenbergina pupa*, *Uvigerina bifurcata*, *Angulogerina angulosa*, *Cibicides aknerianus*, *Pullenia subcarinata subcarinata*, *Anomalina vermiculata* and *Discorbis isabelleanus*.

C. Chilean–Peruvian Province

On the western side of South America lies the Chilean–Peruvian Province which, like the Argentine Province, is characterized by a temperate water foraminiferal fauna. The foraminifera in this area have not been as extensively studied as in other South American areas but can, nevertheless, be mapped zoogeographically.

The foraminiferal faunas on the two sides of the continent are not the same. This observation was first made by d'Orbigny (1839b) and later confirmed by the investigations of Cushman & Parker (1931), Heron-Allen & Earland (1932), Boltovskoy (1957b) and Lena (1966).

According to Boltovskoy & Theyer (1970), the southern limit of the Province lies in the area of Tierra del Fuego. The exact limit is not clearly defined since this area has not been well studied. Our boundary drawn near the western entrance to the Straits of Magellan is tentative. The northern limit is the Bay of Guayaquil where the Panamanian Province begins.

Certain species characterize the entire province: *Buccella peruviana*, s.l., *Cancris inflatus, Nonionella auris, N. chiliensis* and *Uvigerina striata*.

Three subprovinces can be recognized: (a) South Chilean, (b) North Chilean and (c) Peruvian.

The South Chilean subprovince, extending north to 39–40°S, has many species in common with the South Patagonian subprovince, but one of the primary distinctions is the non-dominance of *Buccella peruviana*, s.l. Other distinguishing features are the presence of species unknown elsewhere in South America (*Cassidulina bradyi, Hoeglundina pleurostomata* and others) and the absence of *Elphidium*, a genus very abundant on the Atlantic side.

The North Chilean subprovince is situated between 39–40°S and 15°S. The fauna which sets it apart from its neighbouring subprovinces is characterized by *Bolivina plicata, B. punctata, Bulimina pulchella, Cancris inflatus, Cassidulina auka, Eponides meridionalis, Globobulimina pacifica, Glabratella pileolus, Nonionella chiliensis* and others. *Elphidium* is very rare. The faunal character of this subprovince differs from the South Chilean subprovince by having various species which are warm-temperate whereas the overall aspect of the southern area is cold-temperate.

As one goes further north along the Chilean–Peruvian coast the proportion of warm water species increases. Cushman & Kellett (1929) found *Rotalia rosea, Eponides repandus, Discorbis consobrinus* and *Trochammina kellettae* in Peruvian and equatorial waters. These are all warm temperate species. This change in character of the fauna leads to the creation of another subprovince, the Peruvian, which is bounded by 15°S and the Bay of Guayaquil and is characterized by *Bolivina costata, B. punctata, Cancris inflatus, Discorbis corus, Eponides repandus, Nonion pizarrense, Rotalia rosea* and some others.

The fauna of the Juan Fernandez Islands is somewhat unique. It has elements of a cold-temperate fauna (*Guadryina triangularis, Sigmoidella pacifica, Cassidulina crassa, Angulogerina carinata*) but also has representatives of Indian and Australian waters such as *Spirillina spinigera, Patellina advena, Tretomphalus bulloides* (Cushman & Kellett, 1929). Which zoogeographic province this area belongs to is unknown and the problem requires further investigation.

The foraminiferal assemblages of the islands off the Chilean coast are also unique but have not been well enough studied to warrant description.

D. Panamanian Province

The existence of the Panamanian Province has only recently been established. The geographic position of the region and the prevailing water temperatures indicate that it should possess a well defined tropical fauna. The fauna found in this area, however, lacks the typical warm water genera *Amphistegina, Peneroplis, Marginopora, Amphisorus* and the tropical species of *Quinqueloculina* and *Elphidium*. Data from Boltovskoy and Gualancañay (in press) place the southern boundary of this province near Guayaquil, Ecuador. The province extends north along the Panamanian isthmus and up to the southernmost extremity of Baja California (Lankford & Phleger, 1973).

E. Antarctic Province

The foraminiferal assemblages in the Antarctic Province are typical of cold water regions. The Antarctic region lying south of South America can be divided into 3 subprovinces on the basis of foraminifera, (a) Bellingshausen Sea, (b) Weddell Sea and (c) South Georgia Islands.

If comparisons are made between zoogeographic maps of South America based on molluscan distribution (Carcelles, 1950; Carcelles & Williamson, 1951) and those based on foraminiferal data (Fig. 43), various discrepancies are obvious. The malacologists consider the Magallanian zoogeographical province to extend from 37°S on the Chilean coast around Tierra del Fuego to 43°S on the Argentine coast. As mentioned above, the foraminiferal distribution shows no noticeable change at 43°S, but does exhibit a difference between the two sides of the continent.

Boltovskoy (1957b) explained these discrepancies on the basis of the recent geologic history of South America. The establishment of a free passage of water between the South Atlantic and South Pacific is a relatively recent event, perhaps dating only from the Late Pleistocene. Prior to that time there was a terrestrial connection between Tierra del Fuego and the Palmer Peninsula in Antarctica. The Straits of Magellan did not exist and the faunas of the two coastal areas were isolated from each other. When the Drake Passage was formed the waters of the two oceans mixed and Pacific Ocean water penetrated into the South Atlantic and onto the Argentine continental shelf. The faunas of both sides of the continent began to mix. Molluscs, because their ontogeny includes a planktonic larval stage, are able to migrate much more rapidly than benthonic foraminifera, most of which do not

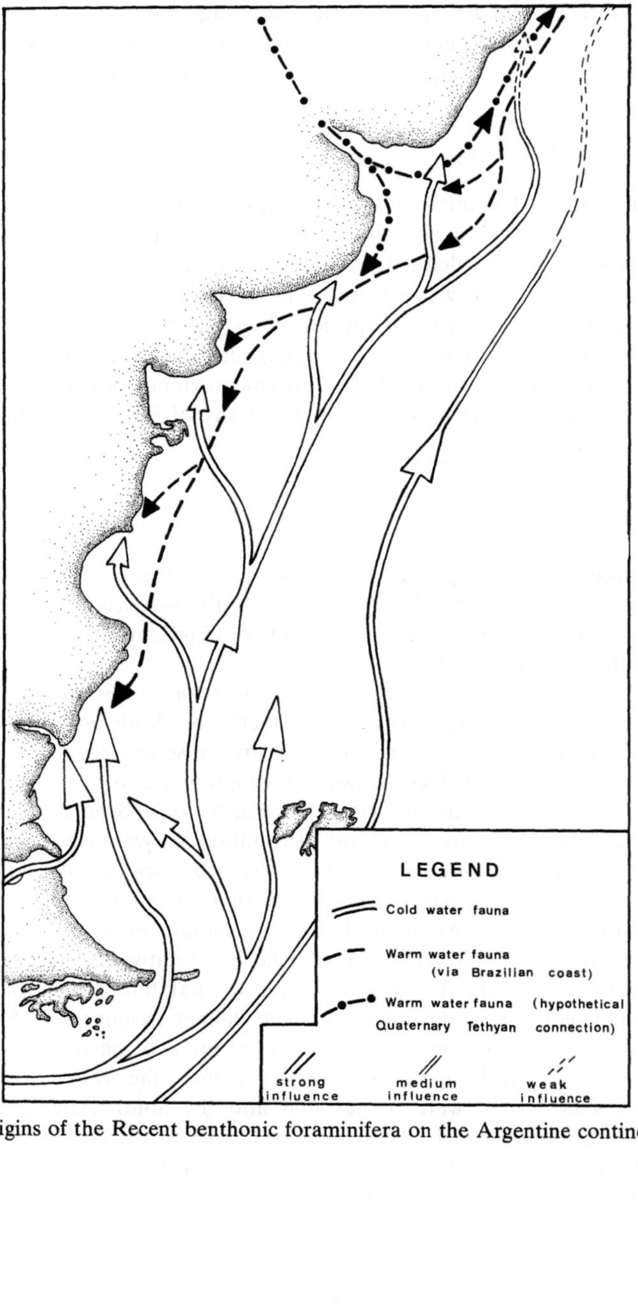

Fig. 45. Origins of the Recent benthonic foraminifera on the Argentine continental shelf.

possess such a stage in their ontogeny. Consequently, the molluscan faunas have had sufficient time to obliterate the Pleistocene provincial differences whereas the foraminifera have not. The geological history of this area may also explain the differences between the foraminiferal faunas of the Bellingshausen and Weddell Seas.

A map showing the hypothetical influences of recent geological history on faunal migrations along the Argentine continental shelf is shown in Fig. 45.

Cold water benthonic foraminiferal faunas inhabit two large areas, the Arctic and Antarctic. The various warm and temperate water provinces are connected with each other but the cold water provinces are separated by great depths and by tropical waters. They are characterized by the presence of few species, a high proportion of agglutinated forms and a pronounced absence of multichambered lagenid forms.

It has long been observed that the number of taxa of all groups of organisms increases toward the equator. For example, some 2000 species of animals are known from the Barents Sea, 7000 from the Mediterranean and about 40 000 from the waters of the Indo-Malaysian Archipelago; this is an increase of 20-fold over the Barents Sea. It has been estimated that the number of animals living in Indonesian waters is 100 times the number living in the Laptev Sea. Tropical waters are characterized by great ecological variability with a consequent large number of ecological niches. This ecological variability coupled with the high rate of metabolism and growth that occurs in warm waters means that species can grow, reproduce and evolve rapidly to fill the great variety of available environments.

The high proportion of agglutinated specimens in Arctic and Antarctic waters may be explained by the high solubility of $CaCO_3$ in cold water. The only calcareous specimens frequently encountered in these waters belong to the Miliolidae, Biliminidae, Nonionidae and Elphidiidae.

The foraminiferal faunas of the Arctic are somewhat better studied than those of the Antarctic, although increasing numbers of recent investigations are concerned with the Antarctic sector. Most of the studies in these areas are of a taxonomic nature. The following is a list of those publications dealing with the Arctic foraminifera (Fig. 46).

1. Parker & Jones, 1857
2. Parker & Jones, 1865
3. Norman, 1876
4. Carpenter, 1876
5. Mereschkowsky, 1878
6. Mereschkowsky, 1879
7. Brady, 1878
8. Brady, 1881a
9. Brady, 1881b
10. Brady, 1882
11. Terquem & Terquem, 1886
12. Schlumberger, 1894
13. Goës, 1894
14. Schaudinn, 1896
15. Kiaer, 1899
16. Kiaer, 1908
17. Kiaer, 1909
18. Awerinzew, 1908
19. Awerinzew, 1911
20. Derjugin, 1915
21. Cushman, 1910–1917
22. Cushman, 1920a
23. Cushman, 1921
24. Cushman, 1933
25. Cushman, 1948b
26. Mesjathev, 1923
27. Mesjathev, 1926a
28. Mesjathev, 1926b

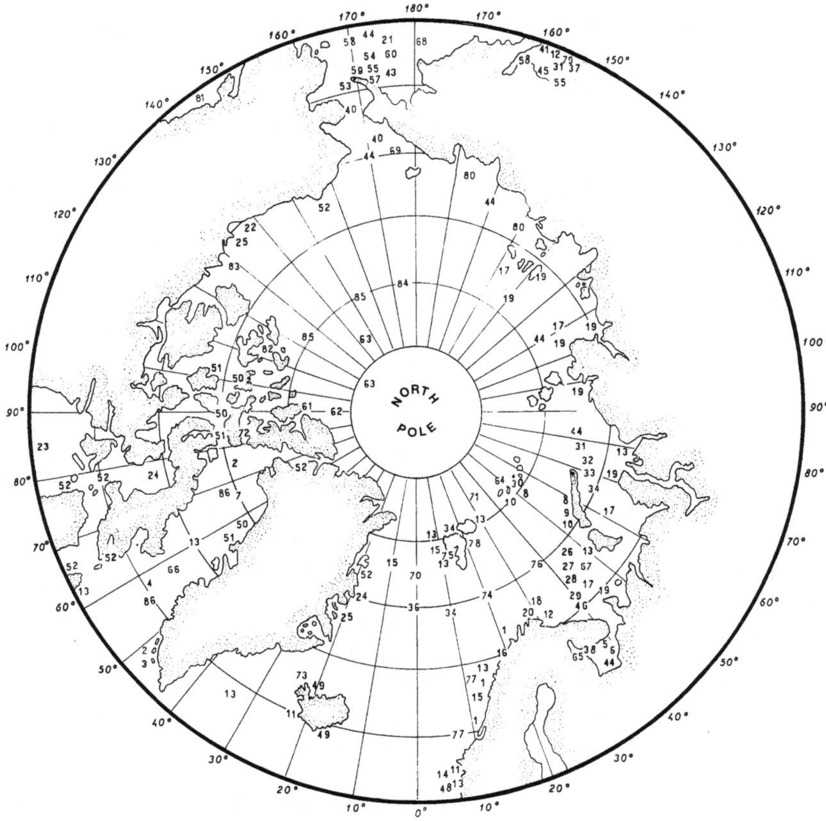

Fig. 46. Arctic foraminiferal studies. The numbers correspond to the listed publications and are located approximately where the studies were conducted.

29. Mesjathev, 1929
30. Spärck, 1933
31. Stschedrina, 1936a
32. Stschedrina, 1936b
33. Stschedrina, 1938
34. Stschedrina, 1939
35. Stschedrina, 1946
36. Stschedrina, 1947
37. Stschedrina, 1950a
38. Stschedrina, 1952a
39. Stschedrina, 1952b
40. Stschedrina, 1952c
41. Stschedrina, 1953a
42. Stschedrina, 1953b
43. Stschedrina, 1955
44. Stschedrina, 1956a
45. Stschedrina, 1956b
46. Stschedrina, 1958b
47. Stschedrina, 1962
48. Nørvang, 1941
49. Nørvang, 1945
50. Phleger, 1951
51. Phleger, 1952b
52. Loeblich & Tappan, 1953
53. Carsola, 1952 Ms
54. Saidova, 1957a
55. Saidova, 1957b
56. Saidova, 1958b
57. Saidova, 1959
58. Saidova, 1960a
59. Saidova, 1960b
61. Ericson & Wollin, 1959
63. Bé, 1960b
65. Majer, 1962
67. Jarke, 1960
69. Cooper, 1964
71. Stschedrina, 1964b
73. Adams & Frampton, 1965
75. Nagy, 1965
77. Tamanova & Stschedrina, 1965
79. Lipps & Warme, 1966
81. Todd & Low, 1967
83. Vilks, 1970b
85. Hunkins *et al.*, 1971

60. Saidova, 1961a
62. Green, 1960
64. Basov, 1961
66. Ehrenberg, 1861
68. Anderson, 1963
70. Stschedrina, 1964a
72. Vilks, 1964
74. Christiansen, 1965b
76. Basov & Slobodin, 1965
78. Rouvillois, 1966
80. Todd & Low, 1966
82. Vilks, 1969
84. Herman, 1970
86. Stehman, 1972

The following studies have been conducted on foraminifera from the Antarctic and adjacent areas (Fig. 47): Pirie, 1913; Fauré-Fremiet, 1914; Pearcey, 1914; Chapman, 1916; Heron-Allen & Earland, 1922, 1932; Wiesner, 1931; Earland, 1933, 1934, 1936; Warthin, 1934; Chapman & Parr, 1937; Cushman, 1945; Parr, 1950; Stschedrina, 1958e; Uchio, 1960b,c; Crespin, 1960; Saidova, 1961b; McKnight,

Fig. 47. Foraminiferal studies in Antarctic and adjacent arears. Authors' names are located in the approximate sites where the studies were conducted.

1962; Pflum, 1963; Shishkevich, 1964 Ms; Blair, 1965; Blanc-Vernet, 1965b; Chen, 1966; Lena, 1966; Kennett, 1967, 1970a; Herb, 1968, 1971; Belyaeva, 1969b; Boltovskoy, 1969c, 1971b; Bé, 1969a; Theyer, 1970, 1971b; Boltovskoy & Boltovskoy, 1970; Echols, 1971; Sissingh, 1973; Stockton, 1973.

Unfortunately, the majority of the investigations of both Arctic and Antarctic foraminifera have dealt with the distribution of empty tests. It is always preferable to examine living specimens but particularly so in the case of polar assemblages because empty tests can be easily frozen in ice and transported to areas remote from their natural habitat. The most common case is the transport of shallow water species to deeper water sediments. This mixing of faunas greatly confuses ecologic and zoogeographic conclusions.

The small numbers of species in cold water Arctic areas was demonstrated for Arctic foraminifera by Stschedrina (1956a) who summarized various Arctic studies. Below are her figures:

Greenland Sea	105	East Siberian Sea	60
Arctic Ocean, s.s.	54	Chukchi Sea	43
Barents Sea	123	Bering Sea	138
White Sea	43	Japan Sea	250
Kara Sea	127	Okhotsk Sea	400
Laptev Sea	55	Northeast Pacific	215

The total number of named species in these Arctic areas is between 700–750. This number is probably an overestimate because of taxonomic problems and the tendency of many workers to split the fauna into too many taxa (see Chapter XVIII).

Stschedrina (personal communication) listed the taxa that are typical of these areas. Within the area from the inner shelf coastal zones to the outer shelf (75 m) the following species are common: *Ammotium inflatum*, *Elphidium gorbunovi*, *Eggerella advena*, *Reophax curtus*, *Hippocrepina indivisa*, *Spiroplectammina biformis*, *Patellina corrugata*, and *Cassidulina* spp.

Among the typical Arctic genera are *Hormosina*, *Arenosphaera*, *Alveolophragmium*, *Ammotium*, *Astrorhiza*, *Reophax* and *Hippocrepina* which are all agglutinated forms, and *Cassidulina*, *Ehrenbergina*, *Buccella*, *Cibicides*, *Astrononion* and various Miliolidae among the calcareous forms.

European Arctic faunas appear to be somewhat different in character than Asiatic Arctic assemblages. The former are dominated by *Hormosina arctica* and *Arenosphaera perforata* whereas the latter are characterized by *Elphidium gorbunovi* and *Ammotium inflatum* and others (Stschedrina, personal communication). This difference may be due to the presence of two separate water circulation systems in the Arctic; one in the Greenland–European sector and the other in the Canada–Alaska–Siberian sector.

Green (1960) also suggested geographic differences among Arctic faunas. He claimed that Arctic faunas from the continental slope in the Western Hemisphere were richer in calcareous forms than were slope assemblages in the Arctic Eastern Hemisphere. Stschedrina criticized this conclusion on the grounds that the samples from which the comparisons were made were not taken from the same depth and were therefore not comparable.

The Antarctic foraminiferal fauna is quite similar at the generic level to the Arctic. For the most part, the same genera are found at the same depths. However, the Antarctic fauna is somewhat richer in terms of number of genera, some of which are unknown in Arctic waters. At the specific level the two areas are generally different.

Saidova (1961) noted that the foraminiferal abundance in Antarctic sediments is considerably greater than that in sediments of the northeast Pacific, although the faunas are generally similar. However, the Antarctic calcareous fauna is some 70 times more abundant than the Pacific fauna but the Antarctic agglutinated fauna is only 20 times greater than that of the Pacific. The number of agglutinated specimens on the slope and in the bathyal zones is relatively low.

Uchio (1960c) said that the characteristic trait of Antarctic benthonic foraminiferal faunas is the relative scarcity of agglutinated species. This observation is contrary to the data of Stschedrina (1972) who investigated high latitude benthonic foraminifera for many years. She studied material from a great many localities in both Arctic and Antarctic waters (Figs. 46 & 47), and summarized data on polar agglutinated forms as follows:

(a) Antarctic waters (5–5300 m) contain 197 agglutinated species, a quantity 3 times that found in Arctic waters where there are 64 species in the depths 12–4000 m. She attributes this difference to the greater amount of free contact and mixing of Antarctic water with the surrounding areas than in the Arctic. The greater variability of bottom conditions in the Antarctic and the presence of very different water masses are also cited as reasons for the differences in diversity.

(b) There are 26 genera in common between the two zones, 20 of which are represented by the same species as well. The other common genera possess species which are endemic to particular local basins.

(c) The species in common between the two areas are not completely identical but can be distinguished from each other morphologically.

Moving away from both poles into temperate water, one finds a relatively large number of species common to both the Northern and Southern Hemispheres. There are also a large number of species in both hemispheres which, while taxonomically distinct, are very similar. These 'bipolar' species include *Psammosphaera fusca, Bathysiphon argenteus, Rhizammina algaeformis, Reophax curtus, Haplophragmoides canariensis, H. glomeratus, Verneuilina advena* and *Elphidium incertum*, among others. This faunal similarity appears strange at first glance since the only connection between these temperate zones is by way of abyssal depths or through tropical water, neither of which are hospitable for these species.

This bipolarity is not restricted to foraminifera but is known in most marine benthonic organisms. Although the occurrence of parallel evolution is not impossible, perhaps the most logical explanation is that the change in water temperature and lowering of sea level during the Pleistocene glacial epochs permitted a greater mixing of northern and southern temperate water forms than can occur today. The remarkable similarity of northern and southern temperate water faunas, noted especially by those investigators who studied both, e.g. Heron-Allen & Earland, Stschedrina, makes this explanation sound credible. Whether or not the drop in temperature occurring during the Pleistocene (5–6 °C) would be sufficient to permit mixing of the faunas is not known.

In addition to these bipolar species which exhibit symmetrical distribution across the equator, there are species groups which show symmetrical patterns in an east–west direction. These species are called anfiboreal and anfiatlantic/anfipacific species.

The anfiboreal species are those sub-arctic species which are found not only in Arctic water masses but also in the North Atlantic and North Pacific. Anfiboreal species are known among benthonic foraminifera as well as in other animal and plant groups. Stschedrina (1958a) lists the following foraminifera as typical anfiboreal species: *Crithionina goesi, Ammodiscus catinus, A. planorbis, Glomospira glomerata, Haplophragmoides fragile, H. jeffreysii, Lagena ornata* and *L. semilineata*. The occurrence of these species in two areas which are separated by a continent and hostile marine environments is best explained by an increase in water temperature at some time in the past (Eemian interglacial?) during which the faunas could have migrated across the Arctic area.

Anfiatlantic and anfipacific species refer to those species living in both Atlantic and Pacific oceans in temperate waters; *Haplophragmoides columbiensis, Cassidulina californica, Ammodiscus minutissimus, Trochammina charlottensis, T. kellettae, T. pacifica, Placopsilina bradyi, Nonionella miocenica, Robertina charlottensis* and others. The distribution of these species can be explained in the same way as that of the anfiboreal species, although the degree of warming necessary to accomplish the necessary migration would have to be rather extensive, perhaps more than is now thought to have occurred. It is also possible that migration might have occurred across the Panamanian isthmus and through the Tethyan region during cooler episodes of the Tertiary.

This latter hypothesis is probably the best explanation for anfiamerican species, those common to both the Atlantic and Pacific coasts of the Americas.

Coral reefs are shallow water marine environments that merit special mention because their foraminiferal fauna is unique. These reefs are confined to tropical waters whose average temperature is 23–24 °C and whose minimum temperature is at least 20 °C. In addition to an elevated temperature these complex wave resistant organic structures require shallow water, normal marine salinity and low turbidity. The number of ecological niches in the coral reef environment is large and the faunas associated with reefs are variable, a trait typical of most tropical water environments. Despite this variability, there are certain benthonic species common to and dominating the foraminiferal fauna of reef environments: *Amphistegina lessonii, A. madagascariensis, Calcarina spengleri, Baculogypsina sphaerulata, Marginopora vertebralis, Borelis* spp., *Cymbaloporetta* spp., *Heterostegina* spp., *Peneroplis* spp. and various robust, thick walled Miliolidae.

The preceding sections have dealt almost exclusively with the distribution of benthonic species in the littoral and shelf zones. These areas have been far more extensively sampled and examined than have the bathyal and abyssal zones of the oceans. One of the few zoogeographic zonations of benthonic foraminiferal species from all depths was prepared by Saidova (1970a) for the Pacific Ocean. She described 6 main provinces and several subprovinces but, unfortunately, did not cite the characteristic foraminifera. Her provinces and subprovinces are therefore very difficult to check, recognize and compare with other areas.

3. Qualitative vertical distribution

The earliest investigators of benthonic foraminifera noted that most species exhibited a tendency to be restricted to certain depths. The most recent detailed investigations of specific areas have provided the data necessary to separate various vertical zones by their characteristic species.

The majority of these more recent studies dealt with shelf and slope faunas. A number of investigators have created specific zones characterized by particular species: Norton (1930), Natland (1933, 1957), Phleger (1942, 1951b, 1956, 1963, 1964, 1965b), Phleger & Walton (1950), Höglund (1947), Parker (1948, 1954, 1958), Lowman (1949), Colom (1950, 1952), Said (1950), Butcher (1951), Carsola (1952 Ms), Crouch (1952, 1954), Bandy (1953, 1954, 1956), Walton (1955), Kruit (1955), Stschedrina (1956a, 1958a), Saidova (1957a, 1960a, 1961a, 1966a), Christiansen (1958), Polski (1959), Waller (1960), Uchio (1959, 1960a), Green (1960), Chierici, Busi & Cita (1962), Bandy & Echols (1964), Bandy & Rodolfo (1964), Vilks (1964, 1969), Wilcoxon (1964), Loep (1965), Bandy & Chierici (1966), Pflum (1963), Eade (1967), Hooper (1968, 1969), Kennett (1968c), Betjeman (1969), Blanc-Vernet (1969), Caralp, Lamy & Pujos (1970), Chiji & Konda (1970), Roettger (1970), Vedantam & Rao (1970), Dupeuble, Mathieu & Momeni (1971), Mathieu (1971), Theyer (1971b), Matoba & Nakagawa (1972), Rosset-Moulinier (1972), Streeter (1972), Lankford & Phleger (1973), Pujos-Lamy (1973), Sellier de Civrieux & Bermúdez (1973) and Fillon (1974).

Stschedrina (1957), Parker (1958), Phleger (1960b, 1965b) and Murray (1973b) prepared generalized depth zonations based on data from the literature (Fig. 48).

Stschedrina (1957), after analyzing data from her own investigations and from those of other workers, concluded that vertical faunal boundaries occurred at the natural physical breaks on the ocean floor. She described shelf (0–200 m), upper bathyal (200–750 m), lower bathyal (750–2000 m), and abyssal (>2000 m) zones.

Parker (1958) created 5 zones which were quite distinctive from those of Stschedrina. She noted marked faunal breaks at 50, 143–205, 500–700 and 1000–1300 metres.

Phleger (1960b) established depth boundaries at 20–30, 50, 70–125, 1000 and 2000 metres.

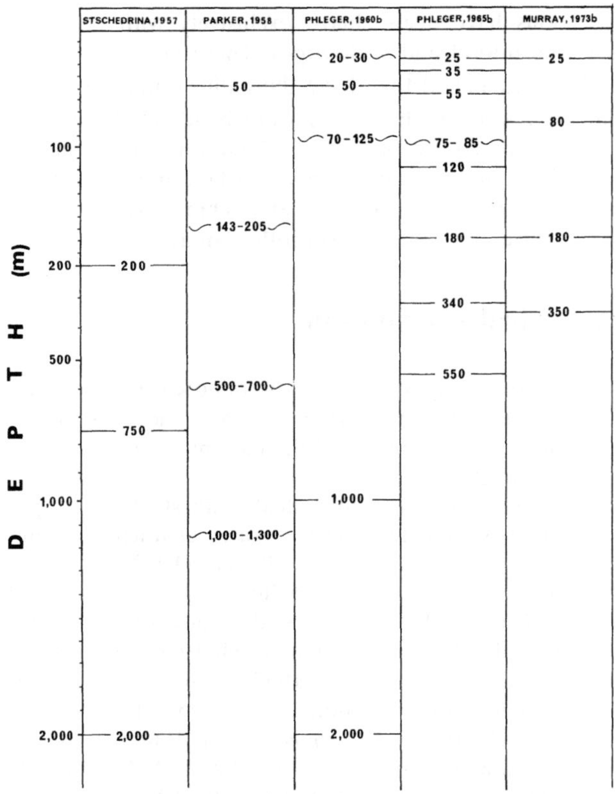

Fig. 48. Vertical benthonic foraminiferal zonation schemes.

He described the 70–125 m break as the most pronounced and perhaps the most important of these boundaries. Its position varies from place to place because it is defined as the lower limit of those waters subject to seasonal variations. Water above this break, especially in the middle latitudes, is subject to considerable fluctuation in environmental factors such as temperature and salinity. This is the zone in which photosynthesis occurs and a rich, renewable phytoplanktonic source of nutrition is available. The effects of turbulence are much more strongly felt here whereas below this boundary conditions are more constant.

The natural boundaries indicated by the other faunal breaks are not as well explained. Phleger said that the 1000 m limit may correspond to the deepest position of the permanent thermocline and that the 2000 m boundary marked the depth below which many bacteria could not survive. As bacteria are indispensable in the food regimen of many foraminiferal species, these species would be unable to live below this depth.

In a later work Phleger (1965b) expressed less optimism about the possibility of establishing universal depth limits for the benthonic foraminifera. He compared the depth boundaries established by many investigators in the northeast Pacific and was impressed by the lack of agreement between zonations. He said that boundaries based on living species distributions were far more reliable than those utilizing the dead fauna because of downslope post-mortem transport. For the same reason he stated that the upper depth limit of a species is more meaningful than the lower. Using these criteria he established an eastern Pacific depth zonation with boundaries at 25, 35, 55, 75–85, 120, 180, 340 and 550 m.

Murray (1973b) synthesized several studies of vertical zonation along the Pacific coast of North America and established faunal breaks at 25, 80, 180 and 350 m in a scheme which agrees rather well with that of Phleger. Murray noted that there is little similarity between different zonations at depths greater than 350 m.

It is necessary to remember that each of these five zonation charts was prepared on the basis of many samples from numerous investigations. The following points may help explain some of the similarities and variations between these zonations:

(a) The species composition of a fauna varies with depth.

(b) The vertical variations in faunal content are controlled to a certain degree by the laws of ecological zonation. The difficulties in establishing zonational limits is in part due to the absence of any well defined criterion with which to define foraminiferal biofacies.

(c) The limits of the established zones correspond, to some extent, to natural physical zones on the ocean bottom. These boundaries are better known and better defined in shallow water than in deep. Consequently, there is more agreement among zonation schemes for shallow depths.

(d) Depth is not the only factor controlling the vertical distribution of a species. The zonation plan established in one area will be appropriate for another region only if all significant environmental conditions are the same. If they are not, the depth boundaries will vary.

The depth distribution of genera shows that many genera are confined to particular zones. The classic example is *Elphidium*, which is characteristic of very shallow water even at quite different latitudes.

An examination of the depth distribution of species is more difficult and reveals a less consistent pattern. If the species lists prepared by various investigators are compared, it becomes clear that there are numerous contradictions. The following observations may clarify these differences:

(a) Vertical distribution in different latitudes, especially for shallow water environments, is controlled more by temperature than by depth. Other environmental factors are also significant locally. The same species may have very distinct vertical ranges in different areas; e.g. *Buliminella elegantissima* is found only to depths of 150 m in the San Diego area (Uchio, 1960a) but to a depth of 820 m in the Gulf of California (Phleger, 1964).

(b) Each area has species which are endemic to that particular area.

(c) The taxonomy of foraminifera is so confused that often different investigators employ synonyms or homonyms which make the comparison of their data difficult.

(d) The reliance of many investigators on dead tests may create confusion because the vertical distribution of the living organism may be different from that of the empty test. Phleger (1956) illustrated this point in the Gulf of Mexico by showing that living specimens of *Rotalia beccarii* were abundant to depths of 40 m and occasionally were found at depths down to 80 m, while empty tests of *R. beccarii* were abundant to 45 m and found to a depth of 110 m. Phleger explains this discrepancy between living and dead species distribution on the basis of downslope post-mortem displacement of tests and/or the presence of a relict Pleistocene fauna representing a time of lower sea level.

McGlasson (1959) noted that depth zonation based on living and dead species did not coincide in the Santa Catalina Basin, California except in very shallow water.

Uchio (1960a), based on a study of 95 species, concluded that the majority of the specimens underwent downslope movement after death.

It is quite clear that the differences between living and dead species distributions can create confusion in the establishment of vertical zonation schemes. Only depth zonations based on the vertical distribution of living specimens merit much confidence and should be used for comparative purposes. On the basis of the distribution of living specimens, we present a general characterization of vertical changes in the generic benthonic foraminiferal fauna from the littoral to the abyssal zones.

A. Littoral Zone (intertidal zone)

This zone, situated between the levels of high and low tide, is subject to very brusque changes in environmental conditions. The only species capable of surviving in this zone are those which can resist rapid water movement, daily changes in water depth, temperature, salinity and other ecologically significant factors. Foraminiferal tests in this zone have the advantage of being flattened and of fitting firmly to the substrate (*Rotorbinella, Discorbis, Trochammina*) where they graze on microflora. Other littoral zone genera whose shape enhances their survival in the surf zone are the attached forms *Cibicides, Acervulina* and *Planorbulina*. Specimens of *Peneroplis* are attached at only one point and turn in the water to present their most streamlined shape to water movements. Species with thick and strong walls (*Elphidium crispum, Quinqueloculina seminulum, Rotalia beccarii*) and those genera with especially strong connections between chambers (*Meandroloculina, Nubecularia*) are well adapted to life in the littoral zone. Specimens small enough to fit in the interstices of the bottom sediment grains are also offered some protection in the littoral zone. There are, however, some species found living in this zone whose particular adaption to high energy conditions is not yet known, e.g. *Buliminella elegantissima, Miliammina fusca, Nonion depressulus, Elphidium* spp. and *Miliolinella* spp.

Because of the great movement of sediment particles in the littoral zone it is particularly important to distinguish living from dead tests when examining the fauna in these zones. Only the living tests are truly representative of the area. But even living tests can be transported beyond their normal place of occurrence by storms and tidal currents. The transported specimens may continue to live in the new environment even though the conditions are not optimal. The most common direction of such transport is from the sublittoral to the littoral zone. This phenomenon of *in vivo* transport makes the preparation of a list of actual littoral species difficult. It also makes difficult the definition of a faunal boundary between the littoral and sublittoral zones.

We cite below some of the observations of investigations dealing specifically with living foraminifera in the littoral zone.

Hedley & Underwood (1957) found 38 species along the north Kent coast of England of which 16 were living. Most of these species belonged to the genera *Elphidium, Nonion, Rotalia, Miliolinella, Trochammina* and *Tholosina*.

Reiter (1959) examined the intertidal foraminifera of Santa Monica Bay, California and found only 17 living species in total fauna of 129 species. During the 7 months of this study a number of species were always found among the living populations (*Rotorbinella lomaensis, R. turbinata, Discorbis monicana, Trochammina pacifica, Cibicides fletcheri* and *Planorbulina mediterranensis*). Among the living specimens which were numerous but not as persistent were *Buliminella elegantissima, Quinqueloculina, Angulogerina* and *Scutuloris*.

Cooper (1961) analysed material taken from the California and Oregon coast. Of the 124 species encountered, 64 were living. The most common genera were *Rotorbinella, Buccella, Cassidulina, Discorbis, Elphidium, Trochammina, Quinqueloculina* and *Miliolinella*.

Boltovskoy (1963a) and Boltovskoy & Lena (1966b, 1970b) found that about 50% of the 130 species comprising the benthonic foraminiferal fauna in the littoral zone at Puerto Deseado, Argentina were alive. The most common of these were *Buccella frigida, Miliolinella subrotunda, Nonion depressulus, Quinqueloculina seminulum, Miliammina fusca* and *Rotalia beccarii.*

Located above the level of the average high tide are tidal pools in which foraminifera can live. Sea water reaches these pools only during exceptionally high tides or storms. Some only receive water from rainfall or surf action. The physiochemical environment in these pools varies considerably with time because of the irregular input of sea water. Salinity decreases when rain provides the main source of water and increases when there is excessive evaporation. Vegetative activity in these ponds results in a variety of pH levels. Despite the differences between the water in these pools and normal sea water, both Boltovskoy and Cooper noted that the specific content of the fauna varies little from the littoral zone fauna. The fauna in these pools is quantitatively poorer and exhibits a greater proportion of deformed specimens as a consequence of the abnormal conditions.

B. Sublittoral Zone

This zone is bounded by the level of low tide and the break in slope at the edge of the continental platform. The depth at which this break in slope occurs varies geographically and is usually between 135–200 m. This zone is commonly divided into 3 regions: a nearshore shelf or turbulent zone, an inner shelf extending out to a depth of 50–80 m, and an outer shelf. The benthonic foraminiferal fauna of each of these is discussed below.

1. *Nearshore shelf* (turbulent zone)

The water in this zone is subject to movements from storms, tides and surges. The average depth is about 20 m but varies from place to place as a result of local conditions. For example, at La Jolla, California a marked turbulence can always be detected at a depth of 9 m but is observed at 30 m only when exceptionally long wavelengths occur.

As is the case in the littoral zone, attached foraminiferal specimens with thick strong walls are common. The fauna is similar taxonomically with that of the littoral zone. The most common genera are *Elphidium*, *Rotalia*, *Quinqueloculina*, *Poroeponides* and to some extent, *Textularia*. Very few planktonic species occur here.

2. *Inner shelf*

The benthonic foraminiferal fauna is much more abundant and diverse in this zone. Agglutinated species constitute but a small fraction of the whole fauna and are represented by relatively simple genera such as *Textularia*, *Trochammina*, *Reophax* and *Saccammina*. Planktonic species are common only in regions where the shelf is narrow and steep. The predominant benthonic genera are *Rotalia*, *Elphidium*, *Quinqueloculina*, *Triloculina*, *Spiroloculina*, *Massilina*, *Discorbis*, *Buliminella* and *Buccella*. The proportion of lagenids and buliminids is greater than in the shallower zones. Tropical waters of this zone are enriched in *Miliolinella*; the shallower tropical waters are characterized by *Peneroplis* and *Archaias*; the deeper by *Sorites*, *Heterostegina* and *Cycloclypeus*. The generic composition of this zone persists from cold to tropical waters but the species change; e.g. *Quinqueloculina seminulum* (temperate) to *Q. vulgaris* (tropical) or *Rotalia beccarii beccarii* (temperate) to *R. beccarii parkinsoniana* (tropical). As a general rule the tropical water fauna of this zone is very rich.

The differences between the inner shelf fauna and the littoral zone/turbulent zone fauna lie not only in the richness of the former but also in morphological differences within a genus. Myers (1943c) noted that specimens of *Elphidium* taken from the lower part of the inner shelf had 40% more chambers and were 60% larger than specimens from the littoral zone. Myers attributed these differences to a more regular source of nutrient material on the shelf.

3. *Outer shelf*

Although this part of the continental shelf has much in common with the preceding zone, there are faunal differences: (a) the number of taxa is about twice as great, (b) the proportion of agglutinated specimens in significantly less, (c) the proportion of planktonic specimens is greater and exceeds the benthonic population at the outer margin of this zone and (d) the proportion of various hyaline calcareous species (lagenids, buliminids, cibicidids, etc.) has increased at the expense of various porcellaneous taxa (except *Pyrgo* which is quite common). The typical coastal species, *Rotalia beccarii, Elphidium crispum* and *E. macellum* have disappeared except for isolated occurrences, usually in water less than 100 m deep.

The predominant genera are: *Cassidulina, Cibicides, Nonion, Nonionella, Uvigerina, Virgulina* and the agglutinated form *Bigenerina*. Other agglutinated forms occur, particularly forms with large complex tests such as *Textularia, Liebusella, Cuneolina*, and *Martinottiella*.

C. Bathyal Zone

This zone begins at the lower limit of the continental shelf and extends to a depth of about 3000 m. Practically all investigators observed a faunal boundary in this zone, but placed it at different depths in different areas. Nevertheless, it is possible to speak about the upper part (upper slope, transitional zone) and lower part (lower slope) of the bathyal zone.

The number of benthonic taxa in the upper part is about the same as in the outer shelf but the number of planktonic specimens is greatly increased and constitutes a majority of the foraminiferal fauna. The agglutinated specimens constitute about 5% of the population and are generally, but not universally, large complex forms. The dominant genera are *Bolivina, Uvigerina, Cassidulina, Pleurostomella, Gyroidina, Chilostomella, Allomorphina, Pullenia, Bulimina*, several *Cibicides* and *Nonion* and some others. Among the porcellaneous forms *Pyrgo* is dominant and is represented by very large specimens which are in contrast to its small ones typical of the inner shelf. Close relatives of *Pyrgo, Pyrgoella* and *Biloculinella* are also present.

Although the shelf and upper slope zones have many genera in common, their species composition is generally different.

The benthonic foraminiferal population of the lower part of the zone is reduced both in diversity and abundance, and is only about one-tenth (or less) that of the planktonic thanatocoenose. Agglutinated species play an increasingly important role in the benthonic population, comprising 10–20% of the fauna in the upper parts and a greater proportion in deeper waters. The most common genera are *Bulimina, Uvigerina, Epistominella, Glomospira, Gyroidina, Pullenia, Laticarinina, Trochammina, Planulina*, some *Cibicides, Nonion*, numerous *Fissurina, Parafissurina* and *Lagena*.

D. Abyssal zone

The benthonic foraminiferal fauna of the upper part of this deep zone is fairly well known and some descriptions of its vertical zonation have been attempted. The fauna of the deeper (ultra-abyssal) zone has not been well studied.

For some time the only data on abyssal foraminifera came from the works of

Brady, Cushman, Schott and a few others. All these earlier investigators noted the significant increase in agglutinated species at great depth. Various Astrorhizidae and the genera *Technitella*, *Cyclammina* and *Thurammina* were considered typical agglutinated species of this zone. An interesting large miliolid whose outer layer was composed of homogeneous silica was found at a depth of 7230 m by the *Challenger* expedition.

In more recent times additional investigations of this zone have been conducted and our knowledge of vertical zonation has increased. The agglutinated species predominate. Some species persist across a wide depth range. Below 8000 m an entirely new group of calcareous genera occurs. The agglutinated species are, for the most part, new species of *Hyperammina* and *Reophax*, or new subspecies of taxa known from lesser depths (*Haplophragmoides nitidus*, *H. glomeratus*, *Trochamminella bullata*, etc.) Some of the species found at these depths are common at a variety of depths (e.g. *Saccammina difflugiformis*); others are characteristic only of deep water, e.g. *Thurammina compressa* and *T. papillata*. As miliolids are commonly associated with shallow water, the discovery of a large, robust *Miliolina* at 7210–7230 m is worth noting. Subsequent discoveries of miliolids in deep cold water have confirmed this anomaly (Herb, 1971, Theyer, 1971a).

An important contribution to our general knowledge of the deep water benthonic fauna was made by Saidova (1959, 1960a,b, 1961a) based on abundant material from the Okhotsk, Bering and Japan seas, and the northwest Pacific. She concluded the following:

(a) Calcareous benthonic specimens live only to depths of 2500–3000 m and are most abundant between 400–1500 m. Only in the open Pacific are they found as deep as 3500 m. Only agglutinated species live below these lower depths.

(b) The principal traits of the benthonic assemblages of the abyssal zone deeper than 3000–3500 m are that the specimens are large but not abundant and are characterized by simply constructed forms (*Rhabdammina*, *Bathysiphon*, *Reophax*, *Hormosina*, *Nodosinella*, *Haplophragmoides*, *Cribrostomoides* and *Cyclammina*).

Saidova noted that these deep water forms are adapted to high pressure, persistently low temperatures and low oxygen content conditions. These conditions also reduce the likelihood of calcium carbonate precipitation. She prepared a generalized table of vertical distribution in bathyal and abyssal waters (Table 7) from which she concluded that there was a strong inverse correlation between depth and the degree of evolution exhibited by the assemblage.

These conclusions of Saidova regarding the agglutinated fauna lead to a curious conclusion when coupled with the observation that in the littoral zone and shelf the agglutinated species tend to be simple, primitive forms. The most evolutionary advanced agglutinated forms appear to be predominant in the middle of the bathyal zone. In deeper or shallower waters, there is an increase in the proportion of primitive species.

Saidova (1961a) also noted (Fig. 49) that in the sublittoral zone calcareous specimens outnumber agglutinated forms 100:1; in the upper part of the bathyal

Table 7. Relation between depth and evolutionary complexity of deep water benthonic foraminifera

Depth (m)	Fauna	General traits
1500	*Nonion* *Cassidulina* *Bolivina*	Highly evolved genera
1500–3000	Agglutinated genera	Complex agglutinated genera
3000–3500	*Rhabdammina* *Reophax* *Nodosinella* *Cribrostomoides*	Primitive non-complex agglutinated genera

zone, about 30 : 1; in the lower part of the same zone the abundance of calcareous forms ranges from equality to 20 times greater than agglutinated specimens; in the abyssal zone the calcareous forms are almost absent.

Saidova (1967b) developed an interesting schematic portrayal of the distribution of genera and families of foraminifera in the Pacific Ocean. Unfortunately, this scheme was based on the study of empty tests. The studies of Lukina (1967a) demonstrated that shallow water species are commonly transported to great depths in this area (*Amphistegina radiata* to 1000 m; *A. madagascarensis* to 3095 m).

Bandy & Rodolfo (1964) studied the living and dead foraminifera in the Peru–Chile Trench and found a rather limited number of living species. The most interesting observation is the presence of living *Stilostomella antillea*, a calcareous species, at a depth of 6011 m.

That calcareous benthonic species can occur at great depths was also shown by Lukina (1967a) who described *Lagena feildeniana, L. glans, Fissurina formosa angusticarinata, Pyrulina acuminata, P. angusta, Gyroidina orbicularis, Nuttallides rugosus* and *Uvigerina peregrina dirupta* at 5000–5670 m. She also described *Sigmoilina obesa, Cibicides pseudoungerianus, Siphouvigerina ampullacea* and *Cassidulina subglobosa* from depths as divergent as 780–6750 m. A rich agglutinated fauna also inhabited these depths in the central Pacific (Lukina, 1967b).

The pertinent characteristics of the benthonic foraminiferal fauna of the deep sea can be summarized as follows:

(a) Agglutinated species comprise the bulk of the fauna and are found as living specimens in the deepest parts of the ocean.

(b) Calcareous specimens are increasingly rare below 3000 m but are found living as deep as 6000 m and sometimes still deeper.

(c) Most of the fauna is composed of cosmopolitan species.

(d) The fauna is relatively well adapted, perhaps because of environmental stability.

(e) The relative proportion of living specimens is low in part because of low productivity and in part because of low clastic sedimentation rates which prevents rapid burial of the empty tests.

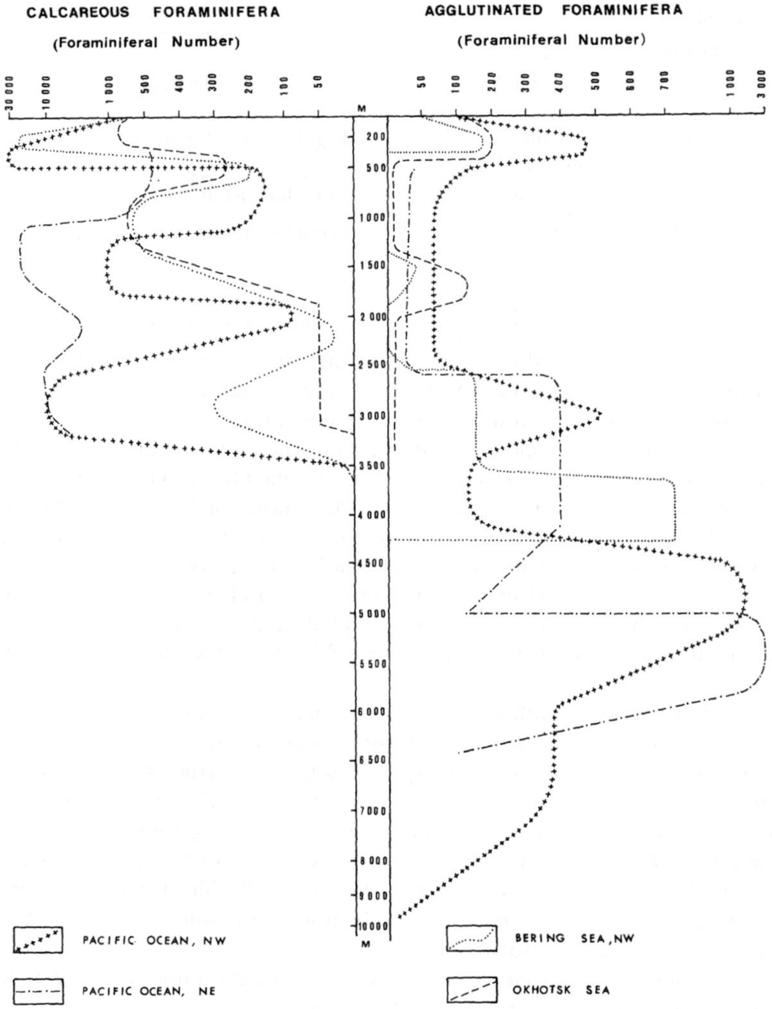

Fig. 49. Quantitative benthonic foraminiferal distribution at different depths (modified after Saidova, 1961a).

4. Quantitative horizontal distribution

Cultured foraminifera tend to disperse evenly throughout the culture chamber if conditions are uniform. This is particularly true of recently born individuals. This phenomenon can be explained by the competition between individuals for resources. Uniform dispersal allows the maximum available food, light, substrate, etc. for each member of the population. These same dispersal and distributional patterns would be observed in nature if environmental conditions were homogenous. They are, however, not uniform, nor is there only one population of organisms competing for the same resources.

Each species has optimum environmental conditions under which it grows and reproduces best. If two or more species have similar or overlapping ranges for one or more environmental conditions, competition for these ideal conditions will determine the particular distributional patterns of species (zoogeography). Because environmental conditions are not uniform, there will also be competition among the members of a species for the optimum locations. This competition produces a distributional pattern which ideally takes the form of concentric circles; the greatest frequency of occurrence is in the centre and decreasing abundance occurs toward the periphery. This type of distribution is modified, sometimes extensively, by variations in local environmental conditions and by competition between species.

Where conditions are optimal a species will occupy all biotopes available to it and it will reach its maximum abundance. On the periphery of these conditions, the species will not occupy all the biotopes, will exhibit an interrupted and irregular distribution and the population will consist of isolated and rare specimens.

This kind of distribution pattern can be observed in the numerous distributional maps that have been prepared by different investigators to show species distributional patterns. As an illustration of this modified concentric pattern see Figs. 50 and

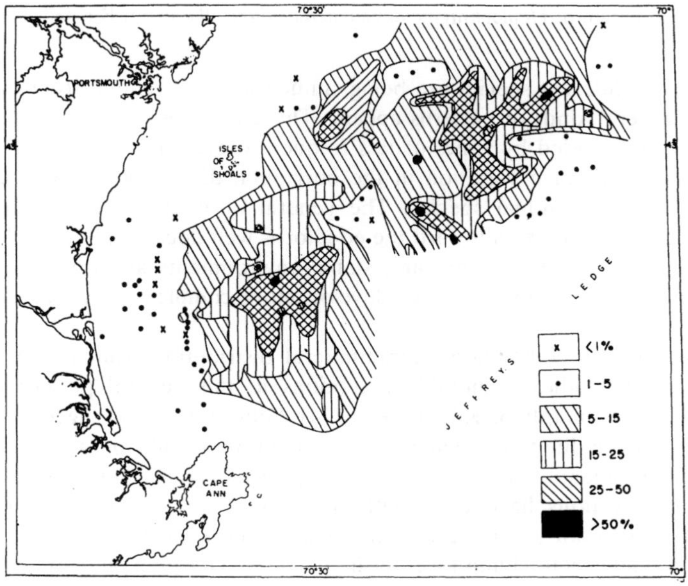

Fig. 50. Distribution of *Reophax curtus* off the New England coast (after Parker, 1952b).

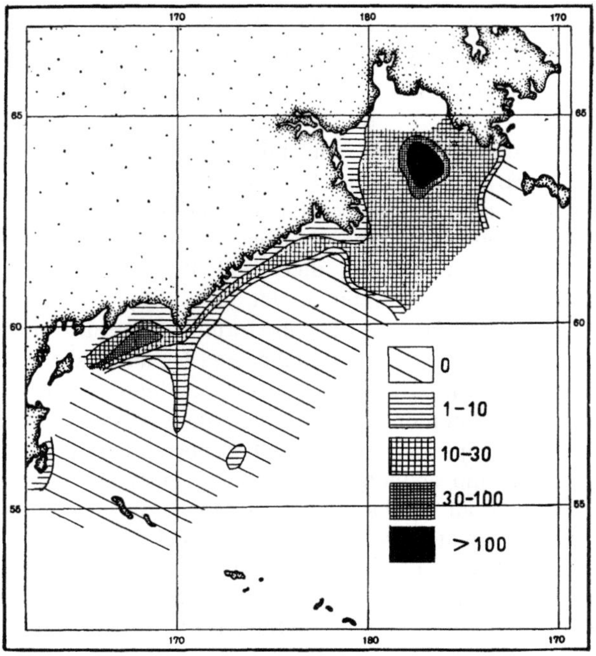

Fig. 51. Distribution of *Haplophragmoides sphaeriloculum* off eastern Russian coast (Belyaeva, 1960).

51. Although different species are being studied in areas remote from one another, the same general kinds of patterns appear. There is a central area or areas of great abundance surrounded by zones of less frequent occurrence.

There is quantitative variation in the horizontal distribution of entire faunas as well as that of individual species. The general observation that the diversity of marine organisms increases toward the equator is well documented. The low latitude zones are characterized by constant, stable environments and offer the organisms living there a better opportunity to develop more complex, highly organized communities.

Detailed studies of benthonic foraminiferal distribution from the Arctic to the Gulf of Mexico indicated notable exceptions to this general pattern (Gibson & Buzas, 1973; Sen Gupta & Kilbourne, 1974). Gibson & Buzas suggested that these anomalies may be due to particular localized environmental conditions. The increase in local environmental stress may decrease species diversity (Sanders, 1969).

These observations have dealt with the quantitative distribution of benthonic foraminifera on a fairly large scale. On a smaller, more localized scale the foraminifera show a tendency to clump together in 'patches'. This phenomenon is discussed in the next section.

5. Microdistribution

Various authors studying benthonic foraminifera in small, restricted areas have noted that two samples taken in close proximity to each other often exhibit different abundances of living specimens and different relative proportions of species. One of the first observations of this type was made by Parker & Athearn (1959). Some data were truly surprising; e.g. Richter (1961) found 14 living specimens in one sample and 293 in a sample only a few metres away; Lutze (1968b) noted two samples one metre apart with 12 and 212 living specimens respectively.

These observations lead to the conclusion that benthonic foraminifera have a non-uniform microdistribution. This mode of distribution has been referred to as 'patchiness' by various authors who have noted the phenomenon (Shifflet, 1961; Seiglie & Bermúdez, 1963; Lynts, 1966b; Troithkaja, 1970; Wright & Hay, 1971; Matera & Lee, 1972, Nyholm & Gertz, 1973; Olsson & Eriksson, 1974).

Despite the widespread recognition of patchiness, it has not been well studied. Only in some of the more recent investigations has the size of the patches been defined (Lutze, 1968b; Schafer, 1968; Boltovskoy & Lena, 1969). Lutze found patches 1–2 m in diameter whereas Boltovskoy & Lena demonstrated that the patches have an irregular shape and a size which is dependent on the sample spacing and on the contour interval used to map them. (Fig. 52). The smaller the contour interval the smaller and more numerous the clumping of high density areas (patches) becomes.

There is no reason to necessarily expect the benthonic foraminifera to be distributed homogeneously on the bottom. Buzas (1970) studied the detailed distribution of four species in a very small area and noted that only one of the four had a homogeneous distribution (defined statistically). In an earlier study, Buzas (1968b) suggested that benthonic foraminifera could be distributed in random, regular or aggregated arrangements on the bottom. He said that aggregates or patches are the most likely because of the clumping of offspring during asexual reproduction and/or aggregation around food clusters. Schafer (1971) noted that if a given species asexually reproduces uniformly throughout the year it will occur in patches. Patchiness has the effect of increasing the diversity of the habitat and reducing the interspecific competition in a local area.

One of the problems with the study of the microdistribution of benthonic foraminifera is the lack of any well-accepted definition of the word 'patch'. A numerical definition should probably incorporate some measure of the rate of change of specimen density as a function of distance. Until such time as a definition is proposed, there will be no full understanding of the phenomenon of patchiness.

6. Quantitative vertical distribution

Until recently the only quantitative parameter considered in the examination of vertical faunal distribution was the number of species. Most, but not all observations

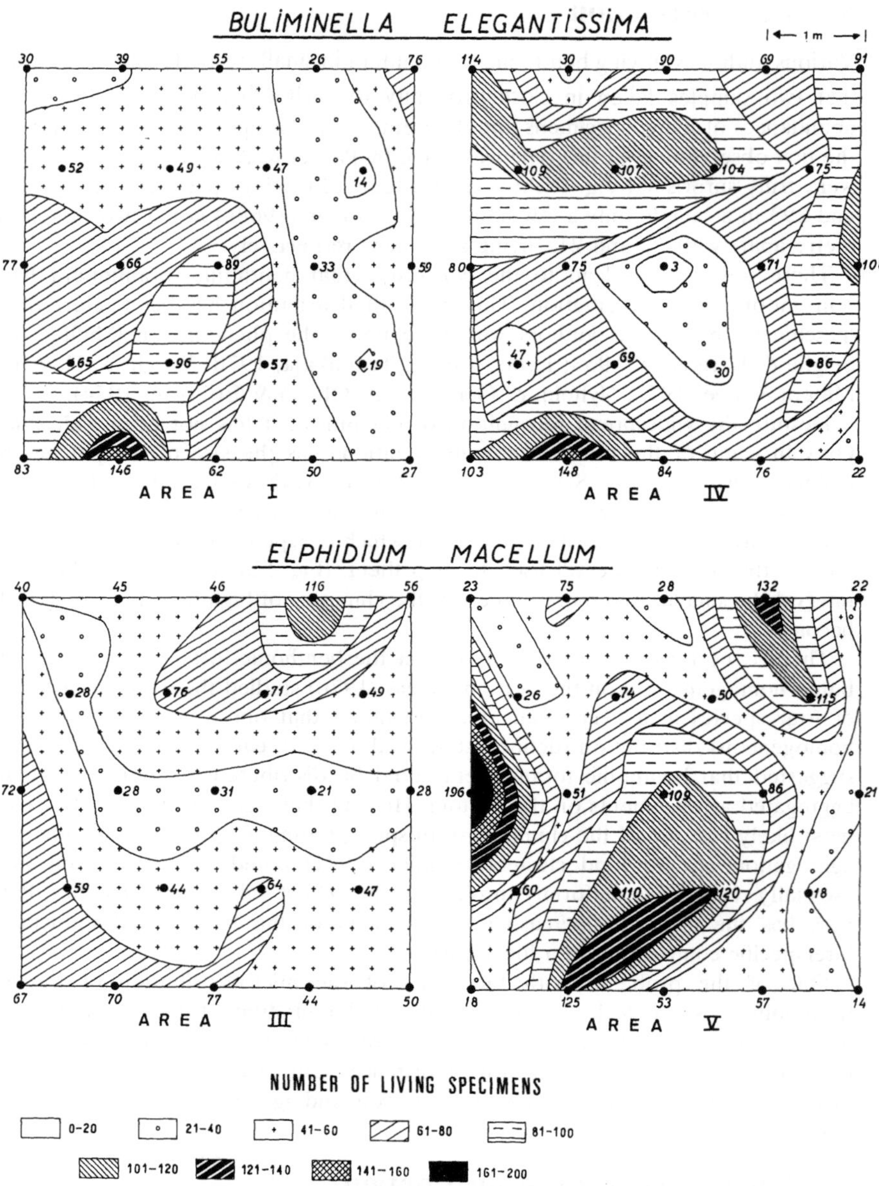

Fig. 52. Microdistribution of *Buliminella elegantissima* and *Elphidium macellum* at Puerto Deseado, Argentina (after Boltovskoy & Lena, 1969).

of species distribution indicated a decrease in the number of species with depth. This decrease was usually explained as a decrease in the number of ecological niches available with depth. More recent utilization of diversity indicies which measure the distribution of individuals within a species as well as the number of species indicate that, as a general rule, deep water environments are more stable than shallow water environments. We cite below some of the data based on the traditional measure of the number of species and follow this with some of the more recent studies whose diversity measures include species dominance as well.

Data from the Challenger expedition (Murray, 1895):

Depth (m)	Number of species	Relative quantities
<183	4248	27.6
183–914	1887	12.3
914–1830	616	4.0
1830–2743	493	3.2
2743–3658	394	2.6
3658–4572	247	1.6
>4572	153	1.0

The distribution of these data is explained by noting that on the shelf, bottom environmental conditions are highly variable and quite favourable for foraminifera. At great depths, there is no light, temperature and salinity conditions are constant, nutrition is scarce and calcium carbonate tends to dissolve. These conditions lead to a decrease in the abundance of benthonic foraminiferal taxa.

Rhumbler (1911) using data of Egger (1893) prepared a table showing depth versus the number of species for benthonic foraminifers which coincides well with the data of Murray:

Depth (m)	Number of species	Relative quantities
0–99	138	7.3
100–499	352	18.5
500–999	132	6.9
1000–1999	147	7.7
2000–2999	53	2.8
3000–3999	79	4.2
4000–4999	38	2.0
5000–6000	19	1.0

Subsequent investigations confirmed the general pattern shown by this data.

Phleger (1951b) suggested that the number of species occurring in depths less than 100 m is about 10 times as great as the number at greater depths. Walton (1955) also considered shallow waters to offer more niches to benthonic foraminifera. He calculated that about 50% of the species in Todos Santos Bay, California inhabited water less than 90 m deep and that 67% occurred above 180 m. Walton also concluded that foraminiferal productivity was a function of depth and in this area reached its maximum between 37–90 m where it was about 6 times greater than at 640–732 m.

Uchio (1960a) found that the greatest number of living specimens in the coastal areas around San Diego, California occurred between 183–274 m (sometimes 100–274 m). In shallower water the population was reduced because the sediments were too coarse and the quantity of available organic nutritive material was consequently reduced. The distribution of the number of species showed a pattern similar to that of specimen abundance: at 37 m there were less than 20 species; the maximum diversity (>50 species) was reached between 110–183 m; the number of species decreased with increasing depth exhibiting a minor increase at 732–1097 m.

Other investigators (Parker, 1948, 1954; Bandy, 1953; Bandy & Arnal, 1957; Chierici, Busi & Cita, 1962; Bandy, Ingle & Resig, 1964c; Vilks, 1964, 1968, 1969; Betjeman, 1969; Buzas & Gibson, 1969; Frerichs, 1970; Ikeya, 1971; Lankford & Phleger, 1973) also noted that the number of species near the coast is relatively small and increases with depth and distance from the coast. A maximum species number is generally reached in the lowest part of the continental shelf and uppermost slope (150–250 m). Beyond this point the number of species begins to decline.

It is very difficult to compare data between different investigators regarding the number of species because, (a) the greater the size of the sample the more likely the investigator is to encounter the less abundant species, and (b) each investigator has a different working definition of the species concept. There is very little that can be done about the latter problem but Hurlbert (1971) suggested a method of standardizing the number of species as a function of sample size:

$$E(S_n) = \sum_{i=1}^{s} \left[1 - \frac{\binom{N-N_i}{n}}{\binom{N}{n}} \right]$$

where

$E(S_n)$ = the expected number of species in a sample of n individuals.
N = number of individuals in the given sample
N_i = number of individuals of the i^0 species
s = number of species in the given sample

If different size samples are normalized to some standardized sample size, n, direct comparisons of species richness can be made.

The preceding data and conclusions were based on littoral, shelf and upper slope faunas. Less is known of the diversity relations in the lower bathyal and abyssal zones, although Stschedrina (1958c) prepared a table showing the quantity of species as a function of depth in the Kurilo-Kamchatka trench:

Depth (m)	Number of species	Species endemic to this depth
950–1255	8	3
2840–2940	33	17
3960–4070	15	6
4850–5570	59	33
6000–7230	16	2
8100–9050	14	11

Diversity measures based on the evenness of species distribution in a sample indicate that equitability among deep sea foraminifera is greater than in shallow water foraminifera (Gibson & Buzas, 1973). They found locally anomalous situations such as the low diversity in the abyssal parts of the northwest Gulf of Mexico difficult to explain.

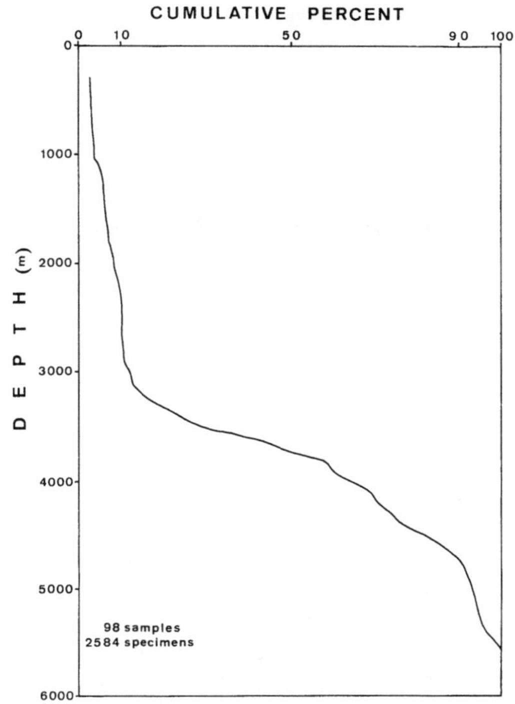

Fig. 53. Cumulative frequency depth distribution of *Cribrostomoides subglobosus* in the Pacific Antarctic (after Theyer, 1971b).

Another change that occurs with depth is the relationship between the number of genera and the number of species. For most organisms the number of species per genus decreases with depth. However, in the San Diego, California area Uchio (1960a) found that at:

0–110 m there were 1.52 species per genus
110–914 m there were 1.54 species per genus
914–1189 m there were 1.72 species per genus

Although this relationship seems contrary to that of other marine organisms, it must be noted that: (a) the differences are not very great, and (b) the data represent only the shelf, slope and upper bathyal zones and are even generalized here. More analysis is needed to draw any conclusions about this phenomenon for foraminifera. This problem also is seriously affected the chaotic taxonomic situation among the foraminifera under which there are so many synonyms and different taxonomic interpretations.

Several investigators have described the quantitative vertical distribution of individual species (*Amphicoryna scalaris* by Scorziello, 1966, and *Quinqueloculina crassimarginata* and *Q. schlumbergeri* by Haake, 1970) but the bulk of the observations were made by Bandy and his colleagues Chierici, Echols, Rodolfo and Theyer who studied these trends in various ocean basins. Theyer (1971b) pointed out that the frequency of occurrence for most species is very low near their upper and lower depth limits (Fig. 53). Normally the shallowest and deepest consistent occurrence is near their 10 and 90% cumulative frequency position.

7. Foraminiferal abundance by weight, volume and surface area

The quantitative abundance of benthonic foraminifera has been calculated in a variety of ways, the most common of which are expressed relative to the weight of sediment, the volume of sediment and the surface area covered by the sample.

A. Weight

The number of foraminiferal tests relative to a given weight of sediment was first utilized by Schott (1935) as a measure of abundance and was called the Foraminiferal Number (FN). Schott calculated the number of tests in 1 gm of sediment which had been dried at 105 °C. His calculations were based only on the 0.2–2.0 mm size fraction of the sediment and included both benthonic and planktonic specimens.

Said (1950) proposed that the size range be extended to 0.1 mm and that benthonic tests be counted separately from planktonic ones. Some later investigators discriminated between agglutinated, hyaline, and porcellaneous tests in their calculations. For ease of communication, these various Foraminiferal Numbers will be referred to as:

FN_t = total number of tests per gram
FN_b = benthonic tests per gram
FN_{cb} = calcareous benthonic tests per gram
FN_{ab} = agglutinated benthonic tests per gram
FN_{hb} = hyaline benthonic tests per gram
FN_{pb} = porcellaneous benthonic tests per gram
FN_p = planktonic tests per gram

Because normal sample preparation techniques (Chapter XI) result in the loss of the very fine fraction ($<62\ \mu$) of the sediment, the FN is generally calculated on the basis of less than the whole sample. To compensate for this error, Barash (1970) proposed that the following calculation be used to calculate FN:

$$FN = \frac{k}{g(100)} n$$

where

k = the weight percentage of that size fraction of the sample which was used in the calculation
g = weight of material examined
n = the quantity of tests found in the sample

This method requires that the investigator perform a size analysis of the sample prior to removal of the foraminiferal tests.

In order to eliminate the increased proportion of error inherent in small samples, it is wise to utilize a relatively large (10–20 gm) sample when calculating FN. The data from each sample should be kept separate in order to obtain information about the geographic distribution of FN in the area under study. Most studies have shown an increase in FN away from the coastal areas. This phenomenon is explained by considering two observations: (a) the clastic sedimentation rate is greater near shore and the relative number of tests in the sediment will be less, and (b) the maximum productivity of benthonic foraminifera occurs some distance away from the littoral zone.

The FN is not a very adequate measure of foraminiferal abundance because it depends on several factors which are difficult to control and compensate for; e.g. sedimentation rate, test solution in deeper water or in low pH zones in shallow water, etc. Consequently, FN is not universally used. Nevertheless, it does give a crude index of the abundance of foraminifera in an area. The variation in values obtained in different investigations is very impressive.

Schott (1935) said that the average FN_t for the Atlantic Ocean was about 2000 with the greatest values (12 700) found just north of the Cape Verde Islands.

Said (1950) calculated the FN_b for the northern part of the Red Sea and found the following variations:

Locale	Average FN_b	Range of values
Suez	498	380–600
Gulf of Acaba	69	8–270
Red Sea	312	22–910

Jarke (1960) found the average FN_t in the central and western parts of the Barents Sea to be 90 ($FN_b = 77$). The maximum value was 900 ($FN_b = 837$) at a depth of 480 m.

Saidova (1960a) in her studies of the Okhotsk Sea calculated the FN_{cb}, FN_{ab} and FN_p separately and prepared maps showing areas of equal FN_t. The maximum value of FN_{cb} (20) occurred near the Kurile Islands and the minimum values (<1) occurred in the deepest zones. The maximum FN_p (6–20) occurred in 600–1300 m of water. These data indicate that the Okhotsk Sea is relatively impoverished in its foraminiferal fauna.

Boltovskoy (1954c, 1957b, 1962b) obtained the following average FN_t values for the inner shelf of South America: Tierra del Fuego, 1400; San Blas Bay, 29; Mouth of the Rio de la Plata, 41; South Brazilian Coast, 316. As a general rule these values give some indication of the quantity of the fauna in these areas. The foraminiferal fauna in the area of Tierra del Fuego is very well developed and that of San Blas Bay is very impoverished. The low value at the mouth of the Rio de la Plata is probably due to the high sedimentation rate there.

The differences in FN between the continental shelves of the northeast Pacific and the Antarctic were discussed by Saidova (1961b). She found 60 times more calcareous and 20 times more agglutinated tests in sediments of the Antarctic than the northeast Pacific. The richness of the Antarctic fauna was also shown by Uchio (1960c) who calculated an average FN_t of 6250 for 10 samples taken at depths from 850–2480 m.

The largest FN_t calculated to date was a value greater than 1 000 000 (Frerichs, 1970). It was calculated from a sample taken in the southern part of the Andaman Sea.

Belyaeva (1969a) prepared a map (Fig. 66) showing the distribution of the number of planktonic tests (FN_p) found in the sediments of the oceans.

At great depths calcareous tests begin to dissolve and this produces a decline in FN_t. That this solution effect is significant was shown by Schott (1935) who demonstrated that the numerical distribution of planktonic tests in the pelagic zone did not correspond to the FN_t distribution in the bottom sediments.

Because the FN commonly is calculated using all the tests in a sample it is only a crude measure of foraminiferal productivity because dead tests are counted along with living tests.

B. Volume

Various investigators (Höglund, 1947; Parker, 1948; Butcher, 1951; Phleger, 1962; Walton, 1955) prefered to calculate the number of tests relative to sample volume. Their results can be transformed to FN_t if the specific gravity of the sediment grains is known and an estimate of the degree of packing (porosity) is made. A rough conversion from volumetric to weight values is accomplished by multiplying the sample volume (in ml) by 2.5.

Different kinds of information are obtained if only the living specimens in a sample are counted. Although there are often very few living specimens in a sample, their numbers give a much better idea of productivity than does the total (living plus dead) fauna. Investigators from Scripps Institution of Oceanography suggested that the number of living specimens in a 10 ml wet sample become the standard for comparison. We could designate these values as FN_l/10 ml. Phleger (1960a,b) said the majority of samples studied in his laboratory had FN_l/10 ml values between 50–200. However, much greater values have been obtained; e.g. 1000–3500 in the delta of the Mississippi River; 2600 in San Antonio Bay, Texas; 1640 in the Laguna Madre, Texas and 1650 in the Ojo de Liebre lagoon, Mexico. All of these regions are characterized by high organic productivity.

C. Surface Area (standing crop)

A much better estimate of the environmental response of the foraminifera is obtained by calculating the number of living specimens per unit of surface area (standing crop). It is important to know the standing crop because areas with large standing crops are areas with high productivity, not only for foraminifera but usually for all organisms. Although the standing crop varies with time, it can be used to estimate productivity and biomass.

The benthonic foraminifera can be utilized to provide estimates of organic productivity because they live in all marine environments (including areas of low salinity) and inhabit all depths. Knowledge of the foraminiferal standing crop is useful for geological purposes as well. An understanding of the rate at which foraminiferal tests accumulate will further our knowledge of the geologic history of calcareous material in oceanic waters.

Although the volumetric calculation of living specimens discussed above is a crude measure of standing crop, it is not necessarily reflective of surface area. Walton (1955) first suggested that the number of living specimens per unit area be calculated. Most of the investigators who have calculated standing crop used only the upper 1–2 cm of the sample on the premise that the benthonic foraminifera live on the surface. However, it has been demonstrated that specimens can live within the sediment at considerable depths (Chapter IX,5; X,2). Benthonic foraminifera also live attached to vegetation (Chapter II,7; IX,5). The presence of living specimens in these habitats are not tallied in conventional methods of calculating standing crop from surface sediment samples.

Various investigators have calculated the standing crop in a variety of areas (Lankford, 1959; Phleger, 1960a, 1964; Richter, 1964a; Bartlett, 1966; Lynts, 1966; Haake, 1967; Saidova, 1967c; Seibold, 1967; Boltovskoy & Lena, 1969b; Rouvillois & Rosset-Moulinier, 1969; Murray, 1970b; Sen Gupta, 1971; Rouvillois, 1971; Lee & Muller, 1973). Most of these studies were conducted in shallow water environments. The greatest reported values for standing crop were on the Grand Banks of Newfoundland where 400 living specimens per square centimetre were found (Sen Gupta, 1971) in one sample. The average value in this area was only 9/cm^2, however. Highest average values were reported in the delta of the Mississippi River where 250/cm^2 were found (Lankford, 1959). In the deep-water study cited above, Saidova (1967c) found benthonic foraminifera specimens containing protoplasm in the greatest depths of the Kurilo Kamchatka trench area. She calculated standing crop values which ranged from 0.035 to 0.1/cm^2 for a zone extending from 1000 to 8000 in depth. The sample from these deep zones were collected in August and September, months that Saidova considered periods of reproduction. She found that from 20–40% of the total foraminiferal fauna consisted of living specimens. In some zones (2500–4500 m) this proportion rose to 50–60%. She calculated the biomass of foraminiferal protoplasm by estimating the average internal volume of the test and multiplying by (a) the standing crop, and (b) the specific gravity of protoplasm (1.027). She estimated that the foraminiferal biomass was 4–10 gm/m^2 in the depth zones 1000–2000, 3000–4000 and 5500–8000 metres. In the less well populated intermediate depths (2000–3000, 4000–5500 m) the biomass was 2–3 gm/m^2. These figures are probably low estimates because only living specimens from the surface sediment were counted.

Zenkevich (1967) claimed that this biomass would be sufficient to satisfy the food requirements of all bottom feeders that ingest foraminifera. Other calculations of biomass were made using similar volumetric estimations (Murray, 1968a, 1969, 1970a,b).

Production is a measure of the total number of organisms formed during a given time span (usually a year) in given area (usually a km^2 or m^2). Although organic production and productivity (rate of production) have been well studied for many planktonic groups, not a great deal is known about benthonic production. What little work that has been done on benthonic foraminifera was summarized by Murray (1967a, 1973b). To calculate productivity it is necessary to know: (a) the initial size of the standing crop, (b) the percentage of individuals that reproduce (generally unknown), (c) the frequency of reproduction (known for some species; Chapter II,3) and (d) the size of the brood (not well known).

Because these 4 factors are imperfectly known, it is only possible to generalize about productivity in benthonic foraminifera. Murray suggested that if the standing crop were very large (>100/cm^2), productivity would be large even if few species reproduced, or they reproduced only annually, and/or few offspring per parent were produced. If the standing crop is very small (<1/cm^2), productivity would be small even if many species reproduced with great frequency. For standing crop values

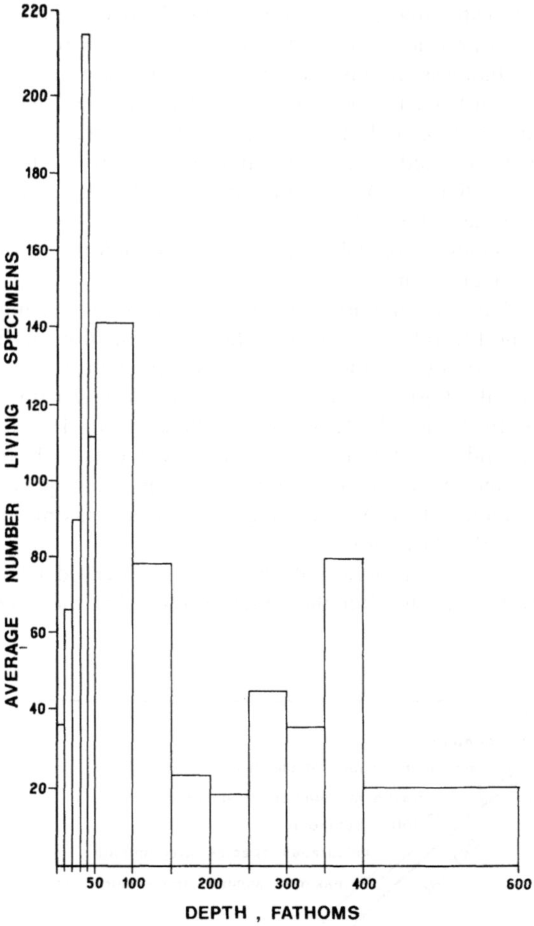

Fig. 54. Depth distribution of total living foraminiferal population, Todos Santos Bay, California (after Walton, 1955).

between these extremes (the typical cases) a variety of productivity values would result from changes in reproductive rate, brood size, etc.

Boltovskoy & Lena (1969b) attempted to estimate productivity from data obtained at Puerto Deseado, Argentina. The average standing crop was 9.24/cm^2. Those species which reproduced once a year comprised about 6% of the living population (0.55/cm^2). The remaining 94% (8.69/cm^2) of the population is composed of species with shorter reproductive cycles. The reproductive frequency is not well known but seems to increase as species size decreases. As the species in this area were small,

Boltovskoy & Lena estimated a one month reproductive period. They then calculated that the annual production is $0.55+(8.69\times 12)$ or approximately 105/cm^2.

To calculate the biomass of this productivity Boltovskoy & Lena weighed 2800 living specimens taken from the bottom in the Deseado area. After being weighed, they were calcined at 500 °C and then reweighed. The resulting weight loss, 12.4 gm, is the dry biomass. Calculated for the estimated productivity in the area, the biomass is 0.465 gm/cm^2. This figure compares favourably with the volumetric estimates of Saidova and Murray described above.

To conclude this chapter we will summarize observations on the distribution of various abundance phenomena.

Various authors have documented the increase in total foraminiferal test abundance away from the littoral zone. Most of these investigations indicate a maximum FN occurring on the outer continental shelf and upper part of the slope (Giunta, 1955; Bandy & Arnal, 1960; Bandy, 1961; Chierici, Busi & Cita, 1962; Anderson, 1963; Bartlett, 1964b; Bandy, Ingle & Resig, 1964c; Ikeya, 1971). This area occurs between 90–100 m and 200–300 m depths. The FN decreases beyond this point.

Walton (1955) found two abundance maxima among living specimens with respect to depth in Todos Santos Bay, Mexico (Fig. 54). These maxima occurred between 30–40 fathoms and 50–100 fathoms.

A few exceptions to this general rule have been observed (Resig, 1958; Jarke, 1960) but they are too few to alter the generalization that the maximum benthonic

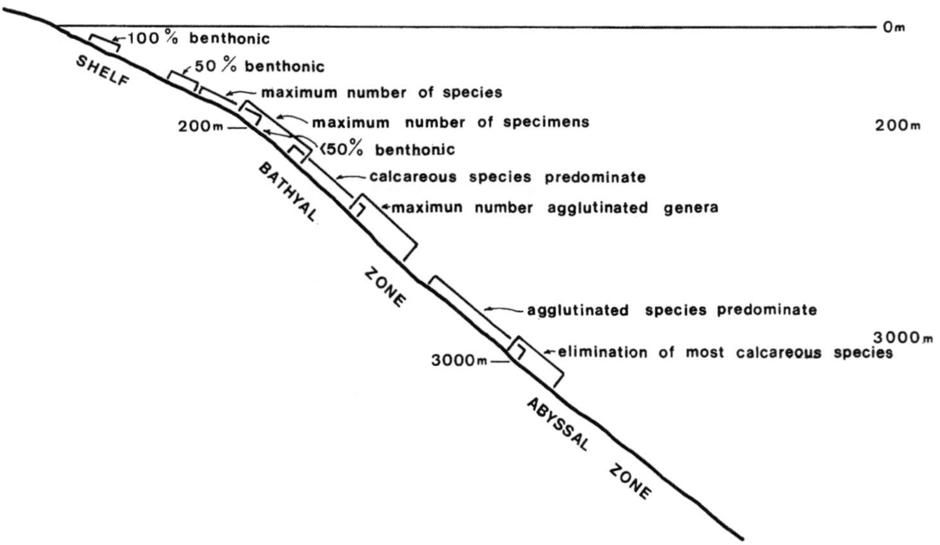

Fig. 55. Approximate location of various benthonic foraminifera traits on the sea floor.

foraminiferal specimen abundance occurs near the outer edge of the continental shelf.

As depth increases, the relationship between the benthonic and planktonic composition of the bottom assemblages also changes. Benthonic specimens dominate in the shallow areas, planktonic ones in deeper zones. The explanation for this change lies not only in the decrease in benthonic specimen abundance noted above, but also in the increase in planktonic specimen abundance in deeper water. As the water column deepens away from the coast, there is progressively more room for planktonic species. Some planktonic species are depth dependent and can not live in shallow waters; others are highly sensitive to salinity changes and can not live in the more unstable near shore shallow waters (Chapter VII,4).

Exceptions to this general trend are found in areas of upwelling where cold, deep, nutrient rich waters reach the surface. Plankton populations bloom in these waters and the resulting increase in productivity may alter the ratio between benthonic and planktonic tests in favour of the latter.

Figure 55 shows summarized observations with respect to the quantitative and, to some degree, qualitative distribution of foraminifera on the oceanic bottom.

V. BENTHONIC FORAMINIFERA OF BRACKISH, HYPERSALINE, ULTRASALINE, AND FRESH WATERS

Aside from temperature, one of the most limiting factors controlling the distribution of marine organisms is salinity. The faunal composition of marine and brackish water areas is different. As a general rule, it is difficult for a marine fauna to penetrate into fresh waters and *vice versa*. Even when there are available migration routes, faunal mixing is impaired by salinity barriers. For thousands of years water from the Mediterranean Sea (salinity 37‰) has mixed with water from the Black Sea (salinity 19‰) and yet the great majority of the marine organisms common to the Mediterranean have been unable to adapt to life in the Black Sea and are completely absent there.

Although adaption of marine forms to waters of lower salinity is accomplished only with difficulty, it is not impossible for various groups, but it does require a long period of time and a low salinity gradient. Under these conditions a special kind of fauna, one adapted to brackish water, is created. Such faunas are usually low diversity faunas. Because animals are generally more sensitive than plants to salinity changes, fewer animal species than plant are found in these mixed environments. Despite their low species diversities, these brackish water environments are quantitatively rich. For example, although the salinity of the Azov Sea is 3 times less than that of the Mediterranean and its marine fauna is 30 times poorer, it is considered one of the richest aquatic areas in the world in terms of its nutrient resources.

There are various schemes in use which classify natural waters on the basis of their salinity. One of the most commonly used classifications is the 'Venice System'. This classification was adopted by the 'Symposium of the Classification of Brackish Waters' which took place in Venice in 1958.

hyperhaline	>40‰
euhaline	30–40‰
mixohaline	0.5–30‰
fresh	<0.5‰

The mixohaline category in this classification has been subdivided into subtypes by various authors. There is no unanimity of opinion over these divisions as different authors place the boundaries at different salinity values in different environments.

Considering only the reactions of the foraminifera, the most appropriate division is:

ultrahaline	>75‰	mixohaline	18–30‰
hyperhaline	40–75‰	brackish	0.5–18‰
euhaline	30–40‰	fresh	<0.5‰

The term hypohaline (hyposaline) is also used in the literature to designate salinity conditions which are below normal. No specific numerical boundaries are set on this term although its upper limit is often considered as 30‰.

All marine organisms, including the foraminifera, can be classified as either stenohaline or euryhaline. The former are those which are very intolerant of changes in salinity. The latter can withstand wide oscillations in salinity.

The following section deals with the foraminifera which dwell in brackish waters. These are generally euryhaline and almost exclusively benthonic, as the planktonic foraminifera are quite intolerant of wide ranges in salinity and of low salinity values.

1. Brackish water

Brackish water can be divided into 3 groups on the basis if its origin: (a) estuarine water of rivers which empty into the sea as well as coastal water strongly influenced by fresh water runoff from the continent, (b) brackish water of relict enclosed basins and semi-enclosed inlets and (c) brackish ground water relict from ancient seas (connate water).

It is difficult to distinguish the foraminiferal differences between waters of the first two types. Brackish ground water foraminifera are more distinct. There the foraminifera have to live in the water that remains in the void space between grains of sediment. The foraminifera of the first two types have been more extensively studied.

Below we will describe separately the foraminifera found in each type of water.

A. Estuaries and Coastal Zone Waters

Numerous studies have been conducted on the foraminifera in these environments. The first of these was conducted by Brady (1870) who described *Elphidium striatopunctatum* and *Nonion depressulus* as species which could live in 'fresh water'. He also described the euryhaline genera *Quinqueloculina*, *Trochammina*, *Lituola*, *Cibicides*, *Rotalia*, *Elphidium* and *Nonion* as being less tolerant to very low salinity conditions. He noted that these genera were associated with the thecamoebian *Difflugia* and that as the salinity decreased the proportion of *Difflugia* increased with respect to the foraminifera. Brady also found brown stained, delicate tests of *Globigerina* and *Orbulina* in brackish zones. These earliest data are of historic interest only, in that Brady was describing dead specimens and did not give exact salinity data.

Walther (1893)[6] synthesized many observations with the conclusion that *Quinqueloculina*, *Cibicides*, *Elphidium*, *Nonion* and *Rotalia* were the genera that could best tolerate brackish waters.

[6] Both Brady and Walther used the denominations *Polystomella*, *Nonionina*, and *Truncatulina* instead of *Elphidium*, *Nonion* and *Cibicides*. They were undoubtedly referring to the genera we now know by the second group of names and for that reason we listed them as such. As we stated in the introduction, we normally conserve the name applied by the original author, but this seemed such an exceptional and non-controversial case that we converted the names to the more modern nomenclature.

Hedberg (1934) studied the foraminifera of Lake Maracaibo which is connected to the sea by a narrow strait and into which various rivers empty. He found that the fauna was different in different areas and was poor in low salinity zones. *Miliammina fusca* was the species most tolerant of low salinity areas with *Elphidium* and *Rotalia beccarii* also exhibiting great tolerances.

Marie (1938) described foraminifera of the Rance River. Specimens of *Elphidium* cf. *excavatum*, *Nonion* cf. *stelligerus*, *Lagena hexagona* and *Rotalia* sp. were found in fresh water zones whose salinity ranged from 0.07–0.12‰. Waters only slightly more saline (0.25–0.9‰) contained specimens of *Trochammina* cf. *nitida*, *Spiroloculina* aff. *grata*, *Bolivina punctata*, *Spirillina vivipara*, *Patellina corrugata*, *Globigerina bulloides* and *Adelosina laevigata*.

Bartenstein (1938) and Bartenstein & Brand (1938) described the foraminifera of brackish waters of northern Germany. They listed *Proteonina difflugiformis*, *Haplophragmoides canariensis*, *Ammobaculites agglutinans*, *Trochammina nitida*, *Nonion depressulus*, *Elphidium excavatum*, *Rotalia beccarii* and *Miliammina fusca*. This last species was the predominant form in Jade Bay just as it was in Lake Maracaibo. In addition to these species, *Jadammina polystoma* was described for the first time in a brackish water environment. This species has proven to be a common one in other similar settings.

This same area was restudied by Richter (1961) who examined only living specimens. He noted the dominance of *Nonion depressulus*, *Elphidium excavatum*, *E. selseyense*, *Trochammina inflata* and to a lesser degree, *Rotalia beccarii*.

The study of Ago Bay, Japan by Morishima (1948b) established the presence of 5 different thanatocoenoses whose existence was explained exclusively by differences in salinity. These zones, from the open ocean toward the coast are: (a) *Globigerina*, (b) *Amphistegina*, *Elphidium crispum*, (c) *Rotalia papillosa*, *Quinqueloculina lamarckiana*, (d) *E. craticulatum*, *Textularia haueri* and (e) *Trochammina*. In other studies (1948a, 1955) the same author established the repreated presence of *Trochammina* (usually *T. globigeriniformis*) in low salinity waters.

Phleger & Walton (1950) described various foraminiferal associations in Barnstable Bay, Massachusetts. The most brackish water environments were characterized by large numbers of *Trochammina* and abundant *Miliammina fusca*.

Post (1951) studied the coastal foraminifera along the south Texas area in a region characterized by great changes in salinity. The genera *Nonion*, *Elphidium* and *Rotalia* exhibited the greatest tolerance to low salinity (down to 9‰). Lesser quantities of *Miliammina* and *Ammobaculites* were also present in low salinity zones.

Röttgardt (1952) created several foraminiferal zones in Kiel Bay, Germany on the basis of salinity changes: (a) lacustrine brackish (1.8–6‰) *Miliammina fusca*, *Proteonina difflugiformis*, (b) brackish (6–9‰) *Miliammina fusca*, *Elphidium granulosum*, *E. asklundi*, *Nonion depressulus*, *Trochammina nitida*, (c) marine brackish (9–20‰) *Reophax subfusiformis*, *R. dentaliniformis*, *Elphidium* sp. and (d) marine (20–30‰) *Eggerella scabra*.

The Rhone delta was divided into two environmental zones by Kruit (1955); a 'terrestrial' and a 'coastal marine' zone. The former was subdivided on the basis of salinity changes with the following sequence of foraminifera occurring from fresh (0.2‰) to normal marine: *Haplophragmoides canariensis provencencis, Trochammina inflata, Rotalia beccarii, Nonion* cf. *depressulus, Elphidium lidoense camarguensis, E. crispum* and *Quinqueloculina* cf. *elegans.*

The following genera were dominant in the brackish waters of small inlets in New York Bay (Ronai, 1955): *Ammobaculites, Ammoastuta, Eggerella, Miliammina, Quinqueloculina, Triloculina, Trochammina, Nonion, Elphidium, Buccella* and *Rotalia.*

Blanc-Vernet (1958) investigated the enclosed coastal ponds near Marseilles and found a poor fauna consisting of only *Rotalia beccarii tepida, Nonion depressulus, Elphidium crispum* and *Quinqueloculina seminulum.*

A general study of the Texas coastal bays (Ladd, Hedgpeth & Post, 1957) revealed that low salinity areas (0–9‰) were characterized by the genera *Rotalia, Elphidium* and *Nonion* and the species *R. beccarii* and *E. galvestonense.*

Ishiwada (1958) analysed the dead foraminiferal tests of Lake Khamana-Ko, Japan which has a connection with the open ocean. Three associations were described: (a) open sea – *Pseudononion japonicum, Elphidium subgranulosum, Rotalia beccarii* (var.), (b) transitional – *Reophax* sp., *Goesella iizukae, Ammobaculites exiguus, Rotalia beccarii* (var.), *Elphidium* sp., etc., (c) lacustrine – *Haplophragmoides canariensis, Trochammina globigeriniformis, Rotalia beccarii* (var.) and some others.

Lankford (1959) studied the living foraminifera of the eastern border of the Mississippi River delta. In addition to the marshes, he described two zones with low salinities. The first, characterized by salinities less than 10‰, exhibited abundant specimens of *Elphidium delicatulum* and frequent specimens of *Ammotium salsum, E. gunteri, Miliammina fusca* and *Streblus beccarii.* The species most characteristic of the second zone (1–32‰) was *Palmerinella gardenislandensis* although specimens of *Bolivina lowmani, Nonionella opima* and *Epistominella vitrea* were common. Lankford concluded that of the species present, *Ammotium salsum* and *Miliammina fusca* were most capable of tolerating the extreme ecological changes which occurred in this environment.

Voorthuysen (1960) studied the living foraminifera of the mouth and lower reaches of the Ems River, Holland. The area where most of the foraminifera were found had salinities of 16.5–30‰ although the salinity was considerably less in upstream areas. The fauna differed from many of those described previously in the low number of agglutinated forms. Only 11 living species were found, the 5 most common of which were *Streblus beccarii, Nonion depressulus, Elphidium selseyense, E. excavatum* and *E. gunteri.*

Water of low salinity (down to 8‰) occurring along the coastal zone of Martha's Vineyard, Massachusetts is characterized by living specimens of *Streblus beccarrii, S. beccarii tepida, Quinqueloculina lata, Q. seminulum, Elphidium galvestonense, E.*

margaritaceum, Protelphidium tisburyense and *Rosalina columbiensis* (Todd & Low, 1961).

The Lagoa dos Patos, a large lagoonal estuary in southern Brazil contains abundant *Miliammina fusca* and various *Elphidium* spp. (Closs, 1962; Closs & Medeiros, 1965; Closs & Madeira, 1968). Other species present in these brackish waters include *Ammotium cassis, Textularia earlandi, Trochammina ochracea* and *Haplophragmoides wilberti.* Smaller nearby estuaries are characterized by *M. fusca, A. salsum, E. discoidale* and *Rotalia beccarii* (Closs & Madeira, 1962, 1967; Closs & Medeiros, 1967).

Ellison, Nichols & Hughes (1965), Nichols & Ellison (1967) and Ellison & Nichols (1970) studied the living foraminifera of the estuary of the Rappahannock River, Virginia and found that the upper reaches of the river (0.5–16‰) were dominated by *Ammobaculites crassus* whereas the lower part of the estuary (>15‰) was characterized by *Elphidium clavatum.* The authors summarized the faunal changes from fresh to normal marine salinity as a transition from thecamoebians to agglutinated foraminifera to chiefly calcareous foraminifera.

Miliammina and *Ammomarginulina* were the genera found by McCrone & Schafer (1966) to be most tolerant of the low salinity waters (5‰) near the head of the Hudson River estuary, New York. Above this zone no living specimens were found. Specimens of *Ammonia* and *Trochammina* found in low salinity water were characterized by a 'chitinous' test rather than by the usual calcareous and agglutinated ones.

The low salinity water of Sabine Lake, an estuary on the Gulf of Mexico, is characterized by *Haplophragmoides, Miliammina* and *Ammobaculites* (Kane, 1967).

Albani (1968a,b) studied the foraminifera in an Australian estuary and was able to divide the fauna into various groups, the brackish water one of which was characterized by *Haplophragmoides canariensis, Trochammina inflata, Miliammina fusca* and *Tritaxis conica.* Unfortunately, the author did not give precise salinity data for the area.

The waters of the Mississippi Sound near Mobile Bay receive extensive fresh water influx and exhibit a highly variable salinity. The low salinity zones of this area are characterized by *Ammobaculites salsus* and *Miliammina fusca* whereas the zones with fluctuating salinity contain *E. gunteri* and *Ammonia beccarii* (Anderson, 1968).

The living foraminifera of the estuary and lower reaches of the Quequén Grande River, Argentina were studied by Boltovskoy & Boltovskoy (1968) and Wright (1968). The latter studied the Miliolidae whereas the former examined the other foraminifera. In addition to faunal samples collected throughout the year, extensive salinity data were collected which allowed close correlation between salinity conditions and species distribution. The upstream zones which received marine waters only during high tides were characterized by *Miliammina fusca, Elphidium excavatum, Rotalia beccarii* ex gr. *parkinsoniana* and *Haplophragmoides wilberti.* The distribution of the other species is shown in Fig. 56.

Murray (1968b) conducted an extensive study of Christchurch Harbour, England,

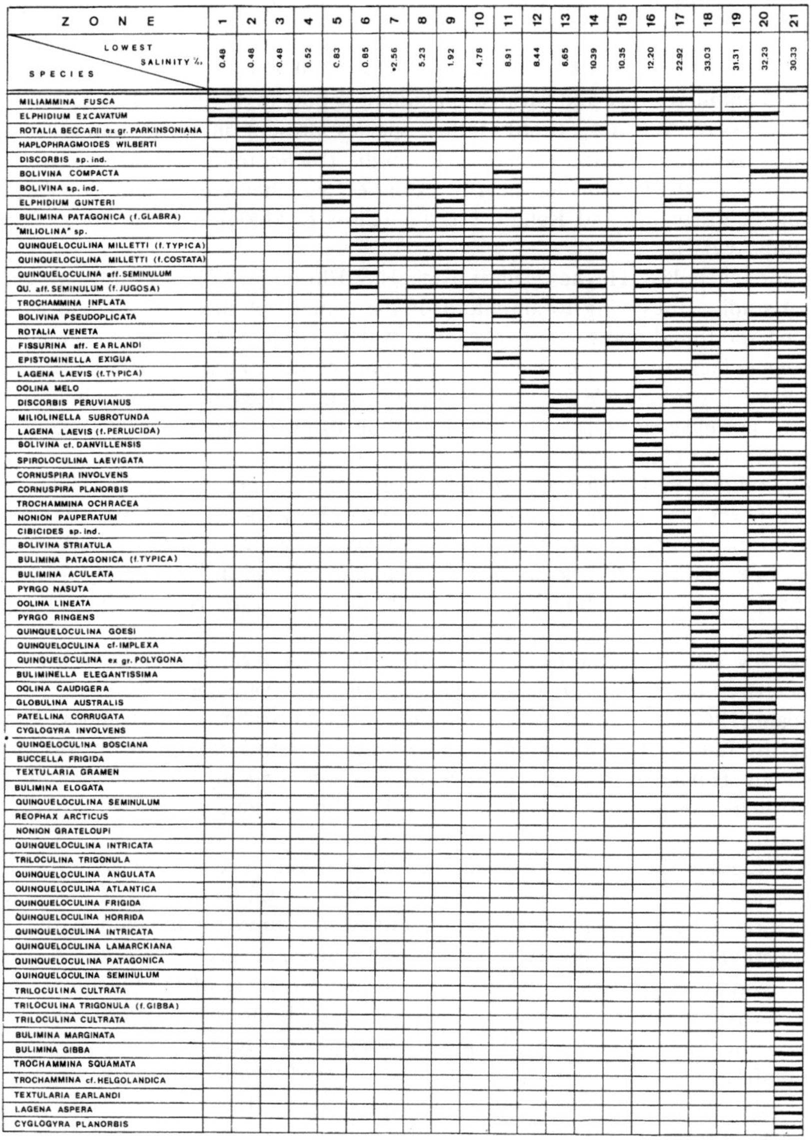

Fig. 56. Distribution of living benthonic foraminifera in the estuary of the Quequén Grande River, Argentina (after Boltovskoy & Boltovskoy, 1968; Wright, 1968).

a shallow water estuary. He described *Miliammina fusca, Elphidium oceanense, E.* sp. and *Protelphidium anglicum* from the low salinity waters of the estuary. He noted that the upper (low salinity) reaches of the estuary were characterized by low foraminiferal diversity. The lower segments exhibited higher diversity values.

The James River Estuary of the North American Atlantic Coast is characterized by *Ammobaculites crassus* in its upper, low salinity, low diversity reaches. The more saline (>14‰) zones nearer the outlet exhibit a more diverse fauna, one dominated by *Elphidium clavatum* (Nichols & Norton, 1969).

Tapley (1969) described the foraminifera of the Miramichi Estuary in Canada. She noted the dominance of *Miliammina fusca* in the low salinity zones and an *Elphidium* fauna downstream.

The fauna of the Arcachon Estuary on the Gulf of Gascogne is dominated in its low salinity zones by agglutinated species: *Miliammina fusca, Trochammina inflata, Ammobaculites agglutinans* and *Jadammina polystoma.* A few calcareous specimens are present: *Nonion depressulus, Elphidium excavatum* and *Ammonia beccarii* (Le Campion, 1970).

Akers (1971) described estuarine and tidal marsh foraminifera from the Beaufort, North Carolina area. He found living specimens of *Ammobaculites, Ammonia beccarii, Arenoparrella mexicana, Miliammina fusca, Siphotrochammina lobata, Tiphotrocha comprimata* and *Trochammina laevigata.*

The artificially created estuaries in the harbour area at St. Malo, France have been invaded by a brackish water fauna dominated by *Eggerella scabra, Cribrononion lidoense, Protoelphidium paralium* and *Cribroelphidium excavatum* (Rouvillois, 1972).

Lamb (1972) described *Miliammina fusca* from the upper part of Mobile Bay and *Ammobaculites salsus* from the middle part of the bay where the salinity was higher.

The brackish water regions of the Gironde Estuary, France are dominated by agglutinated species (*Jadammina polystoma, Labrospira jeffreysii, Trochammina pacifica, T. ochracea* and *Haplophragmoides wilberti*) although some calcareous species are present (*Nonion depressulus, N. orbiculare, Ammonia beccarii tepida* and *Elphidium lidoense*) (Pujos, 1973b).

B. Brackish Waters of Relict Enclosed Basins and Semi-enclosed Inlets

Only a few studies of these areas have been conducted.

The Black Sea serves as a model for a semi-enclosed type of environment. Its average salinity is 19‰ and a limited number of species are distributed among 15–20 genera, the most common of which are *Nonion, Elphidium, Quinqueloculina, Trochammina, Ammobaculites, Eggerella, Rotalia, Hipocrepinella, Massilina, Discorbis* and several others. Somewhat more than 50 species, subspecies, varieties, etc. have been recorded from the area, but as several of these undoubtedly are synonyms, the actual number of taxa which inhabit the sea is about 25.

The Caspian Sea (12–13‰) and the Aral Sea (11‰) are typical of enclosed basins.

The foraminiferal fauna in these environments is greatly reduced in diversity. Only 9 species are known from the Caspian Sea including *Rotalia beccarii*, *Elphidium granulosum*, *Nonion depressulus* and *Ammobaculites pseudospirale*. Only 4 species have been found in the Aral Sea. The reduced number of species in the latter area is explained less by the reduced salinity (which is only slightly less) than by the chemical nature of the dissolved salts (the water is very high in sulfates), the low water temperature and a recent geological history which includes rapid salinity fluctuations.

These areas are relicts of the Tethyan Sea which was much more extensive during the Tertiary and has now diminished in size and been separated into distinct zones. There are numerous other similar relict Tethyan lakes in Eurasia and North Africa. The foraminifera of some of these have been studied.

Daday (1884) described a foraminifer in the lakes of Rumania, *Entzia tetrastomella*, which is very similar to *Trochammina*. Similar forms were described by Krapin (1929) in the mouths of the rivers which feed Lake Elton. Gauthier-Lièvre (1935) found that the potable water from wells in l'Oued Rhir, Sahara, south of Bikstra, contained specimens of Miliolidae, *Trochammina* and *Anomalina*. The brackish waters of the same area contained *Ammodiscus*, *Nonion*, *Cibicides* and *Opthalmidium*.

Specimens of Haplophragmoides canariensis were found in the swamps around Erfurt, Germany (Bartenstein, 1939).

Le Calvez & Le Calvez (1951) examined the fauna of two ponds along the Mediterranean coast whose salinity is highly variable (up to 70‰ but averaging about 20–25‰). They found specimens of *Rotalia beccarii*, *R. beccarii tepida*, *Nonion depressulus*, *N. pompilioides*, *Elphidium excavatum*, *E. littorale* and *Quinqueloculina costata*.

Agglutinated specimens, probably belonging to *Haplophragmoides*, were found in Lake Bol'shoe, Kulikovskoe, Soviet Union (Rezvoj, 1951) and in small lakes and marshes in Germany (Rieth, 1958).

The examples mentioned above refer to Recent foraminifera. Investigations of the fossil faunas found in similar types of environments reveal similar faunal assemblages. The Miocene seas of Austria, Rumania and southern Russia became disconnected from or only partially connected with the open ocean. The best studied of these environments is the Sarmatian Sea whose foraminiferal fauna contains isolated representatives of *Elphidium*, *Nonion* and some Miliolidae.

C. Brackish Ground Water

The foraminifera of this biotope have not been as well studied as those of others.

The first discovery of this type was made in wells located about 250 km east of the Caspian Sea in the Kara-Kum Desert (Brodsky, 1928, 1929). The salinity varied between 5–11‰ and the temperature of the water between 17–20 °C. Specimens of *Lagena*, *Nodosaria*, *Discorbis*, *Textularia* and *Globigerina* were found. All the tests were very small (<160 μ) and completely chitinous. The sediment pore space was about 15–20 times greater than the specimen diameter. Although there is some doubt about the generic determinations made by Brodsky, there is not doubt that the specimens are foraminifera. Since these specimens did not contain protoplasm, it is possible that they were partially dissolved relict marine fossils.

Nevertheless, 27 other wells in the same Kara-Kum Desert later yielded about 300 foraminiferal specimens that did contain protoplasm (Nikoljuk, 1947, 1948). These specimens were described as *Miliolina oblonga arenacea*, *Spiroloculina turcomanica* and *Fischerina* sp. All were megalospheric, small and had delicate walls. The character of the water in which these specimens were found is very similar to Mediterranean Sea water in terms of its salinity and oxygen content.

In the same central Asian region, in Balpash-Sor Lake, Schmalhausen (1950) found 80 specimens of living foraminifera which were identified as *Borovina zernovi*, a species very similar to *Jadammina polystoma* and *Entzia tetrastomella* both of which are common to brackish waters in other localities. These specimens were interpreted as having been part of an interstitial fauna which was carried by ground water into the lake where it died due to the elevated salinity. The 3 faunas described above are all relict faunas of the Tethyan seaway.

A comparison of the foraminiferal faunas of the 3 biotopes described in this section reveals several common elements. Above all, the fauna is not diverse. There are no planktonic species with the exception of a few *Globigerina*, but these may not live in the environments. The number of specimens is, in general, not great. Taxa with chitinous and agglutinated walls usually predominate (*Trochammina*, *Entzia* and *Borovina*). *Miliammina fusca*, also agglutinated, is common. Calcareous species seem to be less able than agglutinated forms to tolerate the lowest salinities found in these environments. Of the calcareous species found, *Rotalia beccarii*, various *Elphidium* (*E. excavatum*, *E. selseyense*), *Nonion depressulus* and *Nonion tisburyense* are the most abundant. Various Miliolidae (*Quinqueloculina*, *Cornuspira*, *Spiroloculina*, etc.) are also capable of adapting to brackish water, although their tolerance is somewhat less than the hyaline perforate and chitino-agglutinated forms. Under severe reductions in salinity, these species become very small and lose their calcareous test, exhibiting only their organic lining. This same loss of test has also been observed in other calcareous species, namely in *Rotalia beccarii* in low salinity bays along the North Sea (Pratje, 1931). The lituolid genera, *Haplophragmoides*, *Ammoscalaria* and *Ammotium* are also frequently found in brackish waters.

The differences between the brackish faunal assemblages found in different sites can be due to differences in quantities and kinds of dissolved salt and to differences in local geologic histories.

2. Hypersaline and ultrasaline water

Hypersaline water (40–75‰) commonly occurs in sites that have been isolated or cut off from normal marine water. The upper limit for marine animals is about

200–220‰ and is somewhat greater for plants. More detailed information on salinity tolerance limits of individual species is given in Chapter IX,2.

Information on the foraminifera of hypersaline waters is not abundant. Said (1950) found that the specimens in the northern part of the Red Sea where salinity values are 40–41‰ were no different from those living in waters of normal marine salinity.

Specimens of *Rotalia beccarii* and *Elphidium poeyanum* live in parts of the Caspian Sea where the salinity reaches 50‰ (Furssenko, 1959).

Tinoco (1958) described specimens of *Streblus beccarii catesbyana* from a Brazilian lagoon whose salinity ranged from 55–71‰. Although the specimens were not living when sampled, Tinoco, after studying various complimentary factors, concluded that the species does live in these waters.

Ayala-Castañares & Segura (1968) encountered the following foraminifera in the hypersaline Laguna Madre, Mexico: *Miliammina fusca, Ammobaculites dilatatus, Quinqueloculina rhodiensis, Q. seminulum, Triloculina* sp., *Eponidella gardenislandensis, Palmerinella palmerae, Ammonia beccarii* (var.), *Elphidium* cf. *fimbriatum, E. galvestonense, E. gunteri, E. incertum mexicanum, E.* cf. *koeboeense, E. matagordanum, E. poeyanum, E.* sp. Unfortunately, these authors did not give any exact salinity data.

Murray (1970a) found that *Peneroplis planatus* was the dominant foraminifera in the internal part of the Abu Dhabi Lagoon, Persian Gulf, where the salinity was between 41–53‰. There were also numerous specimens of *Elphidium* cf. *advenum.*

Sellier de Civrieux (1968) studied foraminifera of the Unare and Píritu lagoons in central Venezuela. The salinity of these lagoons is subject to great periodical changes and can be as high as 90‰ (ultrahaline water). He stated that only 3 species were found alive and can thus tolerate such high salinity: *Ammonia tepida, Quinqueloculina* sp. and *Cribroelphidium poeyanum.* Sellier de Civrieux (personal communication) later expressed some doubt with respect to the determination of the last species; apparently it belongs to the Elphididae but perhaps to another species.

3. Fresh water

Every textbook on foraminifera states that these organisms are almost exclusively marine since the representatives of only one family, the Allogromiidae (Lagynidae of some authors) live in fresh water. This family has not been well studied and there is a tendency among some protozoologists to exclude it from the Order Foraminiferida. Kudo (1969), for example, put them in the Order Testacida. Galloway (1933, p. 6) stated that 'the removal of the Lagynidae would leave in the foraminifera only salt water forms'. The same idea has been expressed by many authors of well known books (Glaessner, 1945; Cushman, 1948; Sigal, 1952; Pokorny, 1958; Loeblich & Tappan, 1964). However, studies of the last few years have revealed that there are several species of non-allogromid foraminifera that do live in fresh water.

Actually, the first evidence that certain foraminifera can tolerate fresh water environments was published by Schultze (1854) who put Mediterranean mud containing specimens of Rotaliidae in fresh water. He changed the water several times, and after 5 weeks he transferred the mud back to sea water and observed that numerous specimens were living. The failure of Schultze to positively identify his specimens may have been the reason why these observations were ignored by later workers. Succeeding investigators could not duplicate these observations since the species were unknown. Experiments with cultures of other foraminifera have not yielded any species that could tolerate a fresh water environment.

The first foraminifera found living in fresh water under natural conditions were discovered in the vicinity of Buenos Aires in the Rio de la Plata, at a depth of about 2 m in silty and sandy sediments (Boltovskoy, 1958a). The salinity of this area is between 0.1–0.15‰. Many specimens of *Nonion tisburyense* and one specimen of *Trochammina*, all with protoplasm, were found. The latter foraminifer was undersized and badly developed. *N. tisburyense* was first described by Butcher from brackish water in tidal pools of Cape Cod, Massachusetts and later in some other brackish water localities, but not in fresh waters.

The second observation of living foraminifera in fresh water was by Closs & Medeiros (1965), who observed specimens of *Miliammina fusca*, *Reophax arcticus* and *Nonion tisburyense* in the Dos Patos Lagoon, Brazil in an area where fresh waters are present during the main part of the year. All these species have been found previously in brackish waters.

As mentioned above, Boltovskoy & Boltovskoy (1968) encountered living specimens of *Miliammina fusca* in fresh water zones of the Quequén Grande River, Argentina, and very small individuals of *Elphidium excavatum*, *Haplophragmoides wilberti* and *Rotalia beccarii parkinsoniana* in a part of the river where the fresh water environment is interrupted only from time to time by brackish water brought in by high tides.

Boltovskoy & Lena (1971) studied about 1000 samples collected from 1963–1967 in the fluvial part of the Rio de la Plata and several samples from the fresh water lakes in the area south of the river. Their study not only confirmed the observations made by Boltovskoy (1958a), but expanded the list of foraminifera found with protoplasm in these fresh water conditions.

Based on their study and on previous data, Boltovskoy & Lena drew the following conclusions: (a) there are non-allogrominid foraminifera which dwell in fresh waters or in areas where fresh water conditions prevail during most of the year, (b) *Nonion tisburyense* thrives in a pure fresh water environment as well as in low salinity brackish waters, (c) the following foraminifera were found alive as rare or isolated specimens in fresh water habitats: *Miliammina fusca*, *Psammosphaera* sp., and *Trochammina* sp., (d) the following foraminifera were found living in areas where fresh water conditions are interrupted from time to time by short periods of brackish water influx: *Nonion tisburyense*, *Miliammina fusca*, *Rotalia beccarii parkinsoniana*, *Elphidium excavatum*, *Haplophragmoides wilberti*, and *Ovammina* sp. and (e) the pure fresh water milieu as well as that invaded periodically by brackish waters can have different foraminiferal assemblages in distinct areas although these areas are situated near each other and apparently have similar environmental parameters.

An example of the last situation is found in the fluvial zone of the Rio de la Plata where there is a relatively rich population of *Nonion tisburyense* but a marked absence of *Miliammina fusca*. However, no specimens of *N. tisburyense* have been found in the nearby fresh water lakes or in the more southernly Quequén Grande River, although *M. fusca* is present in fresh water zones of the latter.

These and other faunal differences between these areas may be due to unknown ecological factors or to the geological history of the region. Ihering (1927), supported by zoological and paleontological data, proposed that a Miocene arm of the Tethys existed which connected the Gulf of Mexico and the Caribbean with the South Atlantic. Further faunal support for this idea came from the observations of Camp (1952), Szidat (1955) and Boltovskoy (1958a). If true, the Rio de la Plata is of Tethyan origin, but the nearby lakes and the Quequén Grande River are post-Miocene and were not connected with the Tethyan arm.

VI. BENTHONIC FORAMINIFERA OF MARSHES, MANGROVE SWAMPS AND LAGOONS

1. Marshes

Marshes are areas that are partially or completely covered by a thin (no more than 20–30 cm) film of water. At times the water may disappear completely. The basic source of the water in these areas is the tide or storms although some water is of terrestrial origin, both surface and subsurface. The most characteristic trait of marshes is the active vegetative growth occurring on them. The salinity is commonly lower than normal marine values, usually fluctuating around 20‰, although it may occasionally exceed 35‰ under high evaporation and low precipitation conditions. Marsh temperatures are also variable and closely follow surface air temperatures. The pH is usually less than 7. Emery & Stevenson (1957) found that the pH of sediments in Newport Bay, California, averaged between 5.9–7.5. Phleger (1960b) noted pH values of 6.8–7.5 in the Mississippi River delta marshes and 6.6–7.4 in San Antonio Bay, Texas. These are daytime values. During the night the values are even less since there is an increase in available CO_2 which combines with water to form carbonic acid and therefore lowers the pH. The acidity of these waters is explained by the large quantity of decaying vegetation. Marsh flora were described by Hedgpeth (1957) and a detailed analysis of ecological parameters was conducted by Phleger & Bradshaw (1966) and by Bradshaw (1968).

Tropical zones are well known for the presence of mangrove swamps occurring along low lying marine coasts. Because the shallow water, variable salinity and medium to fine grained muds characterizing these mangrove swamps are also characteristic of other marshes, we will consider mangrove swamps as marshes when considering foraminifera.

As a general rule, marsh environments exhibit the greatest degree of variability of any inhabited by foraminifera. The degree of environmental variability generally decreases from the shore zones into deeper water.

In general, the foraminiferal faunas of marshes are not very diverse. The number of specimens occurring at a given locale is highly variable; this is in keeping with the variable nature of the biotope. No planktonic species live in the environment. The benthonic fauna is almost exclusively agglutinated, probably because of the low carbonate content of the water and the low pH values. Most marsh samples contain 4–5 species but up to 16 species per sample occur. These more diverse samples usually contain examples of bay and open ocean species in addition to typical marsh forms. The areal zones occupied by different species are quite separate and exhibit sharp boundaries. There are very few species which live both in the open ocean and in marshes. The marsh is clearly a unique environment. If the distribution of foraminifera were controlled only by temperature and salinity, the majority of the marsh species could live in the open ocean or at least in shallow near shore areas with similar temperature and salinity values. The presence of a unique marsh fauna seems to indicate that some, as yet unknown, factor must be of great ecologic importance.

There have been several investigations dealing exclusively with marsh foraminifera (Walton, 1955; Phleger, 1955, 1965a, 1966, 1967, 1970; Warren, 1957; Behm & Grekulinski, 1957; Lankford, 1959; Parker & Athearn, 1959; Bradshaw, 1968; Murray, 1971; Lee & Muller, 1973) and various studies which include extensive data on foraminifera in marsh environments (Parker, Phleger & Peirson, 1953; Phleger, 1960b; Benda & Puri, 1962; Closs, 1962; Phleger & Ewing, 1962; Bandy, 1963b; Richter, 1964a,b; Ellison, Nichols & Hughes, 1965; Haake, 1967; Lutze, 1968a; Haynes & Dobson, 1969; Le Campion, 1970; Ellison & Nichols, 1970; Akers, 1971). The foraminiferal fauna of mangrove swamps was described by Saunders (1958), Murray (1970a), Phleger (1966, 1970) and Gregory (1973).

Parker, Phleger & Peirson (1953) said that the marsh fauna was very similar in most marshes and differed only in minor details. They discussed the marshes of the Texas coastal zone, most of which had many species in common with each other; only a few species were unique to specific marshes.

The most numerous and extensive investigations of marsh foraminifera have been those of Phleger. He has published numerous studies utilizing material from many areas (North America, Europe, Australia, etc.). He summarized many of these investigations (1970) and prepared a list of species endemic to the many marsh environments he examined: *Ammoastuta inepta, Arenoparrella mexicana, Discorinopsis aguayoi, Haplophragmoides* spp., *Jadammina polystoma, Miliammina fusca, Protoschista findens, Pseudoeponides andersoni, Tiphotrocha comprimata, Trochammina inflata, T. macrescens*. Not all of the species occur in any given marsh or in all geographic regions.

There are other species common in marshes but not unique to them. They are also abundant in adjacent bays, inlets, estuaries, etc. Some, such as *Ammotium salsum*, are primarily marsh forms, whereas others, such as *Rotalia beccarii, Elphidium* spp. and various miliolids, are probably lagoon or near shore species.

The sedimentation rate in marshes is generally higher than in other environments. Consequently, the ratio of living to dead specimens is very high.

Some species occur in almost every marsh area (*Trochammina inflata* and varieties, *Miliammina fusca* and *Jadammina polystoma*). In addition to universally distributed forms, some marsh faunas contain species which appear to be geographically restricted. Some of the North American Pacific Coast marshes contain *Protoschita findens* and *Textularia earlandi*, species not recognized in the Gulf of Mexico marshes. *Ammoastuta inepta, Trochammina comprimata* and *Pseudoeponides andersoni* occur in Gulf of Mexico marshes but not in Pacific Coast marshes. Different species of *Haplophragmoides* occur in different geographic areas.

Murray divided marsh foraminiferal faunas into 3 groups (1973b):

(a) Cosmopolitan species found in all marsh types at all latitudes: *Ammotium salsum, Arenoparrella mexicana, Miliammina fusca, Trochammina inflata, Jadammina macrescens* and *J. polystoma*.

(b) Species confined to hyposaline marshes: *Protelphidium tisburyense, P. anglicum, Tiphotrocha comprimata, Ammobaculites* sp., *Pseudoclavulina* sp., *Haplophragmoides* sp., '*Proteonina*' sp. and *Elphidium articulatum*.

(c) Species confined to hypersaline marshes: *Pseudoeponides andersoni, Discorinopsis aguayoi, Glabratella* sp., *Glomospira* sp. and *Textularia earlandi*.

He also noted that representatives of the hypo- and hypersaline marshes as well as *Miliammina fusca* also dwell in non-marine environments.

Marsh foraminiferal faunas and those of estuaries and enclosed basins show certain degrees of similarity. Many genera and some species are common to all these environments. The greatest similarity lies among the agglutinated taxa, a similarity due, no doubt, to the low salinity conditions that prevail in most of these environments. There are, however, differences between the faunas of these environments. Marshes exhibit lower proportions of calcareous taxa. Among the known influential ecological factors is the low pH of the substrate which brings about the dissolution of the empty calcareous tests and significantly accentuates the differences between death assemblages in different environments. The high acidity of the substrate does not seem to affect the living calcareous specimens. The organisms appear to be capable of secreting more calcium carbonate than is dissolved. An additional factor may be that the pH of the bottom water is not as low as that below the surface where empty tests are buried. However, some investigators (Closs, 1962) found no evidence of dissolution of calcareous tests which were buried in the sediment of lagoons and marshes.

As stated above, the mangrove swamp environment is a variant of the marsh environment. The few studies conducted on the foraminiferal faunas of this biotope reveal the presence of the following common species: *Rotalia beccarii, Ammotium salsum, Trochammina inflata, Miliammina petila, Elphidium* spp. and *Haplophragmoides* sp

2. Lagoons

Another biotope which exhibits great variability in its ecological conditions is the lagoon. These are naturally occurring, shallow water basins partially separated from the open sea by a strip of alluvial material.[7] They are generally characterized by a

[7] There are two different terms, lagoon and laguna. The former signifies 'a shallow sound, pond or lake generally separated from the open sea' (Glossary of Oceanographic Terms: U.S. Naval Oceanographic Office, Special Publication, 1966, 204 pp.). The latter discribes the same sort of water body except that it is connected with the sea (*Ibid.*). However, in the foraminiferal literature only the first term is in general use. As there is no evidence yet that these two environments vary in their foraminiferal assemblages, we will follow tradition and use only the term lagoon.

mixture of fresh water from the land and sea water from the ocean. The particular ecological conditions of a given lagoon depend upon the climate, the nature of the connection to the open sea, the extent of subsurface flow, etc. Salinity values vary between 20–30‰ at the very least, and may descend to 0‰ and rise to 100‰. Temperature values also undergo wide variation.

It is often difficult, if not impossible, to establish differences between the conditions in a lagoon and those described in the previous chapter as coastal waters. The faunas of these two environments are similar.

According to Lowman (1949), *Ammobaculites* exhibits the greatest tolerance to low salinity in lagoons. With a slight increase in salinity *Miliammina* appears followed by *Haplophragmoides* and *Trochammina*. At yet higher values the genera *Rotalia* and *Elphidium* are able to survive.

The foraminifera of the Matsukaya-Ura lagoon in Japan were studied by Takayanagi (1955) who found few species and few specimens in this variable low salinity environment. Most (80‰) of the species in the lagoon were agglutinated forms. The principal species were *Haplophragmoides canariensis*, *Ammobaculites agglutinans*, *Goesella iizkae*, *Miliammina fusca*, *Trochammina globigeriniformis*, *T. inflata*, *Rotalia* cf. *nana*, *Elphidium matsukawauraense* and *Rotalia beccarii*.

An extensive study of the foraminifera of various Japanese lagoons and brackish water lakes was conducted by Hada (1936–1967). A resume given in the last work of this series revealed that only 14 species (9 agglutinated and 5 calcareous) comprised the fauna in these environments: *Haplophragmoides canariensis*, *Ammobaculites agglutinans*, *A. amarus*, *A. cassis*, *Miliammina obliqua*, *M. arenacea*, *Rotalia beccarii*, *Quinqueloculina boueana*, *Buccella frigida*, *Buliminella elegantissima*, *Trochammina globigeriniformis*, *T. nitida*, *Elphidium fabum*. The majority of these species are cosmopolitan and typical of cool waters.

Phleger (1960b) cited the following species as typical of the Gulf of Mexico coastal lagoons: *Elphidium delicatulum*, *E.* aff. *tumidum*. In a later work (Phleger & Ewing, 1962), dealing with lagoons in the Gulf of California, *E. gunteri*, *E. tumidum*, *E.* sp., *Streblus beccarii* (var.), and various miliolids were cited as typical.

The lagoonal environments associated with the Gulf of California and the Southern California Coast are characterized by *Ammonia beccarii tepida*, *Elphidium articulatum* and *Quinqueloculina akneriana* (Bandy, 1963b).

An agglutinated and hyaline perforate fauna characterized the lagoons along the Gulf Coast area of Florida (Benda & Puri, 1962). The dominant species were *Streblus beccarii* and *Elphidium* spp.

The large relatively land locked Baltic Sea is in many respects a giant mixohaline lagoon. The shallow brackish waters are dominated by *Miliammina fusca*, *Cribrononion* cf. *gerthi* and *C. asklundi*. The deep basin colder brackish waters are characterized by *C. excavatum clavatum*, *E. incertum*, *Astrammina sphaerica*, *Reophax dentaliniformis regularis*, and *Hippocrepina flexibilis*.

The mixohaline (27.8–28.4‰) lagoonal waters of Tracadie Bay, Canada (Bartlett, 1966) contain specimens of *Miliammina fusca*, *Eggerella advena*, *Ammotium cassis*,

Elphidium incertum and *E. orbiculare*. This fauna is similar to lagoonal assemblages of more southern latitudes.

Ayala-Castañares & Segura (1968) described the fauna of the Laguna Madre, Mexico. Salinity values in this hypersaline lagoon range between 38–47‰. The dominant living species is *Ammonia beccarii*.

The shallow waters of the Bottsand Lagoon in Northern Germany is characterized by salinities of 12–18‰ and the foraminiferal fauna is dominated by agglutinated and calcareous perforate forms including *Cribrononion articulatum* and *Miliammina fusca*.

The shallow water hyposaline lagoons lying behind carbonate barrier islands along the coast of British Honduras are characterized by a fauna rich in *Ammonia, Elphidium* and agglutinated forms (Wantland, 1969a).

The largest coastal lagoon in Brazil, the Lagoa dos Patos, was studied by Closs (1962) who found salinity variations between 0–28‰ and a fauna characterized by *Elphidium gunteri, E. excavatum, Miliammina fusca. Trilocularena patensis, Textularia earlandi, Ammotium salsum, Arenoparrella mexicana, Haplophragmoides wilberti, Spiroplectammina biformis* and many thecamoebians.

Serova (1961a) described a fossil lagoonal assemblage from the Tertiary sediments of western Kamchatka. She found thin layers of argillaceous lignite which contained *Trochammina vitrea* alternating with normal marine sediments (which contained molluscs and foraminifera). Serova suggested that the lignitic layers represented repeated regressions of the sea with the consequent production of lagoons and swampy vegetated areas in which only *T. vitrea* could survive.

The distinctions between faunas of hypersaline, normal marine, and hyposaline lagoons is shown in Fig. 57. The three end members of the diagram represent the three major wall structures of living foraminifera; porcellaneous (miliolids), calcareous perforate (rotalids) and agglutinated (textularids). These end members can be used to distinguish between various environments (Murray, 1968b, 1973b). The plot here shows that hyposaline lagoons are dominated by a variety of mixtures of agglutinated and calcareous perforate forms whereas hypersaline lagoons are dominated by calcareous forms (both perforate and porcellaneous). The normal marine lagoon is similar to the hypersaline lagoon except that it has fewer porcellaneous and more agglutinated forms.

The characteristics of the foraminiferal faunas that inhabit lagoons can be synthesized by noting the low diversity of the fauna (although it is about twice as rich as the marsh environments). It has a great deal in common with the fauna of the coastal environment. The fauna consists of about 12–15 common genera and about 30–40 species. The proportion of calcareous taxa is greater than in the marsh biotope. The number of specimens is usually not large although there are isolated exceptions to this observation. Planktonic species are completely absent. The most common foraminifera are *Elphidium* spp. *Rotalia beccarii*, various Miliolidae, *Ammobaculites dilatatus, Ammotium salsum, Nonion tisburyense, Reophax* spp. *Textularia* spp., *Trochammina* spp., *Haplophragmoides* spp. In those parts of the lagoon closest to

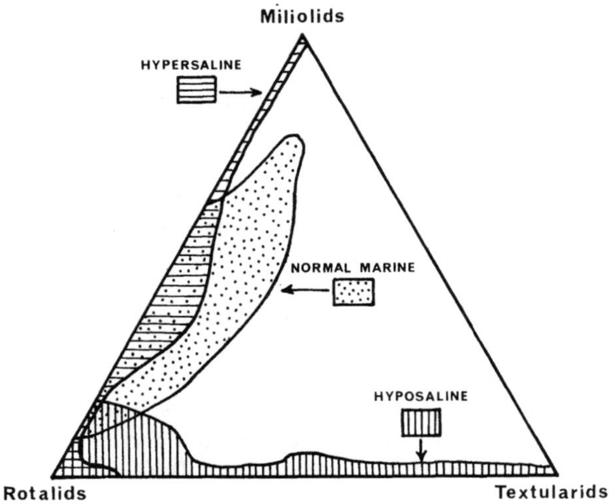

Fig. 57. Triangular diagram delimiting different lagoonal environments (after Murray, 1973b).

Foraminifera	Hyposaline ‹30‰	Normal marine 30‰-40‰	Hypersaline ›40‰	Remarks
Ammotium salsum	——			Cosmopolitan
Bolivina	——	——		Cosmopolitan
Buccella peruviana, s.l.	—	——		Temperate and cold
Buliminella elegantissima	—	——		Temperate and cold
Dahlgrenia	——			Temperate and cold
Discorbis	—	——	——	Temperate and warm
Eggerella advena	—	——		Temperate and cold
Elphidium	——	——	——	Cosmopolitan
Glabratella		——	——	Temperate and warm
Haplophragmoides wilberti	—	—		Cosmopolitan
Miliammina fusca	——			Cosmopolitan
Miliolinella subrotunda	—	——	——	Cosmopolitan
Nonion tisburyense	——			Temperate and warm
Peneroplis		——	——	Warm
Quinqueloculina milletti	——	—		Temperate
Rotalia beccarii, s.l.	——	——	——	Cosmopolitan
Textularia	—	——		Cosmopolitan
Trochammina	——	——	——	Cosmopolitan

Fig. 58. Salinity regimens of selected lagoonal and estuarine benthonic foraminifera.

the open ocean, or in those lagoons whose salinity values are high for other reasons, other taxa appear: *Buliminella elegantissima*, *Buccella*, *Bolivina*, *Cornuspira*, *Discorbis*, etc. As a general rule, there are greater numbers of agglutinated species when the salinity is low and *vice versa*. In those parts of the lagoon most distant from the sea the number of thecamoebians increases.

The fauna of the lagoonal environment is often similar to that of the estuarine environment (Chapter V). The distribution of particular lagoonal and estuarine species by salinity type is shown in Fig. 58.

VII. PLANKTONIC FORAMINIFERA

1. General characteristics

Plankton are the organisms that float in the sea. In general they lack self-locomotion (foraminifera, radiolaria, diatoms, etc.) or are very slightly mobile (dinoflagellates, copepods, etc.). Their principal means of transport, at least in the horizontal direction, is the oceanic current system.

All planktonic organisms are very small. This is understandable when one considers the relationship between the surface area and the volume of a spherical or subspherical body. The ratio between these two parameters:

$$\text{area of a sphere} = 4\pi R^2$$

$$\text{volume of a sphere} = 4\pi R^3/3$$

is known as the 'specific surface' and equals $3/R$. The greater the value of the specific surface, the greater the resistance to the water offered by the organism and the greater the buoyancy. Since the specific surface increases as the size (radius) of the organism decreases, it is not surprising that the overwhelming majority of planktonic organisms are small.

Plankton can be separated into the following types on the basis of size:

megaloplankton	>1 m
macroplankton	5 mm–1 m
mesoplankton	1 mm–5 mm
microplankton	50 μ–1 mm
nannoplankton	<50 μ

Although some planktonic foraminiferal tests are larger than 1 mm, the whole group is usually considered as microplanktonic.

The planktonic foraminifera are relatively recent arrivals into the foraminiferal scenario, having first appeared in the Triassic Period. Benthonic foraminifera are numerous in rocks of the Early Paleozoic.

Planktonic foraminifera inhabit marine waters of normal salinity. They are not found in fresh or hypersaline waters. For the most part they can not tolerate brackish waters either, although *Globigerina bulloides*, *Orbulina universa* and *Globigerinoides ruber* may be able to live in salinities somewhat lower than that of the open ocean.

As noted above, the reduction in size is one of the characteristics which permits the organism to increase its buoyancy. Nevertheless, this single factor is insufficient to create a floating foraminiferal test. One must note that both the test, composed of $CaCO_3$, and the protoplasm with a specific gravity of 1.04, are denser than sea water.

In order to float, the planktonic foraminifera must have certain devices which make them lighter. Either the test or the protoplasm may exhibit these devices.

The organisms increase their buoyancy by: (a) economical use of test material; (b) increasing friction with the water by means of test shape and ornamentation; (c) formation of special flotation devices. We will treat each of these methods separately.

(a) Economical use of test material. Rhumbler (1911) noted that specimens of *Orbulina universa* had walls of different thickness depending on their vertical position in the water column. Those living at the surface had an 18 μ wall whereas the walls of those living at depth were 24 μ thick. In addition to a decrease in weight in specimens found at the surface, there is an associated increase in the size of the primary aperture and the development of supplementary apertures and larger pores. *Globigerinoides trilobus*, forma sacculifera, for example, is a specimen with relatively thick walls. Its test is lightened by means of a large principal and many supplementary apertures.

The principal functions of the pores in planktonic foraminifera are not well understood. They may serve as passageways between the external environment and the cytoplasm. They certainly serve to decrease the weight of the test. Kemna was the first to measure the pore diameter of various species. He noted the following pore diameters: *Globigerina pachyderma:* 2.5 μ, *Pulleniatina obliquiloculata:* 5–6.3 μ, *Globigerinoides conglobatus:* 8.5 μ.

The small diameter of the pores in *Globigerina pachyderma* in conjunction with its thick walls and small aperture led Kemna and Rhumbler to consider this species as benthonic. This opinion was first expressed by Brady (1881) who observed that this species did not often appear in plankton nets but was found in great abundance in bottom samples. This idea with respect to a benthonic habitat for *Globigerina pachyderma* has been revived by modern workers (Uchio, 1960b: Green, 1960; Blanc-Vernet, 1965a). However, it is certain that this species comprises part of the surface plankton. It may be that one *forma* of this species or one stage in its ontogenetic development may live on the botton. Another planktonic species, *Candeina nitida*, has extremely small pores (1 μ) but possesses a line of large pores along the sutures and no one doubts that it is a planktonic species.

The viscosity of water increases as the temperature decreases. The viscosity of water at 0 °C is twice as great as that at 25 °C. This means that tests will sink twice as rapidly in the warmer water. To avoid this, warm water specimens must somehow be lightened. One way that they can compensate for the decrease in water viscosity is to decrease the calcite : protoplasm ratio of the organism. If the pores increase in size and/or number this ratio would be decreased. Studies by Bé (1968), Frerichs *et al.* (1972) and Bé, Harrison & Lott (1973) indicated that test porosity does indeed increase in warm waters.

(b) Increasing friction. Planktonic foraminifera commonly exhibit various ways of increasing their frictional drag with water in order to increase their buoyancy.

Häcker (1908) proposed the following relationships between various physiobiological factors and the maximum buoyancy of a planktonic body. Buoyancy is proportional to the transverse section of the body (horizontal projected area), the transverse section of the appendages, and the salinity of sea water. Buoyancy is inversely proportional to the specific gravity of the organism the volume of the organism and the temperature of sea water. Rhumbler suggested that the transverse section is the most important of these factors.

Various methods of increasing frictional drag are exhibited by the planktonic foraminifera. In some genera (*Globorotalia*, *Globotruncana*) the tests become flattened. Others develop fine, long spines (*Orbulina*). Various genera possess spines and at the same time are flattened to a certain degree (*Globigerina*). It is interesting to note that on most spinose genera the spines are distributed in a random pattern on the test. If the spines serve as buoyancy-increasing devices this would seem to indicate that there is no preferred living orientation. *Hastigerina* has spines located non-randomly around the periphery.

The presence of the spines is related to the spherical form of the chambers. The first planktonic foraminifera probably possessed globular chambers. This assumption is based on the observation that a spherical surface is the most convenient surface for the development and expansion of flotation mechanisms. One may suppose that these early planktonic foraminifera had a hardened outer covering. Because a smooth surface constitutes a poor support, the foraminifera began to develop small papillae between the pores. Later these papillae were transformed into spines. Spines not only increase frictional drag (because of the increase of the transverse cross-section), but also serve as excellent supports for protoplasm. By means of the spines, the protoplasm is able to expand easily outside of the test. This expansion further increases the transverse section of the organism. Some spine systems are so extended that they are 5 times the diameter of the chamber from which they grow. Spine length varies considerably within specimens of the same species. Rhumbler (1911) noted distinct specimens of *Globigerina* taken from a single sample whose spine lengths varied between 0.14 to 0.83 times the diameter of the entire test. He also noted that in dense waters the length of the spines is somewhat less than that observed in lighter waters. These observations mesh with those of Wesenberg-Lund who described the length of spines as being inversely proportional to the ambient specific gravity of the water. The same observations were made with respect to the large stout appendicies on various representatives of *Globorotalia* from different climatic zones (Morozova, 1958).

Spines are usually acicular and have a circular cross-section. However, *Hastigerina* possesses lanceolate spines with three flanges and barbs at the extremities. Rhumbler (1911) proposed that these spines are defensive mechanisms. Hemleben (1969b) suggested, however, that these spines serve a supportive function for the addition of new chambers. The spines of the juvenile interior trochospiral stage of *Orbulina universa* described by Bé, Harrison & Lott (1973) may serve the same supportive function. We mentioned earlier that not all planktonic foraminifera have spines.

Foraminifera with flattened tests such as *Globorotalia* have no spines. Because this species is flattened dorso-ventrally its horizontal surface is relatively large and its buoyancy is increased. Moreover, some species have special configurations which serve to increase their transverse section: *Globorotalia menardii*, forma fimbriata has a peripheral flange; *Sphaeroidinella dehiscens* has a lobate margin.

Of the planktonic foraminifera with globulose chambers, those devoid of spines are most primitive according to Rhumbler (1911). Examples of such species are *Globigerina radians*, *Globigerina ternata* and to a certain extent *Globigerinita glutinata*.

Certain species including representatives of the genera *Globoquadrina*, *Globorotalia*, *Globigerinita*, *Pulleniatina* and *Candeina* do not possess spines at any stage during their lives. There are indications that some species have spines during their juvenile stage but lose them upon maturation. It is very common to find a population in which some specimens of a given species possess spines while others lack them. At the moment no correlation between this phenomenon and age of the specimens or any other factor is evident.

It has been suggested that the spines at times are able to provide a shadow or covering to protect the protoplasm and symbiotic algae from excessive solar radiation.

(c) Special flotation devices. Subbotina (1953) described numerous specimens of the Globigerinidae from Cretaceous sedimentary rocks which have special appendages situated on the spiral axis over the last chamber. These appendages have an oviform ampulla-like shape and are empty. Their size is variable, sometimes smaller than the ultimate chamber, sometimes equal in size to several chambers. In this latter case they usually cover the entire ventral side. The smooth delicate imperforate walls, usually lacking an apertural opening, seem unlikely to have contained protoplasm. Subbotina suggested that they probably were gas filled and acted as flotation devices. She said that other investigators, notably Cushman, upon viewing these appendages, interpreted them as герontic deformations of the test. Subbotina insisted that these appendages are supplementary flotation structures. She gave further support to this hypothesis by noting that these features are found principally on species with large walled tests. No such structures are observed on living specimens. The bullae and tegillae found on some species are only apertural features which have direct contact with the external environment and therefore they could not serve as a true flotation device.

In addition to increasing test buoyancy, the planktonic foraminifera also exhibit various methods of altering the nature of the protoplam to increase buoyancy: (a) increasing friction; (b) decreasing specific gravity; (c) formation of flotation devices.

(a) Increase in friction. As described above, some tests become flattened, develop spines, flanges, etc. in order to increase their transverse cross-section. To a certain extent the same thing occurs in the protoplasm, at least in certain genera. In *Globorotalia* the protoplasm, upon being extruded from the aperture, is distributed along the periphery and attaches itself to the rugose ornamentation of the ventral

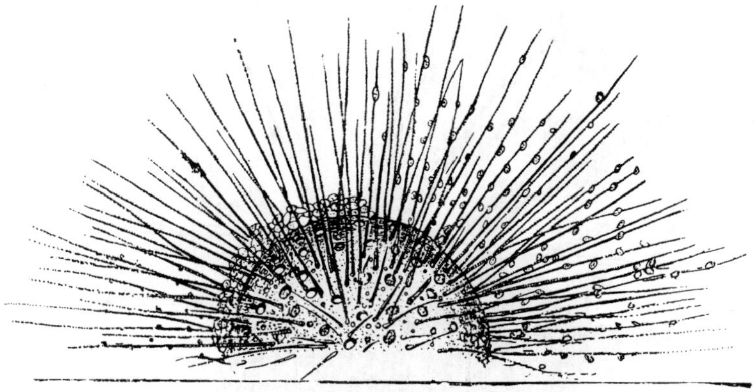

Fig. 59. Orbulina universa with extended pseudopodia (after Rhumbler, 1911); left: zooxanthallae on the test; right: zooxanthallae distributed on the pseudopodia.

side (Rhumbler, 1911). Hofker (1962) mentioned that he was able to see bits of protoplasm arranged in this manner in several preserved specimens. Another method of increasing the friction between the organism and the environment involves inclusions in the protoplasm (Rhumbler, 1911). According to this author the zooxanthellae algae which are symbiotic in some species, e.g. *Orbulina universa*, leave the large pores and attach themselves to the spines, thereby increasing the transverse section of the organism. A similar pheonomenon was observed by Murray (1897) on some specimens of *Globigerina*. The xanthellae also left the test in this case but instead of attaching themselves to the spines, they were located on the test surface between the spines (Fig. 59).

(b) Decrease in specific gravity. The protoplasm of various planktonic organisms contains inclusions such as gas vacuoles, droplets of fatty material, etc. The presence of these inclusions decreases the specific gravity of the body. Inclusions of this type that have been observed in foraminifera are gas bubbles and/or droplets of low salinity water and possible fatty inclusions. Although there is no reason to believe that phytoplankton and zooplankton adapt in the same way, Smayda (1970) was convinced that physiologic adaptions usually prove more suitable for maintaining buoyancy than do morphological ones.

(c) Flotation devices. Rhumbler (1911) described the formation in some *Globigerina* of a gelatinous wrapping. This device was also observed in *Hastigerina pelagica* (Fig. 60) by Murray (1876) and its presence confirmed by Bé (1969b). This gelatinous visicular covering aids greatly in increasing buoyancy. Rhumbler said that this covering is rarely found in foraminifera because it occurs only occasionally or is lost when the specimens are dried.

According to Le Calvez (1953), living specimens of *Orbulina universa* have a great amount of ectoplasm which surrounds the entire test in the manner of the bubbly calymma of some radiolaria.

Fig. 60. Hastigerina pelagica with frothy cytoplasm estraded around test (Bé, 1969b).

Thus, we can see that the planktonic foraminifera have various resources to produce a diminution of specific gravity and a consequent increase in buoyancy.

Apart from the purely planktonic foraminifera there are the 'hemiplanktonic' and 'pseudoplanktonic' foraminifera. The hemiplanktonic species are those which either are predominantly planktonic in habitat, but have a benthonic state, or *vice versa*. *Orbulina universa* is a widely distributed and well known planktonic species. Nevertheless, according to Le Calvez (1936a), it might be more correct to consider it hemiplanktonic. Le Calvez considered those tests of *Orbulina universa* which have within them *Globigerina*-type constructions as the megalospheric examples. This phenomenon occurs only in the planktonic state. He did not find any microspheric examples and suggested that microspheric specimens must be the small thick-walled specimens living on the bottom. Not all investigators share this viewpoint. There are authors (for instance, Hofker, 1959) who suggest that *Orbulina universa* is not a real species but an ontogenetic stage of several planktonic foraminifera. However, according to a recently published study by Vilks & Walker (1974), *Orbulina universa* proves to be a valid taxon.

Another possible example of a hemiplanktonic foraminifera among those commonly considered as planktonic is *Globigerinoides ruber.* Christiansen (1965a) described benthonic forms of this species. According to his studies in the Bay of Naples, these benthonic specimens are high-spired, tightly coiled and represent a microspheric form. In view of the extremely widespread distribution of this species, it is difficult to imagine that its normal ontogenetic development includes a benthonic stage. It is also worth mentioning that this benthonic form was previously considered a separate species, *Globigerinoides pyramidalis* and has been found many times in plankton hauls.

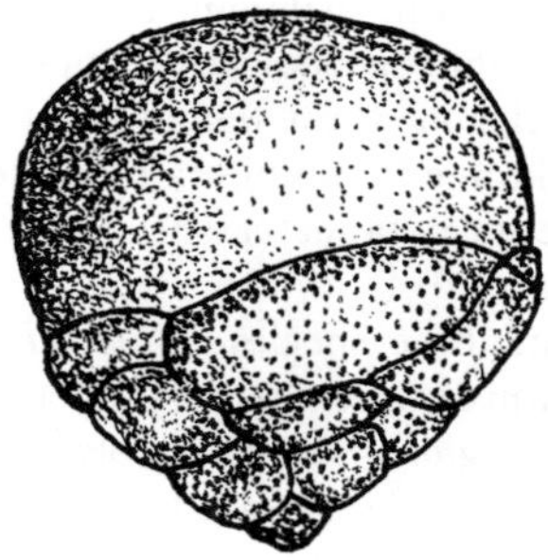

Fig. 61. Tretomphalus, a hemiplanktonic form.

An example of a benthonic species that possesses a planktonic stage is '*Tretomphalus*'. According to Myers (1943b), agamont specimens of this 'genus' rise to the surface for a short period of time (up to 18 hours). The buoyancy for this movement is supplied by means of a special spherical terminal chamber (Fig. 61). The large pores of this chamber serve as openings for the passage of gametes into the water. The specimens fall to the bottom after their brief planktonic venture. Only those that become trapped in algae remain at the surface for much time. Myers (1943b) and Sliter (1965) determined that the float-chambered agamontic state '*Tretomphalus*' is morphologically similar to *Discorbis, Rosalina* and *Cymbaloporetta.* Todd (1971) raised the possibility that several different species may possess this final float stage.

The pseudoplanktonic species are those which become attached to various floating objects, algae, seaweed, pieces of wood, etc. and are carried by currents. These specimens are often representatives of *Cibicides* and *Discorbis.*

Occasionally, examples of typical benthonic species are found in surface waters not attached to any floating material (Murray, 1965a, Boltovskoy, 1966b). These specimens are usually small (100–300 μ), lightweight and free of protoplasm. These kinds of discoveries are usually made in rather shallow waters, often in areas over the continental platform. The discovery of these benthonic specimens is usually explained as the result of storms which erode the tests from the bottom; they are then carried by upwelling water masses and turbulence the short distance to the surface.

However, in the Mar de la Flota (Bransfield Straits), Antarctica, Boltovskoy & Boltovskoy (1970) found living benthonic foraminifera in surface plankton samples collected in water greater than 1000 m deep. The most probable explanation for this is that the living benthonic specimens were transported by icebergs from coastal areas lying farther south. Supporting this hypothesis is the discovery in these samples of specimens of *Elphidium magellanicum*, a typical littoral species.

2. Horizontal qualitative distribution

The species variability of plankton, including the foraminifera, in warm waters is about ten times greater than that in cool and temperate waters. There are 30–40 planktonic foraminiferal species in warm waters and about 3–4 in cold waters. The ease of transport of the planktonic foraminifera by currents and their ability to adapt have established almost identical faunal associations in the Atlantic, Pacific and Indian Oceans.

The only difference is that three species (*Globoquadrina conglomerata*, *G. hexagona* and *Globigerinella adamsi*) which live in the Pacific and Indian Oceans are not found in the Atlantic. *G. conglomerata*, probably disappeared from the Atlantic during the Pliocene. *G. hexagona* disappeared somewhat later, probably during the last interglacial. *G. adamsi* is closely related to *G. aequilateralis* which is known from Middle Miocene to Recent sediments in the Atlantic. *G. adamsi* may be descended from *G. aequilateralis* and may have originated in the post-Pliocene Pacific Ocean after the Atlantic–Pacific Panamanian connection was closed. This species now lives in low latitude Pacific waters and cannot migrate to the Atlantic via high latitude routes.

With the exception of these three species, the planktonic foraminifera follow the general observation that all pelagic species living in the Atlantic have parallel forms in the Pacific and *vice versa*.

The planktonic foraminiferal fauna varies most significantly with latitude. As a general rule, temperature has a greater influence on the distribution of planktonic foraminiferal species than on benthonic ones.

Based on climatic conditions and organic life, the surface waters of the oceans are divided into five zones: (a) Tropical zone (on both sides of the equator): 24–30 °C; (b) Subtropical zones (south and north of the tropical zone), 18–24 °C; (c) Transition zones (warm temperate and cold temperate), 10–18 °C; (d) Subarctic and subantarctic zones; 5–10 °C, (e) Arctic and antarctic zones; 0–5 °C.

Understandably, the boundaries between these zones are not always sharp. Nevertheless, it is possible to group the extant planktonic foraminifera by their preference for warm, cold or intermediate waters. Various preference lists of this kind have been prepared by different investigators. Only some of those appearing in the last 15 years will be mentioned below.

A study made from especially rich material collected from the North and Equatorial Pacific by Bradshaw (1959) revealed the following:

1. Subarctic fauna (southern limit situated approximately at the Subarctic Convergence) *Globigerina* cf. *dutertrei*, *G. pachyderma*, *G. quinqueloba*, *G. bulloides*, *G. eggeri* (dwarfed), *Globigerinoides* cf. *minuta*, *Globigerinita glutinata*.

2. Transition fauna: (a) *Globigerina bulloides*, *G. eggeri* (large), *G. quinqueloba*, *Globigerinoides ruber*, *Orbulina universa*, (b) (in southern part of area): *Hastigerina pelagica*, *Globorotalia truncatulinoides*, *Globigerina inflata*, *Globigerinella aequilateralis*.

3. Warm water fauna: (a) (central zone) *Globigerina inflata*, *Globorotalia truncatulinoides*, *G. scitula*, *Hastigerina pelagica*, *Candeina nitida*, (b) (equatorial zone): *Globigerina conglomerata*, *Globorotalia tumida*, *Pulleniatina obliquiloculata*, *Sphaeroidinella dehiscens*, (c) (both zones): *Globigerinoides conglobatus*, *G. ruber*, *G. sacculifer*, *Globorotalia hirsuta*, *G. menardii*, *Hastigerinella rhumbleri*, *Globigerina hexagona*, *Globigerinella aequilateralis*, *Globigerinita glutinata*.

Parker (1960) analyzed planktonic samples from the Equatorial and Southeast Pacific. She supplied surface salinity and temperature data along with species lists and suggested temperature was a more significant factor than salinity in controlling distribution. Her list is as follows:

1. Widely distributed species (salinity: 33.6–36.5‰: temperature: 9–27.6°C): *Globigerinella aequilateralis*, *Globigerinoides ruber*, *Hastigerina pelagica*, *Orbulina universa*.

2. Equatorial and southeast central species (salinity: 34–36‰: temperature: 16–28 °C): *Globigerinoides conglobatus*, *G. sacculifer*, *Globigerina eggeri*.

3. Equatorial species (salinity: 34–36‰; temperature 24–28 °C): *Globigerina conglomerata*, *G. hexagona*, *Globorotalia menardii*, *Pulleniatina obliquiloculata*.

4. Southeast, central and subantarctic species (salinity: 34–35.7‰; temperature: 9–24 °C): *Globigerina bulloides*, *G. inflata*. *Globorotalia punctulata*, *G. truncatulinoides*.

5. Subantarctic species (salinity: 34‰; temperature 9.2–11.5 °C): *Globorotalia scitula*.

The data presented by Bandy (1960c) supplemented the information of other investigators with one important detail. Various observers had noted that the coiling direction plays an important role in foraminiferal distribution (see Chapter XIV,4). Bandy took coiling into account in his distribution scheme.

Boltovskoy (1962a) noted that it is often necessary to use the taxonomic category of *forma* when dealing with planktonic foraminifera as indicators of different water masses because the category of species is too broad. The small morphological modifications that occur in various species when they change their usual habitat for another are not of a specific character. Nevertheless, they are of great utility in the recognition of various water types. The same species is able to live in different climatic zones or water masses, but will usually be represented by somewhat different morphological forms. Kennett (1968b, 1970a) illustrated these kinds of changes in a detailed study of the latitudinal variation of *Globigerina pachyderma*.

Belyaeva (1964) described the following associations in the epipelagic zone (0–200 m) of the Indian Ocean.

1. Equatorial species: *Globorotalia tumida*, *Globigerina hexagona*, *G. conglomerata*, *Pulleniatina obliquiloculata*, *Sphaeroidinella dehiscens*.

2. Central and equatorial species: *Globigerinoides ruber*, *G. sacculifer*, *G. conglobatus*, *Globorotalia menardii*, *Hastigerina digitata*, *Globigerinella aequilateralis*.

3. Central species: *Globorotalia hirsuta*, *G. punctulata*, *G. truncatulinoides*, *Hastigerina pelagica*, *Candeina nitida*, *Globigerina eggeri*, *G. inflata*.

4. Subantarctic and antarctic: *Globigerina bulloides*, *G. quinqueloba*, *Globigerinita glutinata*.

A study of the Indian Ocean fauna between 0–62°S by Boltovskoy (1969b) revealed the following groups.

1. Warm water species: *Globigerina conglomerata*, *G. hexagona*, *G. dutertrei*, *G. rubescens*, *Globorotalia menardii* (f. tumida), *G. menardii* (f. typica), *G. pumilio*, *Globigerinoides trilobus* (f. sacculifera), *G. conglobatus*, *G. trilobus* (f. typica), *G. ruber* (f. typica), *Hastigerina pelagica*.

2. Warm-temperate water species: *Globorotalia crassaformis*, *G. scitula*, *G. hirsuta*.

3. Cosmopolitan species: *Globigerinita glutinata*.

4. Cold-temperate water species: *Globorotalia truncatulinoides*, *G. inflata*, *G. cavernula*, *Globigerina quinqueloba* (dextral), *G. quinqueloba* (sinistral), *G. bulloides*.

5. Cold water species: *Globigerina pachyderma* (dextral), *G. pachyderma* (sinistral), *Globigerinita uvula*.

Banerji, Schafer & Vine (1971) described the following groups from the Equatorial and North Pacific.

1. Warm water tropical species group (average surface water temperature is approximately 25 °C): *Globigerinella aequilateralis*, *Globigerinoides conglobatus*, *G. ruber*, *G. sacculifer*, *G. trilobus*, *Globoquadrina conglomerata*, *G. hexagona*, *Candeina nitida*, *Globorotalia menardii*, *G. tumida*, *Globigerina calida*, *Globigerinita glutinata*, *Hastigerina pelagica*, *Orbulina universa*, *Pulleniatina obliquiloculata*, *Sphaeroidinella dehiscens*.

2. Warm temperate species (average surface water temperature – 20 °C): *Globorotalia inflata*, *G. hirsuta*, *Globoquadrina dutertrei*, (and some warm water tropical species).

3. Cosmopolitan species: *Globigerinita glutinata*, *Globorotalia inflata* (in subsurface).

4. Cold temperate species (northern boundary approximately at 10 °C isotherm): *Globigerina bulloides*, *G. pachyderma* (dextral), *Globorotalia inflata*, (some cosmopolitan and warm water species).

5. Cold water species (average surface water temperature less than 10 °C):

Globigerina bulloides, *G. pachyderma* (sinistral), *Globoquadrina dutertrei* (small forms), *Globorotalia cavernula*, *Globigerinita glutinata*.

Bé & Tolderlund (1971) established the following distribution of planktonic foraminifera in the Atlantic and Indian Oceans.

1. Arctic and Antarctic zones: *Globigerina pachyderma* (sinistral).
2. Subarctic and Subantarctic zones: *Globigerina pachyderma* (dextral), *G. quinqueloba*, *G. bulloides*, *Globigerinita bradyi*, *Globorotalia scitula*.
3. Transitional zones: *Globorotalia inflata*.
4. Subtropical zones (20°–40° latitude): *Globigerinoides ruber*, *G. conglobatus*, *Hastigerina pelagica*, *Globigerinita glutinata*, *Globigerina rubescens*, *G. falconensis*, *Globorotalia truncatulinoides*, *G. hirsuta*, *G. crassaformis*, *Globigerinella aequilateralis*, *Orbulina universa*, *Globoquadrina dutertrei*.
5. Tropical zone (20°N–20°S): *Globorotalia menardii*, *G. tumida*, *Globoquadrina digitata*, *G. hexagona*, *G. conglomerata*, *Pulleniatina obliquiloculata*, *Globigerinoides sacculifer*, *Candeina nitida*, *Hastigerinella digitata*, *Globigerinella adamsi*, '*Sphaeroidinella dehiscens*'.

Cifelli & Smith (1974) divided the western part of the North Atlantic into 4 latitudinal zones based on the fauna which they contain.

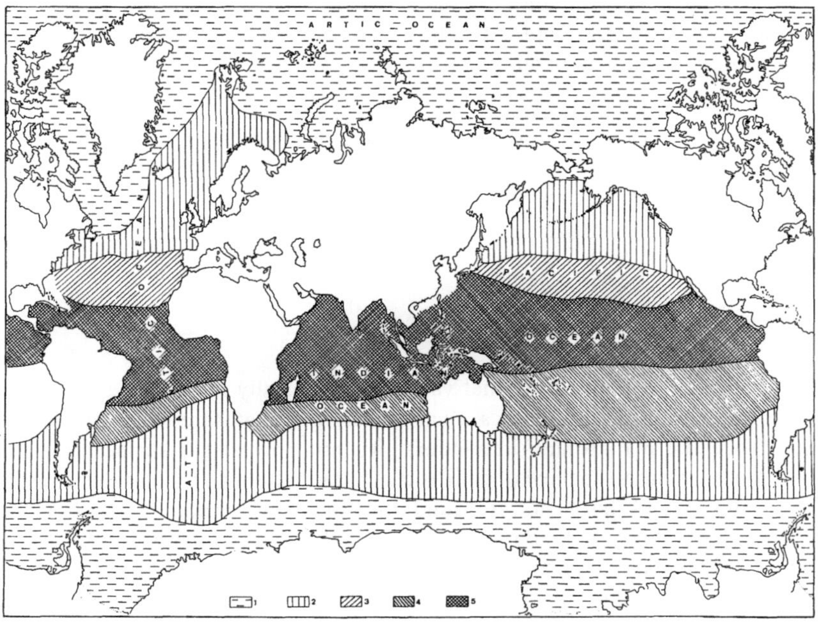

Fig. 62. Planktonic foraminiferal zones of the oceans (after Belyaeva, 1969a); 1. Arctic and Antarctic; 2. Cold Temperature; 3. Northern Subtropical; 4. Southern Subtropical; 5. Tropical.

1. Slope waters (45–38°N): *Globigerina bulloides, G. incompta, G. inflata, G. quinqueloba egelida, Globigerinita glutinata.*

2. Northern Sargasso (38–28°N): *Globigerinoides ruber, G. conglobatus, Globigerinella aequilateralis, Globorotalia truncatulinoides, Pulleniatina obliquiloculata.*

3. Southern Sargasso (28–20°N): *Globigerinoides ruber, G. conglobatus, Globigerinella aequilateralis, Globorotalia truncatulinoides.*

4. Tropical (20–0° latitude): *Globigerinoides ruber, G. trilobus, Globigerinella aequilateralis, Globigerina dutertrei.*

A comparison of all of these data indicates a great similarity between observed zonations. Fig. 62 shows a generalized horizontal zonation for living planktonic foraminifera. A great deal of the differences between schemes is due to different taxonomic determinations; e.g. *Globigerinoides* cf. *minuta* of Bradshaw is the same as *Globigerinita uvula* of Boltovskoy and *G. bradyi* of Bé & Tolderlund, etc.

In addition to differences in taxonomical opinion there are other factors which help explain the deviations among the faunal zonations listed above. Five of these factors are discussed in some detail below.

A. Deviation Induced by Local Changes in Temperature.

It is well known that temperature changes in sea water are much greater in the upper-most waters than at depth. The following data taken from coastal waters near Algeria exhibit this phenomenon.

Depth (m)	Annual oscillation in temperature (°C)
0	10.2
25	6.0
50	2.8
100	2.0
200	1.0
350	0.0

Fig. 63 shows this kind of thermal variation graphically for a locale in the North Atlantic. It is clear that the signficant variations in temperature occur only near the surface. Since the planktonic foraminifera live not only at the surface but also at depth it was commonly believed until recently that thermal oscillations present in surface water would have no profound influence on foraminiferal populations. However, studies done in the Sargasso Sea (Bé, 1960a) dispelled this idea. Between the period of January–May at depths of 0–100 m the following temperate species were found: *Globigerina bulloides, G. inflata, Globorotalia truncatulinoides* and *G. hirsuta*. These species are missing or rare during June–December and are replaced by the tropical species, *Globigerinoides sacculifer, G. conglobatus*, and *Globorotalia*

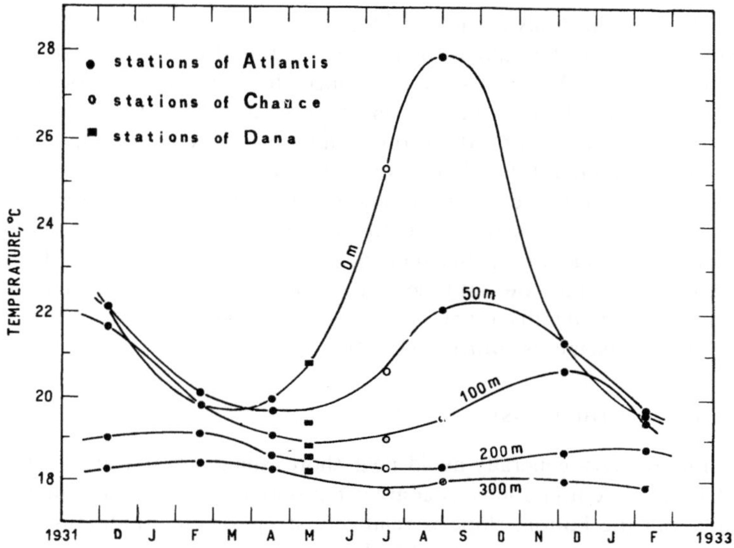

Fig. 63. Seasonal temperature variation at different depths in the North Atlantic (after Iselin, 1936).

menardii. Bé emphasized that this replacement is not due to transport of specimens by different water masses, but due to regular seasonal changes in the physio-chemical properties of the water. It is interesting to note that these changes are so regular that one can estimate rather well the month in which a sample is collected by examining the proportions of different species present.

Bé also defined a third group of species which did not exhibit a periodic maximum development: (*Orbulina universa, Hastigerina pelagica, Globigerinella aequilateralis, Globigerinoides ruber, Pulleniatina obliquiloculata, Globigerinita glutinata, Globorotalia punctulata* and *Globigerina eggeri*).

This oscillation between temperate species and subtropical species influences the composition of the thanatocoenose. Foraminifera with different tolerance limits are found mixed together on the bottom.

Later studies of Bé and others confirmed the observations that planktonic foraminiferal populations are able to change seasonally (Chapter II,3).

B. Currents

Different water masses are characterized by different foraminiferal faunas. These water masses may be transported large distances and consequently the foraminifera characteristic of the water mass may reach locales far distant from their typical area of distribution. For example, *Globorotalia menardii* is carried by the Gulf Stream from low to mid latitudes and *Globigerina pachyderma* is transported from high latitudes to low by the Malvin Current. The changes in direction and velocity of currents are reflected in the thanatocoenose that is found in the zones of mixing of different water masses. We previously mentioned the existence of various types of mixing in the Subtropical–Subantarctic Convergence zone in the South Atlantic. The location of these zones is not fixed but often shifts from one area to another. Thus, the planktonic foraminiferal distribution on the bottom is rather complex.

In some cases currents can create anomalous distribution patterns. In the epipelagic zone of Venezuelan water for instance, typical cold water species *Globigerinita uvula* and *Globigerina quinqueloba* have been recorded. They were evidently brought to the low latitudes by subsurface Subantarctic intermediate currents which occur at depths of 650–800 m, and later carried up to shallower depth by vertical water movements (Miró, 1971 Ms).

C. Distance from the Coast

Planktonic foraminifera generally avoid near-shore, shallow water areas (the continental shelf). They are found in coastal areas but only in regions where there are no river mouths and the bottom slope is relatively steep. These conditions are found along the length of the California coast and numerous planktonic foraminifera have been found there. On the Argentine continental shelf, on the other hand, very few planktonic foraminifera are found in near-shore plankton tows.

Some species are more abundant than others in near-shore zones. Phleger (1954a) showed that *Globigerina bulloides* and *Globigerinoides ruber* are the only planktonic foraminifera found near the northwest coast of the Gulf of Mexico. At greater distances from shore the number of planktonic foraminiferal species increases. Boltovskoy (1970b) established the presence of *Globigerinoides ruber*, *Globigerina bulloides*, *G. quinqueloba*, *G. pachyderma* (f. superficiaria), *Orbulina universa* and *Globigerinita uvula* in the coastal zones on the Argentine continental platform. The occurrence of these specimens was rare and all tests were very small. In the same manner, Adegoke, Dessauvagie & Kogbe (1971) noted that the bottom sediments of the coastal zone of the Gulf of Guinea (0–25 m) contained a few specimens of *Globigerinoides ruber*, *G. trilobus* and *Globoquadrina dutertrei*.

Thus, it appears that *Globigerinoides ruber* and *Globigerina bulloides* are capable of living in the water close to the coast. These species along with *G. pachyderma*, (f. superficiaria), *G. quinqueloba*, and *Globigerinita uvula* are capable of living in salinities of 30.5–31.0‰ which appears to be the minimum salinity for planktonic foraminifera (Boltovskoy, 1970b).

D. Geologic History of a Region

In some water bodies one can observe that various species belonging to different climatic zones can live together and prosper. This phenomenon is noted in the Gulf of Mexico where typical low latitude species such as *Globigerinoides sacculifer* live along with middle latitude species such as *Globorotalia truncatulinoides*. The first of these species is more abundant in the southern Gulf whereas the second is more numerous in the northeast. Phleger (1954a) suggested that *Globorotalia truncatulinoides* is a relict species which invaded the Gulf during the southern shift of cool water faunas during a Pleistocene glacial stage. It is able to survive now only because the winter-water temperature in the northeast Gulf is low enough to allow reproduction. *Globigerinoides sacculifer*, on the other hand, is a tropical species which has migrated from the Caribbean and reproduces only in the summer.

A similar explanation may account for the occurrence of *Globigerina bulloides*, a cold-temperate species, in the Fosa de Cariaco, Venezuela. Apparently it penetrated low latitudes in the Pleistocene glacial period and in some places survived after the glacial period terminated. It is likely that at the same time *Globorotalia inflata* penetrated low latitudes. However it could not survive the increase in temperature and now can be found only in bottom sediments of the Venezuelan waters as empty shells; in other words it is found there as a Pleistocene fossil, but not as a living Recent plankton species.

There are other anomalous distribution patterns; e.g. the presence in the eastern Gulf of Mexico of *Globorotalia inflata* and *Globigerina pachyderma*, and the absence of *Globorotalia crassaformis* and *Pulleniatina obliquiloculata* in the Mediterranean Sea. The explanation of these unusual distributions is often found in an examination of the geological history of the area.

E. General Character of the Region

It has become more and more apparent in recent years that there are ecological factors other than temperature which may control the distribution of planktonic foraminifera in the epipelagic zones of the open ocean. Ericson *et al.* (1961) noted that the distribution of *Globorotalia menardii* in the Sargasso Sea could not be explained solely by the temperature distribution.

Cifelli (1971) presented a good review of the temperature requirements of various planktonic foraminifera in different areas. He concluded that the foraminifera were sensitive to several other ecological parameters in addition to temperature.

Bé & Tolderlund (1971) demonstrated that the distribution of *Globigerinoides ruber* and *G. sacculifer* in the warm waters of various regions is not dependent on the surface water temperature but on the salinity. The latter species has its maximum abundance in salinities of 34.5–36.0‰. *G. ruber* reaches its maximum in salinities greater than 36‰ or less than 34.5‰. Boltovskoy suggested that various biological factors other than temperature and salinity are more useful for defining water masses. His ideas are described in more detail in Chapter XIV,14.

Thus, it is apparent that although the principal cause of the qualitative horizontal distribution of planktonic foraminifera is temperature, there are several different conditions which can modify the general distributional aspects resulting from this principal cause.

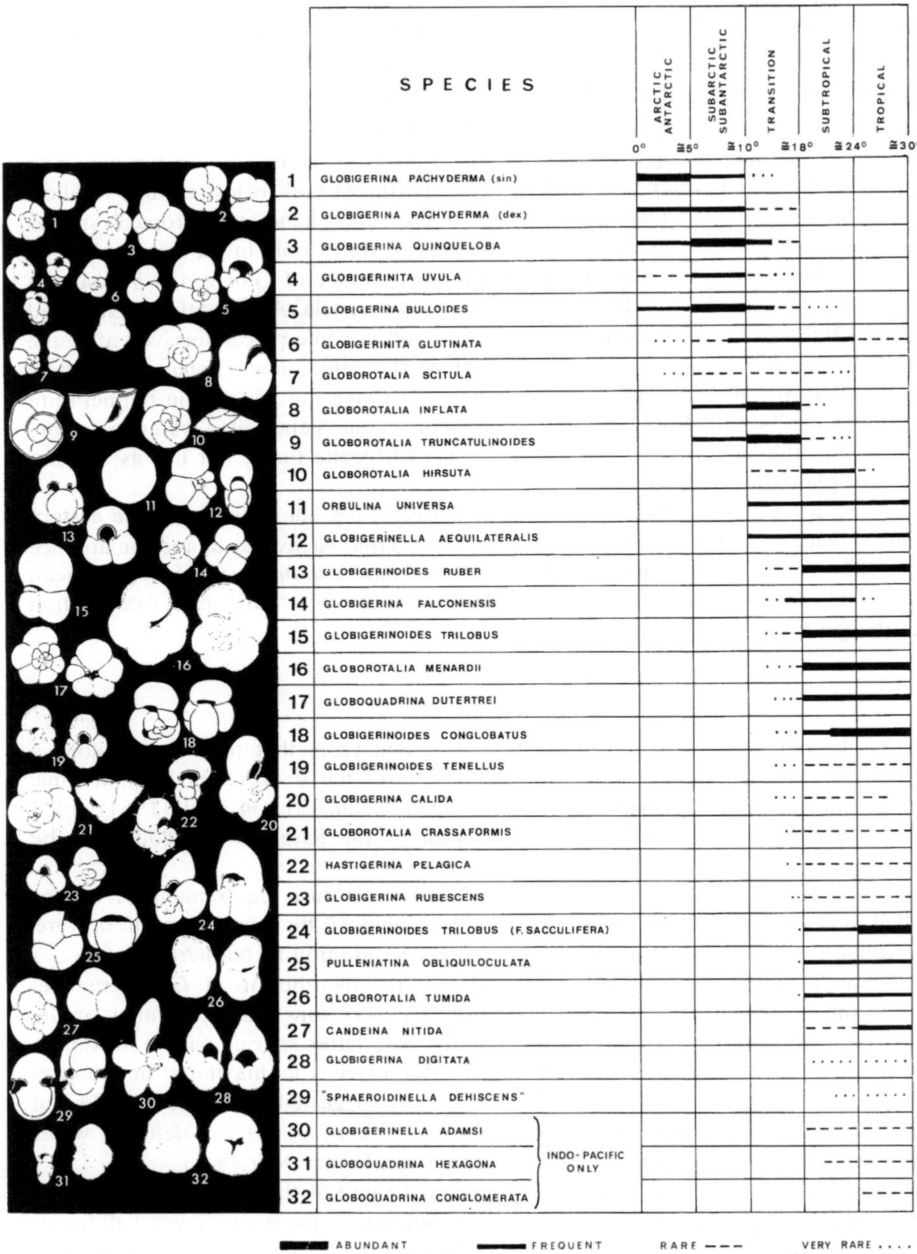

Fig. 64. Principal planktonic foraminifera and their horizontal zonation.

A summary of the observations made on the horizontal distribution of planktonic foraminifera is given in Fig. 64. This figure is based on data taken from the first edition of this book, an excellent table prepared by Bé & Tolderlund (1971), other data from the literature, and recent observations of one of the authors (EB). Included in the table are only the most important species and those for which few taxonomic problems exist. The table is a generalization and like most generalizations has certain limitations.

3. Horizontal quantitative distribution

When we compare the number of specimens comprising the different zooplankton taxa we see that the foraminifera are an important group. According to King & Demond (1953) the planktonic foraminifera represent 1.3–9.9% of the zooplankton. In the winter this percentage decreases, and in the summer the foraminiferal contribution to the zooplankton increases. In the winter, the planktonic foraminifera are the sixth most abundant group after the copepods, chaetognaths, tunicates, krill and siphonophora. In the summer, they are in third place after the copepods and chaetognaths. In some summer months the foraminifera are able to surpass the copepods and account for 60% of the zooplankton. Hida & King (1955) noted that the quantity of planktonic foraminifera is very high in the summer in the equatorial Pacific where they are the second most abundant group of zooplankton.

Based on their biomass, the contribution of the planktonic foraminifera would be much less than their numerical contribution indicates. To determine the biomass of a particular component it is necessary to calculate the average weight of a specimen and multiply by the average number of specimens per unit volume of water. Despite their abundance, their biomass is low because of their small size. The average weight of a *Globigerina* specimen captured in the plankton in the Okhotsk and Bering Sea was only 0.015 mg. (Lubny-Gerthyk, 1953). The water in these areas is very cold and the specimens were probably *G. bulloides* and *G. pachyderma* whose tests are generally rather small. The planktonic foraminifera of warm water are larger and their average weight is greater.

The quantitative distribution of foraminifera can be described in two ways: (a) abundances relative to different areas and (b) absolute abundances relative to some unit of volume or weight.

It was once thought that the warm regions of the oceans contained greater numbers of foraminifera than the cold areas because the process of biologic extraction of calcium carbonate from sea water is more intensive in warm water than in cool water. However, this prediction does not completely correspond to reality. Bradshaw (1959) demonstrated that the subarctic area is characterized by a large number of specimens (although few species). As one goes towards the Equator the quantity of specimens decreases and reaches a minimum in the central mid-latitude areas, increasing again toward the Equator.

It is probable that the concentration of foraminifera in surface waters is very similar to the total zooplankton concentration. The concentration of foraminifera depends on the distribution of phytoplankton, which in turn is a function of the distribution of nutrient salts, particularly phosphate (Parker, 1960).

An analysis of the horizontal distribution of planktonic foraminifera on a small scale indicates that although the area occupied by a species may be continuous, the populations are not uniformly distributed. There seem to be patches of high and low concentrations of planktonic foraminifera just as there are for other plankton (Wiebe & Holland, 1968). Murray & Renard (1891) mentioned that during one day on the voyage of the Challenger they encountered an enormous quantity of one planktonic genus, whereas the following day almost no specimens of the same genus were found. Lohmann (1920) and Hentschel (1933) confirmed these observations with numerical data. Unfortunately they did not describe the distance between peak abundances and so could not establish the real character and size of the patches.

Just as the distribution of the benthonic foraminifera seems to occur in clumps several investigators of planktonic foraminifera have noted that living foraminifera occur in what they refer to as patches (Bé, 1959a; Bradshaw, 1959; Phleger, 1960b; A. Smith, 1963; Boltovskoy, 1964a, 1965c; Bé & Hamlin, 1967; Ujiié, 1968; Boltovskoy & Boltovskoy, 1970). Some of these investigators merely mentioned patchiness without giving data. Cifelli (1962a, 1965), Bé & Hamlin (1967) and Ujiié

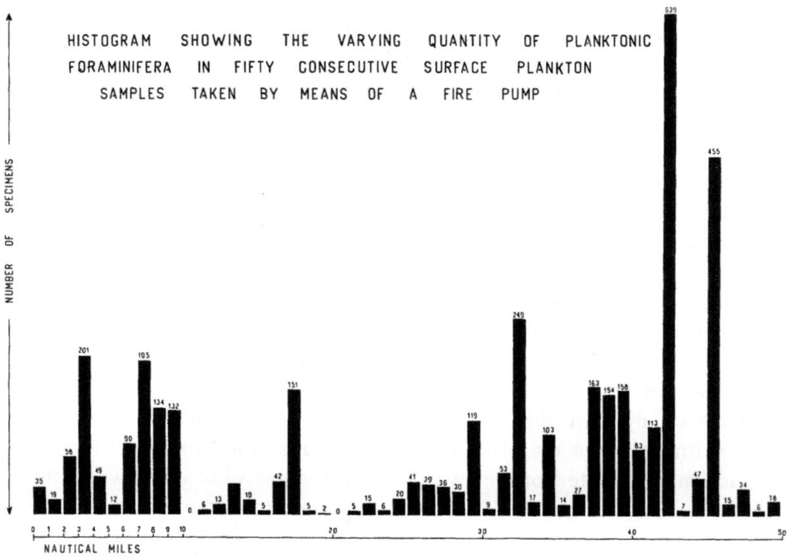

Fig. 65. Lateral variations in abundance of surface planktonic foraminifera (after Boltovskoy, 1971c).

(1968) did not publish data regarding horizontal quantitative distributional changes but did give tabular and map data which indicate local patchiness.

Phleger (1960b) gave an example of the uneven quantitative distribution of living planktonic foraminifera within a short distance. He described a situation in which at two stations 25 miles apart, plankton samples taken at 40 m yielded 2 and 73 specimens per m^3 respectively. The temperature and salinity were similar. A similar example, even more convincing, was provided by Boltovskoy (1964a) who collected two samples from areas of similar hydrologic conditions 35 miles apart and found 0 and 1830 species of planktonic foraminifera from equal volumes of water.

Boltovskoy (1971c) collected surface zooplankton samples at a sampling interval of about one sample per nautical mile (Fig. 65). Each sample consisted of about 3220 litres. All species in the planktonic foraminiferal fauna in these samples exhibited a well pronounced patchiness in distribution, especially in the quantitatively rich areas. The greatest variation between adjacent stations consisted of 198.2 and 2.2 specimens per cubic metre of filtered water; i.e. a 90 fold increase in abundance was recorded within a distance of one nautical mile.

The best method to express the absolute quantity of planktonic foraminifera is to calculate the number of specimens in a unit volume of water, usually one cubic metre. This number is similar to the Foraminiferal Number which we described in Chapter IV,7. This parameter may be expressed by the symbol FN_p/m^3 (planktonic Foraminiferal Number per cubic metre of sea water). Boltovskoy, in the first edition of this book, and Miró (1971Ms) presented extensive bibliographical data with respect to this problem. Many investigators calculated FNp/m^3 in areas of their studies. Some of these data are presented in Table 8.

The greatest FN_p/m^3 reported in the literature was 100 000 (Berger, 1969b) but no further details were given.

Bé & Tolderlund (1971) prepared a map of the Atlantic and Indian Ocean showing zones of different densities of surface planktonic foraminiferal populations. This map indicates that the largest populations live in areas which correspond to the surface current system and that the central areas of the North Atlantic and South Atlantic are areas low in foraminifera.

A review of Table 8 reveals that there is little data on planktonic foraminiferal density considering the extent of the oceans. Moreover, the data are, at times, contradictory. A tentative evaluation of the data shows an unequal distribution of the number of specimens and an extremely wide range in FN_p/m^3 values. These differences not only reflect the fact that spatial distribution of planktonic foraminifera is very uneven and changeable (which is the result of a number of factors including the ecology and geologic history of the area) but is also a consequence of a fact that plankton nets of different mesh size were used by different researchers. Berger (1969b) emphasizes that the sampling technique is a very significant contributer to this wide range. He devised a mathematical method to partially compensate for the variations in sampling procedures (Chapter X,3). When plankton tow data were examined in the light of Berger's corrections, the Indo-Pacific Ocean values

Table 8. Planktonic foraminiferal density in the ocean

Author	Area	Water layer (metres below surface)	FN_p/m^3 Average	FN_p/m^3 Maximum
Lohmann, 1920	Atlantic Equatorial Current	75		63 000
Lohmann, 1920	Atlantic Equatorial Current	400	2000	
Phleger, 1951b, 1954b	NE Gulf of Mexico	25–50	5–6	73
Bradshaw, 1959	North Pacific	surface		560
Bé, 1959a	Atlantic central water mass	210–355	2.5	
Bé, 1959a	Gulf Stream	210–355	6.2	12.5
Bé, 1959a	Atlantic, continental slope	210–355	0.7	
Bé, 1960b	Arctic	surface	0.07	
Bé, 1960b	Caribbean	surface	7.7	
Bé, 1960b	North Atlantic	surface	3.4	
Parker, 1960	Equatorial Pacific	surface	0.1–1.0	
Parker, 1960	South Pacific, S of 40°S	surface	<0.1	
Belyaeva, 1962a	Indian Ocean	0–25	1.07	
Belyaeva, 1962a	Indian Ocean	25–50	1.09	
Boltovskoy, 1966b	SW Atlantic	surface	2–10	591
Cifelli & Sachs, 1966	North Atlantic, continental slope	surface	16–39	
Cifelli & Sachs, 1966	Sargasso Sea	surface	0.8–20	
Bé, 1969a	Antarctic	surface	>10	
Beers & Stewart, 1970	Pacific, California coast	upper layer		31 000
Boltovskoy & Boltovskoy, 1970	Antarctic, Mar de la Flota	surface	0.4	
Banerji, Schafer & Vine, 1971	Equatorial and North Pacific	surface	0.07–8.2	
Boltovskoy, 1971b	Antarctic water, Drake Passage	surface	14.4	
Boltovskoy, 1971b	Subantarctic water, Drake Passage	surface	34	
Stehman, 1972	Baffin Bay	surface		10
Parker, 1973	Gulf of California	surface		882
Boltovskoy, 1973a	South Atlantic	surface	6–405	1332

were higher than those of the Atlantic. Maximum concentrations occurred in the North Pacific and lowest values were recorded in the Sargasso Sea. As a general rule, planktonic foraminiferal density seemed to be least in central mid-latitude regions and greatest in equatorial zones and fertile high-latitude coastal areas.

One may judge planktonic foraminiferal populations not only on the basis of plankton samples but also from an investigation of the thanatocoenoses. Studies of the dead specimens in the sediments reveal quantitative as well as species distribution data (Fig. 66). In order to judge to what degree these data correspond to the actual distribution of planktonic foraminifera a number of additional factors must be considered in detail. During its life a planktonic foraminiferal specimen floats and is carried by currents in different directions. Murray (1897) was the first to call attention to the simple fact that the composition of the thanatocoenose corresponds in some way to the composition of the foraminiferal populations that live in the waters above the bottom. He concluded that the dead tests sank so rapidly that currents had insufficient time to transport them to another site. Murray's observations were later echoed by Lohmann (1903), Schott (1935) and others.

The first calculation of planktonic foraminiferal settling rates was made by Krümmel (1907) who estimated a settling velocity of about 2 cm/sec. Zenkevich (1951) calculated that the velocity of sedimentation of 0.4 mm diameter tests is 2.11 cm/sec. At this velocity a test could fall 1000 m in 14 hours; the deepest parts of the ocean could be reached in only a few days. During such a short time no great amount of lateral transport could occur. On the basis of these calculations many authors assumed that the specimens found on the bottom are reasonably accurate representatives of the fauna living in the water column.

Later calculations of sinking velocity by Berthois & Le Calvez (1960) revealed that although the velocity at the surface was approximately 2.0 cm/sec, this value decreased with depth due to changes in water density and viscosity. Their values indicate a slight increase in the time required for a test to reach the bottom. Nevertheless, the settling time would only be a few days at the most.

Several researchers, studying diatoms (Jousé & Semina, 1955) and foraminifera (Hamilton, 1957; Belyaeva, 1962b, 1963; and some others), noted strong similarity between the living and dead assemblages in the areas of their studies. They concluded from this observation that the tests sank very rapidly.

However, this agreement is not unanimous. Phleger (1954) noted that many of the tests collected in planktonic tows were empty. He concluded from this that the fall velocity must not be very great. Bé (1960a), Boltovskoy, (1964a, 1969b), Miró & Marval (1967), and Saidova (1968) have also observed empty tests in planktonic samples. The number of these tests increases markedly with depth (Fig. 67).

It is interesting here to note the work of Vinogradov (1961) who calculated the sinking velocity of different planktonic organisms as well as the rate of decay of the organic material of these organisms. He concluded that small pteropods could reach abyssal depths in 2–3 days without undergoing significant decomposition. Nevertheless, he noted that an examination of pteropod shells from Indian Ocean plankton samples taken at various depths did not support his calculations. At depths greater than 1000 m (and often below 500 m) only 10–20% of the pteropod shells contained recognizable soft parts. These organisms live in the upper 200 m. Below 2000 m practically all the shells were empty. Below 3000 m no specimens with soft parts were encountered. From these observations Vinogradov concluded that the process of sinking is evidently much slower than supposed. He suggested that this retardation

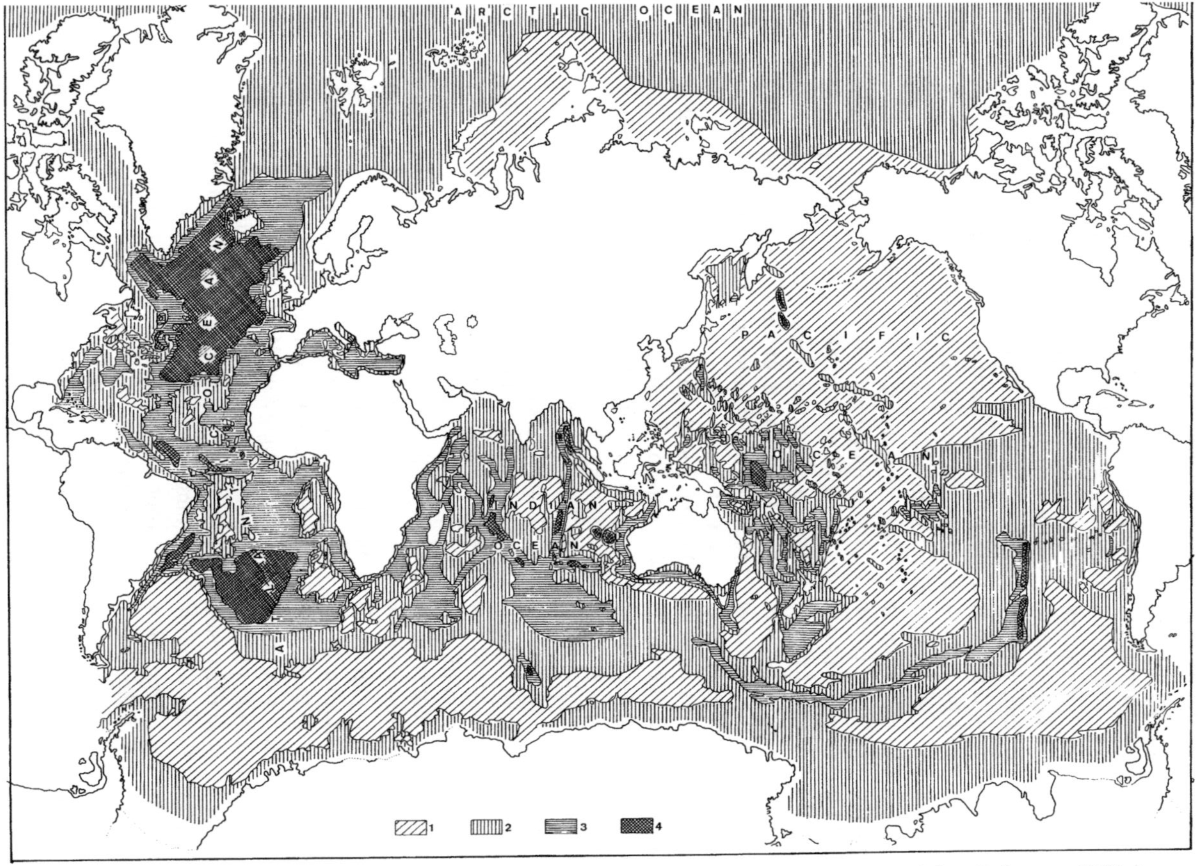

Fig. 66. Quantitative distribution of planktonic foraminiferal test in oceanic sediments (after Belyaeva, 1969a).

Code	Number of specimens/gm
1	<1
2	1–1000
3	1000–10 000
4	>10 000

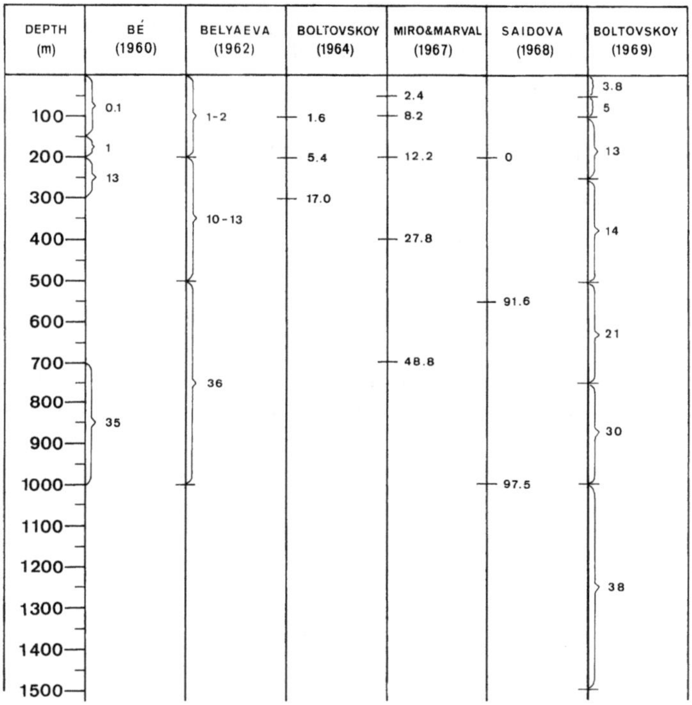

Fig. 67. Percentage of empty planktonic foraminiferal tests found at different depths.

could be explained by the influence of the thermocline and turbulence of the water.

Berger (1969b) criticized the results of earlier laboratory experiments on settling velocity as being up to ten times too high because they were conducted on thick heavy tests in a static water column whereas most sinking planktonic foraminifera have small light tests and fall through a relatively turbulent water column.

In a study of the decomposition of foraminiferal protoplasm Boltovskoy & Lena (1970a) concluded that the fall velocity of tests is much less than is commonly supposed. They suggested that it takes weeks or even months for a test to travel the distance from the upper layer to great depths.

Berger & Piper (1972) conducted the most recent experiments on settling velocity of planktonic foraminifera. They used tests from bottom sediments and from the water column and found that on the average the tests behaved like quartz spheres whose diameters were 2.4 times smaller than the test. Different species behaved differently based on their shape and wall thickness; the thicker specimens sink the fastest. Ranked from rapidly sinking to slow sinking the species are:

1. *Globorotalia tumida*
2. *Pulleniatina obliquiloculata*
3. *Sphaeroidinella dehiscens*
4. *Globoquadrina conglomerata*
5. *Globorotalia inflata*
6. *G. crassaformis*
7. *Globigerina pachyderma*
8. *G. falconensis*
9. *Globoquadrina dutertrei*
10. *Globigerinoides conglobatus*
11. *Globorotalia menardii*
12. *G. truncatulinoides*
13. *G. hirsuta*
14. *Globigerinoides tenellus*
15. *Globigerinoides sacculifer*
16. *G. ruber*
17. *Globigerinella siphonifera*
18. *Globigerina calida*
19. *G. bulloides*
20. *G. rubescens*
21. *Globoquadrina hexagona*
22. *Globigerinita iota*
23. *G. glutinata*
24. *Orbulina universa*
25. *Globigerina digitata*
26. *Candeina nitida*
27. *Hastigerina pelagica*

The modal values for test settling velocities (Fig. 68) confirmed the predictions that the sinking rate was up to an order of magnitude less than had been calculated earlier.

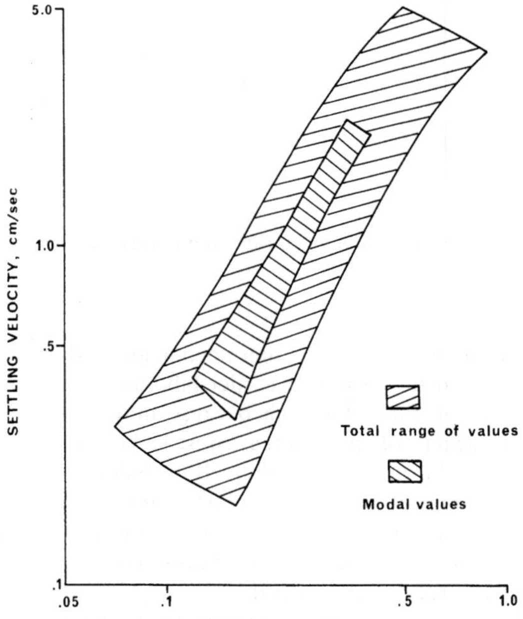

Fig. 68. Experimentally determined settling velocities of planktonic foraminifera (after Berger & Piper, 1972).

How can we reconcile the correspondence of biocoenose and thanatocoenose with the observations indicating a slow settling velocity of dead tests? It may be that the test during its downward trajectory passes through different layers affected by currents operating in different directions. Because these movements are in different directions, they tend to cancel each other and the test is deposited at a point nearly below its life habitat. Thus, the bottom population of empty tests would be quantitatively and qualitatively similar to the living population of the upper water layers.

However, this similarity is only evident at certain depths. In abyssal depths this correspondence disappears due to the increase in solubility of the tests with depth. This phenomenon (Chapter IX,3) plays a significant role in the quantitative and qualitative distribution of planktonic foraminifera on the bottom and warrants a detailed discussion.

According to data of Murray & Renard (1891), calcium carbonate disappears rapidly in bottom sediments as the depth increases. An analysis of 231 samples from the Challenger Expedition showed that the calcium carbonate content of the samples was:

50%	at	3000–4000 m
2%	at	5000 m
<1%	at	6000 m

Based on these data Rhumbler (1911) said that dead foraminiferal tests comprise the *Globigerina* ooze primarily in oceanic depths less than 4000 m. Correns (1939) affirmed the dissolution of calcareous tests in depths between 3600–5500 m. Saidova (1960a) observed that planktonic foraminiferal tests are only found at depths less than 3000–3500 m. It is interesting that the same author later described undamaged planktonic foraminiferal tests in the intestines of bottom feeding benthonic echinoderms (sea urchins, starfish, ophiuroids and holothurians) collected at depths of 7000 m (Saidova, 1968). The implication is that these tests reached this depth without undergoing dissolution and were protected from damage by incorporation into the digestive system of other organisms. Belyaeva (1962) noted that planktonic foraminiferal tests began to show the effects of dissolution in the Indian Ocean at depths of 4500–4700 m. Kennett (1966) found that the dissolution effect is very strong at depths as shallow as 500 m in Antarctic waters (Ross Sea).

Belyaeva (1968) and Saidova (1962, 1970b) presented data regarding dissolution phenomena; their data generally support the observations made above that the vertical limit of calcareous planktonic foraminifera lies between 3500–5000 m. As a general rule, the compensation depth decreases with increasing latitude.

That some planktonic species may be more resistant to solution than others was first mentioned by Walther (1893) who described the lower depth limit in bottom sediments of *Globigerina bulloides*, *Globigerinoides ruber* and *Orbulina universa* at 5760 m. Stubbings (1939) claimed that *Sphaeroidinella dehiscens* was one of the most resistant species because of its thick walled construction. Arrhenius (1952), Phleger, Parker & Peirson (1953), Ericson *et al.* (1961), Belyaeva (1962b) and Oba (1969) also listed observations regarding the differential solution of various planktonic species but the greatest amount of work on this problem was conducted by Berger (Berger, 1967, 1968, 1970a, 1971a; Berger & Soutar, 1967; Berger & Parker, 1970; Parker & Berger, 1971). He created a new term – lysocline – to describe the level at which maximum solution change occurs. According to Berger, dissolution occurs to a greater or lesser extent at all depths, but becomes more intense in water over 1000 m deep and is almost total below 3000–5000 m.

Parker & Berger (1971) prepared a list of planktonic species according to their resistance to dissolution. Their list, from least resistant to most, is:

1. *Hastigerina pelagica*
2. *Globorotalia anfracta*
3. *Globigerinoides ruber*
4. *Globigerina rubescens*
5. *Globigerinoides tenellus*
6. *Globigerinella siphonifera*
7. *G. adamsi*
8. *Globigerinoides sacculifer*
9. *G. conglobatus*
10. *Globigerina quinqueloba*
11. *G. bulloides*
12. *G. calida*
13. *Globorotalia fimbriata*
14. *Globigerinita uvula*
15. *G. glutinata*
16. *G. iota*
17. *Globorotalia* sp. A
18. *G.* sp. B
19. *Globigerina falconensis*
20. *Candeina nitida*
21. *Orbulina universa*
22. *Globorotalia scitula*
23. *Globoquadrina hexagona*
24. *Globigerina digitata*
25. *Globoquadrina conglomerata*
26. *Globorotalia hirsuta*
27. *G. truncatulinoides*
28. *G. inflata*
29. *G. crassaformis*
30. *G. cultrata*
31. *Globoquadrina dutertrei*
32. *Globigerina pachyderma*
33. *Globorotalia pumilio*
34. *Pulleniatina obliquiloculata*
35. *Sphaeroidinella dehiscens*
36. *Globorotalia tumida*
37. *Turborotalia humilis*

Although this ranking is the most complete and accurate at the moment, it should not be considered definitive. According to some authors, the position of certain species should be somewhat different. The case of *Hastigerina pelagica* is an interesting one. Many of the investigators cited above, notably Ericson, Belyaeva and Oba, agree with the position of *H. pelagica* as the least resistant species. However, Boltovskoy (in press) observed that specimens of *Hastigerina pelagica* preserved in a low pH liquid from equatorial Pacific samples are in much better condition than specimens of *Globigerinoides* and *Globorotalia*. He suggests that this phenomenon may be explained as follows:

(a) *H. pelagica* has small pores and a low pore density. Although *Pulleniatina obliquiloculata* has a similar pore density and pore diameter, its resistance appears to be lower, even though its walls are much thicker. This indicates that some factor other than porosity or wall thickness exists which prevents rapid solution.

(b) Living specimens of *H. pelagica* contain a layer of protoplasmic material which covers the whole test (Fig. 60). When the specimen dies this capsule may cover the test and protect the walls from solution. Boltovskoy & Lena (1970a) showed that protoplasm decomposes over a period of weeks or months, during which time the *H. pelagica* tests would be protected. Thus, *H. pelagica*, at least in plankton samples, appears to be more resistant than other planktonic species. The absence of *H. pelagica* in bottom material may be due to the extreme fragility of its test rather than its high solubility.

The problem of selective dissolution is not totally understood although the studies of Berger have shed considerable light on this process. Certain discrepancies that still exist can be explained not only by the scarcity of observations, but also by the fact that test solubility depends as much on environmental parameters as on test characteristics. Berger & Soutar (1970) suggested that under anaerobic conditions the bottom sediments will be saturated with calcium carbonate and thus very little dissolution of foraminiferal tests will occur. Under aerobic conditions burrowing organisms mix the saturated interstitial water with undersaturated bottom water resulting in the undersaturation of interstitial water. This oxygenation of the bottom sediment may also increase the CO_2 production which in turn increases calcite solubility. If the sedimentation velocity is great the water layer situated directly above the bottom will be saturated with calcium and consequently, the majority of the tests will remain undissolved (Berger, 1967). The pH and water temperature will also partially control solution rates. A detailed discussion of the relationships between different species, temperature and solution is given in Parker & Berger (1971). Specimens of the same species from different localities will often exhibit distinctly different resistances to dissolution.

The rate at which dissolution occurs and the way the test is attacked varies according to the test and the environment. Generally, the ornamentation is dissolved first. The walls and other parts of the test are removed later. It is interesting to note that the spines are more resistant than the test walls. The high concentration of organic matter in spines (Rhumbler, 1911; Hemleben, 1969b) may account for this observation.

The depth of 5000 m appears to be a critical one for planktonic foraminiferal tests in low and medium latitudes, whereas the critical depth is much less in high latitude waters.

Benthonic foraminifera are considerably more resistant to dissolution than the planktonic species which are deposited on the bottom.

The planktonic foraminifera accumulate and form the deposits known as *Globigerina* ooze. By weight, these foraminiferal tests comprise from 10–53% of the ooze. Most of these tests are species of *Globigerina* and *Globorotalia*. Correns (1939) estimated that the ooze consists of about 6000 foraminiferal tests (>200 μ) per gram of sediment. In addition to foraminifera the ooze is made up of unicellular calcareous algae coccolithophores, volcanic glass, fine clay and a small quantity of other organic remains. This ooze usually makes its first appearance at a depth of about 600 m, although under special conditions it may form in much shallower depths (as low as 100 m in rare cases).

Globigerina oozes cover about 34.5% of the ocean floor and constitute a layer of considerable thickness. The ooze comprises about 36% of the Pacific Ocean floor, 67% of the Atlantic floor, and 54% of the Indian floor.

These percentages would be greater if the planktonic foraminiferal tests did not undergo solution at great depths. Because of this dissolution the Pacific Ocean, which has large areas of great depth, contains proportionately less area covered by ooze.

Schott (1935) calculated that the central part of the Atlantic received an average of 1.2 cm of ooze per 1000 years; his minimum and maximum values were 0.53 and 2.13 cm respectively. In a later work the same investigator widened the range to 0.5–3.3 cm/1000 years (Schott, 1939). Lisitzin (1971), in a study devoted to sedimentation rates, calculated similar values (1–3 cm/1000 years) for the calcareous ooze areas of the Atlantic. A map of sedimentation rates prepared by Lisitzin showed that this rate is most significantly affected by depth due to the dissolution factor.

4. Vertical qualitative distribution (stratification)

Data on species distribution as a function of depth are not abundant. Emiliani (1954), after studying the oxygen isotope composition (see Chapter XIV,5) of various planktonic species, demonstrated that the calcification of different species takes place at different water temperatures. These temperatures correspond to different depths. From this observation Emiliani concluded that although different species are able to move vertically to select optimum ecological conditions, they tend to choose specific depths at which the majority of their lives are spent. He presented the following classification: (a) Surface species: *Globigerinoides conglobatus, G. ruber, G. sacculifer,* (b) Mid-depth species: *Globigerina dubia, Pulleniatina obliquiloculata, Globorotalia menardii,* (c) Deep water species: *Globorotalia tumida, G. truncatulinoides.*

None of the specimens studied by Emiliani were interpreted by him as having lived below 220 m.

Later, the same author (Emiliani, 1971a) qualified some of his earlier observations by noting that according to oxygen isotope data some species construct their tests at a constant depth whereas others may alter their depth habitat while secreting their tests. According to his data, *G. menardii* sinks as it grows in Caribbean waters but remains at a constant depth in Pacific, Indian and Mediterranean waters.

Bandy (1956) and Waller & Polski (1959) suggested that certain planktonic species lived at certain depths. These conclusions were drawn not from studies of plankton samples but on the basis of studies of bottom samples. However, Phleger (1960b) criticized this method saying that a study of bottom samples would indicate a vertical stratification even if there were none because of the decrease in the total number of planktonic foraminiferal species in coastal areas. In nearshore waters only the abundant species would be detected in bottom samples and the rarer species would be detected only in deeper water samples.

Bradshaw (1959) concluded that there is no true stratification in planktonic foraminiferal distribution. He said that the apparent layering is due to the tropical submergence.

According to Parker (1960), there are certain slight differences between plankton samples collected at different water depths in the equatorial Pacific. These differences occur for only a few species; *Globigerinoides sacculifer* seems to exhibit a preference for surface water whereas *Globorotalia menardii* and *Hastigerina pelagica* are encountered more frequently in deeper waters.

Belyaeva (1962a), after studying the vertical species distribution at 14 stations in the tropical Indian Ocean, came to the conclusion that stratification does not take place among planktonic foraminifera. However, in a later paper (Balyaeva, 1964) she cited some observations which prove that some species prefer particular depths.

Banerji, Schafer & Vine (1971) prepared a list ranking 14 planktonic foraminiferal species by their general compatibility to water depth. Beginning at the surface and going downward, the species are: 1. *Globigerinoides sacculifer*, 2. *Globigerinella aequilateralis*, 3. *Hastigerina pelagica*, 4. *Globigerinoides conglobatus*, 5. *G. conglomeratus*, 6. *Pulleniatina obliquiloculata*, *Globigerinoides ruber*, 8. *Globigerinita glutinata*, 9. *Globigerinoides trilobus*, 10. *Globigerina bulloides*, 11. *Globoquadrina dutertrei*, 12. *G. hexagona*, 13. *Globorotalia hirsuta*, 14. *G. inflata*.

The last named of these species was found in equal abundance at all depths.

Parker (1973) proposed a similar but not identical list. In the upper 100 m she found from the surface downward: 1. *Globigerinoides sacculifer*, 2. *Globigerina bulloides*, 3. *Globigerinella siphonifera*, 4. *Globoquadrina dutertrei*, 5. *Pulleniatina obliquiloculata*.

The most detailed study of vertical species distribution was conducted by Bé & Tolderlund (1971) and Bé (in press). The latter author placed the plankton species into three groups based on their depth habitat.

(a) Surface dwellers: The majority of the spinose species; that is, all the *Globigerinoides* (*G. ruber*, *G. sacculifer*, *G. conglobatus*) and some *Globigerina* species (*G. quinqueloba*, *G. rubescens*). It is interesting to note that adult tests of *Globigerinoides* and *Globigerina* species which live in surface waters can be found in sub-surface layers as well but in this zone they have fewer spines or lack them completely.

(b) Intermediate-water dwellers which occur mainly in the layer between 50–100; four non-spinose species (*Pulleniatina obliquiloculata*, *Globoquadrina dutertrei*, *Candeina nitida* and *Globigerinita glutinata*), and five spinose species (*Globigerina bulloides*, *Hastigerina pelagica*, *Orbulina universa*, *Globigerinella aequilateralis* and *Globigerina calida*).

(c) 'Deep-water' dwellers. They live in the euphotic zone during their early stages but on becoming adults they primarily occupy depths below 100 m. There are 12 species in this group; namely all the *Globorotalia*, non-spinose *Globoquadrina pachyderma* and *G. conglomerata*. According to Bé, *Globorotalia scitula* (large form) and *G. crassaformis* are especially 'deep-water' dwellers as the former is found frequently at the depth 5000–1000 m in sub-Antarctic waters and the latter in deep water in sub-tropical areas. The following three spinose foraminifera should be considered as 'deep-water' representatives: *Hastigerina digitata*, *Globigerinella adamsi* and '*Sphaeroidinella dehiscens*'.

Quantitative studies in the eastern Pacific (Berger, 1971b) confirmed these vertical differences in the position of spinose and non-spinose species. Oxygen isotope studies of fossil (Pleistocene) specimens in the Caribbean (Lidz, Kehm & Miller,

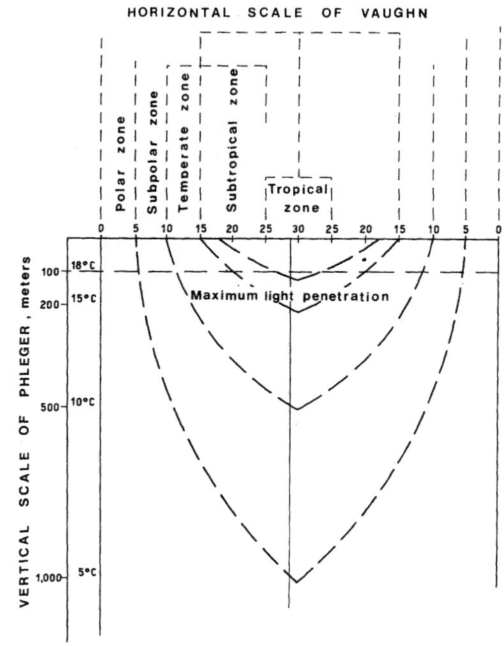

Fig. 69. Relations between horizontal and vertical zonations in the ocean (after Dardenne, 1954).

1968) and the Indian Ocean (Oba, 1969) also showed that *Globigerinoides* spp. (Spinose) lived in near surface waters and *Globorotalia* spp. (non-spinose) in deeper zones.

Summarizing, we can say that apparently a vertical stratification of foraminiferal planktonic species does exist and the correlation of the presence or absence of spines with surface or sub-surface water layers appears to be valid.

As for the tropical-submergence phenomenon, in which cold-water species living at the surface in high latitude areas are encountered at depth in tropical zones, it is well known in several zooplankton groups; e.g. the copepods, polychaets, chaetognaths, etc. Figure 69 illustrates this submergence and the vertical motion that a species may undergo in order to maintain a temperature regime as it moves horizontally.

This submergence phenomenon is also known in foraminifera, but there are relatively few detailed observations on it.

Schott (1935) noted the presence of *Globigerina bulloides* and *G. inflata* in the deep water layer in one area. They were absent in a layer only 100 m higher, a layer inhabited solely by tropical species. The same phenomenon has been observed for *Globorotalia truncatulinoides*. Bradshaw (1959) mentioned similar situations occurring in plankton samples taken at depth in the equatorial and warm temperate zone. He described the occasional presence of *Globigerina bulloides, G. quinqueloba, Globigerinoides* cf. *minuta* and *Globorotalia truncatulinoides* in these samples. He added that one cannot exclude the possibility of these being isolated specimens.

The tropical-submergence phenomenon has also been reported by Belyaeva (1962a), Boltovskoy (1964a) and Jones (1966 Ms).

A good example of this phenomenon is cited by Miró (1971 Ms), who found *Globigerina quinqueloba* and *Globigerinita uvula* in the upper layers in Venezuelan waters. These species were brought there by Subantarctic intermediate currents and lifted by upwelling phenomenon.

5. Vertical quantitative distribution

Carpenter (1875) and Brady (1884) believed that small and young planktonic foraminifera live at the surface while large and adult forms live in deeper water. Walther (1893) thought that the planktonic foraminifera lived at the surface but that most of them reproduced at great depth. He cited as evidence the observations of Carpenter who in 1869 noted the increased turbidity of water at a depth of 1371 m and said that it was due to the increased number of juvenile specimens of *Globigerina*. Murray (1897) suggested that planktonic foraminifera live exclusively at the surface.

Lohmann (1920) was the first to recognize that planktonic foraminifera live at various depths. It would be logical to suppose that the most dense populations would be in surface waters because the phytoplankton material is more abundant there. The following observations of Lohmann confirm this supposition.

Depth (m)	Relative quantity of planktonic foraminifera
0	100
50	67
100	60
200	7
400	3

Schott (1935) confirmed these conclusions with the observation that the quantity of specimens in the upper 100 m was significantly greater than that in water of greater depth. Phleger (1945) displayed the data of Schott in tabular form:

Depth (m)	Number of samples	Average number of planktonic foraminifera per sample
0–100	71	489
100–200	7	8
200–400	13	24
400–600	19	10
600–800	12	11
800–1000	19	28

Phleger (1945) studied the vertical distribution of planktonic foraminifera in the North Atlantic and found the following:

Depth (m)	NF_p/m^3
75–100	0.8
325–500	0.5
650–1000	0.7

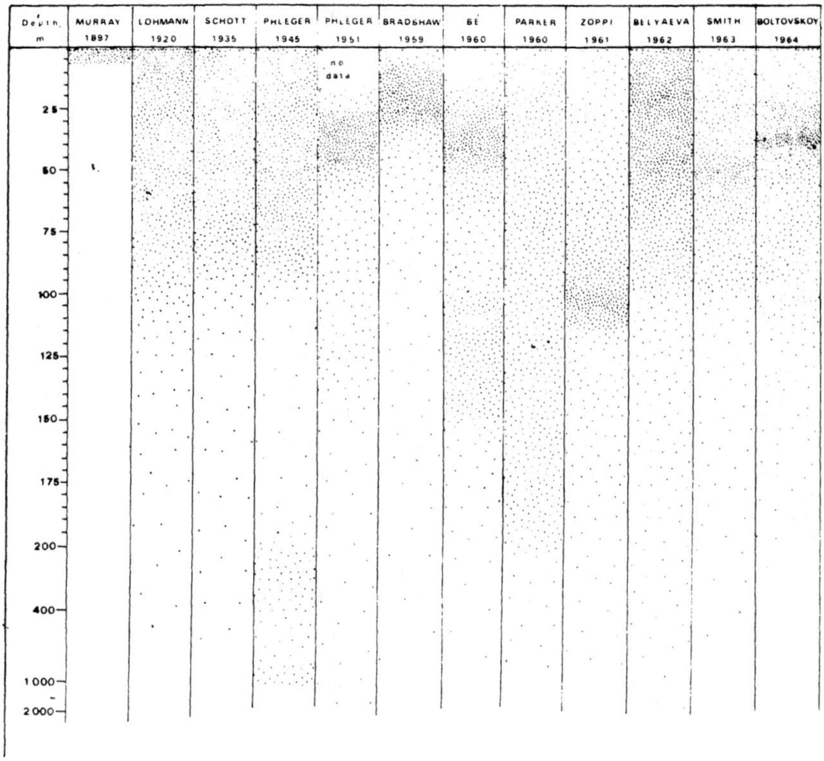

Fig. 70. Vertical quantitative distribution of planktonic foraminifera. Dot density within a column is proportional to abundance but is not comparable between columns.

This same author later examined the number of planktonic foraminifera taken between 15–2000 m (Phleger, 1951). He concluded that living specimens were present at all depths but their quantity varied considerably. Between 20–145 m the FN_p/m^3 was 29, whereas between 175–2000 m the average FN_p/m^3 dropped to 12. The richest layer lay between 25–50 m.

Bradshaw (1959) suggested that the surface-water layer (0–6 m) was not the richest zone but rather that the greatest concentration occurred between 6–30 m. A marked diminution in number of specimens was noted between 50–100 m and few specimens were observed at 200 m and below.

In the Sargasso Sea Bé (1960a) also noted that the maximum planktonic foraminiferal populations did not occur at the surface. He found that the greatest concentration lay in the upper 200 m, specifically between 25–50 m and 100–150 m. Parker (1960), studying the South Pacific, noted the maximum development of planktonic foraminifera in the upper 200 m. Zoppi (1961) observed that the greatest number of species in the Caribbean occurred at 100 m and diminished upward and downward from this depth. Data from the Indian Ocean (Belyaeva, 1962a) revealed the same high concentrations in the upper 50 m where optimum light, temperature, and nutrition conditions prevail. A. Smith (1963) also pointed out the high density of planktonic foraminifera in the upper 200 m in a northeast Pacific study. This investigator said that the upper 100 m were more enriched than the 100–200 m zone and that maximum populations occurred at 50 m. Saidova (1968) found the greatest concentration in the Pacific Ocean, Kurilo-Kamchatka Trench area, between 50–200 m.

According to Boltovskoy (1964a) the maximum abundance occurs at 40 m in the western equatorial Atlantic:

Depth (m)	Average number of specimens per sample	Percent relative to surface sample
0	180.3	100
25	242.3	134
40	454.1	252
50	218.2	121
80	42.4	24
100	44.1	24
200	22.0	12
300	9.8	5

Jones (1971) indicated that the position of maximum abundance of planktonic foraminifera in the Florida Straits migrates slightly with the seasons but lies generally between 60–100 m.

Figure 70 shows graphically the quantitative vertical distribution of planktonic foraminifera as reported by various investigators. These data are, of course, very generalized and are not always consistent. In part these inconsistencies are due to different sampling procedures of the investigators and their different divisions of the top 200 m. Moreover, one must realize that each study represents a different area and is subject to different local conditions.

6. Vertical migrations

This phenomenon has been incompletely studied with respect to planktonic foraminifera and the existing data are somewhat contradictory. Two types of vertical migrations will be discussed: (a) those resulting from changes occurring during the life of the individual (the ontogenetic cycle), and (b) those that have a daily repetition.

A. Ontogenetic Migration

Various planktonic foraminiferal species develop thick walls upon reaching maturity. The outer calcite layer of the walls of various species collected in deep water has been subject to additional calcification and is considerably thicker than the walls of specimens of the same species found in the surface and near-surface plankton. Although Rhumbler (1911) believed that this layer formed organically during the life of the individual, the majority of the later investigators believed that the calcification occurred at depth after the death of the individual.

In a very detailed study Ericson *et al.* (1961) determined that only some species collected at depths greater than 300 m or from bottom samples possessed this crystaline cortax of calcite (*Globorotalia menardii menardii, G. menardii tumida, G. truncatulinoides, Globigerina inflata, G. pachyderma, Globigerinoides sacculifer, G. conglobatus* and *Orbulina universa*). Specimens of *Pulleniatina obliquiloculata* and *Globorotalia scitula* did not exhibit this thickening, nor did the earlier-mentioned species when encountered in the photic zone. The thickness of this cortex is often several times greater than that of the original wall. The thickening is often greatest on the first chamber of the last whorl; it then diminishes toward the later chambers. The ease with which this layer is removed from the test, and the discovery of the original smooth, thin, typical wall beneath, indicated that this layer is not the result of recrystallization of the original wall but is a supplementary addition. It is also not the result of inorganic precipitation of calcite over the test at depth since neither benthonic species nor some other planktonic species possess the layer. Occasionally surface specimens exhibit this crust. Ericson concluded that this deposit must be formed by the living organism. He suggested that this secondary layer is added when the mature individual descends to deeper water to reproduce.

Thus, we see that planktonic foraminifera can increase their weight intentionally. This increase contributes to their vertical migration. Both Murray (1895) and then later Bradshaw (1959) noted that the largest specimens of a given species are most commonly found at the greatest depth. The increase in size decreases their buoyancy.

Whether or not the majority of planktonic foraminifera descend to deeper water to reproduce has been difficult to establish. Le Calvez (1936a) described in detail the vertical migration of *Orbulina universa*. He found that orbuline specimens in the near-surface waters of the Mediterranean have a trochospiral globigerine test within them (Fig. 71). At a depth of 50 m this interior test becomes displaced to one side of the

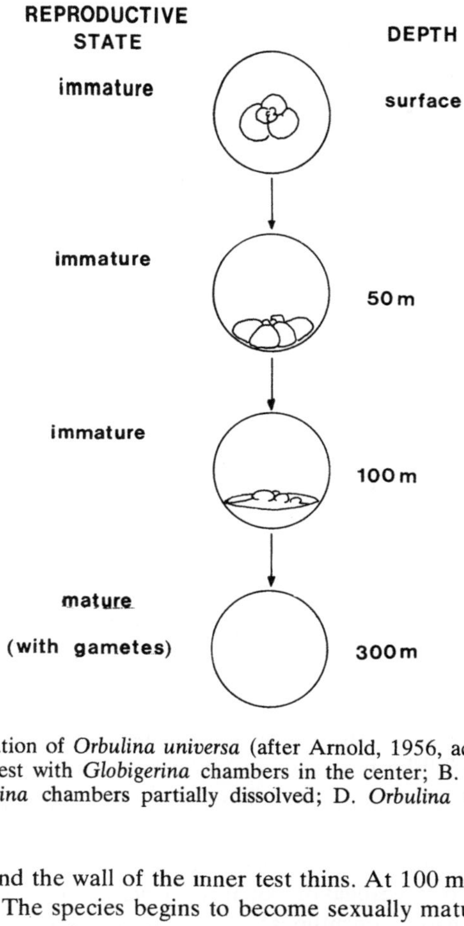

Fig. 71. Vertical migration of *Orbulina universa* (after Arnold, 1956, according to Le Calvez, 1936a); A. *Orbulina* test with *Globigerina* chambers in the center; B. *Globigerina* chambers displaced; C. *Globigerina* chambers partially dissolved; D. *Orbulina* test without traces of *Globigerina* chambers.

outer orbuline test and the wall of the inner test thins. At 100 m the globigerine test partially disappears. The species begins to become sexually mature below this point and at 300 m at full maturity, the internal test is gone. Similar observations on *O. universa* were made at different depths in the Northeast Pacific (A. Smith, 1963) and Indian Ocean (Bé, Harrison & Lott, 1973). Several other investigators have also expressed the idea that the planktonic foraminifera descend to deep water to reproduce; e.g. Boltovskoy (1971a) and Bé & Tolderlund (1971).

At the same time there are observations indicating that foraminifera do not descend to deep water to reproduce. Studies in the Indian Ocean (Belyaeva, 1962a) revealed the greatest number of juvenile specimens were found in the upper 100 m. Berger & Soutar (1967) also stated that reproduction seems to take place mainly in the upper 100 m.

Nevertheless, it seems reasonable at the moment to think that reproduction takes place at depth where conditions are more tranquil and appropriate. The greater percentage of young specimens in the surface water is best explained by the supposition that, immediately after birth, the young specimens ascend to the surface zone where nutritional conditions are better.

We must summarize by stating that the data on ontogenetic migration is insufficient to permit definitive conclusions to be made about its nature.

B. Daily Migrations

In general, planktonic organisms concentrate in surface waters during the night and sink to certain depths during daylight hours (Fig. 72). This phenomenon has been

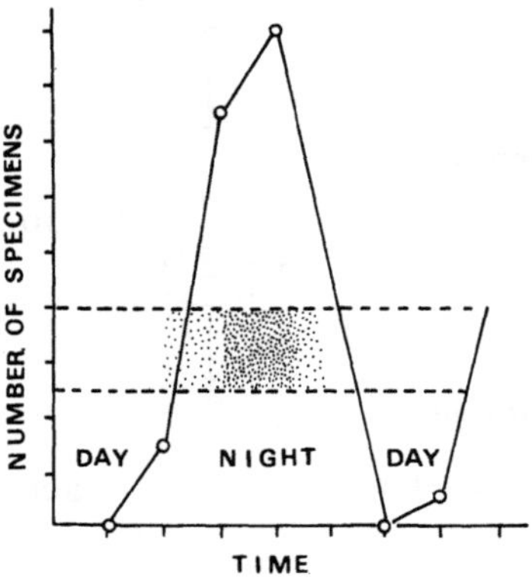

Fig. 72. Vertical plankton migrations, Sea of Japan (after Zenkevich, 1951). Dot density proportional to plankton density.

well studied for various groups of zooplankton and it seemed logical to assume that the planktonic foraminifera would follow this pattern. Ever since Owen (1868) wrote that the greatest surface concentration of planktonic foraminifera occurred during the night a number of investigations of this phenomenon have failed to yield any consistent and satisfactory conclusions.

Some investigators described a migration pattern consistent with that of other zooplankton. A. Smith (1963) said that the onset of daylight brought a vertical migration of the organisms from the surface to a zone about 50 m deep.

Other authors (Rhumbler, 1911; Bradshaw, 1959; Bé, 1960a; Parker, 1960; Bé & Hamlin, 1967; Jones, 1971) noted the opposite effect, affirming that the specimens move toward the surface during daytime.

According to Tsi-chung & Sau-yee (1964), planktonic foraminifera have a very complicated vertical migration pattern. They noted surface concentrations after 4 P.M. and at dusk, followed by a downward movement of the organisms at midnight. The specimens began to rise about 10 A.M. and were concentrated in surface waters at noon, after which time they began to sink again.

Boltovskoy (1964a, 1966b, 1973a), Boltovskoy & Boltovskoy (1970) and Stehman (1972) did not find any significant differences between the numbers of specimens collected in day and night catches and consequently they concluded that no daily vertical migration occurs.

A statistical examination of the daytime and nighttime plankton tow data of many authors by Berger (1969b) showed that many of the variations between total planktonic foraminiferal concentrations during the day and night could be due to chance. However, he also conducted a detailed examination of data, species by species, and noted that some species, notably those that live in shallow, cold to temperate, fertile waters, exhibit significant diurnal migration. Other species, those living in shallow, tropical, low fertility waters, exhibited very little variation in daily concentration.

Various explanations have been offered to explain the phenomenon of vertical migration, Bradshaw (1959) and Parker (1960) suggested that the vertical movement is caused by the production of oxygen by the symbiotic algae inhabiting the foraminiferal test. The rapid production of this gas during the daylight hours would produce low density ampullae which would cause the test to rise. The use of this oxygen for respiration during the night would increase the specific gravity of the test and cause sinking. Since the greatest light penetration and symbiotic algal concentrations in planktonic foraminifera should occur in tropical regions, one would expect the greatest vertical movement due to algal oxygen production to occur in these areas. Yet the data of Berger mentioned above indicates that this is not the case.

Bé (1960a) suggested that the correlation between the low density of foraminifera in surface waters at night and the ascent of other zooplankton at that time is due to the consumption of a large part of the foraminiferal population by the other zooplankton during the early nighttime hours. Bé points out that this explanation does not require any vertical migration but it does pose the problem of how the foraminifera can maintain their total population if large quantities of them are lost daily. There is no significant evidence of partially digested foraminiferal material in the alimentary tracts of the other zooplankton.

It is clear that the problem of the daily vertical migration of the planktonic foraminifera is not solved. There is disagreement as to whether or not vertical diurnal migration occurs.

VIII. CLASSIFICATION

1. Zoological nomenclature

As new lands and seas were being discovered during the great exploration periods of the past, the list of known organisms grew rapidly. These new organisms were generally described by the common names used by people native to the area of discovery. After some time, during which the list of names grew longer and longer, naturalists began to realize that the list was overburdened with many names for the same organisms. Another inconvenience was the designation and description of these organisms in many different languages, further increasing the difficulties of comparison. It became apparent that the use of a common language for biological nomenclature was essential for the normal development of the natural sciences. In Karl Linné (1707–1778) the naturalists found the solution to their problems. His monumental work, *Systema Naturae*, which was devoted to this problem, was first published in 1735. The appearance of the tenth edition of this work in 1758 is traditionally considered as the birthday of modern zoological nomenclature.

Linné proposed that the species be considered as the principal taxonomic unit. Each species should have as its name two Latin words or latinized words, followed by the name of the author who originally described the organism.

This system, called the binomial system of nomenclature, proved to be very convenient, and has been adopted by naturalists all over the world. Its use, however, has only been obligatory since 1905 at which time the Fifth International Zoological Congress studied the ways in which the Linnean system had been used during the previous century and then established the International Code of Zoological Nomenclature (ICZN). The rules established by this congress were published in French, English and German and were made mandatory. In later congresses, the rules were modified and many supplementary rules have been added. A special International Commission was created to act on all the changes and Supplements. New rules, emendations, and decisions about particular names are published in the Bulletin of Zoological Nomenclature, London.

The primary characteristics of zoological nomenclature are:

(a) It is independent of botanical nomenclature. Nevertheless, if some organism previously considered as a plant is discovered to be an animal, its botanical name is conserved.

(b) The rules, are obligatory for all authors and failure to follow the rules invalidates any new name. Exceptions may be made to this stipulation, but only with the approval of the Special Commission, elected during an International Congress; changes and modifications may also be effected by a Congress.

(c) All names must be Latin or latinized words. Generic and higher level denominations must be capitalized. The date of establishment is recorded for each name, and a 'law of priority' then exists for that name. This priority is valid for names established since 1758.

It is neither possible nor necessary to describe all the provisions of the ICZN here, but certain aspects of the Code which are used daily by the zoologist will be discussed.

A. Genera

A new genus is established when its description and diagnosis are published and its type species (*typus generis*, genotype or genoholotype) is designated. The type species is the species which, according to the original author, is a member of that genus and is typical or characteristic of that genus. The genoholotype may have been described earlier, even under a different generic name.

According to the Code, the genoholotype, once established, cannot be annulled. Consequently, if later studies reveal additional or different traits among the populations previously considered as a particular genus, a new genus may be created with a new type. Nevertheless, the old generic name continues to exist and only loses those species moved to the new genus.

To establish a new genus, the author gives it a new name, even though this name may vary by only a single letter from existing generic and subgeneric names. This name should be singular and in the form of a noun in the nominative case. If the generic name is written alone, it is followed by the name of the original author, a comma and the year of original publication; e.g. *Quinqueloculina* d'Orbigny, 1826.

If the generic name is the same as a pre-existing name, even if the pre-existing one is applied to a different type of animal, the name is a homonym and should be changed. The new name is followed by the designation nom. nov. (*nomen novum*) and the original author and data as before.

Often the same genus (or the same species) receives two different names which are then synonyms. According to the principle of priority, the more recent of these names, the junior synonym, should be annulled. However, the Code of Zoological Nomenclature also states that the application of priority may be moderated to preserve a long-accepted name in its accustomed meaning. When conflicts arise between long-established usage and strict application of the rules, in many cases it is better to act permitting the Principle of Conservation to outweigh the Principle of Priority. Several interesting examples of good and bad usage of both principles are discussed by Todd (1963) and Le Calvez (1969).

If a genus is separated into two independent genera, the old generic name is applied to that group of species containing the type species. The other group receives a new generic denomination.

If the original interpretation of some genus has been altered by subsequent investigations, the notation emend. (*emendo*) is placed after the author's name and separated from it by a comma; e.g. *Ammodiscus* Reuss, emend. Loeblich & Tappan, 1954, indicates that the concept of the genus *Ammodiscus*, as originally conceived and established by Reuss, has been interpreted by Loeblich & Tappan in 1954 with modifications. If a genus is described by two authors it is recommended that their names be united by '&' or 'et'. If the original interpretation of a genus has been broadened or diminished the notation s.l. (*sensu lato*) or s. str. (*sensu stricto*) is added.

Subgeneric names are written in parenthesis directly after the generic denomination; e.g. *Gaudryina* (*Siphogaudryina*) *siphonifera*. The subgeneric name is subject to the same rules listed above for generic names.

B. Species

The procedures that are followed to establish a new species are similar to those for erecting a new genus. Each new species, however, must be typified by one example, the holotype, a specimen which is representative or typical of the population which the new species represents. The holotype ought to exhibit all the characteristic traits of the species. In the original publication the location of the holotype must be mentioned along with the description, diagnosis and illustration of the new species. The designation of a holotype is an important aspect of nomenclature in that it conserves the concept of a species even when the species is studied by other methods which may yield a new description. The designation of a new species should include a description not only of the holotype but of other original specimens (paratypes) which give an idea of the degree of variability present in the new species. The greater the number of paratypes studied and described the better will be the concept of the new species.

The idea of allowing one specimen to represent an entire population has provoked some controversy and there have even been suggestions to eliminate the holotype (Emiliani, 1969) and substitute an idealized concept (a true 'Platonic Form') based on the author's study of the paratypes.

Holotypes and paratypes are referred to as primary types. There are numerous other primary as well as secondary and tertiary types (see Frizzell, 1933, for a good description of types). We will mention only topotypes, which are specimens collected from the same locality as those used to describe a new species.

The name given to each organism is tripartite: generic, specific and trivial. The generic name is that of the genus; e.g. *Bulimina*. The trivial name is that which is added to the generic name, e.g. *marginata*. The specific name, or binomial name, is the combination of generic and trivial names; e.g. *Bulimina marginata*.

Having recognized a new species, the author gives it a trivial name, adding this to the appropriate generic name. The trivial name must consist of a single word. It can be noun or adjective; if the latter it should be modified to agree with the gender of the generic name. Masculine terminations are 'us', femine 'a' and 'is' and neuter 'um'. For instance: Masculine – *Cibicides lobatulus, Ammodiscus rugosus;* Feminine – *Cassidulina crassa, Tetrataxis conica;* Neuter – *Elphidium macellum, Loxostomun aculeatum.*

If the investigator has difficulties with the name of the genus it is best to refer to a Latin dictionary. Some cases are difficult, especially when the generic name is not originally Latin and/or the original author did not interpret its gender. As an example, *Nonion* and *Reophax* mean nothing in Latin; they are merely combinations of letters. Montfort, who introduced these genera said nothing about their gender. Since he used them initially as masculine nouns, we should follow his example; e.g. *Nonion depressulus* and *Reophax curtus.* There are yet more difficult cases which require yet more searching. For example the feminine gender of *Pyrgo* was established only when the original source of the name, a nurse, mentioned in a work of Homer, was discovered.

If the trivial name is derived from a geographic source it is terminated with 'ensis' when masculine or feminine, and 'ense' when neuter; e.g. *Eponides columbiensis, Canepaia brasiliensis, Elphidium oregonense.* Trivial names whose source is a proper name end in 'i' if masculine (*Globigerina eggeri*) and in 'ae' if feminine (*Discorbis parkerae*).

There are numerous works dedicated to the problem of the correct composition of scientific zoologic names. Unfortunately, they do not always agree on certain points. This is a rather complicated problem and requires a great deal of work. The work of Macfadyen & Kenny (1934) is dedicated especially to the problem of foraminiferal names. The paper by Emiliani (1952) is also very useful since he utilized dictionaries of ancient languages to try to clarify scientific names. Various suggestions for the foraminiferal names are given in Frizzell (1954) and Ratnovskaja (1967).

A new trivial name may not be the same as a name used for another species belonging to the same genus. If it does it is a homonym and, just as for generic homonyms, must be replaced. In both cases the notation nom. nov. should be used to prevent the later use of the same name. Hofker (1951a) described and named a new species, *Allomorphina pacifica*, not knowing that this name had been used previously by Cushman & Todd (1949) who established *A. pacifica* n.sp. Thus, the name designated by Hofker was a homonym. Thalmann (1952) renamed Hofker's species *A. fragilis* nom. nov. Those that use this name in the future will write it as *A. fragilis* Thalmann. Cases such as this are frequent.

If an investigator moves a species to a new genus it is necessary to place the name of the original author in parenthesis. In 1839 d'Orbigny described *Bulimina elegantissima* d'Orbigny. In 1911, Cushman established a new genus, *Buliminella*, and placed d'Orbigny's species in the new genus. The designation is now written *Buliminella elegantissima* (d'Orbigny).

Subspecific names are written after the specific name and are accompanied by the name of the author and year of designation; e.g. *Spirillina vivipara revertens* Rhumbler, 1907. The gender of the subspecific name should agree with the generic name.

C. Open Nomenclature (*Nomenclatura aperta*)

Occasionally the investigator will have either a limited number of specimens to examine or poorly preserved material. Bibliographic reference material or comparative specimens may also be lacking. Under these circumstances the investigator should indicate his determinations with a certain degree of uncertainty. Unfortunately, the ICZN does not provide any provisions for the reflection of this uncertainty. In actual zoological nomenclatural practice various procedures for treating uncertainties regarding taxonomic determinations have developed. Although these procedures are not legalized, they have become generally accepted by tacit agreement. These 'rules of open nomenclature' are not rules in the formal sense and are open to continuous alteration and discussion. Several of the more important of these are discussed in the following paragraphs.

Those specimens which are completely indeterminate systematically are designated as *incertae sedis* (unknown place).

If the generic position of a specimen is indeterminate the designation gen. indet. (*indeterminata*: indeterminate) is used; if the species is indeterminate, the phrase sp. indet. is used: e.g. *Bolivina* sp. indet. indicates that this specimen can be identified only to the generic level.

The uncertainty of a systematic position is indicated by the use of a question mark; e.g. *Bulimina?* indicates that there is uncertainty about the designation of the genus and that the species is unknown. '*Bulimina*' *marginata* d'Orbigny signifies that the name requires revision. If the question mark is placed before the name of the species, e.g. ?*Rupertia stabilis* Wallich, the entire determination is placed in doubt. *Bulimina? marginata* d'Orbigny is an indication that the specimen belongs to this species but that the genus is uncertain. *Bulimina marginata* d'Orbigny? signifies that the specimen may belong to this species but the exact identification cannot be determined.

There is ample use of the abbreviations cf. and aff. The designation cf. is placed in front of the trivial name to signify that the author does not doubt the generic designation but is less sure about the species. The specimens do not correspond in all respects to the type but do demonstrate a great similarity. As an example, *Nonionella cf. atlantica*[8] indicates that the species examined has much in common with *N. atlantica* but is not exactly the same, or because of inadequate numbers of specimens, the author is unwilling to declare the unconditional attribution of *N. atlantica* to the species under study. Cf. is the abbreviation of *confer* (similar). Aff. is the abbreviation of *affinis* (affinity, relative) and is also used between the generic and trivial names. *Nonionella* aff. *atlantica* indicates that the specimens studied by the

[8] Strictly speaking, it would be more correct to write *Nonionella* cf. *N. atlantica* since the ICZN does not allow the insertion of any word between the generic and trivial name of a species except a subgeneric designation (in parenthesis). Nevertheless, the common usage is to place cf. and aff. between the two names.

investigator were not undeniably members of the species *N. atlantica* but are very similar to it, and, at the very least, are close relatives of the species.

There is no regularity or consistency in the actual use of the terms cf. and aff. This is understandable considering the lack of rules governing their use.

The abbreviation ex gr. (*ex grege*; of the flock or herd) is also used to signify that the studied specimens have many traits characteristic of a particular species but are not completely identical. The traits observed lie around the fringe of the known species. As an example, *Cibicides* ex gr. *lobatulus* means that the specimens so designated are related to and have similar traits to the well known cosmopolitan species *C. lobatulus*.

If the investigator does not wish to give a specific determination, and does not wish to show how the species is related to any known species, he may write only the generic name and add the designation sp.; e.g. *Oolina* sp.

There are other terms used to indicate the kind and degres of uncertainty of a particular taxonomic determination. These are discussed in Richter (1948).

2. Biological significance of taxonomic categories

The principal taxonomic category is the species. Discussion over the species concept has been extensively and intensively conducted for many years. Some suggest that the species has no actual basis in nature, that it is a mental abstraction created for reasons of convenience. There are suggestions that zoologic (biospecies) and paleontologic (chronospecies) species are different because of the introduction of lineages evolving through time in the latter case. However, most zoologists and paleontologists accept the concept of the species, at least in practice, as a reality which exists in time and space and can be identified in naturally occurring zoological and paleontological populations.

The species is generally considered as a population which satisifies the following criteria: (a) physiologic – those individuals belonging to the same species can successfully breed and produce viable offspring; (b) morphologic – although the individual members of a species may be very different, they all exhibit certain morphological similarities which are not shared by populations of other species; (c) geographic – two species, even those closely related to each other, can occupy the same area without mixing.

Nevertheless, there are exceptions to these considerations: (a) there are cases of individuals, considered as separate species, which are able to produce viable offspring, (b) the morphological distinctions between some members of two species (the end members) are so slight as to make delineation difficult, (c) related species tend not to coexist geographically and if they do, tend to occupy different ecological habitats.

The subspecies is the only taxonomic category inferior to the species. The subspecies is generally considered as a geographic isolate that is still part of the species and shares most of the traits of the species, differing only morphologically. Under repeated and long lasting geographic isolation a subspecies may develop into a new species.

Subspecies belonging to the same species (a) can breed freely producing viable offspring, (b) usually exhibit transitional forms among themselves and (c) can not coexist in the same area for any length of time because the resulting interbreeding would eliminate the distinctive traits of each subspecies.

The zoologist as well as the paleontologist occasionally encounters cases in which the areas of distribution of two subspecies are superposed. In such cases the two subspecies as well as their transitional forms will be found together. Vertebrate faunas often exhibit this characteristic. There are numerous examples of this mixing among foraminifera.

It is often difficult to recognize and distinguish subspecies. They may often be distinguished statistically. As a general rule, the typical specimens of a particular subspecies ought to comprise no less than 75% of the population of the species.

The subspecies is the lowest taxonomic category recognized by and governed by the rules of the ICZN. There are, however, commonly used categories inferior to the subspecies, the most common of which in foraminiferal usage are *varietas* and *forma*. There are no rules governing their use and no formal definitions of these categories. We will explain them by example and comparison with the subspecies.

Varietas is the category just inferior to subspecies. A variety generally has 2–3 distinctive traits which are not as constant nor as visible as those of the subspecies. Its geographic distribution is poorly known. Several biologists utilize this category during the outset of an investigation when it is not known if a particular set of organisms represents a subspecies or a local deviation of the *forma*-type. Later studies usually resolve this problem and the variety name can be discarded. Since it never had legal status, it need not be annuled and it does not burden the nomenclature.

Forma is a taxonomic category still lower in the hierarchial scale, yet above the individual variation level. A *forma* generally has only one distinctive trait, which, although usually pronounced and visible, is not very important and can be easily lost. The *forma* does not have a well defined geographic distribution and its members are found interspersed among the typical examples of a subspecies or species. The trait which distinguishes forms is usually a distinctive morphological feature, often a detail of the ornamentation. *Formae* are usually named after their peculiar trait; e.g. f. elongata (elongate), f. costata (with costae), f. parva (small), etc. These names, as is the case with varietal names, have no defense with regard to the law of priority and can be duplicated and eliminated with ease.

The use of the category *forma* is very convenient because it (a) allows a more precise presentation of the morphological relationship among the specimens which are being studied, (b) reduces the number of synonyms and (c) preserves all the small differences which, although they are of an intraspecific nature, are interesting and revealing from an ecological point of view. This latter point is perhaps the most important because almost all the morphological details peculiar to a particular *forma* (loss or increase in ornamentation, change of size, etc.) are a consequence of environmental responses.

As the use of *varietas* and *forma* is not regulated, these categories are not official, but only supplementary. Because of this they are separated from the species name by a comma and followed by the name of the author of the species. As an alternative, the *varietas* or *forma* name may be placed in parenthesis in the same position. They should not be written in italics.

Unfortunately, many investigators have interpreted a *forma* as a new variety and then in turn as a new subspecies, and finally as a new species. The case of *Bulimina patagonica* illustrates this point. Some specimens of *B. patagonica* lack spines and were described by Cushman & Wickenden as *B. patagonica*, var. glabra n. var. On the Argentine Continental Shelf the same forms are found mixed with other specimens showing only some or no spinose ornamentation. The ornamentation is not a constant trait and depends on the peculiar environmental conditions and on the age of the individual. For these specimens the use of the category *forma* would have been much more correct. They should have been named *B. patagonica* d'Orbigny, forma glabra Cushman & Wickenden.

The taxonomic category immediately superior to the species is the genus. Classification by genera is not as critical as the classification into a species because a species, even if placed in several different genera, can be compared from one study to another. The generic category is less exact and well defined than the species. The placement of a species into a genus, because of the large number of genera already described, is unfortunately often based primarily on the 'taste' of the author. Generic names are subject to the ICZN and should be erected with care and seriousness.

The family is the taxonomic category superior to the genus. The genera which are united to form a family should possess certain common traits which should be of greater importance than those which are used to separate genera.

Families are united into orders, and these into classes, etc. In addition, there are transitional categories, the subfamily, suborder, etc. and superfamily, superorder, etc.

The standardized suffix for subfamily names is "inae" and for family names, 'idae;' e.g. subfamily Lenticulininae Sigal, 1952, and family Lagenidae Schultze, 1854. The taxonomic categories superior to the family do not have fixed endings but many authors prefer to use:

superfamily	'oidea' (or 'acea' or 'icea')
suborder	'ina'
order	'ida'
superorder	'ica'
subclass	'ia'
class	'ea'

The bibliography dealing with taxonomic categories and associated problems is enormous. We will mention only a few of the more modern works which deal with foraminifera (Boltovskoy, 1958b; Butler, 1963; Vialli, 1963; Sigal, 1966). Some taxonomic problems are dealt with in great detail by Charlton de Rivero Palacio & Bermúdez (1963) who also have a very ample bibliography of works dealing with such problems.

3. Development of the idea of a natural foraminiferal classification

The first accounts of the discovery of foraminifera are found in the works of Strabo (63 BC–20 AD) who described specimens of *Nummulites* from the pyramids of Egypt. Strabo considered them as fossilized lentils which had been used as food by the slaves who constructed the pyramids. Specimens of *Nummulites* were also mentioned in various other works published prior to the Middle Ages. They, along with other paleontological remains, were considered as freaks of nature.

Various natural scientists of the following centuries, among them, Gesner, Hooke, M. Lister, Beccarius, Plancus, Gualtieri and others, published drawings and descriptions of foraminifera. Beccarius was the first, in 1731, to produce relatively good illustrations of small foraminifera. All these authors considered the foraminiferal specimens as diminutive molluscs.

Linné in the 12th edition of Systema Naturae (1767), included foraminifera under the generic names *Nautilis* and *Serpula*.

The end of the 18th Century and beginning of the 19th Century saw an influx of papers dedicated to foraminifera. These works contained illustrations and descriptions of numerous genera and new species (Walker, 1784; Batsch, 1791; Soldani, 1789–1798; Walker & Jacob, 1798; Fichtel & Moll, 1798; 1803; Lamarck, 1801; 1822; Montegu, 1803, 1808; Montfort, 1808).

The most influential and pioneering investigator of foraminifera during this period was the renowned French naturalist Alcide d'Orbigny whose work (1826, 1839a–c, 1846, 1849, 1852) contained the first detailed descriptions of the entire group. It was d'Orbigny who first placed the foraminifera into the general scheme of the organic world, although he considered them to be Cephalopods (1826). He placed them in a special order, 'Foraminifères', and contrasted them with the 'Siphonifères' whose chambers were connected by a siphuncle. The latinized word foraminifera was first used by Eichwald (1829).

In 1835, Dujardin corrected the error of d'Orbigny and placed the foraminifera in the Protozoa.

D'Orbigny (1826) created the first scientific classification of the foraminifera. This classification was based on visible morphological traits such as the disposition of the chambers and configuration of the test. His classification consisted of 5 families, 52 genera and 552 species. Later amplifications of this classification resulted in the erection of 7 foraminiferal orders (1852). This early classification, pioneering in its

Fig. 73. Eminent 19th Century scientists who studied the biology and distribution of Recent foraminifera.

time, has lost its significance since it was purely morphological and artificial. D'Orbigny did not consider phylogenetic or ontogenetic relationships as these were essentially unknown at the time. Consequently, various growth stages of the same species were interpreted as representatives of different orders.

In a natural classification, insofar as possible, the natural kinships between the units of classification, i.e. the phylogeny, ought to be considered. Unfortunately, the phylogeny can not be verified directly but only by examining both the ontogeny of individuals and the detailed stratigraphic relationships among fossil lineages. Despite the numerous studies of these sorts conducted in the last 100 years, our knowledge of the phylogenetic relationships in foraminifera is still sparse.

After the initial classification by d'Orbigny, numerous other classification schemes were proposed (Schultze, 1854; Reuss, 1861; Carpenter, Parker & Jones, 1862; Zittel, 1876; Schwager, 1877; Brady, 1884; Neumayr, 1887; Rhumbler, 1895, 1923; Eimer & Fickert, 1899; Lister, 1903; Douvillé, 1906; Schubert, 1907; Wedekind, 1913; Galloway, 1926, 1933; Cushman, 1927, 1948a; Chapman & Parr, 1936; Glaessner, 1945; Hofker, 1951b, 1956; Sigal, 1952; Le Calvez, 1953; Pokorny, 1958; Furssenko, 1959; Loeblich & Tappan, 1961, 1964). The continuing development and specialization of foraminiferal studies has led in the past few decades to the specialized taxonomic study of certain foraminiferal families or groups; e.g. Bermúdez (1952) on Rotaliidae; Stainforth (1952a) on uniserial calcareous forms; Avnimelech (1952) on unilocular tubular forms; Reiss (1957) on bilamellar forms, etc.

These classification schemes vary in terms of scientific merit. It would not be feasible nor meaningful to discuss each of them, but some of the more original and important deserve mention.

Reuss (1861) classified the foraminifera on the basis of the composition of the wall (chitinous, agglutinated, and calcareous) and the type of test construction. Several years earlier, Schultze utilized similar criteria but in a less harmonious form. Williamson also mentioned a three-fold division of wall type, hyaline, porcellaneous and arenaceous (agglutinated), in a short communication published in 1854. The classification of Reuss is simple and easily utilized and was generally accepted and used for many years.

The classification of Brady (1884), like that of Reuss, was also simple. He used test form, the number of chambers, and wall structure and composition as bases for his system. The order was divided into 11 families.

Neumayr (1887) was the first to attempt a phylogenetic classification.

Rhumbler (1895) also classified the foraminifera phylogenetically basing the categories on the physical firmness of the test.

Cushman was the first to publish a 'natural' classification. He based his scheme on his own extensive observations and those of the preceding investigators. The ontogeny of specimens was the principal trait on which Cushman based his classification. In his 1928 classification, he described 45 families and 413 genera. In the final edition of his work (1948), 50 families and 769 genera were described. Despite considerable controversy over certain parts of this classification, it stood for many years as the most advanced and most widely accepted one for foraminifera.

In recent years there have been two new significant classification schemes Furssenko, 1959; Loeblich & Tappan, 1964). Both of these are based on very extensive studies of existing collections; Furssenko concentrated primarily on material housed in the Soviet Union whereas Loeblich & Tappan concentrated primarily on material found in the United States and European institutions. These two new schemes have replaced that of Cushman as the most widely used and accepted.

According to Furssenko, foraminifera comprise a subclass of the Class Sarcodina of the phyllum Protozoa. He divided them on the basis of the following.

(a) For the living specimens: protoplasmic composition and characteristics of the life cycle were considered; e.g. the ameboidal character of the gametes of *Spirillina* resulted in the erection of the family Spirillinidae.

(b) For the fossil taxa, the chitinous test was regarded as the most primitive. The secreted and agglutinated tests were considered as parallel lines of evolutionary development. The most highly developed test type is the calcareous perforate form whose crystals show a preferred orientation. Septa are the result of rhythmic growth and are considered a progressive trait which appear simultaneously in several different phylogenetic lineages. The development of a regular spiral also occurs in different evolutionary branches. Consequently, Furssenko did not consider the spiral foraminifera as a natural group. Apertural configuration is considered a significant evolutionary feature and particular apertural types (radiate, toothed, etc.) are considered as indicative of natural groups. Supplementary skeletal features such as internal canals are considered as a characteristic possessed only by very advanced forms but nevertheless a trait independent of the different phylogenetic lineages. According to Furssenko, the immediate ancestors of foraminifera are the chitinous rhizopoda. The first foraminifera are representatives of Order Allogromiida. The hardening of the outer chitinous tegmen created the first rigid skeleton from which the Astrorhizida and Lagenida evolved. From these all the remaining orders of foraminifera developed. Furssenko distinguished 12 orders.

The views of Loeblich & Tappan on the systematic position and division of the foraminifera varied from that of Furssenko. They considered the foraminifera as an order of the subclass Granuloreticulosia, class Rhizopoda, phyllum Protozoa. The Rhizopoda were divided into 3 subclasses (Lobosia, Filosia and Granuloreticulosia) on the basis of the character of the pseudopodial protoplasm. Finer subdivision was made on the basis of the presence or absence of a test. Among the naked rhizopoda, division was based on the character of the protoplasm, ontogenesis, etc. Among the rhizopods with tests within the 3 subclasses, division was based on the composition and construction of the test walls. Family divisions were made on the basis of the disposition of chambers, the character of the aperture, ontogeny, test development, etc. The character of the gametes played an important role, but this trait is not well known among the foraminifera. Loeblich & Tappan (1964) distinguished 17 superfamilies and 79 families and increased this to 19 superfamilies and 119 families in a revised classification (1974).

There is no doubt that the foraminiferal classification will be further modified as new physiologic, evolutionary and detailed morphologic information is obtained. The creation of a natural classification is difficult, especially when many of the traits used in classification are morphological. Comparison of external similarities may lead to erroneous conclusions about phylogenetic relationships and the result may be the grouping together of forms which are only distantly related (isomorphs). Until recently *Elphidium* and *Nonion* were considered as closely related genera (within the same family) because of their similar morphology. Recent detailed studies of wall construction revealed that they are completely distinct and separated by a greater taxonomic distance than previously thought. The genera *Bolivina* and *Textularia* are well known, widely distributed taxa which, although isomorphous, are not closely related. They represent distinct evolutionary lineages, i.e. they are polyphyletic.

As a conclusion to this section on the history of classification it is interesting to note that the lack of certainty in classification has led some recent authors to list species alphabetically rather than according to their systematic position. In addition, it is interesting to see that more and more recent investigators are placing foraminifera in the categories established by Reuss (calcareous, agglutinated and chitinous).

All the preceeding classifications of foraminifera, from Linné to Loeblich & Tappan, have been based on what Mayr (1969) call an essentialist viewpoint in which the classification is an attempt to discover or reveal the true nature of things. All members of a taxon possess and reflect this essential nature. The recent departure from this taxonomic philosophy is seen in the numerical taxonomic approach (Sokal & Sneath, 1963; Sneath & Sokal, 1973) which assumes a nominalistic viewpoint in which only individuals exist and all classes are artificial. Two individuals are placed in the same taxon because they are similar, not because they have phyletic relationship.

4. Conditional classification

Aside from the classifications described above, which were established to treat and identify the natural relationships between all foraminifera, there is also an artificial, completely conditional classification which isolates foraminifera on the basis of their size: (a) 'large' foraminifera; (b) 'small' foraminifera; (c) 'microforaminifera'. The size boundaries of these three groups are not well defined although the majority of investigators who utilize these divisions generally consider 'small' foraminifera as 0.3–5 mm and 'large' foraminifera as 3–12 mm. These boundaries are not based on any detectable morphological difference, nor on any natural break in size-frequency curves. These size divisions are conditional and made for reasons of convenience.

The 'large' foraminifera are primarily defined as those possessing a complicated internal structure and are studied and described by special methods. They also tend to have a limited stratigraphic distribution. On the whole, they are larger than the 'small' foraminifera, but size is not the trait on which they are primarily defined. There are various genera and species of 'small' foraminifera which are larger than species considered as 'large' foraminifera.

The designation 'large' has no taxonomic significance at any level higher than species. Some genera belonging to the same family may be 'large' whereas others may be 'small'.

The best known 'large' foraminifera are the Fusulinidae, Miogypsinidae, Orbitoididae and Nummulitidae. Most are known only in fossil form; the Fusulinidae from the Pennsylvanian and Permian, and the Nummulitidae from the Paleogene, where they each play important biostratigraphic roles. There are 'large' foraminifera in Recent seas (*Borelis, Sorites, Marginopora, Heterostegina*, etc), but they are not particularly abundant.

The first work pertaining specifically to 'microforaminifera' did not appear until recently (Wilson & Hoffmeister, 1952), but there are isolated references to similar kinds of specimens in the works of the 19th Century, especially in the publications of Ehrenberg.

These small (20–200 μ) tests whose form is usually like that of the 'small' foraminifera are too small to be caught in the sieve sizes normally used in sample preparation. Consequently, they are often lost from a sample prior to its examination. They occur in samples ranging from Cretaceous to Recent.

The taxonomic status of the 'microforaminifera' is unclear. Wilson & Hoffmeister (1952) thought that the majority of them were megalospheric forms of existing genera. They did suggest that some forms which lack a large proloculus were new genera. The idea that the 'microforaminifera' included new taxa was a reasonably popular one at that time. However, three years later, Hoffmeister (1955) declared that these forms were not taxonomically distinct but were only megalospheric forms of pre-existing taxa. The same conclusion was reached by Echols & Schaeffer (1960). A detailed statistical study led Popper (1958 Ms) to conclude that the 'microforaminifera' are immature forms of a variety of species which have suffered a very high mortality rate.

Most 'microforaminifera' were discovered to be chitinous forms and most descriptions of them were based on samples which had been treated with hydrochloric or hydrofluoric acid. It is probable that the 'chitinous' tests are in fact the internal organic membranes of calcareous or agglutinated forms. Such forms were frequently encountered in material treated with acids by many investigators (Deflandre, Eisenack, Pastiels, Wetzel and others). A varety of fossil calcareous species were deliberately treated with hydrochloric acid by Taugourdeau-Lantz & Poignant (1964) and numerous 'chitinous microforaminifera' were produced. Cohen & Guber (1958) produced 'pollen-sized microforaminifera' from 'normal' sized *Ammonia limnetes* by dissolving the calcareous test in dilute hydrochloric acid. The implication of all of these studies is that the 'microforaminifera' may be nothing more than immature specimens of ordinary sized species or the organic membrane of species whose test has been dissolved.

5. Morphological traits and their taxonomic value

The particular kinds of criteria which are used to distinguish the different taxonomic categories will be discussed in this section. The principal criteria are (a) geochronologic, (b) geographic, (c) ecologic and (d) morphologic.

A. Geochronologic criterion

The geochronologic criterion is of great value because it allows the establishment of the correct phylogenetic succession. Without an understanding of the temporal position of species serious taxonomic errors will result as a consequence of assigning morphologically similar but phylogenetically unrelated (convergent) groups to the same taxon. In addition, paleontologic studies yield interesting and useful data about the past geographic and ecologic habitats of living species.

B. Geographic criterion

The geographic criterion plays a significant role in that those organisms which develop in different geographic regions may be similar or very distinct. Specimens of the same species may have different morphological traits in distinct locales. Zoologically this isolation leads to the development of two 'geographic races;' taxonomically this divergence is expressed as two subspecies. Under conditions of continued isolation and development these two subspecies may evolve into two independent species.

C. Ecologic criterion

The ecologic criterion is important since the distribution of organisms and their morphologic variation are often due to ecologic conditions. As taxonomic problems become resolved it becomes useful to understand the various environments and their different ecological factors in order to better understand the conditions under which the phylogenetic evolution of the group occurred. When the ecologic conditions are well understood, it becomes possible to explain the fine detail in the studied fauna. As shown in Chapter IX, the ecology of foraminifera is complicated, and in many cases neither well studied nor well understood.

D. Morphologic Criterion

The morphologic criterion is the principal one from a taxonomic point of view. The morphology of an individual is always present and visible; data on the other criteria may not be. Morphologic characters may easily serve as traits to distinguish between taxonomic units. The morphology of an individual may yield information about the other criteria. In fact, the morphology of an individual is a consequence of the influence of the phylogenetic history (genetics), the geographic isolation, and the environmental conditions. It is necessary to study not only adult forms but also the morphology of juvenile stages for their ontogenetic information. All other biologic information available should also be used in conjunction with morphology. Unfortunately, in the case of foraminifera, this additional data is somewhat sparse and the morphologic criterion bears most of the burden in taxonomic discrimination.

Any morphologic characteristic of a test which might be used to distinguish the test may be used as a taxonomic trait. The number of such traits is very great but not all have the same value and significance.

Morozova (1956) calculated that a specimen of *Truncorotalia aragonensis* possesses 200 such morphologic traits, 130 of which are common among 50 specimens of this species from the same sample. About 50 traits are common among specimens of this species taken from the same stratum (10 samples). Only 20 traits are common among all specimens of this species taken from different zoogeographic provinces. Thus, of the 200 traits only 20 proved useful in the taxonomic discrimination of this species. These 20 traits are the 'taxonomic traits', i.e. the ones which are useful in systematics.

The number of taxonomic traits which can be used at a particular taxonomic level decreases at the lower levels. In the example cited above, 20 traits could be used to distinguish *T. aragonensis* from other species of the same genus. Fewer traits, perhaps 10, could be used to distinguish subspecies, and only one or two are used to distinguish *forma*.

Different taxonomic traits have different values among the various groups of foraminifera. The importance of a particular trait may vary from group to group. It is not possible to correlate a particular change in trait to a particular taxonomic level. There are, however, some general traits that are used at certain taxonomic levels.

family traits: mineralogical composition of test; general plan of chamber disposition;
generic traits: general form of test; wall construction; location and character of the aperture;
species (subspecies, variety) traits: number of chambers; character of supplementary skeleton; character of periphery;
form traits: general ornamentation.

Attention has been given recently to the size and spacing of pores and to the nature of the tooth plate. Nevertheless, there is no general agreement among investigators with respect to these features and there is some confusion as to their proper interpretation. For example, some investigators consider pore diameter and distribution to be a high level taxonomic trait; Hofker is perhaps the most fervent advocate of this trait. He created a porosity index which expresses numerically the size of the pores and the number per unit surface area. This index was employed by Hofker in his classification of the Globigerinidae. The taxonomic value of this index becomes suspect when it is observed that pores of different diameter can be found on adjacent chambers in the same specimen. Pore diameter seems to change in many species as a function of age. The porosity (diameter and density considered together) of some planktonic foraminifera appears to be dependent on water temperature (Chapter XIV). Pore spacing increases with water depth among some benthonic taxa (Stapleton, 1973). Zobel (1968) used pore diameter as a means of separating juvenile specimens of *Globigerina pachyderma* from *Globigerina dutertrei* but found that this tool only worked in certain areas. It may be that pore diameter is a taxonomic trait, but a great many more observations must be made before its importance can be fully determined.

Traits such as the type of construction of the supplementary skeleton, the nature of the septa, and the details of the umbilical region all have a relative importance. In some foraminifera, they are constant and singular traits; e.g. as the chomata in the Fusulinidae, but they may also be identical in unrelated groups, as are the pillars of the Nummulitidae and Discocyclinidae.

Many exceptions exist in the list of taxonomic traits and their levels given above. The wall composition is interpreted as having great taxonomic value. However, under certain ecological conditions this trait may vary within a species. The Miliolidae which live in low salinity situations often have chitinous walls rather than calcareous ones. The microstructure of the walls of different specimens of *Cibicides* were found to be different; radiate calcite in the majority of cases, but microgranular in some (Wood & Haynes, 1957). Hansen (1972a) noted that two species of *Turrilina* whose morphology is quite similar had a granular wall and a radiate wall respectively. For these cases the trait proved to be useful at less than the generic level.

Many features may vary considerably as a consequence of environmental conditions. The disposition of the chambers can not be considered a trait useful at a high taxonomic level in all species since the chamber position in many attached benthonic forms is controlled by the substrate. There are also many examples of convergent evolution in chamber disposition; e.g. the biserial arrangement in *Bolivina*, *Textularia* and *Gumbelina*, and the trochospiral arrangement of *Rotalia* and *Trochammina*.

The phylogenetic development of a group seems to have an influence on the value of the traits used to identify it. The taxonomic significance of a trait appears to increase as the group evolves. At the outset of development the features are more plastic, more variable and less distinctive; they later stabilize. The investigations of Miklukho-Maclay, Rauser-Chernoussova & Rozovskaja (1958) on Fusulinidae and of Reitlinger (1958) on Endothyridae demonstrated how the taxonomic value of some characteristics varied in the course of evolution. Lipina (1950) and Cifelli (1960) also noted the instability of morphologic traits during the first stages of evolution of the foraminifera.

In summary, we wish to suggest that a particular taxonomic trait may prove very useful in certain groups at particular stages in their evolution but may not be equally valuable in all groups or at all times. When using the list of traits given above, each case must be treated separately and must be examined for the various exceptions that may exist.

6. Problems of variability

The problem of morphological variability in the foraminifera is directly related to the problems of classification and to the value which is given to the various taxonomic traits.

The study of variability is important as it is essential to an understanding of the boundaries between species. Moreover, understanding variability may lead to a better knowledge of the origins of species and evolutionary pathways.

Variability may be classified by its cause. Among the foraminifera, variability is: (a) dependent on the reproductive cycle; (b) dependent on age; (c) dependent on environmental changes; (d) dependent on individual genetic variability.

Each of these will be treated separately.

(a) The morphological variability that results from alternation between sexual and asexual reproductive modes is discussed in Chapter II,2.

(b) Age dependent variability is that which occurs as an individual ages. To a greater or lesser extent it occurs in all species. In some taxa the differences between juvenile, adult and gerontic individuals is clear whereas in others only detailed studies reveal the differences. Some foraminifera demonstrate great morphological changes during their ontogeny; e.g. *Pyrgo* passes through a quinqueloculine stage; the cosmopolitan *Quinqueloculina seminulum* exhibits a tendency toward triloculine chamber arrangement when fully mature; specimens of *Q. artica*, upon ageing, become greatly rounded; young specimens of *Pullenia* possess depressed sutures whereas in older specimens the sutures have become raised to the same level as the wall; specimens of *Uvigerina* demonstrate a change in ornamentation as they age; and the limbate and curved sutures of a mature *Eponides repandus* become simple and straight with old age. Various species change the configuration of their aperture, loose the ornamentation on the final chamber, etc. The manifestations of age changes are quite diverse, but from a taxonomic point of view, usually easily recognized.

(c) Variations caused by environmental conditions are much more significant. A large part of Chapter IX is devoted to morphological changes controlled by environmental conditions and so we will not repeat these observations here, except to say that these kinds of variations have resulted in the erection of many invalid taxa.

(d) In order to understand individual variation it is necessary to understand the limits of variability of the different morphological traits which an organism posesses under normal life conditions. In practice this is difficult, if not impossible, to do. We know so little about the life history and behaviour of the foraminifera that it is difficult to know if conditions are normal or not.

Very slight changes in the environment may provoke modifications in test morphology. These environmental changes may be so small as to be undetectable by normal measurement or analysis. Thus, when we describe the variability of various traits, we speak less of the origin or cause of that variation than of its amplitude. Prior to describing the particular variations which exist, it is appropriate to give a brief resume of some of the ideas that have been expressed about variability.

According to Keijzer (1936), Batsch was the first to note and take into account the great morphological variability among foraminifera. Among the great investigators of the 19th Century, Brady was an adherent of the same point of view. As the author of numerous works on foraminifera, both fossil and Recent, and of the classic monograph on the *Challenger* Expedition foraminifera, he had the opportunity to study a great quantity of material from all parts of the world, and he made numerous valid observations about morphological variability. He observed that some forms, e.g. *Peneroplis*, *Spiroloculina*, etc., are characterized by great variation whereas others are much more stable.

Of the later investigators who noted extensive variability among foraminifera, it is necessary to mention Schaudinn (1895c) who described *Calcituba polymorpha* as a species lacking a determined form, and Dreyer (1898) who called attention to the great variability in *Peneroplis*, variability reconfirmed by the studies of Heron-Allen & Earland (1915). The list of those species exhibiting considerable variability was greatly enlarged during this period because the foraminifera were being studied primarily by zoologists, whose natural tendency is to look for, identify and place within the species a wide variety of morphologies. They were helped in their studies by the abundance of material at their disposition. According to the North American viewpoint, these 19th Century investigators would be called 'lumpers' as they tended to allow the presence of a great deal of variation within a given taxon.

This same broad viewpoint dominated during the first quarter of the 20th Century. The best known adherents of this perspective were Vaughan (1923) and Diener (1925) both of whom felt that foraminifera, because of their great variability, could never be used for the determination of fine biostratigraphic diversions. Up to this time foraminifera were being studied by a relatively small number of persons most of whom considered the foraminifera as highly variable organisms.

Nevertheless, beginning in the second quarter of this century, thanks to the growth of the petroleum industry, the study of foraminifera began to assume great importance. Numerous investigators discovered that morphological differences between forms could be used with excellent results in biostratigraphy. The foraminifera were studied extensively by stratigraphers whose interpretation of the species was much more strict. Instead of viewing the group as one with great variability, the stratigrapher's purpose of correlation created a viewpoint in which the greater the number of taxa, the finer the possible stratigraphic subdivision. If the zoologists of the 19th Century and early 20th Century were 'lumpers', the new breed of taxonomists were 'splitters'. Slight changes in morphology were considered sufficient for the erection of a new taxon. Using these new taxa the stratigraphers were able to subdivide geologic strata very precisely. A great deal of taxonomic damage was done by this strict interpretation as an enormous number of invalid names were created. The increase in synonyms was so great that if new names continue to be added at the same rate as during this period, the foraminiferal nomenclature will be overwhelmed with synonyms. This in turn would deprive the foraminifera of their significance in the natural sciences. Various investigators began to call attention to this 'splitting' and pointed out that the living foraminifera are more variable than the current nomenclature indicates and that the species ought to be interpreted more amply. Arnold (1968), after summarizing many culture studies, stated that the species variation potential is generally much greater than that recognized by paleontological taxonomists. Many authors began to accept the idea that some morphologic traits can vary greatly without exceeding the limits of a species. Typical of these works was that of Ovey (1938) on the variability of coiling and Bartenstein & Brand (1938) who described the variable aperture in *Jadammina polystoma*.

After World War II, observations which utilized the great variability of foraminifera began to appear in considerable quantity. The interpretation of the species had again changed and was again considered in its more ample form.

The most important of these investigations, arranged by particular morphological features, will be briefly discussed below.

A. Size

As a distinguishing taxonomic trait, size is of limited value. Adult specimens of the same species from the same place often vary in size by a factor of two, three or more. For instance, LeRoy & Levinson (1974) found that *Hoeglundina elegans* exhibited a size variation from 0.35–1.83 mm. These size differences may be even greater if specimens from different sites are compared. Resig (1958) described specimens of *Eggerella advena* ranging in size from 60–340 μ. Hofker (1932) found various specimens of Astrorhizidae to be 20 times larger in the Bay of Naples than in Arctic waters.

B. Aperture

The long held view that the aperture is an invariate trait and that its position and character are constant, is no longer accepted. The work of Bartenstein & Brand (1938), mentioned above, demonstrated that the aperture may vary considerably, often in response to environmental conditions. These authors described the main aperture of *Jadammina polystoma* as having the form of a curved slit and situated at the base of the apertural face. In addition to this opening, there are from 3–7 additional ones distributed on the test in an irregular manner.

The aperture of *Cribrostomoides* was described by Frizzell & Schwartz (1950) as varying from a single fissure to a series of separate circular orifices.

Arnold (1953b) described two different types of aperture occurring in the Miocene genus *Uvigerinella*. Different types of apertural variations were described for *Ataxogyroidina variabilis* by Barnard & Banner (1953) and for *Hantkenina alabamensis* by Barnard (1954).

Bowen (1957) discussed the aperture as a taxonomic trait and described the confusion that results when the aperture is considered as a significant means of discriminating species. He described several cases of apertural variation within a species.

The multiple and variously situated apertures of the brackish water genera *Ammoastuta*, *Haplophragmium* and *Trochamminita* of Trinidad were described by Saunders (1957) and those in another brackish water form, *Nonion tisburyensis*, by Boltovskoy (1958a).

Krasheninnikov (1958b) did not ascribe much importance to the aperture, particularly in certain groups (Nonionidae and Elphidiidae). After examining specimens from the Miocene of Podolia, from the area north of the Caucasus, and from Recent sediments of the Barents and Okhotsk Seas, he concluded that the aperture in various species of these families was quite variable; cribrate, uniserial or consisting of a simple slit at the base of the apertural face. He asserted that the aperture has no taxonomic importance, but only reflects ecological conditions, particularly the degree of contact required by the organism with the environment; when less is needed, the apertural opening is diminished, and the slit becomes a line of small circular openings.

Kuznethova (1960) cited the *Lenticulina–Robulus* group as an interesting example of apertural variability. The slit-like aperture in Jurassic forms exhibited little variation whereas in Tertiary and Recent forms it showed variability and diversity.

Many of the observations of increased variability in aperture are in brackish water specimens. The individual specimens may be responding morphologically to the lower salinities by altering the character of the aperture.

Studies of cultured specimens have revealed some interesting observations on apertural variability. Arnold (1953a) cultured *Allogromia laticollaris* and noted that not only the typical single aperture form was present but that specimens with multiple apertures were produced. The number of apertures did not seem to depend on the environmental conditions nor did the lineages of single aperture forms or multiple aperture forms persist. Cultured specimens of *Nemogullmia* did not show any consistency of apertural form (Nyholm, 1956).

Not all taxa exhibit apertural variation. Some species are characterized by an apertural position and shape which varies little from specimen to specimen. Representatives of the genus *Bulimina* exemplify this stability.

C. Shape

1. *Coiled forms*

Dreyer (1898) demonstrated that *Peneroplis pertusus* exhibits specimens which are tightly coiled, partially uncoiled, or uncoiled. Similar continuous intergraduations of form were described by Ovey (1938). He observed that *Cristellaria laevigata* (tightly coiled) graded into *Marginulina* whose juvenile stage is coiled but whose mature form is uncoiled.

According to the observations of Hussey & McNulty (1950), specimens of *Planularia planotrochiformis* show all gradations from perfectly planispiral to strong trochospiral coiling. These differences in coiling are often considered generic taxonomic traits.

Dihdkov'skij (1959) observed the extraordinary diversity in coiling forms among the fossil Peneroplidae. He correlated the degree of coiling with water depth and turbulence, suggesting that uniserial tests occupied deeper, less dynamic environments. Studies of the coiling and apertural diversity among Recent *Peneroplis* (Sellier de Civrieux, 1970a) support the observations that this genus is extremely plastic morphologically.

2. *Non-coiled forms*

Studies of cultured foraminifera have often revealed a great diversity in their shape. The studies of *Allogromia laticollaris* (Arnold, 1954b) and *Gromia oviformis* (Hedley, 1962a, see Fig. 74) reveal that these species possess enormous variation potential.

Hendrix (1958) suggested that the test form was affected by sedimentation conditions. He said that the wall thickness, the shape of the cross-section, and the character of the peripheral margin depended on environmental conditions. If water

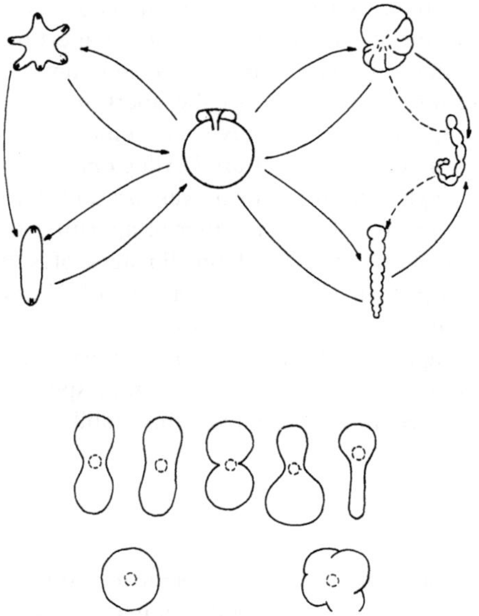

Fig. 74. Morphological variations in *Allogromia laticollaris* (after Arnold, 1954b) and *Gromia oviformis* (after Hedley, 1962a).

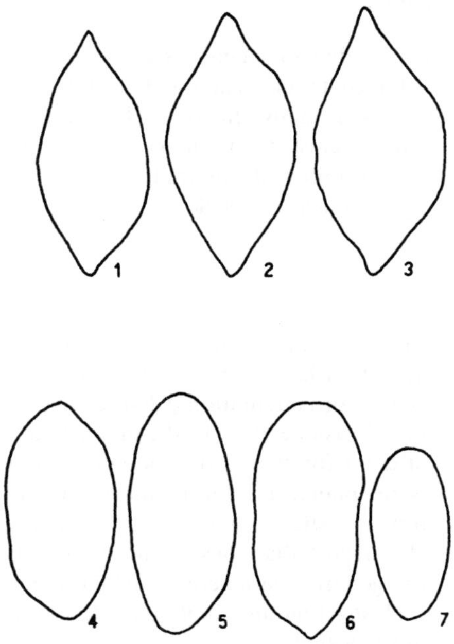

Fig. 75. Variations in the longitudinal profile of *Elphidium macellum* (after Krasheninnikov, 1958b).

density is low and there is little agitation, the test will take on a shape which will increase its buoyancy; with elevated densities and/or water agitation the tests will be more solid and compact.

Krasheninnikov (1958b) demonstrated the variation in the outline of *Elphidium macellum* in a series of transverse sections (Fig. 75). The morphological variation under natural conditions of a variety of taxa has been examined. Particularly useful are studies on the following foraminifera: *Tinoporus* (Keijzer, 1936), Cibicidinae (Dupeuble, 1962; Cooper, 1965), Peneroplidae (Mateu, 1969b) and *Technitella* (Haman, 1971b). Brodniewicz (1965) described changes in morphology in several unilocular species that were the result of salinity and sediment control.

D. Bilateral Symmetry

The asymmetric character of some tests is certainly due to individual variation resulting from environmental conditions. These asymmetrical shapes are quite common in specimens of *Nonion, Robulus, Elphidium,* etc. and are clearly of no taxonomic significance. Various *Nodosaria* are asymmetrically curved and consequently placed in the 'genus' *Dentalina.* Nevertheless, asymmetry may be a fixed, stable trait in some cases and therefore be of taxonomic value, e.g. in *Nonionella.*

E. Number of Chambers

The significance of the number of chambers can not be considered independently from the disposition of the chambers. The number of chambers is not fixed in a given taxon, especially in those with many chambers. It is difficult to determine the limits of variation for this trait; each taxon must be considered separately. Various Rotaliidae illustrate the variability of this trait; in them the number of chambers in the last whorl may vary as much as 100%.

F. Ornamentation

Ornamentation varies greatly and is usually not considered a high level taxonomic trait. Although ornamentation has been used to distinguish new genera, studies of the environmental effects on ornamentation (Chapter IX) indicate that even its use at the species level may not always be wise. In the majority of cases, ornamentation can be used to separate different forms and occasionally subspecies.

Although the spines on planktonic foraminifera have often been considered as ornamentational features, investigations by Bė (1969b), Walker & Vilks (1973) and Vilks & Walker (1974) suggest that these features may be controlled less by the environment than by the generic characteristics of the taxon. Taxonomic discrimination between *Orbulina*, *Hastigerina* and *Globigerinella* on the basis of spine morphology has therefore been suggested.

G. Thickness and General Aspects of the Test Wall

The thickness of the test wall varies considerably in some taxa; e.g. in *Gavelinella intermedia* or in *Orbulina universa*. According to Bé, Harrison & Lott (1973), the wall thickness of the latter species varies from less than 2 μ to over 40 μ. The relative transparency of the test wall also appears to differ among specimens of the same species. This phenomenon may be due to etching of tests in the substrate by acidic solutions (Murray, 1967b; Murray & Wright, 1970) and thus is of no taxonomic importance.

H. Variability of Several Traits Taken Together

The preceding review has dealt primarily with investigations which were concerned with the variations in a single trait. There are also investigations dealing with the totality of morphologic variation. The work of Friese (1951) demonstrated the complete intergradation of the morphological traits of different fossil species of the genera *Elphidium*, *Nonion* and *Robulus*. Numerous other studies dealt with morphological variation in fossil (Bhatia, 1965b; Krasheninnikov, 1958b; Saidova, 1960b; Cifelli, 1960; Adams, 1962; Burdett *et al.*, 1963; Sigal, 1966), Recent (Resig,

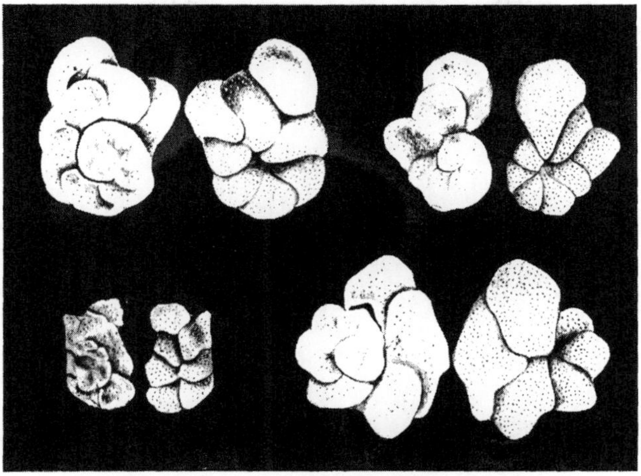

Fig. 76. Morphological variations of *Cibicides lobatulus* (after Nyholm, 1952).

1962; Le Calvez, 1963; Bartlett, 1964a; Lutze, 1964b; Moulinier, 1966; Rosset-Moulinier, 1971; Murray, 1966a; Brooks, 1967; Closs & Madeira, 1968; Sliter, 1970a; Feyling-Hanssen, 1972) and cultured species (Arnold, 1954a, 1964, 1967a,b; Schnitker, 1967).

Noteworthy among the culture studies are the investigations of Nyholm (1961, 1962, 1973) who cultured specimens of *Cibicides lobatulus* (Fig. 76), *Gypsina* and *Acervulina*. He was able to simulate natural conditions much more closely than were previous investigators. The individual variation of specimens of the same species were so extreme that Nyholm concluded that many genera, e.g. *Planorbulina, Cymbalopora, Dyocibicides, Cyclocibicides* and *Webbinella* are ontogenetic stages (or substrate controlled adaptions) of the same species, *Cibicides lobatulus*. He concluded that *Gypsina* and *Acervulina* were also stages of *Cibicides*.

From all these studies it becomes evident that the foraminifera are, for the most part, highly variable. Many species should be interpreted more broadly than they have been by the 'splitters'. There are, however, various species which have demonstrated an extraordinary stability in their morphology over a long period of geologic time. Boltovskoy (1958b) demonstrated that Recent specimens of *Nonion affine* do not differ morphologically from Eocene specimens of the same species. Although these species may conserve their morphologic traits across millions of years, they do often exhibit great geographic variations. These variations in *N. affine*, for example, are a result of variability in environmental conditions. The majority of these variations are of an intraspecific nature and a change of environmental conditions from one state to another results in the production of a different morphological state.

IX. ECOLOGY

Ecology is the study of the reciprocal action between organisms and their surroundings; the influence of the environment on the physiologic activity of the organism and *vice versa*. Ecology thus embraces the study of the environmental limits within which a species exists. These limits are often initially identified as areal limits of distribution; areas in which the species is able to survive; smaller areas in which the species can live and grow; and yet smaller areas in which the species can live, grow and reproduce.

A large number of interrelated factors influence the vital activity of foraminifera and their distribution. Factors such as temperature, salinity and water depth have been extensively studied; others such as nutritive material and the nature of the substrate are less well understood. A variety of other factors, among which are light intensity, turbidity, pH and the concentration of oxygen, calcium carbonate and trace elements, appear to be important but the precise nature of their influence has not been greatly studied nor understood.

The areal limits of a species can be defined in different ways. For example, if species A has temperature reproductive limits of 20–25 °C and salinity reproductive limits of 30–36‰, the reproductive activities of the species will be confined to areas which meet these two constraints (see dotted area in Fig. 77). However, there are

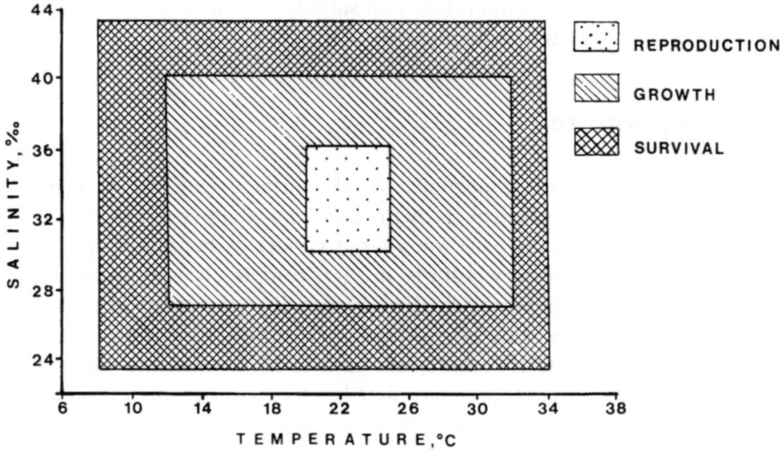

Fig. 77. Hypothetical survival, growth, and reproductive regions for a two-factor, non-interacting ecological system.

broader tolerance limits controlling the growth of an individual (for our hypothetical example, let us say 12–32 °C and 27–40‰) which define a wider area of habitation in which there are certain restrictions on vital activity. There are, for most species, yet wider tolerance limits for mere survival (without growth and reproduction). If these are found to be 8–34 °C and 23–43‰ we can define yet another area in which the species can be found, although with even greater vital restrictions.

There are two reasons why this hypothetical model is unreal. First, there are more than two ecological factors which determine the distributional limits of a species. For this reason the diagram should be drawn in n-dimensional space where n is the number of factors that influence foraminiferal distribution. This sort of diagram is impossible to imagine geometrically, but, in fact, much truer to reality. Second, the ecological factors are not acting independently. Experiments conducted on marine invertebrates (Kinne, 1963) indicated that variation in one ecological factor (temperature, salinity, oxygen, trace elements, etc.) may alter the effects of others. The available data on tolerance limits is usually derived from culture studies where these limits are determined for a fixed set of conditions. The information that species A can reproduce between 20–25 °C was determined under a given set of salinity, pH, substrate, illumination, dissolved oxygen and nutritional conditions. Were any of these changed, the temperature limits might also change. This means that the rectangular shaped boxes in Fig. 77 should probably be replaced by ellipsoids to more nearly represent the interaction that takes place between factors. The data shown in Fig. 125 (Chapter XIV,15) for the survival limits of three planktonic species is more realistic of the kinds of tolerance limit interaction encountered in nature.

Although it is quite clear that various ecological factors are not independent of one another; i.e. light and temperature are related to water depth, and calcium carbonate solubility is related to temperature and salinity, we shall, in the interest of simplicity, treat the various factors individually.

1. Temperature

Temperature is of great importance for all marine organisms. By studying foraminifera in cultures, the survival, growth, and reproductive temperature limits can be determined with relative ease for a large variety of benthonic species.

Conditions in laboratory cultures are not completely normal but there are factors which partially compensate for this abnormality, the main one of which is that a given ecologic factor can be isolated and varied as desired. Culture studies of foraminifera are relatively scarce, but have yielded some interesting results. Table 9 shows the various tolerance limits that have been determined by several investigations on benthonic species. Significant factors which are not included in this table are the length of time a given species can endure a particular temperature and the value of other ecological parameters. The temperatures cited by Bradshaw (1961) are the

Table 9. Temperature tolerance limits for cultured benthonic foraminifera

		Temperature (°C)				
Species	Author	Survival	Growth	Reproduction	Growth	Survival
Allogromia laticollaris	Lee & McEnery, 1970		15		30	
Allogromia laticollaris	Arnold, 1955a			14–32		
Allogromia sp.	Phleger, 1960b	2				38
Bolivina compacta	Bradshaw, 1961					41
Bolivina doniezi	Sliter, 1970b		10	18–22	26	
Bolivina vaughani	Bradshaw, 1961					37.4
Bulimina sp.	Myers, 1935a			<18		
Discobotellina biperforata	Stephenson & Rees, 1965					31.5
Discorbis sp.	Myers, 1935a			<18		
Massilina sp.	Bradshaw, 1961					40
Patellina corrugata	Myers, 1935b			<18–21		
Pyrgo sp.	Myers, 1935a			<18		
Ammonia beccarii tepida	Bradshaw, 1961	–2	15	20–32	35	45*
Rotaliella heterocaryotica	Bradshaw, 1955			14.5–24.5		26
Rotaliella sp.	Bradshaw, 1955			14.5–22.2		26
Rosalina columbiensis	Bradshaw, 1961					38.5
Rosalina globularis	Sliter, 1965			16–25		
Spirillina vivipara	Myers, 1936			18–21	26	
Spirillina vivipara	Bradshaw, 1961					39*
Spiroloculina hyalina	Arnold, 1964			14–24		
Tretomphalus sp.	Myers, 1943c			18		
Triloculina sp.	Myers, 1935a			18		

* see Fig. 78.

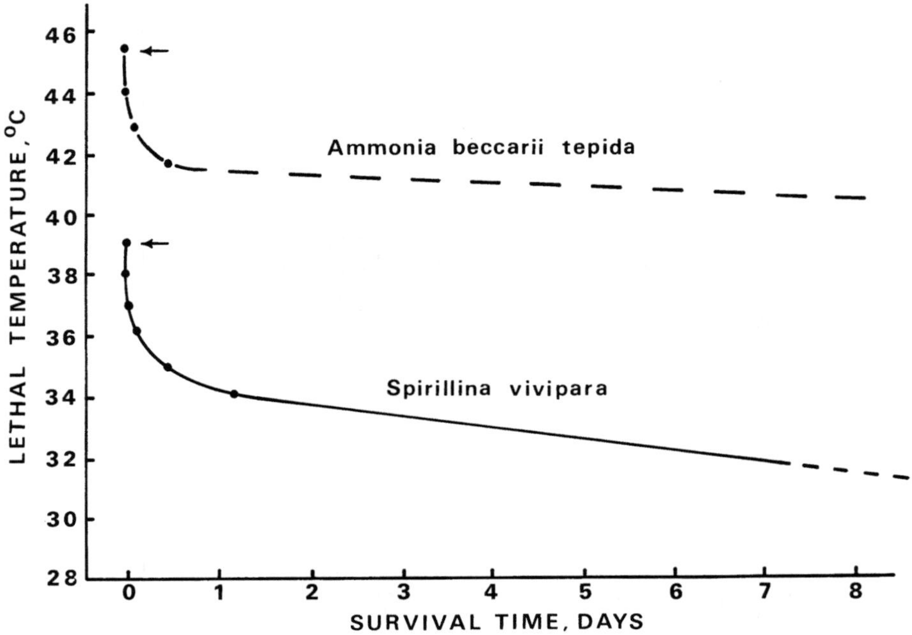

Fig. 78. Survival as function of temperature (after Bradshaw, 1961). The curve indicates the conditions under which 50% of population survived. Arrows indicate instantaneous death of entire population.

maximum that can be tolerated for any length of time. He noted that at slightly lower temperatures the specimens could survive for a longer period but would eventually die (Fig. 78). Schnitker (1974a) indicated that the temperature limits of reproduction may not be fixed for a given species but may vary as a consequence of local thermal conditions. He noted that reproducing specimens of *Ammonia beccarii* kept under varying thermal conditions in cultures respond by shifting the temperature at which they begin to reproduce. These observations suggest a thermal acclimation during the early stages of growth of the individual.

Although most temperature data has been obtained from cultures, some field studies have been conducted which indicate the ranges under actual conditions. One of the major problems with data collected from the field is that one does not know if the temperature ranges occurring where living specimens are found represent estimates of survival, growth or reproductive limits. For that reason the data shown in Table 10 should be interpreted conservatively as representing survival limits.

From the limited amount of data shown in Tables 9 and 10 it is evident that our knowledge of the thermal characteristics of benthonic foraminifera is sparse and

Table 10. Temperature tolerance limits for *in situ* benthonic foraminifera

Species	Author	Temperature (°C)	
Alveolinella quoii	Banner, 1971	26.0	
Ammonia beccarii batava	Murray, 1968a	15.0	20.0*
Buccella frigida	Murray, 1968a	16.0*	
Buliminella elegantissima	Boltovskoy & Lena, 1969b	3.8	14.5†
Eggerella scabra	Murray, 1968a	15.0	20.0*
Elphidium articulatum	Boltovskoy & Lena, 1969b	3.8	20.0
E. articulatum	Lutze, 1968a	0.0	20.0
E. excavatum	Murray, 1968a	10.0	15.0*
E. excavatum	Voorthuysen, 1960	−1.8	24.0
E. excavatum	Boltovskoy & Boltovskoy, 1968	7.0	25.0
E. gunteri	Voorthuysen, 1960	−1.8	24.0
E. gunteri	Boltovskoy & Lena, 1969b	3.8	14.5†
E. oceanense	Murray, 1968a	10.0	20.0*
E. selseyense	Voorthuysen, 1960	−1.8	24.0
Epistominella exigua	Boltovskoy & Lena, 1969b	3.8	14.5†
Miliammina fusca	Murray, 1968a	10.0	20.0*
Miliammina fusca	Wright, 1968	7.0	25.0
Nonion depressulus	Voorthuysen, 1960	−1.8	24.0
Nonion depressulus	Murray, 1968a	5.0	10.0*
Quinqueloculina dimidiata	Murray, 1968a	17.0	20.0*
Q. milletti	Wright, 1968	7.0	25.0
Q. seminulum	Wright, 1968	7.0	25.0
Reophax moniliformis	Murray, 1968a	15.0	20.0*
Rotalia beccarii	Boltovskoy & Lena, 1969b	3.8	14.5†
Rotalia beccarii ex gr. *parkinsoniana*	Boltovskoy & Boltovskoy, 1968	7.0	25.0
Streblus batavus	Voorthuysen, 1960	−1.8	24.0

* cited as optimum range; may represent growth or reproductive limits
† probably the reproductive limits

incomplete. Considering that there are probably more than 1000 living species, the data available on these few dozen looks very insignificant.

The temperature tolerance limits of planktonic foraminifera are better known. Although there are no laboratory observations on the relation between temperature and vital activity, our knowledge of this relationship under natural conditions is rather good. The detailed work of Bé & Tolderlund (1971) and Tolderlund & Bé (1971) provided a great deal of data on surface water temperature tolerance limits and on optimum temperatures. Table 11 shows data from these investigators for the Atlantic and Indian Oceans and additional data from the South Atlantic and its Antarctic sector, and the South Pacific (Boltovskoy, 1966d, 1969a, 1970b).

The role of temperature appears to be twofold: affecting the distribution and the morphology of the organism.

Table 11. Temperature tolerance limits for planktonic foraminifera (surface water temperature, °C)

Species	Atlantic and Indian Ocean (Bé & Tolderlund, 1971; Tolderlund & Bé, 1971)	S. Atlantic and its Antarctic sector (Boltovskoy, 1969b, 1970b, and unpublished data)	S. Pacific and its Antarctic sector, minimum temp. (Boltovskoy, 1966d)
Globigerina pachyderma (sin.)	0–24	−1.4–23.5	
G. pachyderma (dex.)	0–24		
G. pachyderma, f. superficiaria			−1.38
G. quinqueloba	1–21	−1–25	
Globigerinita uvula		−0.7–25	−0.58
Globigerina bulloides	0–27	−0.7–25.1	−0.48
Globorotalia inflata	1–27	−0.4–25	1.05
Globigerinita glutinata	2.5–30	0.3–29	0.69
Globorotalia truncatulinoides (sin.)	4–27	5–23.4	2.9
G. truncatulinoides (dex.)	4–27	5–22.2	2.9
G. crassaformis	6–27		
G. scitula		8–22.5	8.1
Globoquadrina dutertrei	9–30	18–29	
Globorotalia hirsuta	10.5–26.9	9.3–27	
Orbulina universa	10–30	10.5–29	
Globigerinella aequilateralis	12–30	14.5–25	
Globigerinoides ruber	14–30	10.4–29	
Hastigerina pelagica	14.4–29.5	15–29	
Globigerinoides trilobus, f. typica	15–30	13–29	
G. trilobus, f. sacculifera	15–30	18–29	
G. conglobatus	15–30	20–29	
Globorotalia menardii	16–30	10.8–29	
Globigerina rubescens		17–29	
Globoquadrina conglomerata	17–29		
Candeina nitida	20–30		
Globoquadrina hexagona	25–30		

A. Influence on Distribution

Because different species have different tolerance limits, the geographic control of foraminiferal distribution is evident. The zonation into biogeographic provinces by temperature is discussed in Chapter IV,2 (Fig. 42) and Chapter VII,2.

It is worth noting that not only the absolute temperature but the range and rate of change of temperature may be instrumental in determining distribution. Both Natland (1957) and Cockbain (1963), studying the West Coast of North America, were able to faunally discriminate between areas which had different seasonal ranges of temperatures. Benda & Puri (1962) noted that the greatest populations of foraminifera and ostracoda in the Eastern Gulf of Mexico occurred where the temperature and salinity fluctuation was high.

A far more complicated situation is encountered in the relationship between temperature and vertical distribution. It is difficult to distinguish whether the decrease in temperature with depth is the principal cause of bathymetric zonation or whether other depth controlled factors are predominant (see section 3). An examination of the literature reveals a considerable difference of opinion although very few of the arguments seem particularly well supported. It seems clear that in a natural environment it is difficult, if not impossible, to determine the precise causes of bathymetric zonation.

Norton (1930) was among the first to discuss depth zonation and temperature. In a study of the waters of Florida and the West Indies he described 4 faunal zones based on temperature and depth. This zonation was quite general, and it treated families rather than genera or species. Unfortunately, the author considered only empty tests rather than those living at the time of collection. Since the appearance of Norton's work a large number of investigators have studied the vertical distribution of benthonic foraminifera in a great many geographic areas. Among those authors who have produced a large quantity of bathymetric and temperature data are Bandy, Butcher, Caralp, Crouch, Hooper, Lankford, Loep, McGlasson, Natland, F. Parker, Phleger, Pujos, Streeter, Stschedrina and Walton. Phleger and Parker were the first to introduce annual variation of temperature into their graphic displays of zonation.

The bulk of the work of these authors seems to suggest that temperature is one of the principal factors controlling depth zonation. McGlasson (1959), in studying areas along the southern California coast that had been previously studied by Natland (1933) and Bandy (1953), found a similar depth zonation. He noted, however, that the bathymetric zonation of living forms was distinct from that of the empty tests. He established 3 depth zones:

0–37 m	11.1–20.6 °C
38–74 m	9.6–14.2 °C
75–180 m	8.0–11.1 °C

which contained different living assemblages, although he found that some species lived in more than one zone. McGlasson attributed the zonation to both temperature and depth.

There are cases in which depth appears to have a greater influence than temperature. In the Java Sea, Myers (1942c) noted that a distinct vertical zonation existed from surface waters to deeper waters (30–50 m) although the water temperature varied by less than $\frac{1}{2}$ °C. Sublittoral Arctic and Antarctic waters have the same temperature ranges as do deeper temperate and tropical waters but the faunas are distinct, especially at the specific level.

The temperature factor cannot be solely responsible for the distribution of genera since many genera live in waters of greatly differing temperature but always at the same depth; e.g. *Elphidium* lives in shallow water at all latitudes.

It seems clear that vertical foraminiferal zonation, while it may be partially controlled by temperature, is also subject to some other factors or combinations of factors which are related to water depth.

B. Influence on Test Morphology

Carpenter (1856), speaking about the variable morphology of several species, suggested that temperature probably played an important role in the development of the variations. More than a century of observation and investigation has proven him correct.

At the Subtropical–Subantarctic Convergence in the South Atlantic morphological variants of the same species are found in two water masses whose only known hydrological difference is temperature. These water masses can be separated on the basis of these morphological differences (Boltovskoy, 1962a). Lutze (1962) observed that specimens of *Bolivina spissa* in cold water at depths greater than 1000 m are narrow compared to the broader and more highly ornamented specimens found in warmer, shallower water. Which of the two factors, depth or temperature, is of more importance is unknown.

The correlation between coiling and water temperature in some planktonic forms is discussed in Chapters III,3 and XIV,4.

Our understanding of the effect of temperature on specimen size is somewhat confused by contradictory data (Chapter III,12). According to Rhumbler (1911), Schaudinn and Römer observed that *Triloculina tricarinata* and various members of the Astrorhizidae had larger tests in Arctic waters than in lower latitudes. Rhumbler concurred with this observation, noting the size of specimens collected from the Gauss Antarctic expedition:

Species	Antarctic size (mm)	Largest specimen known elsewhere (mm)
Astrorhiza angulosa granulosa	14.7	6.0
Jaculella acuta	57.0	25.0
Tholosina laevis	2.1	0.7
Reophax cylindricus	10.0	3.6

He concluded that cosmopolitan species, as a rule, grow to larger sizes in cold water. He extended the rule to apply not only to latitudinal changes but also to temperature changes due to depth. Rhumbler also noted that this rule has more value for those species which do not require a large quantity of calcium for their tests, although he cited certain calcareous exceptions, notably *Cornuspira*, the Miliolidae and various Nodosariacea. Those which do require a great deal of calcium (*Orbitolites*, *Amphistegina*, *Heterostegina*) are commonly larger in warm waters. A similar observation was made by Bandy (1963a) on certain species (*Cyclammina cancellata*, *Martinottiella* spp., *Pyrgo* spp. and *Robulus thalmanni*) off the Southern California coast, although there was also an increase in depth and oxygen content coincident with the size increase. Lewis & Jenkins (1969) found *Nonionellina flemingi* to be twice as large near the cooler end of its latitudinal range. Study of fossil *Globigerina pachyderma* in the Arctic showed that specimens were smaller during times of warmer temperature (Steuerwald & Clark, 1972). However, a statistical study of this same species in the Southern Hemisphere revealed no significant correlation between latitude and median test size (Malmgren & Kennett, 1972).

Calcituba polymorpha maintained at 18–19 °C in laboratory cultures produced larger chambers than those grown at 21–25 °C. The specimens were collected in waters of 28–30 °C (Arnold, 1967a).

Awerinzev (1903), studying rhizopods, and Häcker (1908), studying radiolaria, also concluded that larger specimens are produced in colder water.

Not all observations conform to the pattern of increasing size in cooler water. Observations on Recent planktonic species by Stone (1956), Boltovskoy (1962a, 1966b), Bé (1972), Bé, Harrison & Lott (1973), Emiliani (1974) and Hecht (1974), on fossil planktonic species by Myers & Cole (1957) and Lidz (1972), and on fossil benthonic species by Phleger & Hamilton (1946) indicate that there is tendency for specimen size to increase as water temperature increases.

The morphology of planktonic foraminifera is influenced by temperature but it is sometimes difficult to isolate the temperature effect from those of other environmental parameters. Various observations have been made on morphological changes occurring in different latitudes or in different water masses.

Bé (1968) noted a correlation between water temperature and the size and spacing of pores on planktonic species. His measurements and calculations revealed that specimens living in warm tropical waters exhibit 20% porosity whereas those found in cold, polar water have a 5% porosity. An explanation of this correlation was offered by Frerichs *et al.* (1972) who suggested that the test porosity is related to buoyancy. Those specimens living in warm water must in some way compensate for the low density of this water in order to maintain their buoyancy. Perhaps this compensation occurs in the form of a greater pore diameter or pore density, i.e. greater porosity, which reduces the specific gravity of the whole animal. Those specimens living in cold and therefore denser water have less need for this compensation and are less porous.

Boltovskoy (1962a, 1966b) noted that *Globigerinoides trilobus* and *G. ruber* lose their secondary apertures when they live in water cooler than normal. He also noted (1966d) that the aperture of *Globigerina bulloides* decreases in size in high latitude (cooler water) zones. Under the same latitude conditions specimens of *Globigerinita glutinata* have unusually large primary apertures and smaller supplementary ones.

The shape changes of several planktonic species appear to be temperature related. The ratio of asymmetrical to symmetrical forms of *Globigerinella siphonifera* increases in cooler waters (Hecht & Savin, 1972). As this species is a subtropical–tropical water species it may be that the ecological stress applied near its temperature tolerance limits affects its growth processes. The same authors also suggest that growth in an environment with a temperature above the optimum temperature range of a species may account for the diminutive final chamber (kummerform) shown by several planktonic species (Chapter III,11).

Boltovskoy (1959b) noted changes in the shape of *Globorotalia truncatulinoides* in different water masses. These changes were described more fully by Kennett (1968a) who showed that the specimens found in cold waters are more biconvex and have lower spires than those in warmer waters (which are planoconvex and have high spires). The presence of umbilical teeth was shown to be related to increasing water temperature for this species (Takayanagi, Niitsuma & Sakai, 1968).

Although not strictly a morphological feature, the occurrence of the pink colour in some forms of *Globigerinoides ruber* appears to be positively correlated with water temperature (Boltovskoy, 1966b). The direct cause of the coloration may be related to symbiotic zooxanthellae in the test whose activity and abundance is directly controlled by temperature. The species changes from orange in May to ruby red in the warmer months of August–October in the North Atlantic (Bé & Tolderlund, 1971).

It is difficult to determine whether temperature is the primary controlling factor or a secondary one that brings about most morphological changes in planktonic foraminifera. It is even possible that temperature may correlate with morphological change without being a causal factor.

In summary it appears that temperature: 1. Has a great deal of control over the vital activities of foraminifera. Each species has certain tolerance limits as well as a temperature optimum. A species can only tolerate temperature extremes for a short period. 2. Is the primary factor which controls the geographic distribution of species, and an important factor affecting vertical distribution. The range of temperature fluctuation is probably as important as the absolute temperature. 3. Influences the test size and morphology. Although the majority of observations indicate that increased size is a result of lowered temperature, this relationship is not perfectly understood.

2. Salinity

As is the case with temperature, the salinity tolerance limits for survival, growth and reproduction of the foraminifera can be determined. The tolerance limits of the benthonic foraminifera are very wide and quite distinct for different species. Planktonic species are much more sensitive to salinity changes.

Salinities within the normal range (30–40‰) seem to have little effect on the survival of benthonic forms. Even in mixohaline waters (18–30‰) most benthonic species are able to live. In a study of Christchurch Harbour, England, Murray (1968a) noted that an abundant living benthonic foraminiferal population could flourish in waters whose salinity ranged from 0–34‰ although the maximum populations (presumably representing reproduction) occurred during the warmer (and higher salinity) months of summer and early autumn. During one summer in which there was unusually high dilution from stream runoff the maximum salinity reached was only 4.3‰ and very few living specimens were found. It appears that although the specimens were able to survive a wide range in salinity their reproductive limits were much narrower.

Certain forms, notably the agglutinated genera *Miliammina, Trochammina,* and *Jadammina,* the calcareous genera *Rotalia, Elphidium, Nonion* and various Miliolidae are much more tolerant to low salinity conditions than other forms. McCrone & Schafer (1966) found *Miliammina fusca* and *Ammomarginulina fluvialis* in waters of 5‰ salinity. Schafer & Sen Gupta (1969) reported *M. fusca* and *T. inflata* in water of 5‰ salinity. The lower salinity tolerance limit in the James River estuary, Virginia, of *Ammobacuolites* sp. is 0.5‰ and of *Ammonia beccarii tepida* is 6‰ (Nichols & Norton, 1969). Boltovskoy (1958b) and Boltovskoy & Lena (1971) reported living specimens of *Nonion tisburyensis, Psammosphaera* sp. and *Miliammina fusca* in waters with a salinity of 0.1–0.15‰. They regard these species as special adaptions of forms which normally live in more saline water (Chapter V).

The variation in sensitivity of these tolerant euryhaline genera is shown in a comparison of two related species: *Elphidium excavatum* which can tolerate a salinity of 7.25‰ (Voorthuysen, 1960) and *E. crispum* which Myers (1943c) showed to be very numerous at salinities of 19.5‰ but which is seldom, if ever, found in brackish waters. *E. crispum* is a thick walled form requiring more $CaCO_3$ than *E. excavatum.* As the solubility of $CaCO_3$ is proportional to salinity, there may not be enough $CaCO_3$ available in the lower salinity waters for this form.

Many species common in normal marine waters are also found under low salinity conditions. Furssenko (1959) found *Elphidium poeyanum* and *Rotalia beccarii* in water of 7.38‰; and Voorthuysen (1960) established the presence of *Streblus batavus, Nonion depressulus, Elphidium selseyense, E. excavatum* and *E. gunteri* in areas where the salinity was 7.25‰. An interesting occurrence of *Marginopora vertebralis,* a normal marine species, was reported from an intertidal flat where the specimens are exposed to direct sunlight and rainfall during times of low tide (R. Smith, 1968). Of the species enumerated above, those most able to tolerate wide ranges of salinities are probably *E. poeyanum* and *R. beccarii.* These two species have been found in water of very high salinity – 50‰ (Furssenko, 1959). A subspecies of *Streblus beccarii, S. beccarii catesbyana* has been found in salinities of 70.6‰ (Tinoco, 1958), but its lower tolerance limit is unknown. *Cribroelphidium poeyanum,* a synonym of *E. poeyanum,* was found living in a hypersaline (90‰) lagoon in Venezuela (Sellier de Civrieux, 1968).

The problem of whether or not there are true fresh water foraminifera is a difficult one. Are those forms found in fresh water typical of this environment or are they only adapted marine forms? This problem has been partially 'solved' by taxonomic means; the fresh water members of the Foraminifera have been reclassified as Thecamoebians.

Culture studies have provided useful, although scarce data about the relationships

Table 12. Salinity tolerance limits for cultured benthonic foraminifera

Species	Author	Salinity, ‰				
		Survival	Growth	Reproduction	Growth	Survival
Allogromia laticollaris	Lee & McEnery, 1970	5				50
Allogromia sp.	Lee & Pierce, 1963			20–45*		
Discobotellina biperforata	Stephenson & Rees, 1965b	25.5				37.5
Quinqueloculina lata	Muller & Lee, 1969			5–95		
Massilina sp.	Bradshaw, 1961	7–10†				
Peneroplis planatus	Murray, 1970a					53
Bolivina vaughani	Bradshaw, 1961	7†				
Discorinopsis aguayoi	Arnold, 1954a	20				57
Spirillina vivipara	Bradshaw, 1961	7†				
Rotaliella heterocaryotica	Bradshaw, 1955		16.8	26–33	37	
Ammonia beccarii tepida	Bradshaw, 1957, 1961	2†	8	15–40	67	
Elphidium crispum	Myers, 1943c	19.5				48
Elphidium spp.	Murray, 1970a					53
Heterostegina depressa	Lutz, Grabert & Seibold 1971			36–40‡		90

* after 5 weeks, 5–55%
† 12 hour limit
‡ cited as optimum range

between salinity and tolerance limits. Schultze (1854) collected mud from the Mediterranean and preserved it in several changes of fresh water. After five weeks he noted that the Rotaliidae in the sample were still living. He then added marine water to the sample and noted that the specimens continued to live after several weeks. Data from subsequent cultures studies are shown in Table 12.

Bradshaw (1961) noted that *Rotalia beccarii tepida* could reproduce between salinities of 15–40‰, but the time required for a new generation varied with the salinity:

Salinity (‰)	Time between generations (days)
7	47
15	56
25	82
34	35

The growth rate of the specimens was greatest at 34‰. Bradshaw concluded that the salinity and temperature limits of survival are not so important for this species because the survival limits are not often reached in nature. However, the limits of growth and reproduction are often reached in natural settings and are therefore important to the distribution of this species.

Cultured specimens of *Elphidium crispum* displayed feeding and survival characteristics that were dependent on both salinity and temperature conditions (Murray, 1963). Specimens maintained at salinities less than the lower limit for growth were better able to survive at lower temperatures than at high ones. In both cases the temperatures were within the survival range of the species. This example points out the interrelationship of the various ecological factors.

The influence of salinity on benthonic foraminiferal distribution has been amply demonstrated in numerous studies. Coastal zones characterized by considerable mixing of fresh and marine water contain qualitatively and quantitatively different populations from those areas characterized by normal marine waters. The details of these distributions are given in Chapter V.

For the planktonic foraminifera, conditions are much more uniform. They are usually surrounded by water with a salinity of 34–36‰ although they appear to be able to tolerate salinity ranging from 33–40‰. As is true with benthonic forms, individual species have different tolerance ranges. Unfortunately, laboratory cultures of planktonic species have not been as successful as those involving benthonic forms and consequently, our knowledge of tolerance limits is derived from specimens collected in the open ocean. Waller & Polski (1959) suggest that planktonic species can not survive at salinities less than 27‰. However, more recent studies seem to indicate this limit may be too low. Tsi-chung & Sau-yee (1964) cited the 33‰ isohaline as the limit of distribution of planktonic foraminifera in the northern South China Sea. Near the mouth of the estuary of the Rio de la Plata, Boltovskoy (1966b) encountered small living specimens of *Globigerina bulloides*, *G. pachyderma* (f. superficiaria), *G. quinqueloba*, *Globigerinita uvula* and *Globigerinoides ruber* (f. alba and f. rosea) in waters whose lowest salinity was 30.5–31‰.

The vertical and horizontal salinity changes in the open ocean are so small that they have much less effect on the geographic distribution of planktonic species than do other ecologic factors, notably temperature. However, under certain conditions, small differences in salinity may exert strong influences in the geographic distribution of planktonic foraminifera. Bé & Tolderlund (1971) noted that *Globigerinoides sacculifer* is the dominant species in warm, tropical water of 34.5–36‰ salinity, but that *G. ruber* becomes dominant where the salinity is less than 34.5‰ or greater than 36‰. These two tropical water competitive species have their distribution controlled by salinity. Zobel (1971) noted similar salinity control of species distribution in the Arabian Sea where high salinity water masses emanate from the Red Sea and the Persian Gulf.

Many foraminiferal specimens are reduced in size, lose ornamentation, acquire thinner walls or are transformed from calcareous to chitinous as a result of reduction in salinity (Brodsky, 1928; Pratje, 1931; Dolgopol'skaja & Pauli, 1932; Nikoljuk, 1948; Boltovskoy, 1957b; Kurc, 1961) (see also Chapter III). Schaudinn (1895b) noted that *Patellina*, after living a long time in a test tube, became transparent. Le Calvez & Le Calvez (1951), studying the fauna from two lagoons near Banyuls-sur-Mer, France, discovered that at reduced salinity the size of various Rotaliidae and Nonionidae decreased. The same phenomenon was observed by Brodniewicz (1965) in the Southern Baltic Sea for the agglutinated species *Armorella sphaera* and *Hippocrepinella remanei*. Wright (1968) noted that as the salinity decreased in the estuary of the Quequén Grande River, Argentina, the living Miliolidae became less diverse, thinner walled, more variable morphologically and changed their ornamentation patterns.

Furssenko (1959) mentioned an interesting comparison between specimens of *Rotalia beccarii* taken from the Mediterranean (salinity 37–39‰) and from the Black Sea (18–22‰). The tests of the latter were smooth, contained fewer chambers and lacked ornamentation and umbilical plugs. These traits appear to be morphological adaptations due to reduced salinity. The same kinds of observations were made by Blanc-Vernet (1958) and Tufescu (1967) who observed that the brackish environment produces morphological modifications, principally the reduction of ornamentation and wall thickness. In the Dos Patos Lagoon, Brazil, those specimens of *Elphidium gunteri*, *E. incertum* and *E. excavatum* found during the winter months (low temperature and low salinity) had unusually thin walls (Forti & Roettger, 1967). Murray (1966a) noted a correlation between high salinities in the Persian Gulf and forms with great morphological variability. Cultured specimens of *Heterostegina depressa* from this area (Lutze, Grabert & Seibold, 1971) whose optimum salinity was 36–40‰ grew more slowly and exhibited a greater number of abnormalities when the salinity was gradually reduced to 20‰.

These observations deal with the reduction of salinity below normal or 'optimum' values. Although the data is less abundant, there appear to be corresponding alterations when salinity is increased above normal values. Shokhina (in Furssenko, 1959) found that tests of *Rotalia* sp. are largest when salinity is normal and decrease in size at higher (40‰) salinities. *Cribroelphidium poeyanum* exhibits small deformed chambers and an irregular periphery when found in hypersaline lagoons.

Bradshaw's observations (1961) suggest a salinity reaction opposite that already mentioned. He noted that the largest specimens of *Ammonia beccarii tepida* were found in relatively low salinity waters and 'small' specimens in waters of normal salinity.

Some variations in chamber size in a given species are accredited to oscillations in salinity by Rhumbler (1911), who cited changes in *Elphidium* and *Nonion.*

In summary it seems as though salinity: 1 Is directly related to the vital activity of foraminifera. Each species has certain critical limits as well as an optimum salinity for various physiological functions. 2 Affects the geographic distribution of foraminifera. Because of the relatively uniform distribution of salinity in most of the open ocean, this effect is much more pronounced in areas where there is considerable mixing with fresh water. Not only the amount of dissolved salts but the daily and seasonal variation in salinity affects the distribution and occurrence of foraminifera. 3 Can affect test morphology, primarily size and ornamentation. Most observations indicate a reduction in size and loss of ornamentation near the salinity tolerance limits.

3. Depth

The ecological effects of depth are difficult to assess because other ecological factors vary with depth (temperature, light intensity, pressure, oxygen and calcium carbonate concentration, etc.). In natural environments it is impossible to determine the isolated effects of this factor. Nevertheless, various morphological and distributional attributes of the foraminifera indicate that there is a depth effect. Brady (1884) noted that at great depths the genus *Miliolina* has a siliceous test but in shallower higher latitude waters of the same temperature the tests are calcareous. Another example is the case of *Elphidium* which always lives in shallow water zones regardless of the wide range of temperature or various other environmental factors at these sites. Many species of *Bolivina* are similar in that they habitually live within a certain depth range regardless of other factors. The composition of the fauna in bathyal zones in tropical waters is not the same as that of shallower temperate or polar waters. Many benthonic foraminifera have the same upper depth limit in various areas of the world even though the temperature conditions of these areas are diverse. For example, *Adercotryma glomerata* occurs in both the Antarctic and Gulf of Mexico but has an upper depth limit of 200 m in each area despite a 11–15 °C temperature difference between the two areas. These kinds of species are referred to as isobathyal by Bandy & Echols (1964). The depth zonation of these species can not be explained by variations in temperature. One must note that although the foraminiferal faunas of the Arctic and Antarctic areas have many traits in common with those of deeper tropical waters, these traits are only of a general nature; e.g. the low number of species and the abundance of agglutinated forms.

Just as there are genera and species whose vertical distribution is narrow, there appear to be some foraminifera which are found in depths so diverse that one gets the impression that depth has no effect at all. *Bathysiphon filiformis* has been found in 2500 m (Brady, 1884) and in 150–250 m (Le Calvez, 1938a). *Cyclammina orbicularis, C. pusilla, Hormosina robusta, Recurvoides contortus, Martinottiella antarctica* and some other species have a depth range from 200–400 m down to 5000–6000 m (Theyer, 1971b). The genera *Reophax, Rhabdammina* and *Cribrostomoides* have also been found in water whose depth varied greatly.

In a summary of numerous studies of different groups of benthonic animals, Beklemishev *et al.* (1973) state that the average upper limit of the deep water benthonic fauna occurs at a depth of 2000 m which corresponds to the middle part of the slope. The authors emphasize that this limit does not have any relation to the bottom relief and apparently depends on the upper limit of deep homogeneous waters.

The generalized depth distribution of various benthonic genera is given in Table 13. The preparation of this table presented some serious difficulties for two reasons: (a) the confused state of taxonomy; and (b) the unrecognized presence of displaced faunas (Chapter XIV,13); i.e. the presence of a species in a given sample does not necessarily mean that the species lived there. Because of these factors we can not claim that the table is absolutely correct. It was composed on the basis of data taken from the literature as well as that from our own studies of the deep water benthonic foraminifera of the South Atlantic and Indian Ocean.

To better understand the table we think that it is necessary to make the following comments:

(a) This table would have greater value if prepared at the species level rather than at the generic level. However, if prepared in this way, it would be immense and be beyond the scope of this book. Many genera have very wide bathymetric ranges and are also represented at different depths by different species. If these particular species are very rare at some particular depth, we neglected them in the compilation of the table. If, however, they can be found in appreciable numbers, they were not neglected. For example, we cite *Spiroloculina* as essentially a shelf genus. It can be found in the bathyal zone as well, and even at depths of more than 3000 m (abyssal zone, *S. pusilla*), but is extremely rare at these depths and for this reason we did not cite its presence in the abyssal zone. Another example is *Quinqueloculina*. Although it also is primarily a shelf zone genus, there are several species which are typical of great depths, e.g. *Q. venusta* and *Q. weaveri*, and as they are sometimes frequently found there, we note this occurrence on the table.

(b) Our interpretation of the taxonomic category of genus is somewhat broader than that of some of our colleagues. This partially explains why the bathymetrical

Table 13. Depth distribution of selected benthonic genera

	Shelf		Slope		
	Nearshore zone and inner shelf	Outer shelf	Bathyal zone		Abyssal zone
			Transitional		
	0 m – ~70 m	~70 m – ~200 m	~200 m – 1000 m	1000 m – 3000 m	3000 m
Acervulina	+				
Allomorphina			+	+	
Alveolinella	+				
Alveolophragmium	+	+	+		
Amphistegina	+				
Angulogerina	+	+	+	+	+
Archaias	+				
Asterigerina	+				
Baculogypsina	+				
Bathysiphon		+	+	+	+
Bigenerina		+	+	+	
Bolivina	+	+	+	+	+
Borelis	+				
Buccella	+	+			
Bulimina	+	+	+	+	+
Buliminella	+	+	+	+	+
Calcarina	+				
Cancris	+	+	+		
Cassidulina	+	+	+	+	+
Chilostomella		+	+	+	+
Cibicides	+	+	+	+	+
Cornuspira	+	+	+		
Cribrostomoides	+	+	+	+	+
Cyclammina		+	+	+	+
Dentalina	+	+	+	+	+
Discorbis	+	+	+		
Ehrenbergina	+	+	+	+	+
Elphidium	+				

Table 13. *(Continued)*

	Shelf		Slope		
	Nearshore zone and inner shelf	Outer shelf	Bathyal zone		Abyssal zone
			Transitional		
	0 m – ~70 m	~70 m – ~200 m	~200 m – 1000 m	1000 m – 3000 m	
Epistominella	+	+	+	+	+
Eponides	+	+	+	+	+
Fissurina	+	+	+	+	+
Gaudryina		+	+	+	
Glabratella	+				
Globobulimina	+	+	+	+	
Globulina	+				
Guttulina	+	+	+		
Gyroidina	+	+	+	+	+
Heterostegina	+				
Hoeglundina		+	+	+	
Hyalinea	+	+	+		
Karreriella		+	+	+	
Lagena	+	+	+	+	+
Laticarinina			+	+	+
Marginopora	+				
Marginulina	+	+	+	+	
Martinottiella		+	+	+	+
Massilina	+				
Miliolinella	+	+			
Nodosaria	+	+	+	+	+
Nodosinella				+	+
Nonion	+	+	+	+	
Nonionella	+	+	+	+	
Oolina	+	+	+	+	
Operculina	+				
Parafissurina	+	+	+	+	+
Patellina	+	+			

Table 13. (Continued)

	Shelf		Slope		
	Nearshore zone and inner shelf	Outer shelf	Bathyal zone		Abyssal zone
			Transitional		
	0 m	~70 m	~200 m	1000 m	3000 m
Peneroplis	+				
Planorbulina	+	+			
Planulina	+	+	+	+	
Poroeponides	+				
Pullenia	+	+	+	+	+
Pyrgo	+	+	+	+	+
Pyrulina		+	+	+	+
Quinqueloculina	+	+	+	+	+
Reophax	+	+	+	+	+
Rhabdammina		+	+	+	+
Robulus	+	+	+	+	
Rotalia	+				
Saccammina	+	+	+	+	+
Sigmoilina	+	+	+	+	+
Sigmomorphina	+	+			
Spirillina	+	+	+	+	
Spiroloculina	+	+	+	+	
Textularia	+	+	+	+	
Trifarina	+	+	+	+	
Triloculina	+	+	+	+	
Trochammina	+	+	+	+	
Uvigerina	+	+	+	+	+
Valvulineria	+	+	+	+	
Vertebralina	+				
Virgulina	+	+	+	+	+

ranges of some genera appear wider than in similar tables of some other investigators.

(c) The existing observations with respect to ultra-abyssal depths (>6000 m) are so poor and scarce that we prefer not to include them at all.

Some species have been cited by particular authors as being typical of certain depths. But when all the data are examined it becomes clear that these species occur over a much wider range than is indicated by any one study. For example, *Hoeglundina elegans* was found to have its upper depth limit at 50 m by Kustanowich (1964), at 100 m by Colom (1950) and Bandy & Chierici (1966), at 730 m by Walton (1955), and at 2450 m by Bandy (1953). *Uvigerina auberiana* was cited by Waller (1960) as being indicative of water deeper than 120 m and by Parker (1954) as being indicative of water deeper than 250 m. The upper depth limit of *Glomospira gordialis* was 100–200 m in California waters (Uchio, 1960a) and 600 m in the Mediterranean (Parker, 1958). As these are all well known species, these depth variations are not a consequence of misidentification. One reason why these depth limits vary so much from place to place is due to the effect of the overlying water mass on the bottom fauna, especially in the shelf and upper slope zones (Vinogradov, 1948; Beklemishev *et al.*, 1973 and several others). The nature of the benthonic fauna appears to be controlled by the particular physio-chemical characteristics of the water mass. However, two water masses possessing the same characteristics may exist at different depths, and consequently the 'depth ranges' of particular species will be different in these two areas.

Bandy (1960a) described relationships between test morphology of benthonic forms and depth. Without trying to explain the mechanism of alteration, he noted that many genera when found in depths outside their normal range are altered morphologically. These alterations are usually the same for a given situation. These changes are different for different groups; the most notable are those affecting the calcareous hyaline forms. Bandy gave many examples of changes due to increased depth: *Pullenia* becomes more spherical (the specimens on the shelf are compressed whereas those in bathyal waters are wider); *Cibicides* is more oblate and has more prominent sutures (*C. aknerianus* and *C. pseudoungerianus* of the shelf and *C. wuellerstorfi* of the bathyal zone). *Eponides* decreases in size (*E. antillarum* and *E. repandus* in shallow water compared with *E. tumidulus* and *E. turgidus* of deeper waters). The Rotaliidae generally acquire a more blunted peripheral margin (*Cancris* of the shelf compared to *Valvulineria*, *Baggina*, and *Gyroidina* of the depths). *Bulimina*, *Bolivina*, and *Uvigerina* increase in size and ornamentation. In later articles Bandy (1963a, 1964) noted that *Martinottiella* spp., *Pyrgo* spp., and *Gyroidina* spp. all increased in size with depth. Frerichs (1970) noted that *Bolivina* spp. develops sinuate sutures below 200 m, that the ornamentation of *Angulogerina* decreases with depth, and that the costate species of *Uvigerina* in the outer shelf and upper bathyal zones becomes hispid (covered with short fine hair like spines) at depths greater than 600 m. Changes noted at the specific level are shown in Table 14.

There seems to be a very clear correlation between planktonic test morphology

Table 14. Morphological changes in some benthonic species as a function of increasing depth

Species	Author	Alteration
Angulogerina carinata	P. Smith, 1964	Walls become thick, opaque, lose carinae
Bolivina argentea	Lutze, 1964b	Loss of costae, keels and spine
B. pseudobeyrichi	P. Smith, 1964	Develops a keel
B. seminuda	P. Smith, 1963	Longer, wider, more flaring
B. seminuda	P. Smith, 1964	Greater number of clear spots in wall
B. subadvena	P. Smith, 1964	Smaller, less robust
Buccella frigida	Theyer, 1966	Thicker, fewer chambers, keel
Bulimina denudata	P. Smith, 1964	Loss of spines
Cassidulina laevigata	P. Smith, 1964	Development of carinae
C. tumida	P. Smith, 1964	Development of wide serrate keel
Cyclammina cancellata	Bandy, 1963a	Larger
C. cancellata	Theyer, 1971c	Larger, narrower; then smaller again
Eggerella advena	Resig, 1963	Smaller
Glandulina laevigata	Bandy, 1963a	Apical spine develops
Pyrgoella sphaera	Bandy, 1963a	Larger
Robulus thalmanni	Bandy, 1963a	Larger
Suggrunda eckisi	P. Smith, 1964	Shorter
Uvigerina peregrina	P. Smith, 1964	Spines change to costae

and depth. Several species secrete a secondary calcite crust or cortex as they age and sink in the water column. This phenomenon was first observed in *Globigerina pachyderma* (Bé, 1960b) and later found in other species, namely in *Globigerinoides conglobatus*, *G. sacculifer*, *Globorotalia inflata*, *G. menardii*, *G. tumida*, *G. truncatulinoides* and *G. crassaformis*. Orr (1969) discovered the presence of cortex also in *Globigerinoides ruber*. Investigations of secondary thickening done on the basis of oxygen isotope ratios (Chapter XIV,5) have yielded data both supportive and contradictory of the concept of encrustation as an ontogenetic adaption to meso- or bathypelagic depth habits (Hecht & Savin, 1970; Emiliani, 1971a).

Bé & Tolderlund (1971) concluded that, with the exception of *Hastigerina pelagica*, the spinose planktonic species all live in surface waters whereas most of the non-spinose species are found below 50 m.

The direct effects of an increase in depth are increasing pressure and decreasing light penetration. As light is absorbed very rapidly in the first few meters of the sea, its influence is not felt except in surface waters. The effects of reduced illumination are discussed in section 11 of this chapter. At the surface, atmospheric pressure is a bit more than 1 kg/cm^2. With each meter the pressure increases 0.1 atm; at a depth of 10 000 m the pressure is about 1000 kg/cm^2. Life exists at all depths (pressures) but the characteristics change with depth.

Various investigators have studied the influence of pressure on different types of organisms ranging from bacteria to crustaceans. They found that elevated pressure produced various physiological effects. Some species, especially those collected in shallow water, died when exposed to higher pressures, whereas others survived, indicating a range of pressure tolerance.

The only systematic studies of pressure effects on foraminifera were conducted by Bradshaw (1961) who demonstrated that while *Ammonia beccarii tepida* could survive a pressure of 400 atm for 3 days; a pressure of 1000 atm killed all the specimens. However, *A. beccarii tepida* is a shallow water species and the results of these experiments should not be taken to indicate that all foraminifera react the same way to pressure changes. Living benthonic foraminifera have been collected from the Tonga Trench between 10 415–10 687 m (Belyaev, Vinogradova & Filatova, 1960). Bradshaw's experiments point out the kinds of adaptions that the foraminifera undergo.

Indirect effects due to increasing depth include a decrease in temperature and an increase in the partial pressure of CO_2 with an corresponding increase in $CaCO_3$ solubility. Beneath a certain depth the high partial pressure of CO_2, the low temperature, and the high water pressure create a situation in which the rate of solution of calcium carbonate is greater than its rate of supply from above. This calcite compensation depth lies between 4000–5000 m in the central Pacific. It is deeper in the Atlantic Ocean because of lower CO_2 concentrations. It is shallower at higher latitudes because of lower temperatures, rising as high as 400–500 m in the Ross Sea, Antarctica (Kennett, 1966; Fillon, 1974). Calcium carbonate solution also occurs above the compensation depth (Peterson, 1966) and has noticeable effects on planktonic foraminifera (Berger, 1967; Pytkowicz & Fowler, 1967). Berger (1968) noted the presence of a zone (lysocline) above the compensation depth at which the rate of solution increases rapidly.

The various planktonic species do not react equally to dissolution. The subspherical, massive, non-spinose, incrusted forms seem to be more resistant to solution than the others. Those forms which live in the upper portion of the water column seem to be more easily dissolved. The effects of the solution process on the tests is to fragment them, dissolve the early, thinner walled chambers and corrode the outer walls. Other facets of this topic are dealt with in Chapters VII,3 and XIV.

Calcareous benthonic foraminifera are more resistant to solution than planktonic forms. Certain species are better adapted than others to withstand the corrosive effects of deep water. The number of calcareous benthonic foraminifera that can manage to secrete calcite in deep water is small. Consequently, the proportion of agglutinated species as well as specimens increases.

In summary it seems that depth: 1. Is an important ecological factor, but one whose isolated effects are not well understood. It may influence foraminifera only because of depth related changes that occur in other environmental parameters; e.g. as depth increases the pressure increases, the temperature decreases, and carbonate solubility increases. 2. And depth-related changes have a greater effect on the vertical ranges of calcareous species than on those of agglutinated species. 3. Affects the morphology in a variety of ways; there is (a) a tendency for many species to become more rounded with an increase in depth, (b) a tendency for many species to increase their ornamentation in deeper water (although many do not) and (c) marked change in size with an increase in depth.

4. Nutrition

A variety of phytoplankton, particularly pennate diatoms, serve as nutrient material for foraminifera, although such things as flagellates, algae, algal gametes, seaweed, bacteria, copepods, small echinoderms, nematodes, infusoria, spores, microcrustaceans, ciliates and even other foraminifera have been listed as food sources. For details see Chapter II. This dependence upon plant material for various fundamental compounds, including albumin, means that the photosynthetic process is of great immediate importance to the foraminifera.

Described in a simplified, schematic manner, photosynthesis is a process which utilizes light energy to produce free oxygen and hydrogen. Chlorophyll in the presence of solar energy disassociates the water molecule into a hydrogen ion (H^+) and an unstable hydroxyl radical (OH^-). Chlorophyll also aids in the disassociation of the CO_2 molecule into component parts. Two of the OH^- unite to form a water molecule and a free oxygen ion ($OH^- + OH^- \rightarrow H_2O + O^{-2}$). At the same time the liberated H^+ unites with oxygen and carbon ions to form carbohydrate which in turn combines with various nitrogenous and phosphatic materials to form albumin.

In the marine environment there is sufficient water for these processes, but the quantity of phosphate and nitrates is limited and they therefore are the substances controlling productivity. One would suspect that in those areas where phosphate and nitrate are abundant the greatest phytoplankton development would occur. As the foraminifera depend on phytoplankton for nutrition and the phytoplankton depend upon phosphates and nitrates, foraminiferal abundance is controlled, at least in part, by the PO_4 and NO_4 distribution. This dependence was demonstrated by Schott (1935) whose surface (0–100 m) planktonic foraminiferal abundance maps correlate well with phosphate and total plankton distribution maps (one would not expect a perfect correlation as nutrition is not the only factor influencing the foraminiferal distribution). Bradshaw (1959), Parker (1960) and Berger (1969b) also observed that the most numerous populations of planktonic foraminifera are found where the inorganic phosphate concentration is the highest.

Both under natural conditions and in the laboratory it is difficult to completely isolate the effects of various nutritive materials. The artificial nature of the laboratory environment always creates a bit of doubt about the validity of making analogies between results of experiments and natural situations.

Myers (1943c) described the influence of nutrition on the chamber growth of some species. In high latitudes in the winter, specimens receive little nutrition and the few new chambers added by multichambered forms are differentiated from those added in the summer because they are shorter and smaller. This phenomenon is most notable in those genera whose chambers are added in annular series (*Sorites*, *Amphisorus*). It is very easy to distinguish those specimens which have spent a winter in high latitudes. Myers (1942a) found that specimens of *Operculina* in western Java exhibited episodes of rapid, robust growth corresponding to the two periods of the year in which there was abundant nutritive material.

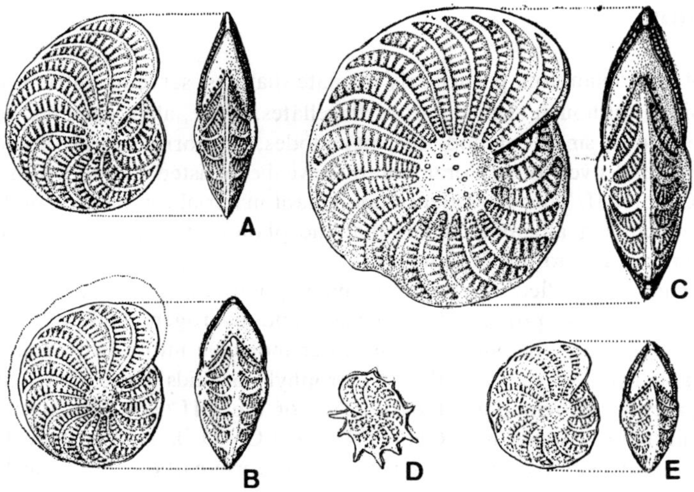

Fig. 79. Elphidium crispum morphology as a function of nutrition (after Myers, 1943c); A. Young specimen with high, narrow chambers; favorable nutrition; B. Young specimen with short, broad chambers; unfavorable nutrition; C. Mature specimen with irregularity produced during a winter with reduced nutritive material; D. Juvenile specimen with spines; E. Mature specimen from a tidal basin where growth is slow.

This same author, studying *Elphidium crispum* from Plymouth Harbour, England, noted that in spring when nutritive materials are abundant the chambers added are large and fat, but those formed in the autumn, when there is less nutritive material available, were relatively reduced in size, and were shorter and narrower. This variability in alimentation produces a sinuous periferial margin in the specimens (Fig. 79). In those areas where nutrition was generally abundant the specimens were large and elongate. In those areas with a deficiency of nutritive materials the specimens grew more slowly, were rounder and more globose with a less pronounced keel, and had a flatter apertural face. These observations agree with laboratory observations that there was diminution in size of specimen and the formation of abnormalities when specimens were grown in cultures having few food organisms (Myers, 1936; Murray, 1963). Myers noted that in tropical regions where growth is somewhat slower, although more constant, the tests of *E. crispum* are considerably more globose and depressed than those from high latitudes.

Another observation of Myers regarding *E. crispum* is that specimens from the sublittoral zone are 60‰ larger and have 40‰ more chambers than specimens of the littoral zone. He attributed these differences to the reduced quantity of nutritive material in the littoral zone.

Seasonal variation in nutrition affects not only morphology but also the abundance of the organisms. Tolderlund & Bé (1971) showed that planktonic foraminiferal abundance in the North Atlantic is clearly seasonal and fairly well correlated with the abundance of other plankton. The same correlation has been observed in benthonic foraminifera (Walton, 1955).

Several authors have noted a positive correlation between the abundance of foraminifera and the abundance of nutritive material as indicated by various indicies of high organic production. Phleger (1960b), after studying quantitative data from several sources, concluded that the number of specimens in a given volume of sediment is especially high in those areas where there is high organic production; in deltas, in areas of upwelling, and in areas with abundant vegetation. Said (1950) made correlations between the foraminiferal concentration and the quantity of nitrogen in the bottom sediment of the Red Sea. In a later work he concluded that in well oxygenated basins, nutrition and substrate texture are the most important factors influencing the distribution of benthonic foraminifera (Said, 1951a). The foraminiferal abundance in San Pedro Bay, California, is high where the phosphorous and nitrogen content of the sediments is high (Bandy, Ingle & Resig, 1964c). Closs (1962) observed a strong correlation between the number of foraminifera and the number of diatoms in Dos Patos Lagoon, Brazil. Shokhina (in Furssenko, 1959) noted that the abundance of *Rotalia beccarii* and *Elphidium poeyanum* is proportional to algal development and the quantity of aerobic bacteria. Belyaeva (1961) observed that in the Indian Ocean the population of planktonic foraminifera is always more abundant in regions of upwelling where nutritive salts, and consequently, phytoplankton, are more numerous. Similar correlations in the Pacific Ocean between nutrient abundance and benthonic foraminiferal abundance were made by Saidova (1963).

Boltovskoy (1954c, 1956b) studying foraminifera from the Patagonian continental shelf concluded that unfavorable nutrition conditions resulted in various deviations in test morphology.

Culture experiments concluded by Arnold (1954c) and Bradshaw (1955) indicated that an overabundance of nutritive material could be unfavorable (see Fig. 130 and Chapter XV). Laboratory experiments conducted by Arnold (1967a) indicated that the nature of the distribution of food on the substrate is a factor influencing test morphology. Specimens of *Calcituba polymorpha* living on a uniform algal mat developed a dendritic form but when the food was confined to clumps, the foraminifera added new chambers in a highly contorted and tightly confined manner. The morphology was directly related to the physical proximity of the food source.

In summary we conclude that nutritive material: 1. Is among the most important factors governing foraminiferal distribution and abundance. The physio-chemical ecologic factors (temperature, salinity, depth, turbidity, illumination, etc.) set the gross limiting conditions for distribution, but within these limits it is often difficult to effect correlations between these factors and quantitative distribution. The nutritional requirements, while not well understood, probably determine the finer distribunal patterns. 2. May influence the size and morphology of the tests. Reduced nutrition tends to result in undersized specimens whose chambers are less regular in shape.

5. Substrate

The textural, chemical and mineralogical nature of the substrate plays a contributory role in the distribution and morphology of benthonic foraminifera. The agglutinated forms, because they derive some of their test material directly from the bottom, are more directly related to substrate conditions than the calcareous forms which secrete their tests chemically.

As the following summary of the literature will show, there is a great deal of what appears to be contradictory evidence regarding the ecological role of various substrate conditions.

A. Influence on Qualitative Distribution

Authors of the last century, notably Brady (1884) and Walther (1893), noted a correlation between the distribution of various benthonic species and the nature of the bottom sediments. During the first 45 years of the present century a few general observations were made to the effect that some foraminifera characterized by large spines (such as *Asterorotalia trispinosa*) preferred soft muddy bottoms, that various genera (*Discorbis* and *Cibicides*) lived attached to algae and other plants, that the Miliolidae were most often found in sandy bottoms whereas the Buliminacea and Nodosariacea seemed to prefer argillaceous substrates, that *Alveolina* preferred a hard sandy or clayey substrate, *Operculina* a soft silty bottom and that *Tinoporus* and *Calcarina* lived attached to densely growing algae.

After the Second World War, studies of foraminifera distribution increased rapidly. Because almost all authors describing Recent forms occurring at a particular site were geologists, they gave information about bottom conditions. Table 15 gives a sample of some of the species whose distribution according to some investigators seems to be affected by the substrate condition.

Lynts (1962) and Schnitker (1971a) have also described marked preferences between certain species and the grain size distribution. Schnitker noted that only those species that live firmly attached seem to be influenced by the grain size.

The preference may not be made on the basis of grain size but on cohesion. Murray (1963) observed that cultured specimens of *Elphidium crispum* preferred a clean hard substrate to a soft clayey one.

There are exceptions to the observations mentioned above; e.g. the Miliolidae have been cited in some areas as preferring sandy substrates and in others as preferring silty sediments.

The relationships between fossil species and the substrate was discussed to some extent by Rauser-Chernoussova & Kulik (1949), Reitlinger (1950), Bogdanowicz (1952), Pozaryska (1957) and Volozhanina (1960), but the most detailed study of the relationship was conducted by Krasheninnikov (1958a, 1959, 1960) who correlated

Gaudryina pseudoturris	Pujos, 1972	Bay of Biscay			×		
Guttulina spp.	H. Schmidt, 1953	Bay of Naples		×			
Hemisphaerammina sp.	Pujos, 1972	Bay of Biscay			×		
Hyalinea balthica	Pujos, 1972	Bay of Biscay			×		
Hyperammina fragilis	Pujos, 1972	Bay of Biscay			×		
Lagena striata	Bouysee & Le Calvez, 1967	French coast	×				
Massilina secans	Atkinson, 1971	Cardigan Bay, Wales				×	×
Melonis zaandamae	Leslie, 1965	Hudson Bay	×	×			
Miliolidae	H. Schmidt, 1953	Bay of Naples			×		
Miliolidae	Giunta, 1955	Ligurian Sea		×			
Miliolidae	Phleger & Lankford, 1957	Texas lagoons			×		
Miliolidae	Drooger & Kaasschieter, 1958	Caribbean			×		
Nonion parkerae	Bouysee & Le Calvez, 1967	French Coast	×				
Nonion spp.	Giunta, 1955	Ligurian Sea			×		
Nonionella turgida	Pujos, 1972	Bay of Biscay			×		
Operculina spp.	H. Schmidt, 1955	Bay of Naples					×
Protelphidium orbiculare	Leslie, 1965	Hudson Bay			×		
Proteonina aff. *micacea*	Rosset-Moulinier, 1972	Breton Coast		×			
Quinqueloculina aspera	Bouysee & Le Calvez, 1967	French Coast		×			
Quinqueloculina seminulum	Pujos, 1972	Bay of Biscay			×		
Quinqueloculina viennensis	Bouysee & Le Calvez, 1967	French Coast		×			
Rectuvigerina phlegeri	Pujos, 1972	Bay of Biscay		×			
Rotalia beccarii	Pratje, 1931	North Sea		×			
Rotalia beccarii	Phleger & Lankford, 1957	Texas lagoons		×			

Table 15. (Continued)

Taxon	Author	Locale	Clay	Silt ('fine grained')	Sand ('coarse grained')	Gravel	Algal substrate
Rotalia japonica	Morishima, 1948b	Japanese bays		×			
Rotalia spp.	H. Schmidt, 1953	Bay of Naples					×
Rotalia spp.	Giunta, 1955	Ligurian Sea			×		
Sigmoilina schlumbergeri	Morishima, 1948b	Japanese bays		×			
Spiroloculina depressa	Morishima, 1948b	Japanese bays		×			
Stainforthia fusiformis	Rosset-Moulinier, 1972	Breton Coast	×				
Textularia bocki	Pujos, 1972	Bay of Biscay			×		
Textularia sagittula	Pujos, 1972	Bay of Biscay			×		
Textularia truncata	Pujos, 1972	Bay of Biscay			×		
Textularia spp.	Sellier de Civrieux & Bonilla, 1971	Venezuelan Coast			×		
Trifarina angulosa	Pujos, 1972	Bay of Biscay			×		
Triloculina bermudezi	Pujos, 1972	Bay of Biscay			×		
Triloculina trigonula	Bouysee & Le Calvez, 1967	French Coast		×			
Tritaxis fusca	Pujos, 1972	Bay of Biscay			×		
Uvigerina peregrina	Bouysee & Le Calvez, 1967	French Coast	×				
Uvigerina spp.	H. Schmidt, 1953	Bay of Naples		×			
Valvulineria spp.	Giunta, 1955	Ligurian Sea			×		

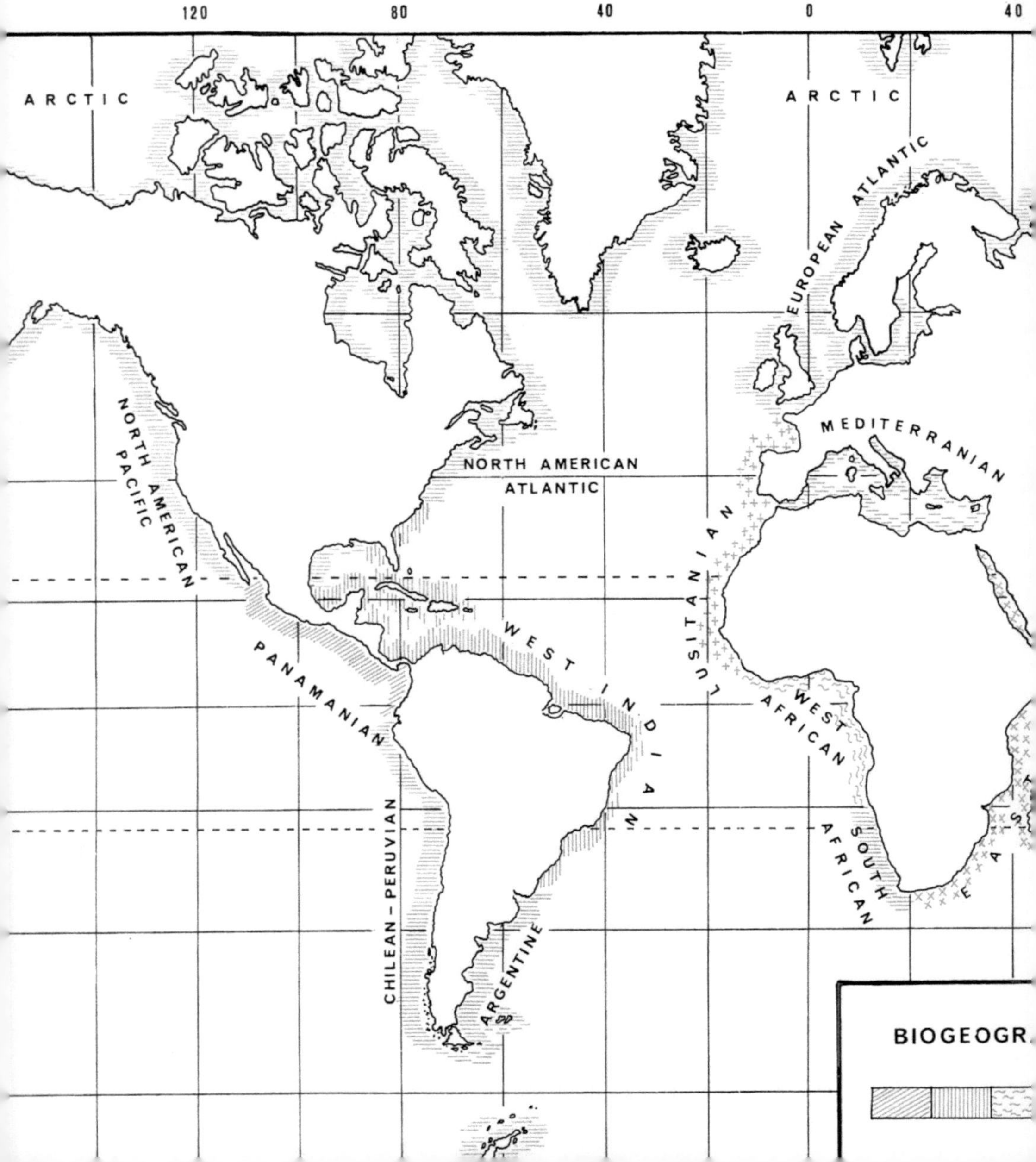
120
80
40
0
40
ARCTIC
ARCTIC
EUROPEAN ATLANTIC
NORTH AMERICAN PACIFIC
NORTH AMERICAN ATLANTIC
MEDITERRANEAN
LUSITANIAN
PANAMANIAN
WEST INDIAN
WEST AFRICAN
SOUTH AFRICAN
CHILEAN - PERUVIAN
ARGENTINE
BIOGEOGR

the occurrence of *Rotalia beccarii, Nonion* and *Elphidium* with the grain size of the Miocene sediments of the Caucausian and Crimean areas. He divided the foraminifera into those he called stenofacial forms (found only in sediments of a narrow textural range) and euryfacial ones (those having a wide range of textural affinities). He mentioned *Elphidium macellum, E. crispum, E. fichtelianum, E. aculeatum, Rotalia beccarii* and *Cibicides lobatulus* as examples of euryfacial species found in the Miocene of Southeast European Russia.

However, other investigators have concluded that the substrate has very little or no influence on the distribution of foraminifera. Studies which support this conclusion were conducted in a variety of environments: open ocean (Saidova, 1960a), continental shelf (Bandy, 1954; Wilcoxon, 1964; Schnitker, 1971a), island shelf (McGlasson, 1959), littoral zone (Dupeuble, 1963; Grossman, 1967), enclosed basins (Christiansen, 1958; Chierici, Busi & Cita, 1962), lagoons (Bartlett, 1964b; Kraft & Margules, 1971; Murray, 1968a), tidal flats (Richter, 1964b) and in general (Stschedrina, 1957). In a laboratory study of *Rosalina globularis*, Sliter (1965) showed that substrate had no effect on the specimens other than restriction of movement by coarse grains.

Sediment grain size may have little effect on other important environmental parameters. Lynts (1966a) discovered that in a shallow water Florida bay there was little or no quantitative relationship between grain size and the distribution of temperature, salinity, depth, pH and oxidation potential. If this is generally true it might account for the lack of correlation between species distribution and substrate type noted by the preceding authors.

Some authors have changed their opinion after a restudy of certain areas. Phleger (1960b), for instance, whose opinions about sediment size and species abundance were cited above, upon examining the detailed works of Parker (1952a) and Stetson (1953) concluded that no clear cut relationship between sediment and foraminifera distribution could be determined. In a later work he reaffirmed this lack of relationship (Phleger & Ewing, 1962).

B. Influence on Quantitative Distribution

Many authors have made comparisons between the abundance of foraminifera and the size of the bottom sediments. Following is a short summary of some of these investigations and an attempt to reconcile the seemingly contradictory conclusions.

Pratje (1930, 1931) found that the quantity of *Rotalia beccarii* specimens in the North Sea increases as the percentage of grains smaller than 0.1 mm increases. Myers (1943c) observed that in the sublittoral zone the greatest number of *Elphidium crispum* are found on compact sandy bottoms or in a mixture of sand and gravel.

Two opposing points of view are held with regard to the relative effect of sandy bottoms and clay bottoms. Several authors (Said, 1950; Miller, 1953; Blanc-Vernet, 1958; Berthois, Le Calvez & Crosnier, 1968, Lukina, 1969; Atkinson, 1971; Kihle, 1971; Schnitker, 1971a) reported a greater total population on sandy bottoms, whereas other authors indicated a greater population on a fine grained substrate

(Freydanck, 1955; Ishiwada, 1958; Hendrix, 1958; Graham & Miltante, 1959; Uchio, 1960a; Closs, 1962; Benda & Puri, 1962; Lidz, 1965; Kaesler, 1966).

Reiter (1959), in studying a sandy coast in California, found that the quantity of living as well as dead specimens is always inversely proportional to the grain size. The greatest quantity occurred where the average diameter of the grains was less than 0.5 mm. This same relationship was noted in the Gulf of St. Lawrence by Vilks (1968) and quantified on the Grand Banks of Newfoundland by Sen Gupta (1967) and Sen Gupta & McMullen (1969). Both Emery & Niino (1963) and Bandy, Ingle & Resig (1964c) noted an increase in organic nitrogen (proportional to nutrition?) as the grain size decreases.

Kurc (1961) cited silty sands covered with vegetation as the optimum site for foraminifera. The influence of vegetative cover on foraminiferal distribution, although known for a century, has only become readily apparent in the last decade as investigators have begun to find that many species are epiphytic. In some environments the number of specimens living on the vegetation exceeds the number making up the bottom populations (Wright & Hay, 1971). The preference of certain genera (*Cibicides*, *Planorbulina*, *Rosalina* and the Miliolidae) for vegetation rather than the bottom was noted by Blanc-Vernet (1969). Some species (*Protelphidium tisburyensis*) appear to be selective with regard to the kind of vegetation they inhabit, whereas others (*Ammonia beccarii* and *Elphidium* spp.) are non-selective (Lee *et al.*, 1969). The study of Lee *et al.* revealed no correlation between the amount of vegetation and the abundance of the foraminifera population.

The attachment of littoral benthonic foraminifera to coraline algae has been noted by Jell, Maxwell & McKellar (1965) and Atkinson (1969). Boltovskoy (1971d) noted that rich populations of littoral foraminifera are found on substrates of fine sand containing molluscan and algal detritus but even richer populations occur on living calcareous algae.

Saidova (1960a), while denying a profound influence of grain size on calcareous foraminifera, did indicate that aggultinated forms are intimately related to the size of the grains on the substrate. Cockbain (1963) noted that some agglutinated species reach their peak abundance on silts (15 μ). In a later article Saidova (1961a) suggested the optimum sites for foraminifera are sands and fine sands mixed with a considerable quantity of mud. The same conclusion was reached by Brodniewicz (1965), Hansen (1965), Kureshy (1969) and Ramanthan (1969). There are several reasons why this combination might be favourable. The coarse grains give physical support to the tests while the fines contain finely disseminated nutritive material. The fines also bind the sand size particles together preventing shifting movements which would tear apart the pseudopodial reticulum with which the organism gathers food.

All of these somewhat contradictory conclusions regarding the character of the substrate (fine or coarse grained) and the fauna (rich or poor) may be explained by noting that, (a) many authors cite examples of a rich or poor fauna without indicating either the number of specimens per unit area or the quantitative content of particular species, (b) the terms coarse and fine are relative and are often not well defined and

(c) many of the observations pertain to the empty test fauna rather than to living specimens although the relationship between grain size and specimen abundance may be greatly influenced by post-mortem transport and selective sorting of empty tests.

Uchio's (1960a) study of living foraminiferal populations on different bottom types resulted in the following data:

Sediment type	Number of specimens per sample
Coarse sand	149
Fine sand	578
Silt	568
Clayey silt	536
Foraminiferal sand or silt	629

Uchio defined foraminiferal sand or silt as a mixture of coarse and fine sediments containing a high percentage of broken shelly material.

Murray (1969) found that living populations from silty sands rich in shelly materials in Vineyard Sound, Massachusetts, were very abundant (440 specimens/30 cm^2) compared to those from adjacent non-shelly coarser sands (20–197 specimens/30 cm^2). This relationship did not hold, however, in samples taken from the Cape Hatteras, North Carolina region. A detailed study conducted by the same author (Murray, 1968b) in Buzzards Bay, Massachusetts showed that muddy silts had an average of 65 living specimens/30 cm^2 compared to muddy sands which had an average of 512/30 cm^2.

From all these (sometimes contradictory) data we can only generalize that based on the majority of observations the optimum bottom for quantitative richness of the foraminiferal fauna is a mixture of sand with shelly detritus and silt. Another very favourable bottom is that covered with algae, especially calcareous algae.

It should be also noted that dead foraminifera behave as sediment grains and several of the observations noted above may be merely the consequence of post-mortem transport and sorting of foraminiferal tests and other detrital grains. Because tests are hollow, they behave hydraulically in the same way as do quartz grains of a smaller diameter. The equivalent diameter depends on the particular shape of the foraminifera but seems to range between $\frac{1}{2}$–$\frac{1}{3}$ the size of the test (Haake, 1962; Grabert, 1971). If post-mortem transport occurs it would tend to associate foraminiferal tests, which normally fall in the medium to coarse sand sizes, with coarse silt to medium sand sized grains. This would make it more difficult to find the relationships which exists between foraminiferal fauna and the substrate.

Several investigators suggested that there is no relationship at all and that apparent correlations are due to environmental factors other than the substrate.

However, that this view cannot be completely valid is attested to by the very obvious relationships between certain attached species and the substrate, and also by the observations of Nyholm (1957) who demonstrated a correlation between the rigidity of pseudopodia and substrate type among the simpler foraminifera. Species with slender fragile pseudopodia are absent from sandy bottoms whereas those with rigid extensions occupy soft sediment and sands. The nature of the substrate is certain to have some physiological or morphological effect on the organism since the penetrating ability of the pseudopodia is related to the sediment.

The intimate relationship between sediment and organism is also indicated by the observation that some foraminifera live buried beneath the sediment–water interface. Some are infaunal and are completely surrounded by sediment (Richter, 1961, 1964b; Buzas, 1965, 1968b; Brooks, 1967) whereas others occupy open burrows of other animals (Daniels, 1970a). Living specimens have been found as deep as 16 cm (Boltovskoy, 1966c) and even 35–40 cm (Akers, 1971). Laboratory experiments (Lee *et al.*, 1969) indicate that some species are capable of vertical motion through a sediment column whereas others are not. The observations of Boltovskoy (1966c) suggest that more subsurface penetration occurs in coarse grains than in fine. Frankel (1972) observed sexual reproduction occurring 3–34 mm below the surface in a pebbly sand substrate.

C. Influence on Test Morphology

The morphology of the multilocular sedentary foraminiferal tests is influenced more by the nature of the substrate than are other forms.

The effect of substrate on test morphology was first noted by Carpenter, Parker & Jones (1862). Hada (1957) indicated that the tests of cosmopolitan agglutinated forms have a coarse surface texture when found on coarse grained sediments, and a fine texture on fine grained substrates. These field observations agree with those made in laboratory experiments conducted by Slama (1954), indicating that some agglutinating species will use whatever materials are available for test construction.

Krasheninnikov (1958a) observed that the 'euryfacial' (see page 253) species have variable forms based on the nature of the substrate, but, unfortunately, this author did not mention the nature of the morphological modifications nor the species examined. Saidova (1953) made correlations between the morphology of *Uvigerina* and sediment character. Myers (1943a) indicated that planispiral, fusiform and discoidal tests tended to be associated with firm sand and sand–mud bottoms. *Nubecularia lucifuga*, which lives on blades of sea grass, has a planispiral form when attached to blades and threads of grass which are narrower than the normal test width of the organism (Arnold, 1967b).

The Miliolidae of the Gulf of Cariaco, Venezula which occupy sandy bottoms are striated and have polygonal and wedge shaped chambers, whereas those living on clay bottoms are unornamented and have smooth rounded chambers (Seiglie & Bermúdez, 1963).

In a study of fossil forms Hendrix (1958) noted thick walled forms in coarse grained sediments and thin walled forms in fine grained sediments. He also noted that finely laminated deposits often contain specimens without ornamentation whereas those deposits with massive bedding yield more highly ornamented forms that have sharp peripheral angles.

It is obvious from the above observations that we know very little about the precise relationship between the substrate and morphology.

In summary: 1. The substrate seems to have some effect on the foraminifera, especially the agglutinated and sedentary calcareous forms, although perhaps not a very important one. While there may be a preferred substrate for a given species, ecological factors other than the sediment type are more instrumental in controlling the distribution and abundance. Even those species which appear to be intimately related to sediment type may in fact be reacting to a yet different environmental factor which is indirectly related to the substrate; e.g. a poorly sorted mixture of sand, silt and clay may provide the optimum supply of nutrient material required by a given species. Despite these qualifying remarks, there are certain species which are found consistently associated with certain types of substrate. 2. There does seem to be a general correlation between the Miliolidae and sandy sediments although this correlation may be a function of a particular hydrologic regime in which $CaCO_3$ concentration is the controlling factor and sand size particles are only a consequence of this hydrology (see section 12). 3. Although the correlations are not very good, the bulk of the observations suggest that fine sand mixed with some shelly fragments and silt or clay support the richest standing crop of benthonic foraminifera. 4. The sedentary forms exhibit the greatest morphological variation as a result of substrate conditions.

6. pH

The alkalinity, acidity or neutrality of water is controlled by the quantity of OH^- and H^+ ions present in it. Normal sea water has a pH of 8.1, but often varies between 7.8–8.3. Under special conditions it is occasionally higher; Moore (1957) reports values as high as 10 in some tidal basins. Some estuaries are acidic (pH 6) because of the quantity of CO_2 liberated by the decomposition of organic materials. Acidic conditions are encountered in some bays because of the influx of H_2SO_4 enriched rivers which drain areas containing decomposed sulfides, e.g. off the Chilean Coast.

There is no doubt that there is a relationship between the pH and the physiology of most marine organisms. The metabolic rate of many fish decreases in an acid environment. Some phytoplankton increase their reproductive rate in an acid setting, whereas an alkaline situation is better for others. At a pH of 6 the fertilization of sea urchin eggs can not take place. Some protozoans cease growing at a pH between 4 and 6.

As is the case with salinity and temperature tolerance, all organisms can be divided on the basis of their tolerance to pH into stenoionic and euryionic forms. Most marine organisms appear to be stenoionic.

The influence of pH on the foraminifera has not been well studied although there is no doubt that there is a significant relationship. This factor appears to play a different role in the protoplasmic mass than in the test. For this reason the two parts will be treated separately.

A. Influence on the Living Organism

There are undoubtedly pH tolerance limits for each species at which certain vital processes slow or cease. More is known about the effects of low pH waters than those with a high pH. The first published account of foraminifera living in low pH waters was published by Gauthier-Lièvre (1935) who found *Trochammina*, *Anomalina*, and various Miliolidae living in fresh water of pH 7.2. Myers (1943c) suggested that small specimens could better survive in acidic waters than could large ones. There does seem to be a correlation between low pH and (a) low foraminiferal abundance and (b) faunas in which agglutinated forms dominate (Phleger, 1970).

Studies of the effect of pH on foraminifera in cultures conducted by Bradshaw (1961) indicate an effect less pronounced than one would expect from observations in nature. Bradshaw observed that *Ammonia beccarii tepida* could survive 25–75 minutes in a pH of 2 although *Spirillina vivipara* lasted but 2–6 minutes in this habitat. *Allogromia laticollaris* survived in pH conditions ranging from 3–9.5 (Lee & McEnery, 1970). Specimens of *Elphidium macellum* and *Quinqueloculina* cultured by one of us (EB) became much less active as the pH was reduced below that of normal sea water and ceased all activity at a pH of 7.

The relationship between substrate type and pH and the effect of this relationship on foraminiferal distribution was first discussed by Miller (1953) who noted that clay sediments rich in decaying organic material (low pH) supported fewer foraminifera than clean sands. Zobell (1946) suggested that well sorted coarse grain sediments have a relative high pH, probably equal to that of the water above, whereas poorly sorted fine grained sediments have a pH lower than that of the superjacent water mass (and would therefore contain a less numerous fauna). Arnal (1961), studying the fauna of the Salton Sea, California, concluded that the most important ecological factor influencing the distribution of the specimens was the pH. The pH of the bottom waters varied between 7.34 and 8.41 whereas that of the bottom sediments varied between 6.74 and 7.96. Arnal noted a correlation between reduced pH and the combination of increased depth and smaller grain size. Under these conditions the quantity of foraminifera decreased.

B. Influence on the Test

At a pH less than 7.8 calcareous tests will begin to dissolve. Such values are often encountered in marshes and some bays which are characterized by a large quantity of decomposing organic material. A thin (1–2 cm) surface layer with a high oxidation potential is underlain by a reducing zone characterized by low pH values (6.7–7.4). Under these conditions calcareous materials begin to dissolve and agglutinated forms become proportionally more abundant.

The effect that a low pH substrate has on foraminiferal tests has been noted by a number of investigators studying marshy enviroments (Parker & Athearn, 1959; Benda & Puri, 1962; Bartlett, 1964b; Phleger, 1966). Schafer (1970) reported living specimens of *Elphidium* surviving in pH conditions of 6.7.

There is a daily variation of pH in the Mission Bay, California marsh (Phleger & Bradshaw, 1966) from 6.8 during the night to 8.5 during the daylight hours. Thus, the foraminifera are subjected to dissolution half of the time. The living organism must spend a great deal of energy recalcifying and one would expect to find a very specialized fauna inhabiting these zones (see Chapter VI,1). Culture experiments on *Rosalina floridana* indicate that the organism is capable of living in a low pH environment and can recalcify easily (Angell, 1967c). Empty tests, however, will not survive long in the low pH, reducing zone.

In summary: 1. Low pH creates a stress situation in which calcareous specimens must spend considerable energy recalcifying their tests. 2. Low pH is detrimental to the foraminiferal organism and thus restricts its distribution. 3. Low pH values in the substrate will cause the dissolution of empty tests. 4. We know little about the effects of highly alkaline (high pH) conditions.

7. Organic content of the substrate

The role played by organic matter in the substrate is varied. Sometimes it can produce the acidic conditions whose effect on foraminiferal distribution was discussed in the previous section. But in other situations the organic material seems to enhance the nutritive content of the water and substrate and thus favour the development of foraminifera.

Exemplifying the latter situation, Phleger (1960b) cites numerous cases along the lagoons and bays of the Texas coast in which there is an increase in the foraminiferal population in finer grained sediment, e.g. those containing a greater quantity of organic materials. The foraminifera of the Ems Estuary also are more numerous in muddy bottoms containing a high organic content than in areas with pure sand (Voorthuysen, 1960). The faunal patterns of foraminifera and ostracoda in the eastern Gulf Coast area correlate well with patterns of organic carbon in the sediment (Benda & Puri, 1962). Populations are high where carbon content is high. Seiglie (1970b) noted that the foraminiferal population is high in Yabucoa Bay, Puerto Rico, wherever the organic content of the substrate is high.

However, the observations of Saidova (1960a) in the Okhotsk Sea do not support these conclusions. She overlaid quantitative and qualitative foraminiferal distribution maps on a map of the hydrocarbon content of the sediment and was unable to find any correlations. In the Wash area, England, the foraminiferal populations were very low in areas of high organic content (Kureshy, 1969).

The nature of the particular organic conditions that would result in increased acidity in one case and increased nutrient content in another are not well known and require more observations than have been made to date.

In summary, organic material in the substrate: 1. Can provide nutritive matter and thus be beneficial to foraminiferal development. 2. Can result in increased acidity and be detrimental to foraminiferal development.

8. Trace elements

Certain chemical elements, e.g. B, Mg, I, Co, Rb, Mo, Zn, Cu, U, Th, Ac and Ra, are amply distributed in the waters, sediments, rocks and soils of the earth but in very small quantities and in a non-uniform manner. Under natural conditions these trace elements are found as inorganic ions and in organic chemical compounds (vitamins, hormones, fermenting agents).

It has been known for more than a century that certain trace elements are necessary for the biological well-being of various organisms, but only in the last few decades have the details of these biological relations been examined. The presence, absence, disequilibria or inter-relations of some of these trace elements in individual organisms can retard or stop normal growth, can provoke abnormal development (monstrosities) and can even induce death. Although most of the detailed studies have been conducted on domestic animals and cultivated plants, it is reasonable to expect that the effects of trace elements on more primitive plants and animals is just as complicated and just as powerful. Vinogradov (1953) concluded that the simple plants are exceptionally variable with respect to the effects caused by the trace elements within them. He noted that some plants previously considered as new species are in reality nothing more than morphological variants of existing species which were created by the presence of certain trace elements. These induced traits are not hereditary and if the trace element imbalance is corrected the plant may return to its normal state. It is possible that some new taxa of foraminifera may be explained in the same way.

Very little has been done with respect to the relationship between trace element distribution and foraminifera, although some analogies with other shell secreting animals may be appropriate, at least to show the kinds of effects which can occur. The gastropod *Helix pomatia* is known to increase in size in copper rich areas. The Pacific coast observations of Schuster (1954) on the gastropod *Olivella* indicate that the presence of very small specimens can be correlated with an excess of ferric ions liberated from weathered Quaternary volvanic rocks there. Tasch (1953) published a detailed summary of those works dealing with dwarfed faunas, especially molluscs. In many cases, he noted the influence of trace elements.

Various authors have enumerated the quantity of trace elements present in the tests of foraminifera (see Chapter III,2). Krinsley (1960) and Barchuk *et al.* (1970) noted a correlation between the presence of Mn in tests and the locations of the specimens. Krinsley observed a latitudinal correlation whereas Barchuk *et al.* noted a bathymetric one. Blackmon & Todd (1959) suggested that the Mg content of calcareous tests depended on temperature conditions, whereas Ponder & Glendinning (1974) indicated no relationship between $MgCO_3$ in miliolids and water temperature or depth. But the relationship between trace elements and ecology is not well studied. Boltovskoy (1956b) tried to make some correlations in a study on the Argentine continental shelf. He noted that in the northern area the foraminiferal fauna exhibited the following evidence of depauperization: (a) few specimens and few species, (b) small size, (c) lack of ornamentation, (d) tendency to asymmetry and monstrostiy and (e) retardation of ontogenetic development in some species. Comparing this fauna with those of other similar regions, he was unable to discover any known environmental factor which was different, but after making spectrographic analyses of the tests of the dwarfed fauna and of normal tests of the same species in other areas he discovered that the former contained trace quantities of Pb whereas the normal specimens did not. Boltovskoy suggested that this element retarded the growth of phytoplankton. The low productivity of diatoms in this area supported this idea. The lack of phytoplankton, a nutrient source for foraminifera, would impair the growth of the individuals. Boltovskoy also suggested that the Pb may have a direct toxic effect on the protoplasm of the foraminifera.

An impoverished fauna in the Hudson River Estuary, New York was correlated with bottom sediments which were depleted in various metallic ions usually important in biological nutrition (McCrone & Schafer, 1966). The preliminary nature of their study and that of Boltovskoy, and the lack of any detailed examination of the problem make it clear that the nature and mechanisms of the effects of the trace element regime on foraminifera is virtually unknown.

In summary, trace elements: 1. May be essential to the existence of some marine organisms (including foraminifera), 2. Can create morphological variants and dwarfed faunas.

9. Turbidity

Excessive turbidity appears to reduce the number of plankton that can live in an area. Since light penetration is inversely proportional to turbidity, it is understandable that in high turbidity (low light penetration) conditions the amount of photosynthetic activity is reduced. Thus, the quantity of foraminiferal nutrient material is reduced. It has been estimated that photosynthetic activity occurs from the surface to depths equivalent to two and one-half times the depth at which a Secchi disc disappears.

It is well known that living planktonic foraminifera require clear and clean water. The reduction of planktonic foraminiferal population near coastal areas is partially explained by the influx of turbid, terruginous water. Tappan (1951) indicated that planktonic foraminifera in the past also required clear, clean water.

Turbidity appears to have similar effects on benthonic foraminifera. Myers (1942c) observed a reduction in the benthonic populations in those areas of the Java Sea where there was rapid sedimentation and high turbidity. Closs (1962), studying the Dos Patos Lagoon, Brazil and Frerichs (1970), examining the Andaman Sea, Burma noted the same phenomenon. However, Stephenson & Rees (1965b) suggested that a mild increase in turbidity in some areas might increase the amount of food available and thus increase the survival of benthonic forms. Anomalous benthonic foraminiferal distribution patterns off the Cameroon coast, Africa, were attributed to the influence of turbid river water which lowered the salinity and increased the amount of suspended material in the water (Berthois, Le Calvez & Crosnier, 1968). Similar observations were made by Uchio (1964) in Japan and Vedantam & Rao (1970) in India.

Some investigators (Stainforth, 1952b; Boltovskoy, 1957b) noted an increase in the abundance of agglutinated forms in turbid waters. It does not necessarily follow that increased turbidity favours the development of these forms, but more likely, that it is selectively unfavourable for the calcareous forms. The agglutinated forms are therefore present in greater relative abundance.

In summary, tubidity: 1. Is a factor which is generally prejudicial to planktonic and calcareous benthonic foraminifera. In an indirect manner it favours the development of agglutinated forms. 2. Has a restricted zone of influence, being effective as a limiting factor only along coastal zones, particularly where there are river mouths.

10. Oxygen

The presence of dissolved oxygen in sea water allows organic material to be oxidized and animal life to exist. Photosynthetic processes produce most of the dissolved oxygen although a small amount comes from direct absorption from the atmosphere. Because of the nature and location of both of these processes, the upper layers of the oceans are the most oxygen enriched. Marine water is most oxygen rich in the spring and early summer because phytoplankton blooms occur at these times. Oxygen content is also related to water temperature. An increase in water temperature decreases the capacity of water to dissolve oxygen. In addition to temperature and phytoplankton blooms the quantity of dissolved oxygen depends also on the quantity of organisms which consume oxygen. The oxygen content of surface sea water usually oscillates between 4.0–6.5 ml/l. Some planktonic organisms can survive in concentrations as low as 0.17 ml/l.

Cold polar surface water charged with oxygen, because of its high density, sinks and migrates toward the Equator. During its travel oxygen is slowly lost, but because of the low temperature some oxygen is retained. However, there are cases of water masses whose oxygen content is so low that it supports little or no life. These masses are usually transitional layers in the ocean and are located beneath the thermocline.

As oxygen is indispensable for animal respiration, those water masses depleted in dissolved oxygen can induce great loss of life if they reach the surface along continental shelves. If there is repeated upwelling in such areas there is a very poor benthonic fauna (Gulf of California, Panamanian Coast).

The specific oxygen requirements of foraminifera are variable. According to Sigal (1952), Pokorny (1958) and Bettenstaedt (1962) genera such as *Haplophragmoides, Trochammina, Cyclammina, Bathysiphon* and *Vaginulina* require little oxygen to survive. None of these authors cited oxygen concentration values nor did they offer explanations for their conclusion. It seems that some benthonic genera (*Elphidium, Buccella, Bulimina, Nonion, Buliminella*) and many planktonic species require a relatively large amount of oxygen and possess various symbiotic algae in their protoplasm to supply this need.

Various studies have illustrated unfavourable effects of low O_2 levels on the diversity and abundance of foraminifera.

In Great Pond, Massachusetts (Said, 1953) and in closed Japanese bays (Nagahama, 1951; Ishiwada, 1958) the number of benthonic foraminifera seems to decrease in areas where there is little aeration of the water just above the bottom. Stschedrina (1957) wrote that oxygen levels at great depth in closed basins may be low enough to affect foraminiferal distribution. The differential tolerances of foraminifera to oxygen levels were shown by Saidova (1960a) who found distinct faunal zones in the Okhotsk Sea which correspond to oxygen concentrations. The foraminiferal tests found in waters containing 4–5 ml/l. were different from those in waters characterized by 2–4 ml/l. In a later work, the same author (Saidova, 1961a) indicated that at oxygen levels less than 1–2 ml/l. the quantity of species is reduced greatly. Harman (1964) noted that low oxygen levels often restrict the distribution of various benthonic species in Santa Barbara Basin, California. A restudy of this basin by Phleger & Soutar (1973) revealed that although there is a low faunal diversity in these oxygen depleted waters the number of living individuals is quite large compared to that of nearby areas of higher oxygen concentration. The oxygen poor areas were dominated by *Globobulimina hoeglundi, Suggrunda eckisi, Bolivina seminuda, Nonionella stella* and *Textularia* sp.

The lower limit of oxygen concentration at which some species can live is quite low. Emery & Hulsemann (1961) suggested that specimens of *Bolivina seminuda* and *Suggrunda eckisi* can live at oxygen levels less than 0.1 ml/l. The studies of P. Smith (1963, 1964) on various species of *Bolivina* found off the Pacific Coast of Central America show that *B. semiperforata, B. minuta, B. subadvena* and *B. pseudobeyrichi* can live in water with 0.30 ml/l. of oxygen; *B. seminuda* at 0.38 ml/l.; and *Suggrunda eckisi* at 0.58 ml/l. All the specimens were well developed and often exhibited their maximum size, flare, chamber height and abundance in the oxygen minimum level which also corresponded to the nitrogen maximum level. Smith was undecided as to whether the robust nature of the *Bolivina* fauna in these conditions was due to a favourable or unfavourable environment. She suggested that low oxygen levels would slow the oxidation of detrital material in the substrate and there would be more nutritive material for individual organisms. Seiglie (1971b) noted the increased abundance of *Ammonia tepida* in Puerto Rican bays where there was low oxygen and abundant organic material. A similar low oxygen, abundant foraminifera zone was noted by Pujos (1970) in the Bay of Biscay, France. Theyer (1971c) noted that specimens of *Cyclammina cancellata* in the Peru–Chile Trench area reached their maximum size in the oxygen minimum zone.

Studies conducted along the northern and western coast of South America (Khusid, 1971; Saidova, 1971; Sellier de Civrieux & Bonilla, 1971; Boltovskoy, 1972; Phleger & Soutar, 1973) revealed low oxygen zones on the bottom (0.14–0.38 ml/l.) and a corresponding depletion in benthonic foraminiferal diversity. No agglutinated forms were present and the calcareous specimens were dominated chiefly by *Bolivina*, *Suggrunda* and *Brizalina* whose size was small and whose ornamentation was poorly developed.

From the above information it appears that *Bolivina* (and its close relatives *Suggrunda* and *Brizalina*) is the most tolerant genus of low oxygen levels. However, despite its tolerance, this genus appears to be morphologically affected by low O_2 content. In contrast to the observations of P. Smith, Lutze (1962, 1964a), Harman

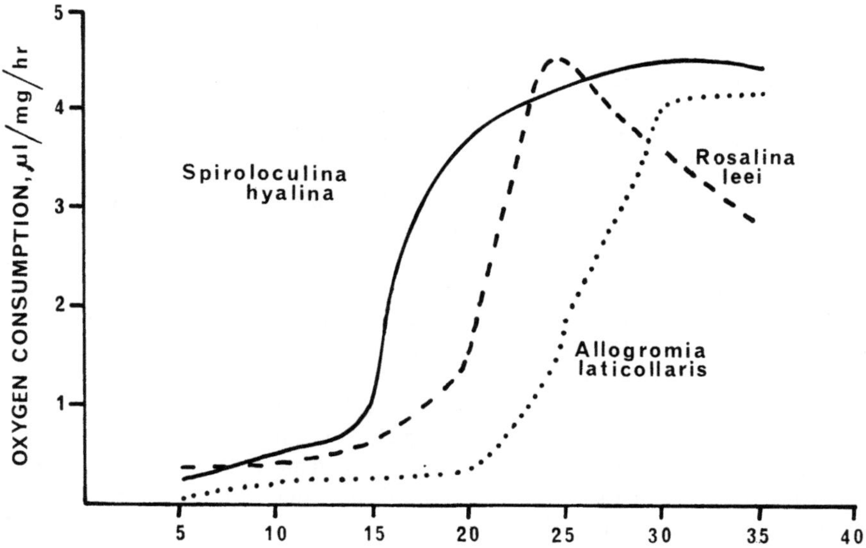

Fig. 80. Oxygen consumption as a function of temperature for three benthonic foraminifera (after Lee & Muller, 1973).

(1964) and Boltovskoy (1972) showed that many *Bolivina* species were narrow, poorly ornamented, and had fragile thin walls when found in poorly oxygenated waters.

The consumption of oxygen by cultured specimens was studied by Bradshaw (1961) who utilized a microspirometer to measure oxygen intake in *Ammonia beccarii tepida*. He noted that individuals of this species consumed an average of 2500–8000 μμl. of oxygen/hr. This consumption was a function of temperature with maximum usage occurring at 35 °C. There was an understandable relation between the size of the specimen and the amount of oxygen consumed. Additional laboratory oxygen consumption data are shown in Fig. 80. Using Bradshaw's data for oxygen consumption, Phleger & Soutar (1973) estimated that the benthonic foraminiferal fauna consumes about 0.26 ml of oxygen per year, only 3.5% of the available oxygen supply. They suggest that despite low oxygen levels in many areas there is sufficient oxygen to conduct metabolic activity at a normal rate considering the small size of the specimens and the low temperatures of most oxygen depleted zones.

In summary, it seems as though low oxygen concentration: 1. Reduces the number of species, 2. Enhances the abundant development of certain species; e.g. mainly the *Bolivina* group, 3. Can alter the morphology of specimens. 4. While affecting foraminiferal populations, is not a limiting factor in their existence.

11. Illumination

The amount of energy transmitted to water by solar illumination varies with depth and with the time of day. The greatest amount of penetration occurs when the incident angle of the sun is lowest (at midday). The different wavelengths of light are absorbed at different rates. Long wavelengths (red) yield the most energy and are also absorbed most rapidly. At a depth of 100 m the red wavelengths are practically non-existent and at 200 m the orange rays disappear. As a consequence of this rapid rate of absorption only 18% of the energy available at the surface is found 10 m beneath the surface.

The development of marine vegetation is intimately connected with the penetration of solar energy. For this reason, it is possible to erect a vertical zonation based on light penetration and life zones. 1. Euphotic zone – from the surface to a depth of 60–120 m; very good illumination; principal masses of phytoplankton and abundant zooplankton. 2. Disphotic zone – from 60–120 m to 200 m; low illumination; site of the so-called 'phytoplankton of the shadows;' some diatoms and green algae present to 350 m; low phytoplankton production; abundant zooplankton. 3. Aphotic zone – >200 m; no light and no photosynthesis; rich and diverse zooplankton in upper part (mostly carnivores and detritus feeders); nocturnal plankton.

We have previously noted (Chapter VII) that foraminifera are found in all of these zones. Light plays an important indirect role in the vertical distribution of the foraminifera. The principal nutritive material of these organisms is phytoplankton whose abundance depends on photosynthesis. Moreover, some foraminifera are inhabited by symbiotic algae which require light (Chapter II,6). For these reasons it is quite understandable that the quantity of planktonic foraminifera decreases markedly at depths greater than 200 m (Fig. 70, Chapter VII). There is no marked change in temperature at this depth; only a decrease in illumination.

The direct effect of light on foraminifera has been noted by several authors. Dujardin (1835), Narchi (1963a), Murray (1963) and Arnold (1964) reported no reaction at all. However, Verworn (1889) found that *Elphidium crispum* was attracted to a diffuse light source. This positive phototropic tendency of *Elphidium* was also observed by Jepps (1942), although she noted that it was not a strong one. A negative phototropic response was noted in *Globigerinoides ruber* by Christiansen (1965a). A retractive response of pseudopodia to intense incident light was first noted by Engelmann (1879) and has been observed often in culture studies.

Rhumbler (1911) proposed the idea that some agglutinated forms such as *Vanhoeffenella gaussi* have transparent 'windows' in the pseudochitin which serve to allow light energy to reach the protoplasm.

There are probably light tolerance limits for foraminifera just as there are temperature, salinity, etc. limits. These have rarely been studied even in cultures. Most investigators who study cultures are concerned with obtaining the optimum light conditions for growth so that they can examine the effects of other ecological factors. For that reason, most of the available data is on the light flux necessary for optimum growth in cultures (see Chapter XV). Only a few data are available regarding the effects of light variation. Röttger & Berger (1972) noted that the rate of growth of *Heterostegina depressa* was greater (one whorl per unit time) at 300 lux (the optimum) than at 600 lux (only one-half whorl in the same period of time). In experiments with the same species, Röttger (1972c) noted that a low light intensity (45 lux) slowed growth.

Ordinary, unfiltered light cast on specimens of *Nemogullmia longevariabilis* by Nyholm (1956) caused retraction of the pseudopodia, partial disintegration of the protoplasm, and cessation of growth. When a green filter was put over the light, digestion and growth returned to normal.

In summary: 1. Light is of direct importance to the nutrition of the shallow water foraminifera because of their dependence on photosynthesizing phytoplankton for food. 2. Light may alter the pseudopodial behaviour of foraminifera. 3. Some, but by no means all, benthonic foraminifera appear to be positively photopropic. 4. Light tolerance limits for the foraminifera are virtually unknown.

12. Calcium carbonate

The amount of $CaCO_3$ that can be dissolved in sea water is a function of water temperature, salinity and pressure. The greater the temperature, the greater the salinity, and the less the hydrostatic pressure, the greater the solubility of the bicarbonate ion. As the majority of foraminifera have calcareous tests and utilize $CaCO_3$ from the water to construct these tests, the role of this material is a significant one. We have already noted the large size and thick walls of specimens found in warm tropical waters. The low calcium concentrations in waters of 50–250 m between 15–20 °C (as discovered by the *Meteor* expedition) may be explained by the rich foraminifera and pteropod fauna found in these zones.

A greater proportion of the foraminifera in polar waters are agglutinated because the quantity of available $CaCO_3$ is reduced in colder waters, prohibiting the construction of thick walled massive calcareous forms. The observations of Le Calvez & Le Calvez (1955) in Villefranche Bay, France are interesting in this respect. They found that the Miliolidae in shallow waters are characterized by strong thick walled species (*Triloculina cuneata, Quinqueloculina vulgaris*) whereas the species in deeper, colder waters are smaller with thinner walls (*Sigmoilina ovata, Biloculina labiata*). The authors explained this observation on the basis of reduced available $CaCO_3$ at depth.

Saidova (1961a) also attributes great importance to the $CaCO_3$ concentration for the quantitative and qualitative distribution of calcareous benthonic foraminifera. In the Northeast Pacific Ocean at depths less than 3000 m in the presence of relatively warm waters (1.7–3.5 °C) which are saturated with $CaCO_3$, the foraminifera are able to easily assimilate the $CaCO_3$. At depths greater than 3000 m, in colder waters (1.0–1.3 °C) depleted in $CaCO_3$, the foraminifera have great difficulty in extracting the chemical from the water, and very few calcareous species exist. The increased solubility of calcareous tests below the lysocline (see section 3) also accounts for the reduction in calcareous species.

Hada (1957) suggested that the distribution of agglutinated foraminifera could be explained by $CaCO_3$ availability. He proposed that their abundance in low salinity and low temperature environments (nearshore and polar regions) was due to the decreased amount of available $CaCO_3$ in these zones. This hypothesis was expanded by Greiner (1969, 1970, 1974). He suggested that agglutinated foraminifera which do not require $CaCO_3$ will dominate in zones of low $CaCO_3$ availability (low temperature, low salinity, high hydrostatic pressure). The calcareous forms require the presence of $CaCO_3$ but the differences in the calcification processes between hyaline forms and porcellaneous forms (Chapter III,4,10) are such that each becomes dominant under different environmental conditions. Forms with hyaline walls have an organic membrane that serves as a template for oriented regular crystal growth. They are consequently more efficient in their use of available $CaCO_3$ than are the porcellaneous forms whose calcification processes appear to be random and spontaneous within the cytoplasmic sheath (Fig. 37). The porcellaneous wall, lacking a well organized nucleating surface, requires waters saturated or supersaturated in $CaCO_3$ (which are found under conditions of high temperature, high salinity and low hydrostatic pressure). The hyaline forms should therefore be dominant in areas of intermediate $CaCO_3$ saturation. Greiner tested this hypothesis by studying the data from about 1100 samples from the Gulf of Mexico. He concluded that the distribution of tests of the agglutinated, hyaline and porcellaneous forms was consistent with the hypothesis.

Seasonal studies of benthonic foraminiferal distribution, abundance, and morphology lend some support to the idea of Greiner. Daniels (1970a) noted that the proportion of agglutinated specimens in the Limski Canal, Yugoslavia, increases in the winter when the temperature drops (and there is less available $CaCO_3$).

The observations of Greiner pertain to whole assemblages and his conclusions reflect general trends. There are, of course, exceptions such as the miliolid (porcellaneous) rich assemblages of deep, cold Antarctic waters.

Theyer (1971a) suggested that the location of deep water concentrations of calcareous specimens may coincide with zones of high salinity. Where temperature is low and less calcium carbonate is available, the proportion of calcareous specimens decreases. Only those species capable of secreting calcite in this hostile, low carbonate environment can survive and, therefore, the diversity of calcareous forms decreases.

Morphological variations have sometimes been attributed to $CaCO_3$ availability. Forti & Roettger (1967) found that the walls of *Elphidium gunteri, E. incertum* and *E. excavatum* in the Dos Patos Lagoon, Brazil become thinner in the winter months when there are lower temperature and lower salinities. A study of fossil forms by Abrard (1928) showed that those Nummultidae found in carbonate rich sediments possessed pillar structures whereas those in non-carbonate environments did not.

In summary, calcium carbonate concentration may: 1. Control the distribution of foraminifera as a consequence of its availability. 2. Influence the size and robustness of calcareous tests.

13. Currents and tides

Currents can directly effect the geographic distribution of foraminifera and can indirectly influence their morphology.

A. Influence on Distribution

Illing (1950) demonstrated that dead tests behave as sediment grains and so are subject to transport by current movements. Current transport of empty tests is one of the main causes of the dissimilarities between the living and the dead benthonic faunas in many bottom samples. While the foraminifer is alive the pseudopodia tend to anchor the test to the substrate, and incidentally, prevent substrate erosion (Ross, 1972). Once dead the test is freed from the bottom, the decaying protoplasm fills the chambers with gas, and transport is easily achieved. Tests between 100–1000 μ are most easily transported as particles of this size require the least fluid velocity to overcome bottom adhesion (Schafer & Prakash, 1968). The smaller tests may even be carried by turbulent transport into the upper layers of water. Murray (1965a), Boltovskoy (1966b) and Loose (1970), have found empty benthonic tests in plankton tows. Murray noted that the surface plankton tow sample in which his specimens were found was taken where the water is 78–107 m deep and that the tests were all small (150–200 μ) and thin walled. Boltovskoy & Boltovskoy (1970) found living benthonic specimens in surface plankton samples in water 1188 m deep. It seems, therefore, that faunal mixing as the result of turbulent current transport is a real possibility.

Current action may directly affect the distribution of living tests in shallow water, high energy zones. Berthois, Le Calvez & Guilcher (1964), Guilcher *et al.* (1965), Le Calvez (1965), Murray (1965b), Kureshy (1969) and Brooks (1973) all noted this influence. The data of Tinoco (1959) on the Amazon delta are interesting for what they show about the effects of large scale currents on foraminiferal distribution. The fauna of the delta itself is a mixed brackish water assemblage but only a short distance away from the delta front, just shoreward from the 100 m isobath the fauna is completely marine. This transition can be explained by the influence of the Equatorial Current which carries marine water coastward and northward.

Daily tidal movement has an effect on benthonic foraminiferal distribution. A number of studies have indicated that where tidal currents are rapid the specimen abundance is low (Richter, 1964a, 1967; Phleger, 1967; Murray, 1968a, 1969; Haynes & Dobson, 1969). According to Richter (1964b) the disruption of the sediments by tidal currents destroys the feeding net of the foraminifer. He demonstrated that in marsh environments the greatest populations occur where the least tidal current movement takes place. This occurs at the low and high tide zones. The assemblages in these two zones are abundant and distinct from one another (Richter, 1967; Phleger, 1970).

Planktonic foraminifera are incapable of self-locomotion and so must rely on current action for movement during their entire life span. Surface currents may carry specimens into areas which are outside their optimum tolerance limits. These movements produce some interesting distributional patterns which are discussed at length in Chapter XIV,14. Bottom currents play a lesser role in the distribution of living planktonic specimens although they may influence empty tests.

B. Influence on Test Morphology

The influence of currents on test morphology and behaviour is less well understood. Le Calvez (1953) suggested that the plastogamic reproductive habits of *Glabratella ornatissima* are an adaption to life in turbulent habitats. The agglutinated form *Haliphysema* possesses a fibrous protein sheath which provides flexibility to the organism and enables it to survive highly agitated conditions (Hedley & Wakefield, 1967a). Delaca & Lipps (1972) proposed that the attachment of *Rosalina globularis* to the substrate by an organic membrane is also an adaption to high energy

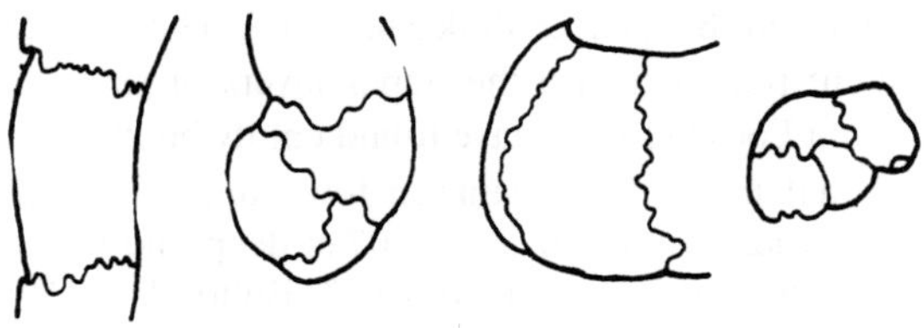

Fig. 81. Fimbrate sutures in *Meandroloculina litoralis* (after Bogdanowicz, 1952).

environments. Hendrix (1958) studied *Cassidulina, Cassidulinella, Uvigerina, Bolivina* and *Virgulina* in the Miocene strata of California and concluded that those specimens living in tranquil areas without strong currents tended to have thick walls with narrow carinate profiles and prominent ornamentation. The specimens found in sediments representing an environment of strong current activity lacked highly ornamented tests and carinae, and were rounded in outline.

Tidal movements and currents have the advantage of stirring up nutritive material and oxygenating the water, but, on the other hand, they tend to increase turbidity and may cause direct physical damage to the test. Bogdanowicz (1952) indicated that Miliolidae living in turbulent waters often exhibit thick walls, a persistent rim and fimbrate sutures (Fig. 81). He interpreted these traits as a response to agitated waters. Lipina (1961) observed similar adaptions in other foraminiferal families and proposed the general rule that foraminifera living under highly turbulent conditions develop tests with some means of reinforcement (supplementary endoskeletons, thicker walls, etc.), but under tranquil conditions small, flattened, thin walled tests predominate. Dihdkov'skij (1959) suggested that loosely coiled tests of *Peneroplis* are typical of quiet (and deep) waters whereas in agitated waters the tests become tightly coiled. He explained these morphological changes as means of increasing the strength of the test under high energy conditions.

In summary, currents and tidal movements: 1. Physically alter the bottom distribution of both living and empty tests, 2. Can transport benthonic forms into the planktonic zone, 3. Can adversely affect the quantity of living specimens in areas of rapid movement (tidal zone), 4. Can create conditions which result in alterations of normal test morphology.

14. Other ecological factors

Factors such as sedimentation rate, interaction with other organisms, the presence of unusual chemical compounds, etc. undoubtedly play a role in the ecology of foraminifera. Some of these additional factors are more influential than others, but few of their effects have been examined in detail.

Various investigators have studied the relationship between the sedimentation rate and the development of foraminifera (e.g. Rauser-Chernoussova & Kulik, 1949; Arrhenius, 1950; Miklukho-Maclay, 1955; Lipina, 1961 and others), but in reality these studies were concerned more with the preservation of the specimens than with their development.

Mateu (1970) cited the wind as a minor environmental factor in shallow water environments because a strong wind could disturb the vegetation on which many benthonic forms live.

The influence of other organisms on the foraminifera is certainly of major importance. The competition for nutritive materials as well as for space, both between and among species, is significant in the distribution and morphology of the foraminifera. Various culture studies have shown that the presence of certain organisms, primarily bacteria, in the cultures are necessary for optimum development (Lee, McEnery & Rubin, 1969). The role of predators is discussed in Chapter II,5 and the effect of human activity on foraminiferal life is treated in Chapter II,8.

The presence of various organic compounds in sea water has been established in the last few decades. These substances, primarily enzymes or external metabolites, are organic secretions which often emanate from flagellates, can influence other organisms in a favourable or unfavourable way. Some groups of algae can not fully develop due to the toxic effect of various organically secreted enzymes in sea water. Some enzymes disappear when the water is heated, although others are very stable and their influence persists even after the secreting individual has died. The diatom, *Nitzschia*, for example, can only grow in water which has been occupied by *Chlorella* if that water has been well filtered. The 'red tides' which kill millions of marine organisms are good examples of the unfavourable effects of toxic enzymes. The lack of correlation between the geographic distribution of phytoplankton with zooplankton is often explained by the influence of these enzymes. These organic compounds surely have an effect on foraminifera but the specific nature of this effect is unknown.

Aside from the role played by organic matter in the sediments, a certain ecological role is played by some minerals found on the bottom. Saidova (1960a) made some interesting correlations between the foraminiferal and iron oxide distribution in the Okhotsk Sea. She demonstrated that there is a relation between the quantity of agglutinated foraminifera and the concentration of iron oxide in the bottom sediments. Stschedrina (1958c) found that the fauna of the Kurilo–Kamchatka basin is quantitatively and qualitatively poor. She suggested that the primary cause of this phenomenon is the mineralogical content of the bottom sediments.

Correlations between various environmental factors and the distribution and/or abundance of individual species or whole faunas do not necessarily indicate a causal relationship. The observation of a correlation requires some explanation on the nature of the effect before we can consider that particular ecologic factor as a significant one for foraminifera. For example, Mangin & Mangin (1969) noted the positive correlation between atmospheric pressure and the number of planktonic foraminiferal specimens captured in a year long study near Villefranche, France. There is yet no explanation for this correlation.

As we conclude this discussion of ecological factors we wish to reemphasize that the effects of the various factors on foraminifera are not isolated from one another. The combinations of effects may produce results which are different from the sum of the effects of the factors if they acted alone.

Muller (as described in Lee, 1974) studied some of these interactions and was able to demonstrate the effect of salinity/pH, temperature/salinity, and temperature/pH taken together on the generation time of 3 cultured species. The results of these experiments clearly revealed the effects of one ecological factor on another. For example, when these factors were considered separately, *Allogromia laticollaris* reproduced in the temperature range 14–32 °C, the salinity range 20–45‰ and the pH range 5–9.5. But when temperature and salinity were considered together the salinity range of reproduction changed from 18–45‰ at 25 °C to a range of 21–41‰ at 20 °C, and to a yet narrower range of 20–30‰ at 10 °C. When salinity and pH were considered together the salinity range of reproduction remained rather constant at about 15–42‰ for pH values beteeen 5–9.5, but time required to produce a new generation increased markedly from alkaline conditions (pH 7.5–9.5) to acidic (pH 5.5) conditions. When temperature and pH were considered together the pH range of reproduction which was 5–9 at 30 °C increased to 5–9.5 at 25 °C, and dropped to 7.5–8.5 at 10 °C.

The effects of intraspecific competition were also considered in this study. Different species seemed to react in different ways. *Allogromia laticollaris* fed and reproduced at a slower rate when crowded than when isolated, whereas *Rosalina leei* fed more rapidly when it was surrounded by other feeding specimens of the same species. The feeding rate of *Spiroloculina hyalina* seemed to be unaffected by crowding although its reproductive rate declined under such conditions.

Muller proposed a simple quantitative relationship to express the intensity of interspecific competition. The relationship was written for the effect of two competitor species on a third and is here expanded to include any number of species:

$$\text{I.S.C.} = \frac{\sum F_i}{n \cdot F_\alpha}$$

where

I.S.C. = interspecific coefficient of competition

F_i = feeding rate when the studied species is in competition with the ith species.

n = number of competitor species

F_α = feeding rate when studied species is grown without competition.

The coefficient varies from 0 (severe competition) to 1 (no competition). This coefficient as calculated considers only the average of the pair by pair reduction in feeding rates (two species in competition with each other) whereas there may be additional interaction when more than two species are competing for a food source.

15. Utility of ecological data

Ecological data on Recent foraminifera is commonly used to draw conclusions about fossil forms. Naturally these conclusions will be more accurate if analogies are made between forms at the lower taxonomic levels. For example, knowing the ecological conditions in which *Bulimina patagonica,* f. glabra lives now and finding this same *forma* as a fossil allows one to draw rather precise conclusions about paleoenvironmental conditions, whereas the citation of this same species without the designation of any *forma* yields less information. Drawing paleoenvironmental conclusions from generic data is yet another step removed from the ideal situation.

Of the many studies which have utilized the analogy approach, we cite two as typical examples. Hornibrook (1968) used the modern distribution of the genus *Amphistegina* to draw conclusions about the New Zealand Tertiary. He noted that this genus is today living in shallow warm water of the Central Pacific and is not present in New Zealand waters. However, this species is present in the Tertiary strata of New Zealand, indicating a warmer climate in the past. He noted that the lower temperature tolerance limit of this genus today is about 20 °C. Finding *Amphistegina* in U. Eocene-M. Miocene New Zealand strata, he proposed that the 20 °C isotherm lay south of New Zealand during that period. The absence of the genus in Plio-Pleistrocene strata suggested a cooling during that period. Wright & Murray (1972) compared an Eocene faunal assemblage with a modern one at the generic level to determine environmental conditions. Although the genera were not very similar, they were able to detect transitions among various shallow water hypersaline lagoonal conditions on the basis of faunal diversity indices and other faunal structures.

Unfortunately it is often difficult to find fossil species which are exactly the same as Recent ones because faunas change with time. Moreover, one must consider that many foraminifera change their habitats with time. The family Lagenidae, for example, is currently found living in waters deeper than those it occupied during the Mesozoic. *Pyrgo* also has gradually moved to deeper water since the Tertiary. The Astrorhizidae of the Paleozoic are found in shallow tropical water sediments, whereas they now live in deep middle latitude waters or in polar coastal zones. *Bulimina inflata* and *Sphaeroidina bulloides* occupied shallower waters in the Miocene than they do today. *Uvigerina* appears to have changed its depth habit with time, as has *Cyclammina* which had a much wider bathymetric range in the Late Miocene-Pliocene than in the Pleistocene and Recent. The agglutinated species *Adercotryma glomerata* is found today on the shelf and in depths greater than 3500 m, but lived on the shallower continental slope during the Pleistocene. Thus, a consideration of possible environmental changes through time must accompany any utilization of taxa in paleoenvironmental reconstructions based on analogies between Recent and fossil taxa.

In those cases where there are not even genera in common between the fossil and Recent faunas, the entire fauna can be used to make comparisons. For example, the designation of a fauna as brackish water, planktonic, primitive, agglutinated, etc., provides a certain type of ecological information even though there are no species or genera in common with Recent faunas from the same environment.

An understanding of foraminiferal ecology is of great importance not only from a theoretical point of view but also because it may help to solve various problems in applied geology and oceanography. Some of the applications of our understanding of foraminiferal ecology are treated in detail in Chapter XIV.

X. COLLECTION OF MATERIAL

1. Fossil material

Although this book is dedicated principally to Recent foraminifera, we wish to treat briefly the methods of collection and preparation of fossil tests. Many of the methods and techniques of collection and preparation are the same for both fossil and Recent material. Most marine biologists and geologists who study the living foraminifera of some region draw their taxonomic conclusions on the basis of the test rather than on the soft parts of the organism. The paleontologist draws his conclusions from the same information.

Moreover, the boundary between the study of Recent and fossil foraminifera is blurred in oceanographic study where samples of living specimens, empty Holocene tests on the sea floor and Neogene tests from cores may be collected and studied simultaneously.

Because of their small size, it is not possible to collect foraminifera as one would the molluscs, arthropods and the majority of other metazoans. In reality, the investigator does not collect foraminifera, but rather collects a sample of sediment or sedimentary rock which contains foraminifera.

Some species are sufficiently large to be seen without the aid of a hand lens or microscope, but these are uncommon. Consequently, a preliminary search of the rock surface with a lens or microscope should reveal the presence or absence of tests. The tests often are visible on a wetted surface more easily than on a dry one.

A large number of samples should be taken from the geologic section. The minimum number of samples would be one per vertical facies change in the outcrop. In a thick continuous section, samples should be taken at some predetermined regular or random interval. The greater the number of samples, the more precise will be the results.

It is usually necessary to clean away weathered and contaminated material from the outcrop prior to sampling. Removal of the outer 40–50 cm is usually sufficient to remove slumped and weathered material. The sample should be as fresh as possible so that the probability of finding unaltered tests is high.

If preliminary examination of the samples or prior knowledge of the section indicates an abundant fauna, samples as small as 5–10 g may be sufficient. In the absence of any prior information it is probably wise to collect at leeast 0.5 kg of material. Parts of the sample can be set aside for different kinds of analysis and for distribution to other investigators.

In addition to taking discrete samples, it is possible to take a sample which includes the entire vertical section of the outcrop. A vertical trench is prepared from top to bottom of the outcrop and samples 30–50 cm long are taken. In this way the entire section is represented in the sample suite; nothing will be overlooked or missed because of isolated sampling. Some precision is lost in the location of exact changes and it may be necessary to resample at smaller intervals to locate exact datum planes.

Field examination of samples is very useful especially if the site is difficult to sample and one visit is all that is possible. Field laboratories designed for the purpose of preliminary sample examination were described by Henson (1934), Rauser-Chernoussova & Furssenko (1937) and Glaessner (1945).

Samples taken from cores are another source of fossil material. The surface of the core must be carefully cleaned to avoid contamination. Cores may be sampled in the same manner as an outcrop, at each facies change or at some predetermined interval. A one centimetre thick slice of the core is commonly taken for examination.

Techniques for the extraction of samples during the drilling of a well are well known in the petroleum industry and are described by Rauser-Chernoussova & Furssenko (1937) and Wicher (1942).

2. Recent benthonic foraminifera

Perhaps the simplest and easiest means of collecting Recent benthonic foraminifera is accomplished along the coast in the littoral zone. Well preserved tests are often cast into this zone as a result of wave action. Collection, separation and cursory examination of the material can be done in the following way. A small scoop of littoral material is placed in a wide mouthed bottle and agitated with some sea water. The sand grains will settle to the bottom slightly faster than the foraminiferal tests which float or remain in suspension. If the upper portion of the water is filtered quickly the floating and suspended material will be separated. Examination of this filtrate with a lens will reveal the presence or absence of tests.

This sort of collecting will result in a death assemblage containing not only Recent but perhaps relict Pleistocene and reworked material from coastal outcrops. Occasionally living specimens are found in these littoral deposits, having been carried into shore by wave action. Living specimens are more valuable than dead in the study of Recent foraminifera and their collection requires that samples be taken not in the littoral zone but from substrates that are permanently covered by water.

Large numbers of living specimens can occasionally be obtained from the agitation of marine algae, either by shaking them in a receptacle or scrubbing them with the hands. Algae found in protected environments without waves and surge action contain considerably richer foraminiferal populations than those occurring along open coastlines. After washing the algae it is helpful to pass the residue containing the tests through a large mesh sieve to eliminate the larger algal fragments and macrofauna. If the residue is then decanted several times in sea water, additional algal debris will be removed. The residue is then bottled and transported to the laboratory for final separation and examination (see Chapter XV,2).

These techniques are convenient and easily accomplished but do not yield great quantities of material and information. It is much better to use devices designed especially for the collection of bottom sediments. These devices can be classified according to the depth of water in which they work best. Some are designed for quantitative studies and others for qualitative work.

A. Qualitative Sampling Devices for Shallow Water

1. Metallic tube

This is a very simple device consisting of a cylindrical metal tube, open at one end with the other closed by a piece of cloth. Attached to a line, it can be thrown from a boat or the shore, pulled across the substrate and will consistently provide a sample.

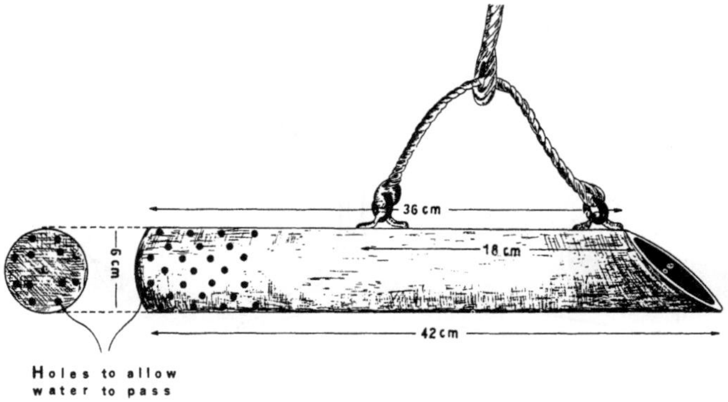

Fig. 82. Metallic tube for bottom samples.

Shupak (1934) described a similar apparatus with some beneficial alterations. The open end is cut obliquely and sharpened. The other end is covered by a metal plate punctured by holes which permit the passage of water (Fig. 82). The sampler works in water up to 10–20 m deep. The tube not only skims off the upper surface of sediment but may also dig deeper into the sediment. Consequently the samples yield only crude information about the Recent fauna. Conclusions about the qualitative foraminiferal composition of an area should be drawn only from large numbers of samples of this type.

2. Hulme surface sediment dredge

This device was designed and successfully used by Hulme (1964) and many others to collect samples at depths up to 20–50 m. One to two centimetre surface sediment samples are scooped up by this device which consists of a steel tube plugged at one end. The tube contains steel fins along the sides to prevent penetration too deeply into the subsurface. The entire apparatus weighs about 9 kg (Fig. 83).

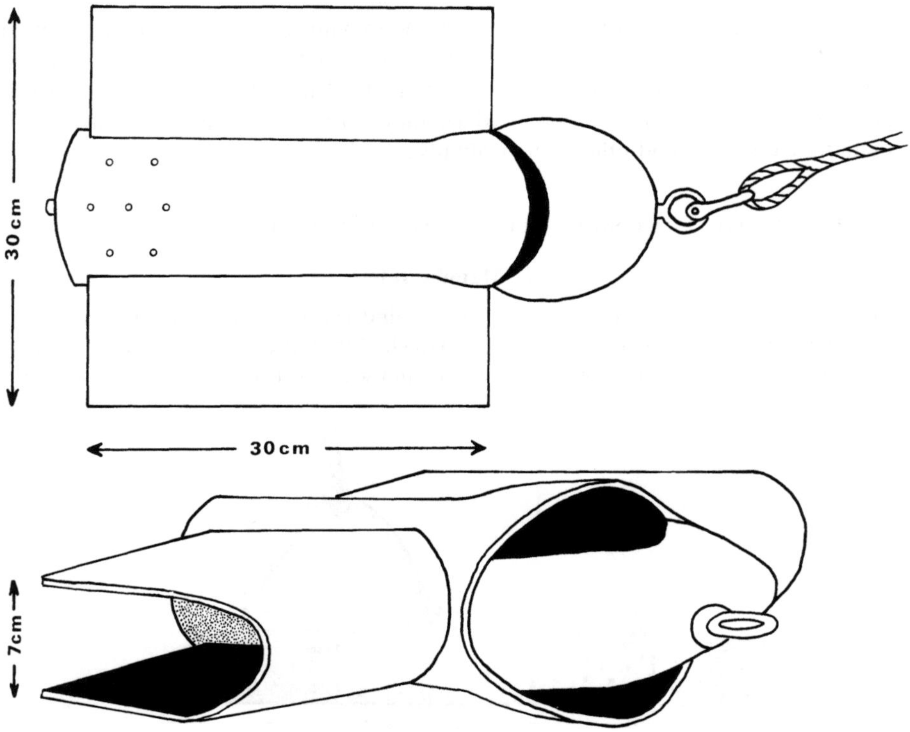

Fig. 83. Hulme surface sediment dredge (after Hulme, 1964).

3. Benson-Kaesler Bolapipe dredge

A lightweight (0.5 kg) shallow water dredge designed to scrape the top 1–2 cm of bottom at distances up to 40 m from the collector was described by Benson & Kaesler (1965). It consists of a 5 cm cylindrical tube, flared at one end and closed at the other with a sample bag. This device is attached by swivels to a line and is spun around the head several times and then thrown.

4. Scraper

In very shallow water with sandy or muddy bottoms, a simple scraping device can provide good qualitative samples. The scraper consists of a metal triangular frame with a 2–3 m pole connected to one of the angles of the frame (Fig. 84). A steel wedge or blade is attached to holes in the frame. The device is used by scraping the sediment off the bottom so that it accumulates in the bag. Although the scraper is designed primarily for qualitative samples, careful use may result in semi-quantitative samples; i.e. 10 passes each 1 m long over an area with a scraper whose frame is 10 cm wide results in a total sample from 1 m^2.

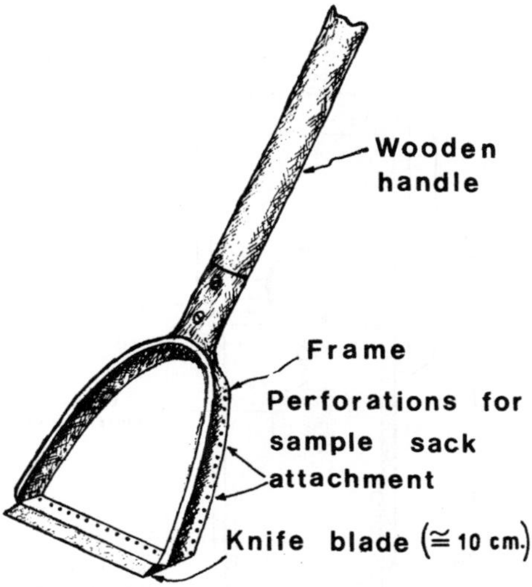

Fig. 84. Bottom scraper.

B. Quantitative Sampling Device for Shallow Water

The Lankford sampler (Figs. 85 and 86) is perhaps the best device for obtaining samples of known surface area and volume in shallow water up to 6 m. The sample, a short core, is taken in a plastic tube (10–15 cm long; 2–3 cm diameter). The core is pushed into the sediment by means of a long pole. Upon extraction the tube is

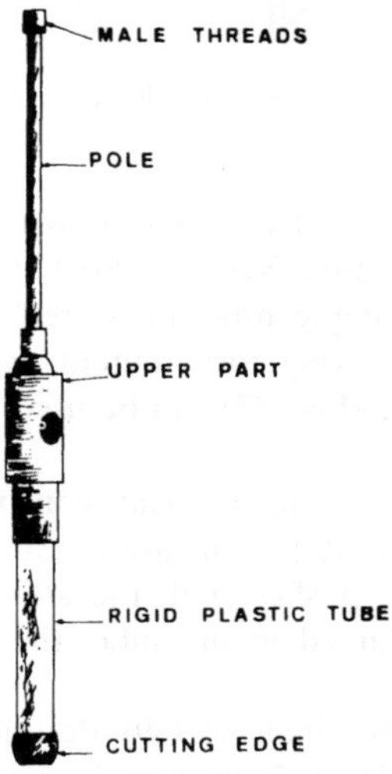

Fig. 85. Lankford sampler, a hand operated coring device.

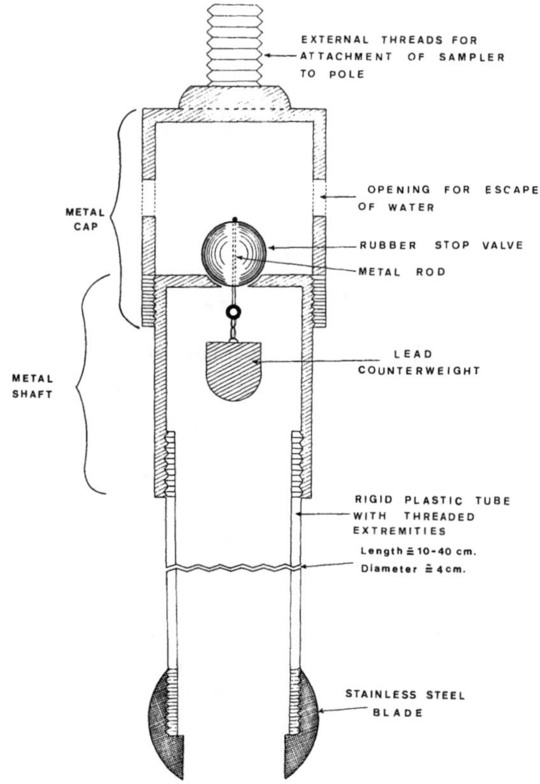

Fig. 86. Construction details of Lankford sampler.

unscrewed from the sampler, the cutting head is removed and the core is plugged. Preservative and stain (Chapter XI) are added at the upper end of the core which is then also plugged. The sample must be stored vertically to avoid mixing the top layers, which contain the living foraminifera, with the lower layers. Once in the laboratory a special device (Fig. 87) can be used to cut off segments of the core for study.

The Lankford sampler is simple, easily constructed and operated and provides samples from which quantitative conclusions can be drawn. It is useful only to the depths to which it can be pushed with the attached pole. Because coarse grained sediments will not be retained in the tube, the device is useful only in relatively cohesive substrates.

Rouvillois & Rosset-Moulinier (1969) described a small coring device which collects undisturbed samples 1–7 cm deep in up to 40 m of water. As the device is

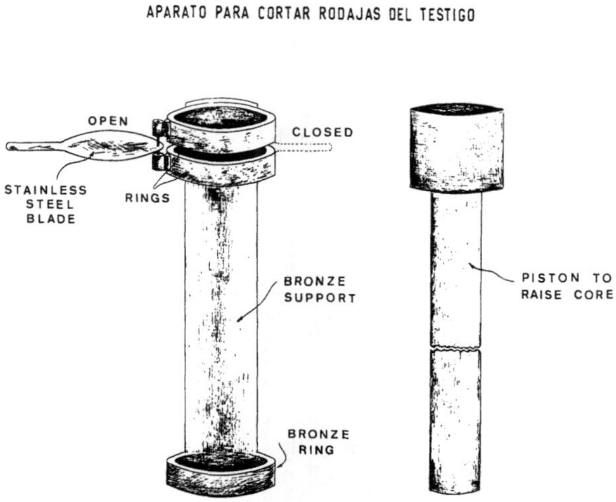

Fig. 87. Device for slicing cores.

designed to close the bottom of the corer when the sample is taken, sediment of any grain size and cohesion can be extracted.

The following two sections deal with sampling devices designed to collect qualitative and quantitative samples in deep water. As these devices are all lowered to the bottom and hauled up again by means of cable and winch, and are usually standard equipment aboard oceanographic vessels, we will not describe their construction in great detail but will only describe the fundamental notion of their operation and the nature of the sample they yield.

C. Qualitative Sampling Devices for Deep Water

1. Small grab samplers

There are several types of deep water sampling devices which yield small bulk samples and are easy to operate. Among them are the orange-peel sampler (Fig. 88), the scoop and others. These devices penetrate the substrate to a certain depth, grab a sample of the sediment and enclose it within the jaws of the device. The mechanics of their operation mixes the sediment sample and destroys the original stratification.

2. Large grab samplers

These are larger versions of the small devices described above which, as a consequence of their weight, are able to penetrate into well indurated materials. Their construction is based on the principle of the shovel with several variations. The Petersen, Gordeev and Okean-50 (Fig. 89) samplers are typical examples of this type of device.

Fig. 88. Orange-peel sampler.

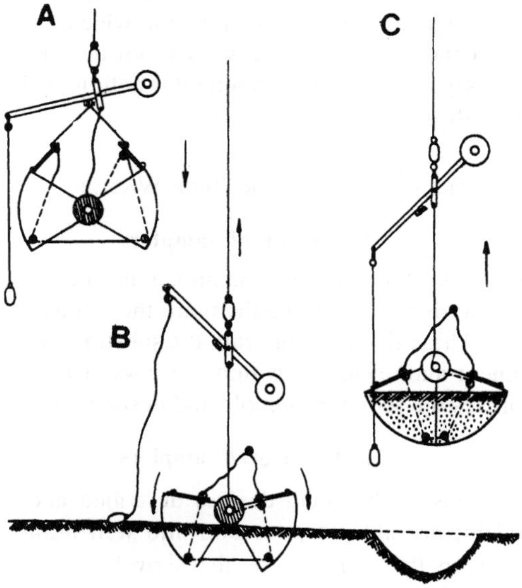

Fig. 89. Method of operation of the Okean-50 bottom sampler (after Zhadin, 1960).

Most rakes and dredges do not work well for foraminiferal samples because the sieve openings are so large that most tests pass through. Only those specimens attached to larger organisms or sediment blocks are retained.

D. Quantitative Sampling Devices for Deep Water

1. Lundquist sampler

This device was described by Höglund (1947) and Christiansen (1958) as being small, easy to operate and as yielding good samples (Fig. 90). It penetrates the sediment due to its own weight. When it is extracted from the bottom the sediment in the tube is pulled downward in the tube producing a partial vacuum in the upper part of the tube. This vacuum impedes the further slippage of the sediment and holds it in place

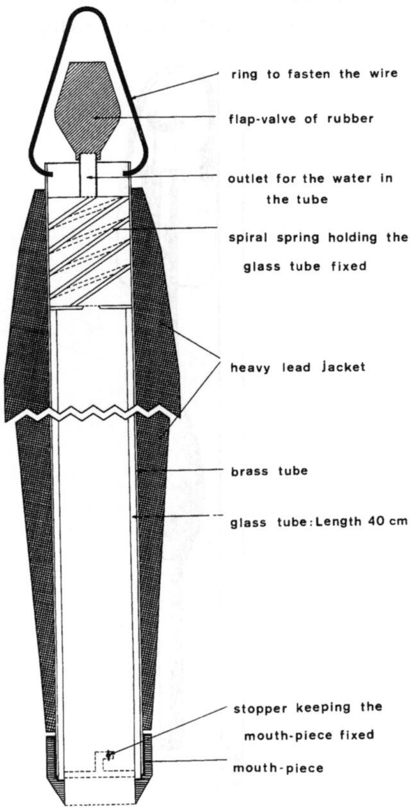

Fig. 90. Construction details of Lundquist sampling tube (after Christiansen, 1958).

as the device is raised off the bottom. In the laboratory, the surface water is removed by a siphon and the sample removed by placing a piston beneath the sample and lifting the core out of the tube. The upper part (1–2 cm) is removed and examined for foraminifera.

The main advantage of this device is its ease of operation and its simple fabrication. It will work in several hundreds of metres of water. The uppermost sediment layers are not disturbed. The main disadvantage is that only fine grained sediment will remain in the tube when it is extracted. Coarse materials lack the cohesion to create the vacuum necessary to hold the sediment in place.

2. Coring devices

All these devices have a main body consisting of a steel tube (with an inner plastic lining) which penetrates the sediment due to it weight. Their great advantage is the removal of a relatively long core of undisturbed material which allows study not only of the living foraminifera in the upper layer but also the pre-Recent material preserved in the lower sections of the core. These coring tubes work well in most

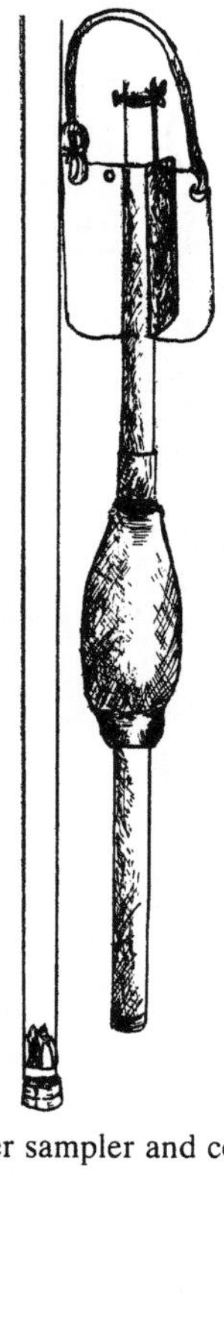

Fig. 91. Phleger sampler and core tube.

sediment types although problems are sometimes encountered in rocky bottoms, in fine sand and in volcanic ash. In the case of the rocky bottom the tube may not be able to penetrate the substrate. Sands and ash generally do not remain in the tube due to their low cohesion.

One of the most widely used coring devices for foraminifera is the Phleger corer (Fig. 91). This device will yield undisturbed cores up to 1.2 m long. Because it is relatively small and light it can be lowered and raised with a small bathythermograph

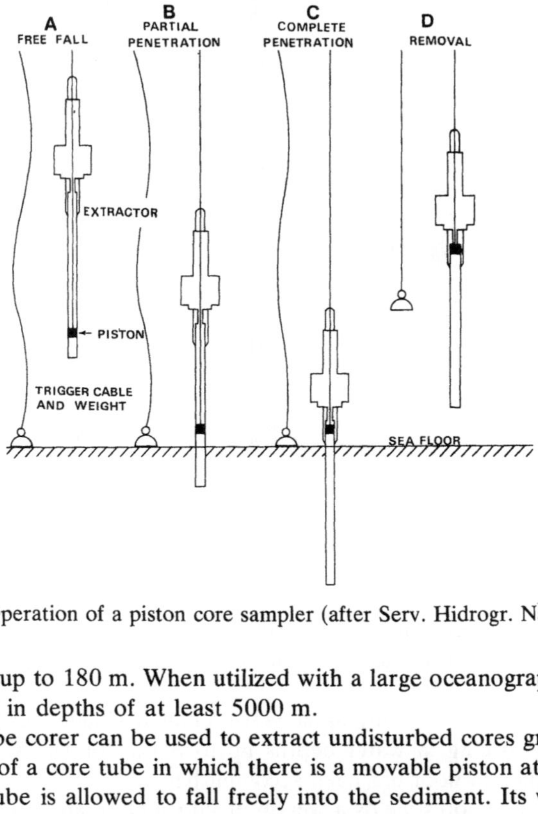

Fig. 92. Operation of a piston core sampler (after Serv. Hidrogr. Naval, 1958).

winch in depths up to 180 m. When utilized with a large oceanographic winch, cores can be obtained in depths of at least 5000 m.

The piston tube corer can be used to extract undisturbed cores greater than 1.2 m long. It consists of a core tube in which there is a movable piston attached to a cable (Fig. 92). The tube is allowed to fall freely into the sediment. Its weight drives the tube into the sediment while the piston remains at the substrate surface. Hydrostatic pressure impedes the formation of an air space between the piston and the core sample. The piston then acts as a suction device against the sample when the apparatus is extracted from the substrate.

The first piston corer was designed by Kullenberg. This device could take cores much longer than simple gravity devices such as the Phleger corer. Ewing made various modifications in the design of this sampler so that it could extract cores of even greater length (Fig. 93).

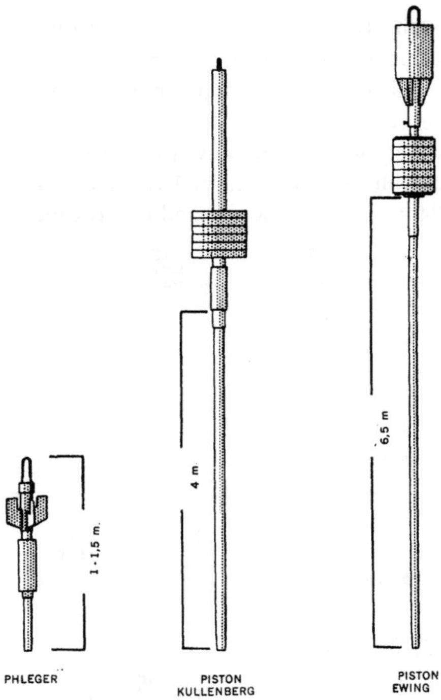

Fig. 93. Coring tubes (after Serv. Hidrogr. Naval, 1958); Phleger corer; Kullenberg piston corer; Ewing piston corer.

Piston coring devices can obtain cores up to 35 m long in water depths up to 3000–4000 m. These cores are generally undisturbed and original bedding can be observed.

In 1968 the deep sea oceanographic vessel Glomar Challenger began cruising and taking cores by drilling in water depths up to 6000–7000 m. Coring to depths of 1400 m below the sea floor has been accomplished.

The study of long cores yields a wealth of information about past faunal changes, evolution, paleoclimates, paleooceanography, etc. (Chapter XIV).

If environmental data is to be taken from a sample it should be measured as soon as possible after removal of the core, as several environmental factors begin to change immediately upon removal from the bottom.

Most samples of Recent foraminifera collected for quantitative study consist of the top 2 cm portion of a core whose surface area represents about 8–12 cm^2 of the substrate. The sample volume is therefore about 16–24 cm^3. Whether or not this sampling area is sufficiently representative of the area from which it was taken is debatable. Studies by Lynts (1966b), Buzas (1968b, 1970), Boltovskoy & Lena (1969a) and Wright & Hay (1971) indicate that there is considerable variation in foraminiferal microdistribution (Chapter IV,6). Therefore it is advisable to even out the patchiness of the foraminiferal distribution by collecting samples from as many sites as possible. It is wiser to have 5 g samples from 10 places than to have one sample of 50 g from one site.

A multiple coring device which collects 5 cores in depths up to 3000 m was designed by Fowler & Kulm (1966). A similar sampler was designed by Willemoes (1964) utilizing 3 cores but the total surface area covered is only 9 cm^2 which is about the same as that covered by conventional single corers.

Several studies have shown that living foraminifera can survive and even thrive abundantly at depths significantly greater than the 2 cm which is the usual sample depth. Boltovskoy (1966c) reported data from cores taken in a variety of sediment types which showed that only 35% of the living specimens were found in the uppermost 2 cm of core. In a study of replicate cores from the Gulf of St. Lawrence, Schafer (1971) found that the top centimetre of core contributed a remarkably small fraction of the living specimens found in the top 9 cm. Data taken from clay and fine silt sediments in a marsh by Akers (1971) was not suitable for quantitative evaluation, but clearly indicated that foraminifera live in moderate abundance down to depths of 40 cm. A statistical study of the distribution of living specimens of *Ammobaculites* indicated that there was no significant change in distribution in the top 10 cm of substrate (Buzas, 1974). Thus, not only do foraminifera live below the surface, but they may live there in abundance, in such abundance that it seems advisable to systematically look for living specimens in the subsurface portions of cores prior to the initiation of any quantitative analysis of the living population.

One of the additional assumptions underlying distributional studies of Recent foraminifera is that the standard core top (8–12 cm^2) is sufficiently large to detect the species present in an area. Dennison & Hay (1967) applied binomial sampling theory to the question of sampling areas and thereby provided a means by which this assumption could be tested. They published a chart which yields information about the probability of being able to detect a particular species if you are given data about its size, abundance on the bottom and if you choose a certain risk of error. Using their chart and assuming that the average foraminifer is about 500 μ in diameter, a standard core tube sample is sufficiently large to detect the presence of a species which constitutes about 0.15% of the sea floor (with a risk of error; $\alpha = 0.01$). This means that if the foraminiferal tests constitute 1% of the sediment one could only be 99% certain of detecting those species which constitute at least 15% of the population. If the tests constitute 10% of the sediment, a species constituting only 1.5% of the foraminiferal population could be detected. We conclude that under most conditions, if one wishes to be certain of sampling those species which constitute at least 1% of the fauna it would be wise to take at least two standard core tubes per sampling location.

3. Recent planktonic foraminifera

Planktonic foraminifera, like all plankton, are sampled by plankton nets. A typical plankton net has a conical shape, is about 0.5 m in diameter at its open end and is made of a fine mesh material. At its apical end there is a glass or metal receptacle (plankton net bucket) whose distal end is covered with the same mesh used in the main net. Glass jars are attached with a cord, metal ones commonly are secured with a metal ring. The anterior open end of the net is towed through the water by a cable about 50 m long. This cable is attached to the metal frame, which forms the mouth of the net, by a 3-way swivel (Fig. 97).

There are many variations on this plan. The net itself may be made of different kinds of material. Before the invention of synthetic fabrics they were usually silk. This material shrinks after the first few tows and the mesh openings become reduced in size. In addition, silk nets are delicate, non-resistant and it is difficult to clean accumulated organic debris from them. In the last decade nylon nets have substituted completely for the silk ones.

An informative discussion of the development of plankton sampling devices and a critique of various net types was prepared by Omaly (1966). Many valuable data with respect to planktonic nets, their construction, usage, maintenance, etc. can be found in UNESCO (1968).

The size of the mesh opening is designated by the following standardized mesh numbers:

Number	Openings/in.	Mesh size (mm)	Number	Openings/in.	Mesh size (mm)
0000	18	1.364	10	109	0.158
000	23	1.024	11	116	0.145
00	29	0.752	12	125	0.119
0	33	0.569	13	129	0.112
1	48	0.416	14	139	0.099
2	54	0.366	15	150	0.094
3	58	0.333	16	157	0.086
4	62	0.318	17	163	0.081
5	66	0.282	18	166	0.079
6	74	0.239	19	169	0.077
7	82	0.224	20	173	0.076
8	86	0.203	21	178	0.069
9	97	0.168	25	200	0.064

When this table was established the smallest mesh size was number 25. However, a mesh size with an opening of 35 μ soon appeared and during the last few years some factories began to produce fabrics whose openings are as small as 10 μ and even 1 μ. Now the majority of producers prefer to name nets not according to their numbers, but on the basis of their mesh diameter. This is more convenient since Japanese and Russian mesh systems and numbers are different than those listed above.

The planktonic foraminifera, members of the microplankton, are commonly between 0.2–1.0 mm in size, although some, *Globigerinita uvula* and *G. iota*, are often smaller. Consequently, the best net for the collection of most planktonic foraminifera is Number 12 or 10 (with openings of 0.119 and 0.158 mm respectively). It is not advisable to employ very fine nets as they will only clog with small phytoplankton and therefore make the separation of the foraminifera more difficult. The additional foraminifera that will be retained on the finer nets will be juvenile specimens which are quite difficult to identify and so are of little use.

It is possible to collect rich planktonic foraminiferal samples even with coarse mesh nets. Boltovskoy (1968a) demonstrated that Number 3 (0.333 mm) and Number 5 (0.282 mm) nets yielded abundant specimens because the foraminiferal specimens smaller than the openings became entangled with other plankton.

There are a number of variables which will affect the numbers of foraminiferal tests recovered in a plankton tow and the resulting calculations of foraminiferal density (number of tests per cubic metre; FN_p/m^3). The method by which water volume is estimated, the depth at which the tow begins, the effects of plankton patchiness, the separation method employed in the laboratory are all such variables. But the size of the net openings is perhaps the greatest single factor in explaining the many discrepancies observed between different investigators. Nets tend to clog and (a) allow less water to pass, and (b) decrease the effective size of the opening. These two effects of net clogging act in opposite directions with respect to the calculation of foraminiferal density.

Berger (1969b) compared concentration data from 18 different investigations and found that the finer the net size the greater the estimate of FN_p/m^3. Empirically analysing the data from these studies he proposed an equation which would allow the calculation of planktonic foraminiferal concentration for an equivalent mesh size.

$$FN_p/m^3\,(\text{equivalent}) = FN_p/m^3\,(\text{actual}) \times (M_a^3/M_e^3)$$

where M_a = actual mesh size used; M_e = equivalent mesh size.

Berger proposed that a mesh size of 158 μ be used and he suggested that data be converted to a 158 μ equivalency. However, the equation should work for any M_a between 30–500 μ.

The most primitive method of obtaining planktonic foraminifera is to place a net over the mouth of the ship's fire pump and draw sea water through this device. There are numerous advantages to this technique: the ship need not reduce its speed, the volume of the filtered water can be calculated very accurately, the pump operates in all weather conditions and samples can be taken in high latitudes even when the ship is surrounded by ice. There are some disadvantages as well. The pump, especially if it is an impeller centrifugal pump, damages meso- and macroplankton. Some ships have piston pumps which move water at a lower discharge rate but cause less damage to the specimens. The rate at which the water is filtered is less than that of

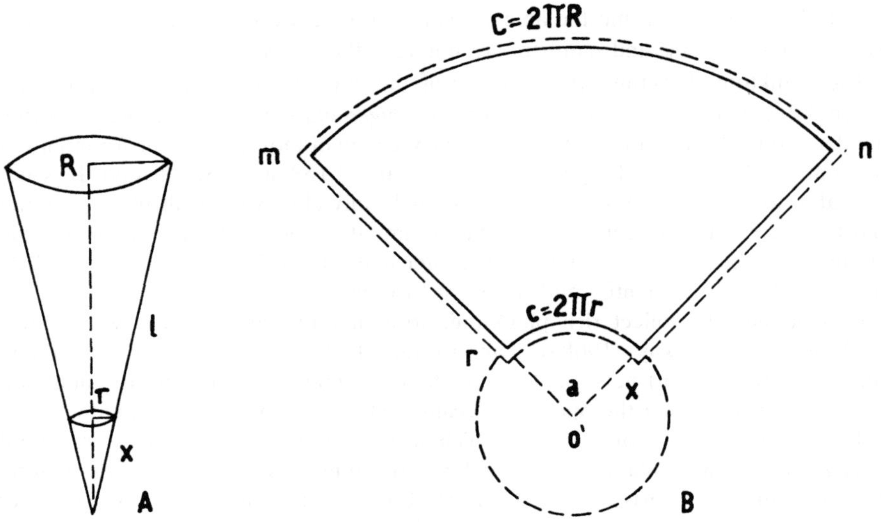

Fig. 94. Pattern design for a plankton net.

conventional towing methods. Samples can be taken as deep as 50 m by using an extension hose from the pump.

Pumps have been designed especially for the purpose of extracting plankton samples (Azor, 1958; Beers, Stewart & Strickland, 1967).

It is possible to prepare a makeshift plankton net if appropriate mesh cloth is available. A cone of material is sewn according to the pattern shown in Fig. 94. The large end of the cone is fitted with a metal ring and the small end cut and fitted with a bucket. The pattern is first laid out on paper. The length of cone and its angle are calculated as follows:

$$x = \frac{(r)(l)}{R - r} \qquad a = \frac{180(r)}{x}$$

where R = radius of the metal ring; r = radius of the bucket; l = length of the edge of the truncated cone (sewn edge); x = length of the edge of the cone from truncation point to apex; a = apical angle of the unrolled cone.

For sewing purposes it is desirable to add an additional 1 cm at the top and sides and 3 cm at the base. This addition should be composed of strong material that can withstand the wear and tear of motion through the water. Without this addition the nylon or silk netting will quickly tear at the receptacle where sewn together or at the metal ring.

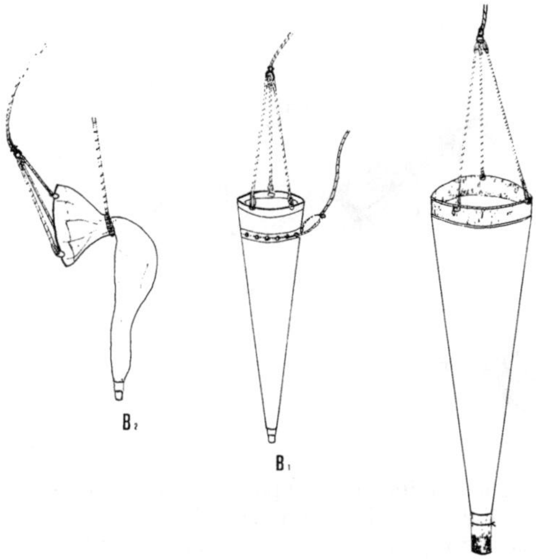

Fig. 95. Plankton nets for qualitative studies (based on Newell & Newell, 1963); A. simple net for vertical tows; B. closing nets; B1. open; B2. closed.

A net of this type is useful only for qualitative analysis because counter currents of water generated by the net's movement through the water remove some of the plankton. To avoid this, an additional short cylinder of heavy cloth is added to the leading edge of the net. Even the presence of this addition does not eliminate all loss and it becomes necessary to calculate a 'coefficient of filtration' which is a function of tow velocity, wear and tear and aperture size. Well made, unclogged nets have a coefficient of about 85%.

Figure 95A illustrates a common qualitative net without the additional cloth cylinder. Fig. 95B shows the same net before closing (B_1) and after closing (B_2). Vertical tows between predetermined depths can be obtained by the use of messenger devices designed to close a net at a particular depth (Fig. 96). Samples from specific water layers can be sampled in this way.

The receptacles attached to the small end of the plankton net are usually made of metal, glass or strong plastic. They contain a 'window' in the wall composed of metallic mesh of the same size as the net proper. Some contain a spigot from which the plankton sample can be extracted. The attachment point of the bucket to the net is a steel ring whose diameter can be adjusted for different size receptacles.

There are numerous designs for plankton nets (Nansen, Hensen, Apstein, Juday, etc.). In spite of their differences, they all consist of a truncated cone of cloth and a bucket to hold the plankton.

One of the most recent net designs is that prepared by the UNESCO Working Group (Bé *et al.*, 1968). They tried to improve the methodology of plankton

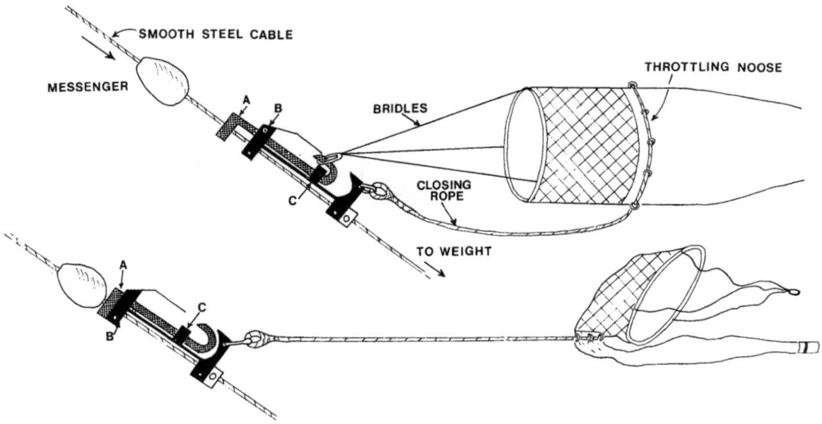

Fig. 96. Simple release mechanism for closing a plankton tow net (based on Hardy, 1970). The messenger strikes part A and drives it down through holes in part B and C thereby freeing the bridles. The pull of the closing rope closes the net by tightening the throttling noose.

sampling by considering all previously designed nets and preparing one that would be the most efficient yet made (Fig. 97).

There are several nets designed to be opened and closed at given depths which also have devices for measuring the volume of water filtered (Clarke-Bumpus, 1950; Niskin & Jones, 1963). Bé, Ewing & Linton (1959) created a vertical tow apparatus which would collect plankton samples simultaneously from three different depths.

If samples are taken at specific depths, the following schedule is suggested: 0, 10, 25, 50, 100, 300, 500, 1000 and 2000 m. If vertical tows are taken in different water layers, the following are suggested: 0–10, 10–25, 25–50, 50–100, 100–200, 200–300, 300–500, 500–1500, 1500–2000, 2000–3000 m. The best advice, however, is to adjust sampling depths or zones in accordance with the water structure at the site. Samples should be taken above and below the thermocline. Sampling usually begins at the lower depths and ends in the upper water layers.

When taking surface plankton samples with the plankton net it is necessary to diminish the velocity of the ship to about 2–2.5 knots to avoid tearing the cloth of the net if it is silk. Modern nylon nets can withstand a velocity of up to 8 knots. Sampling at greater ship velocities usually requires the use of expensive devices and strong winches. Arnold (1962) described a simply made, inexpensive plastic sampler with which all types of plankton could be taken in good condition at velocities up to 17 knots. However, as the volume of water filtered is very low, samples taken with this device give only an indication of the plankton present in the area.

It is necessary to consider the angle of inclination of the towing cable when calculating the sample depth (Fig. 98). The amount of cable which must be extended is a function of the towing angle. For example, if the tow angle is 25° and a sample is

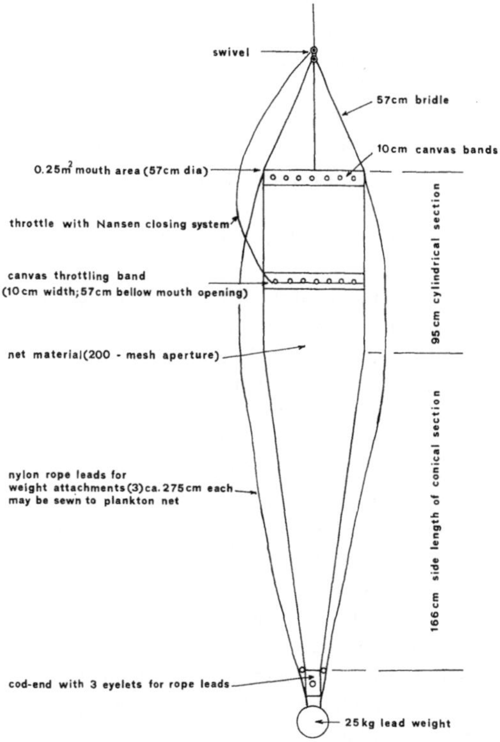

Fig. 97. The WP-2 plankton net for quantitative plankton samples in the upper 200 m of water (after Bé *et al.*, 1968).

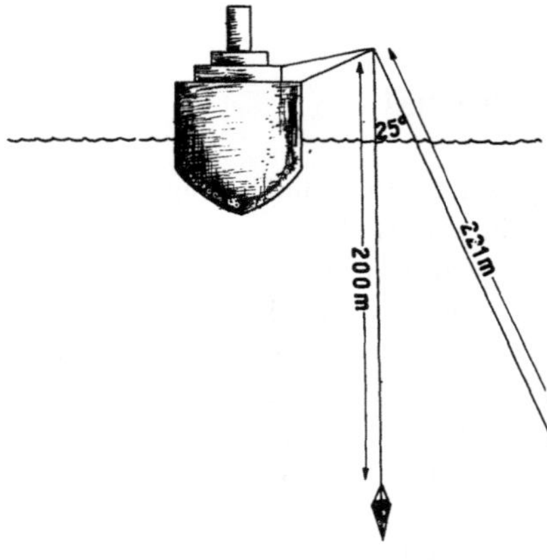

Fig. 98. Cable angle and the difference between true depth and cable length.

to be taken from 200 m it is necessary to use 200/cos 25° = 221 m of cable. If a sample of the water layer between 100–200 m is desired and the cable is inclined at a 20° angle, the net should be opened at 200/cos 20° = 213 m and closed at 100/cos 20° = 106.5 m.

The greater the volume of water filtered the greater the likelihood of finding foraminifera. A water sample of 1000 m^3 usually assures the discovery of a fairly rich assemblage of tests. The quantity of water filtered is calculated by multiplying the cross-sectional area of the net opening by the distance travelled and correcting with the 'coefficient of filtration'. Special volumetric measuring devices can be attached to some nets.

Vertical tows require the ship to stop. The net is lowered with a heavy weight (20–30 kg) to the desired depth and then hauled up at a constant rate (0.5–1.0 m/sec). The weight added to the net should be even greater when oblique tows are conducted because the ship is moving during sampling.

If there is a greater volume of plankton plus liquid in the bucket than will fit in the sample jar it is necessary to eliminate some of the liquid. Unfortunately, the superfluous liquid can not be decanted off because some plankton will be lost. To cope with this problem there is a simple device which we shall call a 'reducer'. It consists of a cup with a handle and a spout and with a bottom made of the same size mesh fabric as the plankton net used to take the sample. All of the sample is poured from the plankton bucket into the 'reducer' and when the volume of plankton plus liquid has been reduced sufficiently (by loss of the liquid through the fabric), the sample is poured into the sample jar.

After each sample is taken it is necessary to wash the net and bucket with fresh water, especially near the constricted end of the net where planktonic organisms may adhere. In the absence of abundant fresh water, well filtered sea water will serve. If the net becomes very clogged or is to be stored, some workers prefer to clean the net with a potassium hydroxide solution or commercial bleach. The latter liquid will dissolve silk but is very effective on nylon.

4. Preservation of material

All benthonic samples may contain foraminifera but the tests may represent dead as well as living individuals. The empty dead tests may represent resedimented fossil specimens as well as recently dead Recent ones. Unless the fossil specimens are filled with sediment, it is usually impossible to distinguish them from the dead Recent ones. The distinction between living and dead planktonic foraminifera is less difficult because the walls of these specimens are generally translucent allowing the investigator to see the yellow-green or yellow-brown protoplasmic mass inside. The problem of reworked fossil material is seldom a problem in planktonic tows except in turbulent zones.

It is important to know which specimens were alive at the time of collection. The presence of protoplasm in the test is an easy and reasonably exact indication of life. In order to make this determination in the laboratory it is necessary to preserve the specimens in some fixative immediately after sampling. The most common fixative is formaldehyde[9] because it hardens protoplasm and is cheap. Its primary defect is its tendency to hydrolize and become acidic with time. The increased pH of the fixative will destroy calcareous foraminiferal tests. The solution must therefore be neutralized or buffered to prevent this hydrolysis. The addition of 10 g of borax ($Na_2B_4O_7$) previously dissolved in hot water, or sodium bicarbonate (Na_2CO_3), or 200–250 g of hexamethyltetramine ($C_6H_{12}N_4$) to a litre of formalin will produce this buffering action. The borax and sodium bicarbonate neutralizers are not completely satisfactory as they retard the drop in pH for only a limited time. If samples are to be kept for long periods they should be examined every 3–4 months and additional neutralizer added if the pH has dropped below 8. We know of cases in which planktonic foraminifera fixed with borax neutralized formalin have shown serious signs of dissolution within a month of collection. Bé & Anderson (1973) suggested that high temperature in the sample storage area may increase the rate of dissolution and they recommended a temperature of 5–15 °C. They also proposed that a high ratio of plankton biomass (and its associated body constituents) to preservative will lower the pH. They suggested that there be at least 9 times as much preservative as plankton in the stored sample. Hexamethyltetramine is a much better buffer than the other two chemicals. Boltovskoy (1969b) described 17 year old planktonic foraminiferal samples in perfect condition which had been treated with formalin-hexamethyltetramine. There are two disadvantages to this method: (a) hexamethyltetramine is expensive, and (b) if the sample is subject to evaporation, the hexamethyltetramine tends to precipitate crystals around the foraminiferal tests.

Another method of controlling pH used by some investigators is the addition of calcite chips to the formalin. As the acidity of the solution increases, the calcite gradually dissolves and restores the pH to neutrality. The chips provide a visual check on the solution.

The neutralized formalin is commonly used in a 4% solution. This is prepared by dissolving one part of buffered formalin in 9 parts water prior to putting the fixative in the sample.

Qualitative benthonic grab samples are treated in the following manner: (a) a half-litre wide-mouthed jar with a tight cap or seal is filled no more than half way with the sample, (b) filtered sea water is added till all but a fifth part of the jar is full, (c) undiluted neutralized formalin is added in an amount equal to 10% of the water in the jar and (d) the jar is sealed and agitated.

For quantitative core samples, the neutralized formalin solution is added gently to the top of the core tube without any agitation.

[9] Formaldehyde (methylaldehyde) is the name of the gas (CH_2O) produced by oxygenation of methyalcohol. A 40% solution of this gas in water produces formalin or formol. Unfortunately, there is great confusion in the literature over the correct usages of these terms.

Samples preserved in this way can be kept for many years without fear of protoplasmic decomposition. Periodic check of acid–alkaline conditions are necessary, however. If the samples are kept away from light, the foraminiferal tests will maintain their original colour.

Planktonic samples are preserved in the same manner as the benthonic ones. Undiluted neutralized formalin is added to the plankton sample in an amount equal to 10% of the volume of the sample and sea water.

Ethyl alcohol is also used as a fixative. Its greatest advantage is that it is neutral. The alcohol also acts as anti-freeze for plankton samples taken in winter for which there is no warm storage place aboard ship. The disadvantage of alcohol is the high cost and tendency to form a precipitate in sea water.

Working Group 23 of SCOR/UNESCO suggested a new method of preserving zooplankton material in 1968. Details of this technique and some background to it were received by personal communication (1970) from H. F. Steedman, Bath University. With many thanks to Dr. Steedman we reproduce excerpts from his letter.

'...formaldehyde, no matter what the buffer, seems to lead to the dissolution of fine calcareous shells. If there is too high a pH within the tissues all the specimens will be softened. This softening may lead to disintegration of small shells, not because the calcium has been dissolved, but because the protein has been either altered or softened in such a way that it no longer holds together the individual pieces of calcium carbonate. If the pH is too low, . . . , then the shells will be dissolved in the usual way. These facts have been amply verified during the course of the SCOR/UNESCO/WG23 work over the last two years, . . . experiments were set up using potassium oxalate to prevent the dissolution of calcium salts. The reaction between calcium chloride and potassium oxalate is a simple one and as a result of the double decomposition which takes place at room temperature in a matter of seconds, insoluble calcium oxalate is produced. It therefore seemed probable that if potassium oxalate was to be utilized in formaldehyde solutions it would prevent dissolution of calcium carbonate. . . . It will be noted that potassium oxalate cannot be used with sea water formaldehyde, because the calcium chloride present in sea water is immediately thrown out as an insoluble powder. Where fine setae, etc., are to be studied the effect of a fine powder scattered over everything would be to make such work almost impossible. Therefore, where potassium oxalate is used it must be in distilled water. May I suggest that you give a trial to the following two formulations:

	I	II
potassium oxalate	2 g	1 g
40% formaldehyde	5 ml	5 ml
propylene glycol	—	5 ml
distilled water	93 ml	89 ml

'To use these fluids take the plankton as it comes from the sea and while still in the net gently wash with distilled water for 20–30 seconds (to remove the sea water) and place the specimens directly into one of the fluids . . . . The sea water precipitation of calcium oxalate should be negligible because the calcium chloride present has been largely removed by the brief wash in distilled water. From that point onwards you should find very little, if any, dissolution of calcium carbonate shells. I have had many different types of molluscs in similar fluids and they have lasted over two years without showing signs of dissolution. The fluids given are suitable for all types of plankton as well as calcareous material.

'The propylene glycol should give flexibility and some translucency to the specimens as well as acting as a fungicide'.

Perhaps the main disadvantage of this technique is the limited supply of distilled water, especially aboard small vessels.

Aves (1959), Marr (1964) and several other investigators suggested that prior to fixation the sample should be treated with some narcotic in order to freeze the protoplasmic material in its position at the time of capture. In the case of foraminifera this may mean the preservation of protoplasm in the aperture or partially outside the test. Aves proposed the use of 5–10 drops of chloroform, a 10% solution of chloral hydrate, or a 95% solution of ethyl alcohol for each 100 cm^3 of sample, but our experience has shown no practical difference between fixed and unfixed specimens.

Some planktologists use 'chinosol' as a fixative (Göke, 1963). Chinosol is the common name of 8-hydroxyquinoline. It is convenient to use because it comes in pill form, but it is not recommended for use with calcareous plankton because, even buffered, it reduces to a pH of 4–5 in a few days.

Frankel (1970) suggested a method for the fixation, staining and preservation *in situ* of benthonic material in the sediment. The specimens are held in place with an epoxy cement which can then be examined and even sectioned to see the spatial relationships of the organisms.

5. Sample registration

The registration of samples is very important. Even if a sample is rich in foraminiferal tests, the absence of information about its collection reduces its scientific value to almost nothing. In one North American laboratory there is a paradoxical but meaningful sign 'Remember that the label is more important than the sample'. Myers wrote that over the desk in his laboratory is a sign which reads 'Label it now or throw it out'.

All samples should contain their own labels and should be recorded in sample registries (see Chapter XVI). Labels should be prepared prior to collection, on durable paper, written with a hard pencil or permanent ink. The kind of information contained on the label depends on the nature of the sample; fossil, benthonic or planktonic.

A. Fossil Material

The label should include the following data: (a) sample number, (b) well data, if appropriate, (c) depth of sample, (d) collection site, (e) lithologic character, (f) stratigraphic zone, (g) other observations, (h) date and (i) name of collector.

When the data is entered in the sample registry, additional data about the well or outcrop should be included; e.g. thickness of unit, paleontology, zonation, etc.

B. Benthonic Material

The label should include the following data: (a) sample number, (b) date, (c) location, (d) water depth, (e) nature of bottom, (f) type of extractor, (g) other observations and (h) name of collector. If the sample was taken during an oceanographic expedition, the name of the ship, the operation and the station should be added.

When the data is entered into the sample registry additional lithological, hydrological and biological information should be included.

C. Planktonic Material

The label should include the following data: (a) operation, (b) ship, (c) date, (d) location, (e) station number, (f) sample number, (g) depth or zone, (h) apparatus used, (i) other observations and (j) name of collector.

When the data is entered into the sample registry additional information should be added such as the duration of the ship on the station, meteorological data, maximum depth at the site, volume of water filtered, hydrological data on temperature, salinity, etc., visual appearance of the sample and how the sample was preserved.

Sample labels should be placed within the sample vial arranged so that its face is visible from the outside.

XI. PREPARATION OF SAMPLES

The job of preparing samples for microscopic study consists of the separation of the tests from the sediment (for benthonic samples) or from other planktonic components (for planktonic samples). This process may involve several steps. In the case of consolidated sediments, the sample must be disaggregated, washed and the foraminiferal components separated from the sediment. For those unconsolidated samples taken directly from the ocean bottom the disaggregation process is eliminated but a new step, coloration, is added between washing and separation. To separate planktonic specimens from a plankton net sample requires special procedures.

1. Bottom samples

A. Disaggregation

Disaggregation is necessary only if the sample consists of consolidated sediments. Weakly cemented sediments can be disaggregated by being soaked in water. Boiling the water sometimes aids in this process. Boiling also serves to partially clean the specimens. If the rock is very well cemented or indurated, the only solution is to prepare thin-sections to view the specimens in cross section. Between these two extremes lie a variety of sedimentary rocks and numerous techniques which may serve to disaggregate them.

If the rock is not particularly well compacted but remains intact after boiling, it is often rewarding to disaggregate the sample by force, although a certain fraction of the specimens may be destroyed in the process. The rock can be broken in small fragments by hitting it with a hammer. The use of a vice is better because specimens often pop out of the sample when pressure is applied. The vice is placed with the grips in a horizontal position. The sample is placed between two plates, a box is placed beneath the vice to catch the disaggregated sample and pressure is applied. Perconig (1966) described the construction of a mechanical apparatus which throws the sample back and forth rapidly in a tube filled with water. There are more complicated devices for disaggregating samples. Descriptions of these can be found in various treatises on micropaleontology (Wicher, 1942; Hiltermann, 1958; Kummel & Raup, 1965).

The foraminifera may be removed from the partially disaggregated material by a variety of different methods. Those which have proven most successful are described below.

1. Soda method

The sample is placed in a non-aluminium container and covered with water to which has been added a spoonful of $Na_2CO_3 \cdot (nH_2O)$. This mixture is then boiled one to three hours.

2. Crystallization method

The sample is placed in a dry container and heated until all water is driven out. Without reducing the heat the sample is covered with crystals of Glauber salt ($Na_2SO_4 \cdot 10H_2O$) and a saturated solution of the same salt is poured over the sample. The sample container is then tightly covered and boiled for a while. After boiling the container is removed from the heat and allowed to cool. The Glauber solution penetrates the pore space and crystallizes. The expansion resulting from this phase change breaks the sample along grain contacts. Several repetitions of the process may be necessary to completely disaggregate the rock.

This same process can be performed without the salt solution. The sample is covered with water and placed in a vacuum chamber. The air pressure is gradually reduced until air bubbles are seen leaving the rock. Returning the rock to normal air pressure, the sample is frozen (down to $-30\,°C$). The ice crystals which form in the pore spaces will disaggregate the specimen. If a vacuum chamber is not available, the air can be eliminated from the pores by boiling for one-half to one hour. The disaggregation of a sample by this method may result in the destruction of hollow foraminiferal tests.

3. Hydrogen peroxide method

The sample is covered with a 15% solution of hydrogen peroxide (H_2O_2). After 10 to 15 minutes the sample begins to 'boil'. This process can be accelerated somewhat by heating the sample. The hydrogen peroxide oxidizes any organic material in the sample producing CO_2 gas which fills the pore spaces and tends to separate the individual grains. This process works best on those samples containing abundant organic matter. After an hour in this solution, foraminiferal tests begin to show signs of etching.

4. Gasoline method

Samples of shale can be heated to eliminate pore water, cooled and then covered with gasoline. After the gasoline has penetrated the pore space (about half an hour) the gasoline is decanted and the sample covered with water. Great tension is produced in the pores when the water reacts with the gasoline and the rock breaks apart.

Aside from these four principal methods, there are others which combine several techniques. Bolli (1952) described a method in which a sample is dried in an oven at 80–100 °C, and then placed in gasoline for 15 to 20 minutes. The gasoline is decanted and a 50% solution of soda ($Na_2CO_3 \cdot nH_2O$) is added. This mixture is boiled until the sample is disaggregated (usually one half to one hour). A mixture of sodium pyrophosphate and 15% hydrogen peroxide was used by different researchers aboard the Glomar Challenger to break down semi-consolidated sediments.

Chamney (1957) gives a good detailed description of various disaggregation methods and indicates the best methods for different types of sediment.

If non-calcareous specimens are found in a calcareous rock or one with a calcareous cement, a weak solution of hydrochloric or acetic acid can be used to separate the foraminifera from the matrix. The acid solution should be changed when it becomes neutralized. The solution should be weak because too rapid a dissolution may destroy the fine structure of some specimens. Samples which are indurated with siliceous material can be disaggregated by a hydrofluoric acid technique (Magné & Defaure, 1964). The sample is broken into small (125 mm^3) fragments. About 100 gm of sample is placed in a one litre plastic beaker to which is added 300 cc H_2O_2 and 150 cc HF. The mixture will effervesce violently. Water can be added to prevent overflow. The sample should be broken down in 5–7 minutes. The reaction is stopped by rinsing the sample in water. There is a tendency for this method to corrode the specimens.

B. Washing and Sieving

The object of washing and sieving is to clean the specimens and remove all particles which are too small to be foraminifera. Juvenile foraminiferal forms whose taxonomic position is difficult, if not impossible, to determine are also removed in this process. Washing and sieving will aid the investigator in the microscopic examination of the sample. It is much easier to microscopically view a field containing particles of the same size than to view a poorly sorted aggregate of grains.

Ordinary U.S. Standard or Tyler sieves are commonly used for the sieving process. For normal or routine observation of most samples the finest sieve size used is 74 μ (Tyler No. 200) and the largest is 400 μ (Tyler No. 40). The number of intermediate sieves used depends entirely on the discretion of the investigator. If the fauna is composed of exceptionally small specimens it may be necessary to use a 61 μ (Tyler No. 250) screen as well.

Wet washing of the sample is performed by placing one sieve on top of the other (smaller mesh on the bottom) and adding the sample from above. The particles are aided in their passage through the sieves by a stream of water. Thus the material is simultaneously washed and divided into different size fractions. It is necessary to be sure that the sieves do not become plugged with sediment and overflow. To avoid this problem one can use only one sieve, the finest, for the washing and then later sieve the dried sample.

There are devices which allow the rapid washing of samples over a single sieve. These generally consist of a container which has a high pressure water inlet on the top and a reinforced sieve screen on the bottom. The sample is placed in the container and the water supply turned on. The particles finer than the sieve openings pass through the screen and the sample is cleaned very rapidly.

It is abolutely necessary that the sieves be thoroughly cleaned after each use to prevent the introduction of spurious specimens in a later sample. One method of avoiding contamination from dirty sieves is to dip the sieve in a chemical stain each time it is used. The stain alters the colour of all particles that remain in the sieve openings. If examination of a sample reveals a specimen stained in this fashion one can assume that it came from a sample other than the one under study. Methylene blue is a very effective colouring agent although it has no effect on pyritized specimens.

When sieving by hand the sample should not be forced through the sieve openings because such pressure, (a) will often destroy delicate tests, and (b) may alter the size of the sieve opening rendering the sieve inaccurate for future use. A hand-operated sprayer (sold with commercial detergent) when filled with water is useful to disaggregate clays without having to touch them and risk damage to the specimens.

Brotzen (1950) suggested an improvised sieve for those without access to standard sieves. He utilized a length of celluloid 20–30 cm × 4 cm × 0.5–1.5 cm. The ends are glued together with acetone to make a circle. A gauze milling gauge is glued to one end with an acetone–celluloid mixture. This jury-rigged sieve is easy to clean and works in both hot and cold water.

If the sample is very argillaceous, the clay sized particles clog the sieve openings very easily. For these samples a decantation method works better. The sample is placed in a beaker, water is added and the mixture is agitated. After standing a few moments the suspended clay–water mixture is poured through the finest sieve. This process is repeated until most of the very fine fraction has been eliminated. The sample can then be wet sieved in the normal fashion. This process is time-consuming but one must consider that many specimens contain air in their chambers and may float in the water. Thus it is desirable to pass the clay suspension through the sieve with each agitation. The slowness of this process can be speeded up a bit by agitating and decanting several clay-rich samples at once.

The washed and sieved sample is placed in a container (a porcelain dish works well) and left to dry in the air, in an oven, under heat lamps or in an improvised double boiler (Fig. 99). It must be noted that several delicate foraminifera, principally small agglutinated forms with very fine walls, cannot withstand forced heating. Numerous experiments have shown that such specimens found in washed but undried samples are not found in washed and dried samples. Experiments conducted by Brodniewicz (1965) showed that calcareous species are much more resistant than agglutinated ones to the washing and drying procedure. Three agglutinated species, *Myxotheca arenilega*, *Hippocrepina cylindrica* and *Saccodendron limosum* present in wet untreated samples were not found after washing and drying. Thus, it is always desirable to verify the contents of a sample by examining a portion of the undried sample. This warning is particularly apt when dealing with samples from low salinity waters which often contain delicate agglutinated forms.

Organic material in samples often tends to shrink and cake when dried. This shrinkage often destroys specimens, especially agglutinated forms. Vilks (1967)

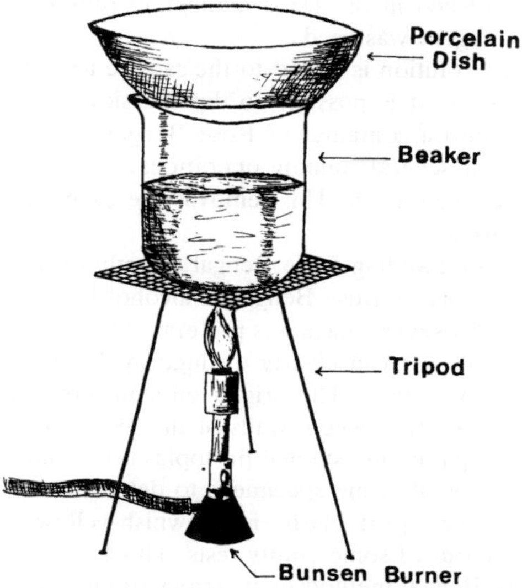

Fig. 99. Improvised method for drying samples.

noted a 69% loss of specimens in 12 such samples after drying. He suggested that the sample be wet washed on sieves under a continuous flow of water for about 10 minutes. The residue is then washed in a 250 ml beaker with 100 ml water and then collected on a millipore filter where it can be examined.

C. Coloration

Various chemical substances can be applied to the sample to determine if the specimens contained protoplasm at the time of capture. Some of these substances react chemically with a complicated protoplasmic protein molecule whereas others are merely absorbed on the surface of the protoplasm as dyes. These agents, by colouring any protoplasm present, indicate if there is protoplasm in the test.

The most commonly used stain is Rose Bengal ($C_2OH_2O_5T_4Cl_4Na_2$) which is of the type absorbed on the surface of the protoplasm. The use of this agent was made popular by Walton (1952). The stain is used in the following manner:

(a) About one gram of Rose Bengal is dissolved in one litre of distilled water. Distilled water is advisable because some of the dissolved salts in undistilled water have a tendency to precipitate the colouring agent. The quantity of Rose Bengal used to make the solution will affect the intensity of the colour shown by the protoplasm. It is therefore advisable to always use the same quantity of Rose Bengal. Rose Bengal has a solubility in distilled water of 36.25 g/l. Ellison & Nichols (1970) suggest that a solution of 1 gm of Rose Bengal, 5 ml phenol and 100 ml distilled water is very effective.

(b) The sample is washed in the usual manner to remove the finer fraction and whatever preservative agent was used.

(c) The Rose Bengal solution is added to the sample and allowed to sit for 10–20 minutes. For convenience, it is possible to dip the sieve on which the sample has been washed directly into a container of Rose Bengal solution. In this manner the solution can be used for several staining operations.

(d) The sample is washed again. This removes the excess colouring agent.

(e) The sample is dried.

Lutze (1964a) proposed adding Rose Bengal directly to the preservative (alcohol or formalin). The solubility of Rose Bengal in alcohol is 7.53 g/l. He also said that staining the specimens for several days was preferable to a 10–20 minute immersion.

Under the microscope one can clearly distinguish those tests containing stained protoplasm from the empty tests. The bright red stain produced by Rose Bengal is usually visible through the translucent walls of the test. On those tests whose walls are unusually thick or opaque the stained protoplasm is visible around the aperture. It may be necessary to break some specimens to determine if stained protoplasm is present because unstained protoplasm is brownish-yellow, a colour difficult to distinguish from the colour of some empty tests. This has been considered a serious drawback to the Rose Bengal method by its critics (Benda & Puri, 1962; Dupeuble, 1964a; Atkinson, 1971; Martin & Steinker, 1973).

There are other defects as well. Above all, one must learn through experience to distinguish between those specimens which contain stained protoplasm and those whose tests have been tinted by the dye. In the case of transparent multilocular tests (*Bulimina*, *Buliminella*, *Bolivina*, *Elphidium*, *Epistominella*, *Lenticulina*, *Nonion*, etc.) the coloration will often be visible only across the walls of the last few chambers. The earlier chambers will seldom show any colour. This is understandable since the protoplasm in the last few chambers fills the test and impedes the penetration of the Rose Bengal into the other chambers and because the only other openings in the test, the pores, are closed off by an organic membrane. In the case of *Quinqueloculina*, coloured protoplasm often extends out of the aperture and a yellowish untinted segment is visible around the apertural rim. The coloured protoplasm of the unilocular calcareous forms (*Lagena*, *Fissurina*, *Oolina*, *Parafisurina*) appears as a small irregular globular mass, commonly adhering to the internal wall of the test.

Although the tests of dead foraminifera lack coloured protoplasm, they occasionally display a red colour or exhibit small spots of red material on the walls. This coloration is due to the presence of other organic material (bacteria, marine worms, etc.) in the empty tests. Insufficient washing may result in red colour remaining in the tests.

The stained protoplasm often shrinks when dried and is therefore more difficult to detect. If the specimens can not be examined while still wet, Ayala-Castañares (1963) proposed the addition of an alcoholic aerosal detergent to the sample. After

24 hours this detergent expands the shrunken protoplasm and therefore increases its visibility.

There are various cases of coloration occurring in specimens which were obviously dead. Bandy observed coloration in broken tests (in Green, 1960); Reiter (1959) mentioned a similar phenomenon occurring in Miocene specimens from California. Le Calvez & Cesana (1972) noted examples of fossil foraminifera from the Pliocene of Nizza and the Eocene of the Paris Basin which stained red when treated with Rose Bengal.

On the other hand, there are cases in which Rose Bengal treatment failed to stain specimens that were undoubtedly alive. Lutze (1964a) observed that protoplasm-filled specimens of *Cribrononion* turned green when Rose Bengal was added to the sample. In some cases this green colour was temporary and the specimens turned red after remaining in the solution for a while. Lutze preferred to interpret all green specimens as alive.

The strongest critique of the Walton method is that of Le Calvez & Cesana (1972). They performed various experiments on living and dead specimens and arrived at the conclusion that Rose Bengal does not always stain living specimens but may give colour to empty tests. They explain the first phenomenon by the presence of a thin organic film which remains after the death of the organism and prevents the penetration of the colorant into the test interior. They explain the latter phenomenon (coloration of fossil forms) by the presence of spurious organic material in the empty test. They conclude that the only valid method of distinguishing between living specimens (with protoplasm) and dead specimens (lacking protoplasm) is the direct observation of protoplasm in the test immediately after collection.

The recent discovery that the chemical Sudan Black B will stain living protoplasm a dark blue-black (Walker, Linton & Schafer, 1974) may prove very important in studies of living foraminifera. This stain has many advantages when compared with Rose Bengal. It is able to penetrate past the first few chambers of the test, it does not stain the organic membrane, it is easily removed from the surface of the test by washing and most importantly, it is very accurate.

Sudan Black B works best when heated and acetylated but the latter step is somewhat complicated. A simple saturated heated solution of the compound gives good results. It is prepared as follows; (a) Sudan Black B is mixed with 70% ethanol to make a saturated solution. (b) The foraminifera which have been preserved in fixative are washed in distilled water. (c) The saturated solution of Sudan Black B, heated to 40 °C, is added to the specimens and maintained at 40° for 30 minutes. (d) Excess stain is removed by washing in a 70% ethanol solution.

D. Preliminary Separation

Flotation methods utilizing differences in specific gravity of the minerals in the sediment are simple and sometimes very efficient methods of separating the foraminifera from the remaining sediment. Both Recent and fossil foraminifera can be separated in this way.

The most primitive method consists of heating a washed and dried sample and then dropping it into cold water. Those tests whose chambers are unfilled will float because of the hot air in the empty chambers.

Decloitre (1966) proposed a separation method for thecamoebians which might prove useful for foraminifera as well if they are small. His method is the following: (a) add water to a sample, (b) pass CO_2 (10–15 minutes) through the sample, (c) let the sediment sink, (d) pour the liquid with the floating foraminiferal tests into another container and (e) when the CO_2 is evaporated from the water, the foraminiferal tests will settle to the bottom.

The use of heavy liquids is a much more efficient means of separation of foraminifera. Traditionally, carbon tetrachloride (CCl_4; sp. gr. 1.59) was used. Hollow calcareous tests float in this liquid whereas quartz, solid pieces of calcite, and most other commonly occurring minerals will sink. Various kinds of funneling and filtering devices can be set up to effect this flotation. Carbon tetrachloride has the advantage of being inexpensive and reusable. Its vapor is toxic, especially for the liver, and it should be used only in a well-ventilated area under a hood. Tetrachlorethene (or perchlorethylene; CCl_2; sp. gr. 1.62) has been used as a less toxic substitute for CCl_4.

Because the specific gravity of calcite is 2.72 those tests which are filled with debris or have very thick walls may not float. For this reason, if CCl_4 is used one must examine the residue as well as the flotant for foraminifera. Multiple treatment with CCl_4 may increase the efficiency of this method.

In addition to foraminiferal tests a variety of other materials will float to the surface of CCl_4. These may include organic debris, microgastropods, micropelecypods, ostracodes, calcareous algae, mica flakes, etc. Thus, one must consider this flotation as only a first step in the separation process and further separation must be done with the microscope.

If the fraction that floats is very small it is sometimes advisable to add to it a bit of the heavier material. This addition, although diluting the concentrate, may prevent the loss of a very small sample.

According to Gibson & Walker (1967) the results obtained by the CCl_4 flotation method are not very satisfactory because many thick-walled and partially filled tests remain with the residue instead of floating. These authors suggest that a liquid with a specific gravity of 2.2 works better. A mixture of 10 parts bromoform (tribromomethane; $CHBr_3$; sp. gr. 2.89) and four parts acetone (sp. gr. 0.792) can be adjusted to the proper specific gravity by testing the buoyancy of a piece of opal (sp. gr. 2.2). Experiments conducted by Gibson & Walker indicated that the CCl_4 method yields only 7–71% of the number of specimens that were found in the floated fraction of a sample treated with the bromoform–acetone mixture. Only 3–4% of the total population remained in the residue after treatment with the bromoform–acetone mixture. According to Sen Gupta & McMullen (1969), the CCl_4

flotation method recovered about 85% of the total number of tests present in a sample. A second flotation using bromoform–acetone generally yielded about 99% of the tests.

The Danish Geological Survey prefers to use a mixture of bromoform (65 parts by volume) and alcohol (25 parts), which has a specific gravity equal to 2.0 (Buch, personal communication).

The use of bromoform is not only better from a test recovery point of view but also because it is considerably less toxic and less volatile than CCl_4. The disadvantage is that bromoform is about ten times as expensive as CCl_4.

A cadmium-based heavy liquid separation method was described by Petelin & Fotiev (1952) and used successfully on foraminifera by Saidova (1956). This liquid is non-toxic and requires no ventilation system.

The use of a vibratory apparatus, the syntron, to concentrate Recent foraminifera was described by Anderson & Hoffman (1963). Morozova (1956) also discussed several different devices used to separate and sort microfossils.

The separated material should be stored, preferably in a small glass tube fitted with a cap or stopper. A slip of paper containing sufficient data to identify the sample should be placed inside the tube with the sample and another glued to the outside for easy identification.

2. Planktonic samples

The use of colorants for Recent planktonic specimens taken from the upper water layers only prolongs the entire preparation process since practically all specimens found in these zones are living. Various studies have shown that the great majority of specimens in surface tows contain protoplasm. If, despite this general observation, the investigator wishes to observe which specimens contain protoplasm he is able to do this easily without the use of a colorant because the walls of the planktonic foraminifera are commonly translucent and transmit the yellow or green colour of the protoplasm.

After collecting a planktonic sample and fixing it, the macro- and microplankton begin to sink to the bottom of the container. It is relatively easy to pipette this material into a Petri dish or watch glass and then pick out the foraminifera under a microscope with a pipette or small (00 or 000) brush. This is the classic method, having been employed since the onset of planktonic foraminiferal studies. It is simple and gives good results, but is time-consuming. The investigator is subject to fatigue and might overlook some tests. He must also overcome a natural bias toward selecting only the large and/or unusually-shaped tests.

It is often desirable and necessary to make taxonomic determinations while yet aboard ship. Such preliminary screenings can be accomplished by using a simple ×30 binocular microscope. The foraminifera are picked out of the sample and placed on a cardboard slide (see Fig. 102). If possible they can be arranged on the numbered squares of the slide by taxonomic category. The hand picking of the planktonic foraminifera from the sample by means of a pipette or brush is a time-consuming task, especially if the sample is rich in phytoplankton. Hamlin & Bé (1963) proposed an apparatus (plankton picker) to facilitate this work.

Bé (1959b) proposed another procedure to rapidly remove tests from the sample. Following is a description of this procedure: (a) To a cylindrical funnel (capacity 1000 ml) with a stopcock add 600 ml of water saturated with NaCl (310 gm NaCl in 1000 ml distilled water at room temperature). This liquid has a specific gravity of 1.20. (b) Decant off the supernate liquid from the sample allowing only the solid part of the sample to remain: (c) Put enough sample into the funnel so that it forms a thin upper cap on the NaCl solution. The plankton with tests (foraminifera, radiolaria, pteropods, etc.) have a specific gravity of about 1.4 and will sink in the solution. The other plankton (copepods, chaetognaths and dinoflagellates, etc.) have a specific gravity of 1.00–1.09 and will float. (d) Open the stopcock so that the heavier plankton pass through the funnel into a collecting beaker. The remaining fluid is agitated and the procedure repeated. When no more material settles, additional sample material can be added. The excess liquid that passes through the stopcock can be returned to the funnel. (e) Both fractions of the sample can then be examined. This method works best for quantitatively poor samples because in rich samples the specimens, especially spinose ones, are retained in the net of phytoplanktonic material.

The very efficient method described below was proposed by Sachs, Cifelli & Bowen (1964): (a) The sample is washed in a 200 or 250 mesh sieve, (b) filtered with a Buchner funnel, (c) dried on filter paper, (d) heated to 500 °C for about two hours in an electric oven, (e) washed again in the same size sieve and (f) dried.

For the most part, only calcareous shells remain in the sample. From these it is quite easy to extract the foraminifera tests.

This ignition method was modified and somewhat improved by Sachs (1965) and Smith (1967) by the addition of several steps described below: (a) Split the sample to preserve some of the non-shelled organisms for study. (b) Wash the sample on filter paper to remove all preservative. (c) Place the sample in sodium hypochlorite or hydrogen peroxide to digest the soft organic matter. (d) Wash again on filter paper to remove the digesting agent. (e) Dry the sample. (f) Heat the sample to 500 °C for approximately two hours. (g) Wash the sample on a 200 or 250 mesh sieve. (h) Dry the sample and begin microscopic examination.

The principal advantage of this method is that it results in a dry residue from which the foraminifera may be picked with much greater speed than in the classic methods.

The disadvantages of the ignition method are that the various steps are time consuming, that specimen loss may occur during final washing, and the principal disadvantage, that this method cannot be used if living-to-dead ratios are desired since all the protoplasm is consumed by heat. Such ratios are extremely important in the study of plankton samples below the surface water layer because there are usually significant numbers of empty tests in these zones (Boltovskoy, 1966a).

In addition, Boltovskoy (1971b) also noted that the ignition method destroys the colour in the test walls. This prohibits the distinction between *Globigerinoides ruber* (forma rosea) from *G. ruber* (f. alba), two forms which occupy slightly different habitats. He also noted that some planktonic species can be distinguished on the basis of the colour of their protoplasm, which is destroyed in the ignition method.

Boltovskoy (1966a) proposed a filter method of separation which consists of the following steps: (a) An aliquot of sample is filtered through filter paper. (b) This filter paper with the sample is dried at a low temperature (25–70 °C). (c) The dried residue is removed from the filter paper and placed in an extraction tray and the foraminiferal tests are picked out. This author made comparisons of the classic method, the saturated liquid method, the ignition method and the filter method and concluded that the ignition method is best for plankton samples with large quantities of soft organic material whereas the filter method is the most efficient and precise if the original samples contain little beside the tests desired for study. In conclusion Boltovskoy (1971b) suggested that all samples be first filtered and examined to determine if large quantities of soft organic material are present; if a small quantity is present, the sample is immediately picked; if a large quantity is present, the sample is burned, washed, filtered and then picked. It should be emphasized, however, that in both cases only aliquots should be treated so that some original material is preserved.

A separation method proposed by Vilks (1970a) involves the use of a low-temperature ashing device, the TRACERLAB LTA-600. The sample is exposed to a flow of oxygen in a radio frequency electromagnetic field under low pressure (0.1–1.0 mm Hg). In this state the oxygen reacts readily with organic material in the sample, although the temperature does not exceed 150 °C. According to Vilks, this method is not injurious to the specimens, is fast and clean, but does require special equipment.

3. Cytological techniques

The protoplasm of foraminifera can be studied by means of the standard techniques used for optical and electron microscopy; i.e. fixation, decalcification, dehydration, embedding, sectioning and staining. The protoplasm may also be studied utilizing the scanning electron microscope.

A. Methods for Optical Microscope Study

1. Fixation

The choice of a fixative depends on the staining agent to be used which in turn depends on the structures to be observed. The size of the specimen is also important as some good fixatives, e.g. osmium tetroxide (OsO_4), penetrate the specimen very slowly. Because foraminifera are generally small, the size problem usually is not critical.

The common fixatives are Bouin, formalin, Carnoy, Clarke, Serra and Zenker.

Bouin works well for studying the cytoplasm and nucleus. It consists of a 15 : 5 : 1 mixture of saturated solution of picric acid, formalin and glacial acetic acid with the last component added at the moment of use. Bouin has been used on *Allogromia laticollaris* by Arnold (1955a), on *Ovammina opaca* by Dahlgren (1964), on *A. flexibilis* by Lena (1972), and in modified form (without acetic acid) on *Globigerina bulloides* and *Globigerinoides ruber* by Lee *et al.* (1965). The time of fixation is a function of specimen size but under no circumstances should it be for more than one hour.

Formalin is a very common and practical fixative. The concentration and fixation time are variable. The most commonly used concentration is 10% formalin in sea water or with a phosphate buffer (pH 7.2). The fixation time used by most investigators varies between 0.5–6 hours.

The fixatives Carnoy (a 6 : 3 : 1 mixture of ethyl alcohol, chloroform and glacial acetic acid) and Clarke (a 3 : 1 mixture of ethyl alcohol and glacial acetic acid) were used by Dahlgren (1964) in a study of the nucleic acids of *Ovammina opaca*. These acids (Carnoy, pH 2.7; Clarke, pH 1.5) hydrolize the nucleoproteids into nucleic acids and proteins and are desirable fixatives when digestive techniques are used. For *O. opaca* the optimum fixation time was 45 minutes. Clarke was used in a phosphate buffer (pH 7.2) for 6 hours at 7 °C in a study of *Hippocrepinella alba* and *Globobulimina turgida* (Dahlgren, 1967b).

Serra's fixative (a 6 : 3 : 1 mixture of 80% ethanol, formalin and glacial acetic acid) gives good results if the Feulgen reaction is to be applied later in the cytological examination (Dahlgren, 1964).

Zenker also can be used if a Feulgen reaction is planned (Lee *et al.*, 1965). It is prepared by mixing mercuric chloride ($HgCl_2$) with potassium bichromate, sodium sulphate, distilled water and glacial acetic acid in the proportions 5:2:1:100:5% with the acetic acid added at the moment of use. The disadvantage of this fixative is that the fixed specimens must be left in running water for 24 hours.

2. Decalcification

This step is necessary only when dealing with calcareous foraminifera. Decalcificants used to date include 0.2 N HCl on *Rosalina floridana* (Angell, 1967b), formic acid (Bronte-Gatenby & Beams, 1950), and the chelate EDTA (ethylendiaminotetracetic acid, versene) in a phosphate buffer on *Globobulimina turgida* (Dahlgren, 1967b).

3. Inclusions in agar

When working with very small specimens it is often easier to manipulate the specimen if it is embedded in a 1% agar solution after fixation (Haller, Ehret & Naef, 1961). Once the agar has hardened it is cut into small blocks which can be dehydrated and later reembedded.

4. Dehydration

The most commonly used method of dehydrating the specimen for cytological study is bathing the specimen in a series of alcohol–water solutions whose alcohol concentration increases.

5. Embedding

Paraffin is most often used as an embedding agent. Another embedding medium used for foraminifera is polyester-was and ester in the proportions 4:1 with an embedding temperature of 37–40 °C (Dahlgren, 1967b).

6. Sectioning

The embedded specimens are sectioned using standard microtome techniques.

7. Staining

The choice of stains for a foraminiferal specimen generally depend on the structures to be studied.

The use of haematoxylin-eosin is widespread because it photographs well. The eosin may be used in an alcoholic or water solution. As the methods of preparation of the eosin and haematoxylin are well described in the classic works on microscope techniques (Darlington & La Cour, 1962; Humason, 1962; Pearse, 1960) they will not be described here. There are various types of haematoxylin used on foraminifera, the most common of which are Heidenhain and Carazzi.

To study nucleic acids the Feulgen reaction (Darlington & La Cour, 1962) is most commonly used. The most difficult step in this method is determining the time of hydrolysis of the sectioned material. This time depends on the kind of material and on the fixatives. The following chart shows the time of hydrolysis in 1 N HCl at 60 °C for various foraminifera and fixatives.

Author	Fixative	Hydrolysis time (minutes)	Species
Arnold, 1955a	?	12	*Allogromia laticollaris*
Dahlgren, 1964	Serra	10	*Ovammina opaca*
Dahlgren, 1964	Carnoy	7	*O. opaca*
Dahlgren, 1964	Clarke	7	*O. opaca*
Lee *et al.*, 1965	Zenker	12	*Globigerina bulloides*
Lee *et al.*, 1965	Zenker	12	*Globigerinoides ruber*

In addition to the application of digestive techniques utilizing ribonuclease and desoxyribonuclease to study nucleic acids (Dahlgren, 1964, 1967b) it is possible to utilize the following stains: gallocyamini–chromalum, methyl green–pyronin (Brachet, 1953; Kurnick, 1955; Pearse, 1960), phenyltrihydroxyfluorone (Wingstrand & Melander, 1952), metachromatic stains toluidine blue (Feder & Wolf, 1965), Azur B (Ruthman, 1966), fluorochromes acridine orange (Schümmelfeder, 1957) and pseudoisocyanine chloride (Sterba, 1963).

To detect the presence of carbohydrates the Schiff reaction is commonly used. Dahlgren (1964) prepared it using thionyl–chloride as a sulpher dioxide source. Lee *et al.* (1965) and Angell (1967b) also used this reaction. In addition, Lee *et al.* utilized lead tetraacetate techniques for carbohydrate detection.

The following stains have also been used on foraminifera: Himes–Moriber triple stain (Himes & Moriber, 1956), Masson techniques, Heidenhain azan and Gomeri's trichrome techniques (Humason, 1962), 1% aqueous fast green, Acetic orcein (Darlington & La Cour, 1962), May Grunwald–Giemsa and Unna blue.

B. Methods for Transmission Electron Microscopy (TEM)

The fixatives currently used for TEM preparation are glutaraldehyde and osmium tetroxide both at pH 7.2. The concentration of these fixatives and the application times are variable for foraminifera. For glutaraldehyde the following concentrations and times have been used: 6% in sea water for one hour (Angell, 1967b, 1971), 4% in buffered phosphate for two hours (Dahlgren, 1967a,b), 5% in 0.1 M buffered cacodylate in 0.0224 M buffered Na–vernalacetate for 8–12 minutes (Lengsfeld, 1969b), 3% in sea water buffered with Na–cacodylate (Marszalek, 1969), 2% in sea water at pH 6.8–7.0 (Febvre-Chevalier, 1971) and 6.5% in 0.1 M phosphate buffer for one hour (Lena & Freire, 1974). For osmium tetroxide the following time and concentrations have been used: 1% in sea water for 60 minutes (Angell, 1967b, 1971), 1% in 0.0667 M Sorensen phosphate buffer for 20–60 minutes (Schwab, 1969), 1% in sea water buffered with Na–cacodylate (Marszalek, 1969), 2% in sea water at pH 6.8–7.0 (Febvre-Chevalier, 1971) and 1% in 0.1 M phosphate buffer for one hour (Lena & Freire, 1974). Additional fixatives have been used, among them: 1% $KMnO_4$ in sea water (Angell, 1967b), Os–Cr mixture (Wohlfarth-Botterman, 1957), and 4% formalin with phosphate buffer (Pease, 1964).

The most highly recommended procedure is the use of a combination of 6.0–6.5% glutaraldehyde buffered at pH 7.2 (phosphate) followed by a second fixation in 1% osmium tetroxide at the same pH (Angell, 1967b, 1971; Marszalek, 1969; Lena & Freire, 1974). Bé & Anderson (1973) described in explicit detail the various steps and procedures for the preparation of a phosphate buffer, a glutaraldehyde fixative and an osmium tetroxide fixative.

Dehydration is accomplished by a sequential series of ethanol solutions similar to those described for light microscopy techniques.

Various embedding materials have been used, including metacrylate (Dahlgren, 1967a,b), vestopal W (Lengsfeld, 1969b; Schwab, 1969), propylene oxide (Gorycki, 1971) and Epon 812 (Angell, 1967b; Dahlgren, 1967a,b; Marszalek, 1969; Febvre-Chevalier, 1971; Lena & Freire, 1974). This last medium has the advantage in that coarse sections (1 μ in thickness) for the optical microscope as well as ultrathin sections (600–700 Å) for the TEM can be produced.

Various stains have been employed for TEM examination of foraminifera including lead citrate, uranyl acetate, toluidine blue and Unna blue.

C. Methods for Scanning Electron Microscopy (SEM)

To study foraminiferal protoplasm by means of the SEM, Marszalek (1969) fixed specimens with 3% glutaraldehyde in sea water followed by 1% osmium tetroxide in sea water (both buffered to pH 7.2 with Na–cacodylate) or with Parducz's fixative (Parducz, 1966). The specimens were then frozen in liquid nitrogen and freeze dried in Pearse tissue (Small & Marszalek. 1969). After sublimation a thin coat of gold–palladium was evaporated on the specimen surface.

XII. SEPARATION OF THE FORAMINIFERA FROM THE SEDIMENT

Having performed the previously described operations, we now have a relatively concentrated sample of foraminiferal tests. The process of separating the tests from non-foraminiferal material, mounting the tests on slides and making taxonomical determinations are all done under the microscope.

A binocular stereoscopic microscope with ×8, ×12.5 and ×18 oculars and ×4, ×8 and ×12 objectives is very convenient for these purposes. The first step of separation can usually be conducted under a magnification of ×32–×50. For taxonomic determination the larger magnifications will often prove more useful. For detailed examination it may be necessary to utilize a common biologic microscope or petrographic microscope with provision for immersion oil study.

To extract the tests, the sample is spread very thinly over an extraction tray (Fig. 100). After placing the tray on the microscope stage the specimens are picked out with a brush or needle. Stinemeyr (1965) described a vacuum needle device which is very useful for use on samples with large numbers of specimens. The apparatus allows the specimens to be removed from the tray and stored with a minimum of lost time. The extraction tray is made of metal (brass works well), is about 6×10 cm, with inclined borders about 1 cm high. It is best painted matt black or green and divided into 1 cm squares. These squares are about the size of the field of view at low magnification. The division of the tray into squares aids in a rational, systematic search of the sample. A jury rigged tray can be made from a photographic plate or piece of glass to which is pasted a piece of subdivided black or green paper. Kornicker (1957) designed a circular tray with a spiral pattern. When the tray is rotated the spiral pattern insures complete examination of the sample even if the investigator is interrupted.

For the greatest ease in picking the sample, the sample should be sieved and each size fraction viewed separately. This process allows the investigator to view only particles of one size and aids in the process of specimen recognition and in removing the tests from the tray.

The foraminiferal tests are removed from the tray and placed on a slide where they can be viewed later more conveniently. The slides can be purchased commercially or fabricated. To make a slide take a cardboard rectangle 75×25×1–1.5 cm and paste a piece of white paper on one side. Do the same to another piece of cardboard using black or green paper. In the centre of the white slide drill or punch a hole of 10–15 mm diameter. Paste the white slide to the black slide so that the white side is up and the black/green paper shows through the hole. The slide can be covered with a piece of glass cut to size and a thin sheet of aluminium which is bent to form a

Fig. 100. Extraction tray.

casing for the slide and glass (Fig. 101). Various kinds of slides can be prepared according to the investigator's wishes (Fig. 102).

There are a variety of ways to remove the tests from the separation tray. A fine (00 or 000) brush can be used either wet or dry. A mouth pipette (Fig. 129) can be used to create a vacuum with which to pick up the specimen. A fine needle lightly coated with beeswax will easily pick a specimen out of the tray. Which of these devices works best is a matter of personal preference as each investigator usually swears by the method he utilizes.

The experienced worker can often recognize individual taxa in the tray and can place an identified specimen in its proper slide or square on a slide. If the investigator wishes only to remove the foraminiferal tests from the non-foraminiferal debris and place them on a slide for later study, the separation process described above is time consuming as the worker must lift his head from the microscope for

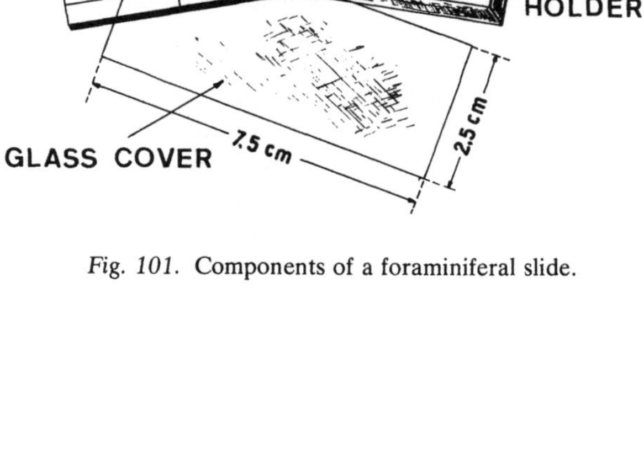

Fig. 101. Components of a foraminiferal slide.

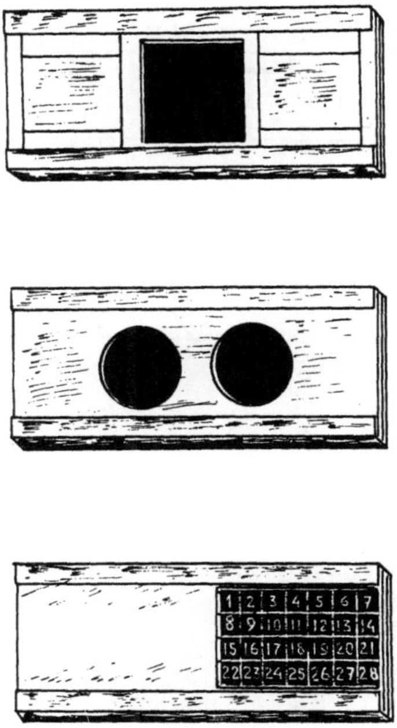

Fig. 102. Different types of foraminiferal slides.

each specimen and see that it is placed on a nearby slide. A rather ingenious arrangement has been devised to reduce this head movement (Triebel, 1947; Hiltermann, 1958). The separation tray is perforated from the bottom in the manner shown in Fig. 103a. The holes must be larger than the largest test that will be encountered. The special microscope stage adaptation shown in Fig. 103b is substituted for the normal stage or placed on top of it. This stage (wooden or sheet metal) has a 2 cm diameter hole on the centre which must be located directly below the objective of the microscope. Beneath the hole a small platform is constructed which will accommodate a slide. A convenient way of constructing this platform is to cut a 2 cm hole in an aluminium slide carrier and attach it upside down to the bottom of the substitute stage. The foraminiferal tests are poured into the tray as usual. They do not fall through the holes in the tray because of the raised rim. The tray is scanned as usual by moving it back and forth across the new stage. When a specimen is found which the investigator wishes to save he picks it up and drops it into the nearest hole (there is at least one hole at a corner of each square on the tray). The specimen falls through the hole in the tray and stage and into the slide beneath. In this manner the investigator can continue to examine, pick and separate the fauna with minimum lost time. Craig & Hogg (1962) described an ingenious suction and sorting apparatus which would save considerable time.

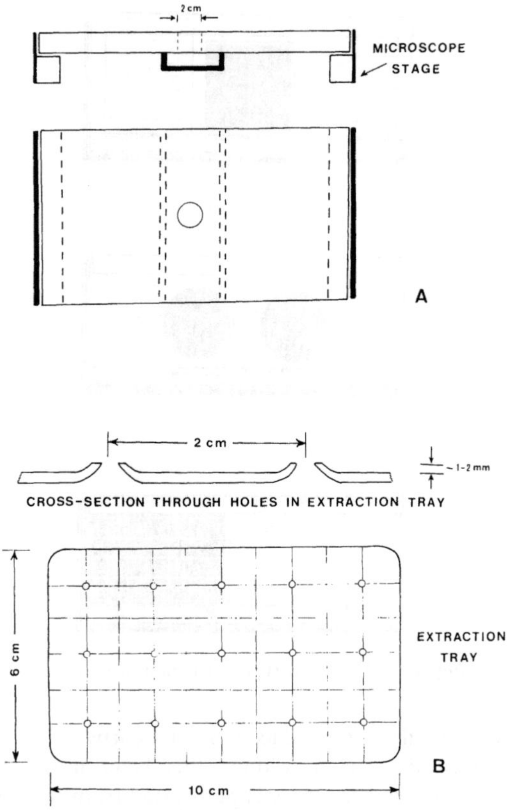

Fig. 103. Schematic diagram of a device to facilitate specimen extraction.

The separated specimens can be attached to the slides with adhesive to prevent their movement or to insure a particular test orientation. The adhesive most commonly used is gum tragacanth which is mixed with distilled water and alcohol to form a paste. A thin film of this paste can be smeared over the slide prior to use so that when a wetted specimen is placed on the slide the adhesive will dissolve and then harden to hold the specimen in place. To prevent the growth of bacteria in the tragacanth a drop or two of some extract; e.g. oil of cloves, can be added to the paste.

XIII. TAXONOMIC DETERMINATION

1. Difficulties

The taxonomic determination of a protozoan species is generally more difficult than for metazoan species because the protozoans are microscopic and possess fewer morphological traits. Moreover, it is often necessary to know the entire life cycle of a protozoan before proper identification can be made. These difficulties apply to foraminifera. We have already mentioned (Chapter VIII) some of the taxonomic problems that one encounters with these organisms: (a) the phenomenon of dimorphism or trimorphism with different micro- and macrospheric generations; (b) the phenomenon of pathologic and monstrous specimens; (c) the phenomenon of polymorphism as a function of ontogeny; e.g. *Quinqueloculina*, *Triloculina* and *Pyrgo*; the juvenile forms of *Gaudryina rugosa* described as *Verneuilina browni*, etc; (d) the large number of phenotypically plastic species; (e) the post-mortem diagenetic effects including differential solution, recrystallization and deformation.

However, perhaps the most difficult problem faced by the foraminiferal taxonomist is the heavy burden of an existing nomenclature full of synonyms. Those who interpret taxonomic categories very narrowly ('splitters' in the common parlance) have established new taxa on the basis of morphologic differences which in all too many cases prove insufficient when examined in the light of the aforementioned phenomena. There are, unfortunately, cases in which new taxa have been erected on the basis of a single specimen. In those cases there is no knowledge of the phenotypic variability of the new taxon.

One of the reasons for the widespread increase in synonyms is the absence of ample bibliographic resources in many laboratories. Many new species descriptions also lack satisfactory illustration and detail in their morphologic, diagnostic and stratigraphic analysis. Although this book does not deal with fossil foraminifera in any systematic way, it is interesting to note the calculation that precise stratigraphic information for approximately 60% of the new taxa erected between 1933 and 1965 (all phyla) was not given in the original description of the taxon. (Shaw, 1971). It is difficult indeed to understand this lack of information if, as so widely stated, two of the primary functions of paleontologic description are to aid in stratigraphic correlation and to establish and clarify evolutionary trends. Nevertheless, the principal cause of the enormous increase in taxa of doubtful validity is the failure to consider the specimens as living organisms and to follow the International Rules of Zoological Nomenclature.

To better understand the rapid increase in taxa we will cite some interesting figures. Thalmann (1952) reported that between 1931 and 1950, 523 new genera and 7758 new species of foraminifera (fossil and Recent) were established. This is an average rate of about 8 per week. The rate appears to have increased since then because Howe (1959) reported that new species were being erected at a rate of 2 per day.

It is interesting to note an investigation undertaken by Boltovskoy (1958b) in the U.S. National Museum, Washington, D.C. This institution probably houses the largest foraminiferal collection in the world. Working with some species of the genus *Nonion*, this investigator discovered that in this collection alone there were 6 holotypes of recently described species, all of which belong to a single species established 100 years before, *Nonion affine* (Reuss). Moreover, other examples of the same species found in the collection had been assigned to yet another 10 different species. There is no doubt that the revision of other collections would reveal even more synonyms. If Boltovskoy had chosen a more morphologically variable species he would have found considerably more synonyms. It is not unreasonable to think that of the approximately 38 000 species that have been described to date perhaps only 3000–4000 have zoological value; the others are synonyms. Such excessive ‘splitting’ has ramifications beyond a chaotic nomenclature. Boltovskoy (1965a) emphasized that the overloading of the foraminiferal nomenclature with an enormous quantity of synonyms would inevitably lead to a decline in the importance of foraminifera as stratigraphic, paleoecologic and oceanographic indicators. Shaw (1969) also pointed out the difficulties that currently exist in comparing one investigator’s work to that of another. He has suggested that the non-objectivity of paleontologic taxonomy makes it impossible for communication and even proposed that traditional zoological nomenclature be discarded in favour of a method which describes the stratigraphy of morphological traits. All the factors mentioned above make the correct taxonomic determination of foraminiferal faunas difficult.

2. Process of determination

The first step in the taxonomic determination of a fauna is a thorough knowledge of the literature pertaining to the area under study and to neighbouring areas. A simple comparison of the published figures with the unknown specimens allows a rapid sorting of the fauna into major groups. Of course, more detailed work must be conducted later to verify these identifications by comparing them with the original publications.

If in the bibliography concerning the area studied a particular species is not found, it is necessary to work from higher to lower taxonomic categories in the following manner. Using the general traits of the specimen and the various foraminiferal treatises, establish the family and the proper genus. As specific descriptions are not contained in most treatises it is necessary to use the various monographs dedicated to certain groups or to certain regions in order to designate a species.

Working with foraminifera, the finest publication dealing with taxa at the specific level is the *Catalogue of Foraminifera*, begun in 1940 by Brook Ellis & Angelina Messina of the American Museum of Natural History, New York. This work contains all the descriptions of new genera, species, subspecies, varieties and forms established throughout the world. The descriptions are accompanied by the original illustrations and are arranged alphabetically by genus and then by species. This work is constantly edited and a supplement containing all the new species and genera is issued annually.

This catalogue is of great value and without it the investigator must study a multitude of publications before deciding which taxon he is examining. The time problem and the likelihood that many publications will not be available give even greater value to the *Catalogue*. The presence of the *Catalogue* does not, however, eliminate the taxonomic problems mentioned in the previous section. The decision as to which extant species, if any, corresponds best to the unknown specimens depends upon the knowledge, experience and to a certain degree, the 'taste' of the investigator.

It is recommended that the investigator, prior to making a determination, prepare a schematic drawing of the specimens and measure them. We have dedicated a later section of this chapter to measurement, as our experience indicates that some investigators, although they are knowledgeable in things much more complicated, often do not know the simple yet necessary procedures of measurement.

The greater the number of specimens studied the better the taxonomic determination, for it is only with a large sample that one can obtain a notion of the variability of the morphological traits, the phenomena of cyclic dimorphism, polymorphism as a function of age, ecological variations, etc.

Juvenile forms are the most difficult to determine since the immature forms of many species are very similar. It is possible to gain some knowledge of the juvenile form of a species by successively breaking the last few chambers of an adult form.

Berger (1966) suggested that it is possible to distinguish between juvenile planktonic species on the basis of coloration with methylene blue, an organic thiazine dye. According to his observations, different species exhibit different colours, although closely related species stain somewhat similarly; *Globigerinita glutinata* and *G. iota* turned violet while *Globoquadrina eggeri* and *G. hexagona* stained green. Numerous attempts on the part of one of us (EB) to repeat these experiments have, unfortunately, yielded no coloration differences between species. Nevertheless, the original colour of the protoplasm (without any stain) may help in taxonomic determination. Boltovskoy (1971b) observed that specimens of *Globigerina pachyderma* and *G. quinqueloba* from the Drake Passage could be distinguished from one another on the basis of protoplasmic colour. The former shows a greenish hue whereas the latter is brownish. Using this criterion the author was able to distinguish juvenile forms. This may be a phenomenon limited to a particular geographic area as it has not as yet been reported for other areas.

In taxonomic determination one must also take into account the details of the internal structure, the examination of which often requires special procedures (the use of clarifying liquids, thin-sections, X-rays, electron microscopy, etc.). These techniques will be described in a later section.

Moreover, one must realize that descriptions, drawings and photographs, although done with great care and skill, never can replace direct comparison. It is always preferable to make a comparison with other specimens. If it is impossible to make such a comparison, or to have an extensive bibliography available, it is best to reserve judgement on the taxonomic determination by putting the specimens in the *nomenclatura aperta* or, at least, to indicate some doubt by the use of corresponding signs (see Chapter VII,1).

Various kinds of statistical techniques have been devised to aid in specific definition. Some of these are discussed in Chapter XIV,1.

An interesting and helpful aid to faunal determination was described by Schnitker (1971b). He noted that the use of a tape recorder attached to the microscope and operated by a foot switch allows the rapid counting and description of specimens that are being observed. One of us (RW) has used this method extensively and found it to be of great help both for determining faunal abundances and in making measurements and verbal descriptions of morphology. Constant attention can be given to the specimens without having to take time to make notations.

The correct taxonomic determination of a fauna is a job of great importance because it represents the keystone on which any subsequent analysis and conclusions rest. Time spent on correct, careful and impartial taxonomy is *always* time well spent.

3. Establishment of a new species

If a new taxonomic entity is erected, it falls under the protection of the law of priority of the International Rules of Zoological Nomenclature and must be respected as a valid name by all subsequent investigators. For this reason, one must be especially careful not to create a new synonym.

Hiltermann (1954) proposed ten rules or guidelines which should be followed when a new taxon is established. Although he proposed these rules for fossil foraminifera they are valuable for Recent species as well. We present them below with some modifications.

(a) A new name should be published only when its author is able to examine specimens of closely related species. Figures and descriptions should never replace specimens in the comparison process. If holotypes are unavailable, the neotypes, paratypes or even topotypes should be examined.

(b) The description should be detailed, the comparisons with related species must be well defined and the species definition clear and irreproachable.

(c) The largest possible number of morphological traits should be included in the definition.

(d) The intraspecific variations should be well defined for each morphological trait.

(e) Comparisons with related species should be presented graphically and/or statistically.

(f) The number of examples studied must be mentioned. For a new species it is advisable to have at least 10 specimens.

(g) The International Rules of Zoological Nomenclature should be strictly followed.

(h) At the same time, one must respect the principles of biology.

(i) In addition to giving an illustration of the holotype, it is very important to figure as many paratypes as possible.

(j) If any doubt, however small, exists as the validity of the new name, it is best to include the species in the *nomenclatura aperta* category, postponing a final decision until the doubts are clarified.

4. Measurements

Measurements, relative as well as absolute, of the different parts of the foraminiferal test yield data of great importance in taxonomical distinction. In addition, these data can give clues to different ecological conditions.

An excellent review article on the role of shell biometry in taxonomic and evolutionary studies was prepared by Scott (1974). He gave numerous examples of different studies involving measurements and offered several suggestions for particular measurements which can be made on different architectural forms of foraminifera.

To make measurements, it is necessary to replace the normal ocular in the microscope with an ocular micrometer. The ocular micrometer is calibrated in the following manner. A scale of known dimensions, usually an engraved millimeter rule graduated into 100 parts, is placed on the microscope stage and the length of each division on the ocular micrometer scale is calculated for a given magnification. For example, when using a ×8 ocular and ×3 objective (×24 magnification), 2 mm on the stage scale is equivalent to 94 units on the ocular micrometer (Fig. 104a), and each ocular micrometer division is therefore equal to $2/94 = 0.0213$ mm or 21.3 μ. If the oculars are changed to ×12 (total magnification, ×36), 1.4 mm on the stage scale is equivalent to 98.5 units on the ocular micrometer (Fig. 104b). In this case, each ocular micrometer division in equal to $1.4/98.5 = 0.0142$ mm or 14.2 μ.

To avoid recalculation or consequences of a poor memory, a table of magnifications and calibrations should be prepared and placed inside the microscope case.

5. External test examination

At times various details of the test construction are not clearly visible. This situation is quite common in fossil or partially decomposed tests. Often the surface cavities

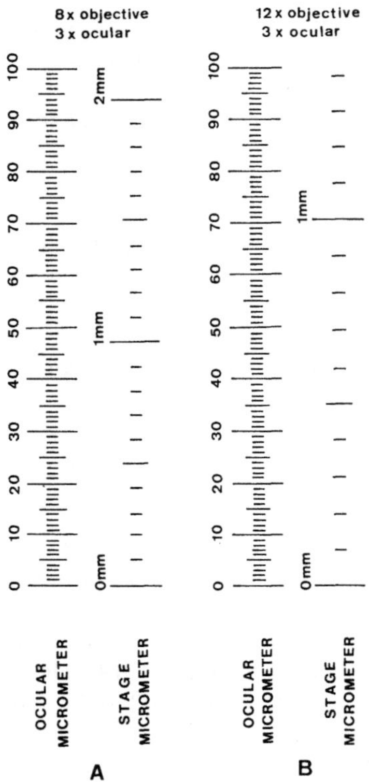

Fig. 104. Comparison of graduated scales at different magnifications.

and depressions on the test are filled with sediment and/or cementing agents. Nevertheless, with special treatment, it is possible not only to eliminate these bits of camouflage but even to use them to an advantage. Tests are often composed of materials which differ from the enclosing sediment. If this is the case, these two materials will often react in different manners to staining agents. Argillaceous sediments often react by becoming darker when stained, thus emphasizing the surface topography of the test even more.

There are a variety of colouring agents available which will bring out surface detail. We will cite two that have proven particularly useful, namely methylene blue and malachite green. The tests are placed in a watch glass with distilled water. When the air has been displaced from the tests, a few drops of concentrated methylene blue (1:150) or malachite green are added to the water. After a few minutes all argillaceous and siliceous particles will show a blue or green colour. The calcareous material requires a greater period of time to stain so that washing the tests after a few minutes will produce a test that exhibits accentuated detail. In the absence of these chemicals, washable green, blue or brown ink placed on the test surface will enhance detail. Some ink contains fine particulate matter which may adhere to the test and give misleading ideas about wall construction. If the ink is filtered or centrifuged it gives much better results.

The surface details of specimens can sometimes be accentuated by heating the specimens to a very high temperature. The specimens are placed on a thin metal plate (a razor blade works well) and held over a flame until they become incandescent. After 15–20 seconds at this temperature the nodes, spines, tubercules and other raised portions of the test become white whereas the porous structures become dark. If the specimens are wetted after heating the contrast between the topographic highs and lows is yet more marked. The specimens become weaker as a result of this process and some delicate structures do not survive it.

A simple mechanical cleaning with a soft metal needle is often very rewarding. To accomplish this cleaning it is necessary to anchor the specimen to a stable surface. A drop of gum tragacanth, a piece of paraffin, wax, putty or Wood metal can be used. Wood metal is prepared by mixing together 7–8 parts bismuth, 4 parts lead, two parts zinc and one to two parts cadmium. A small piece of this mixture is placed on a copper or bronze plate and heated until soft. The test can then be placed in the desired position for cleaning. The cooled and hardened material holds the specimen very firmly. Reheating the plate allows the specimen to be moved or removed. If wax, paraffin or putty is used it is often desirable to add carbon black to the material to give a dark colour against which the lighter coloured test will contrast.

Placing dirty specimens in an ultrasonic bath for a few minutes will often clean debris off the test but may damage delicate tests.

6. Internal test examination

Knowledge of the internal structure of many species is essential for correct taxonomic determination. Various methods are utilized to reveal these details (clarification, chemical etching, rupture, impregnation, radiography and thin sectioning).

A. Clarification

A drop of water placed on the test is sometimes sufficient to reveal internal details of interest when viewed with transmitted light. Glycerine is better as it does not evaporate so readily in the heat from the lamp. It can be removed with alcohol. Xylol, castor oil or aniseed oil also serve as clarifiers.

Prior to using a clarifier it is advisable to dehydrate the specimens by placing them in a 50–60% alcohol solution.

These methods are helpful when examining most agglutinated tests. Water or glycerine often give good results when used on such tests especially if they contain some siliceous cement. These tests occasionally appear to be an agglomerate of grains. With the use of the clarifier their walls become transparent and entrapped air bubbles look silver in colour. The greatest amount of detail can be seen in these specimens by using a combination of reflected and transmitted light.

A method somewhat more complicated, but which yields good results, is fluoridization. Calcareous tests are attacked by hydrofluoric acid and the calcium carbonate is replaced by calcium fluoride which gives the test a transparent appearance ($CaCO_3 + 2HF \rightarrow CaF_2 + CO_2 + H_2O$). The tests are placed in a polyethylene container to which hydrofluoric acid is added. Concentrated acid is used for small tests; dilute acid for large tests. After several minutes the tests are removed and washed over filter paper to remove the acid.

Specimens with thick walls, filled interiors or very complicated internal structures are not good candidates for clarification.

B. Etching and Rupture

The specimens must be held firmly in place (see section 5 for description of techniques). The procedure followed depends on the nature of the test.

For a calcareous test, fix the specimen in paraffin and clean a portion of the test. Then brush it with a weak solution of HCl. As the wall gradually dissolves, the internal structure can be viewed under the microscope. In dealing with members of the Orbitoididae a cold 0.2% solution of HCl applied over a period of 1–3 minutes works quite well.

Another effective method involves the mixture of a weak solution of HCl with a paste of gum tragacanth. This acidic paste can be applied to the test wherever desired to rupture the wall. After the opening is made, the paste can be removed with a brush and water. The disadvantage of this method is that the heat from the viewing lamp may dry the paste before the acid has had time to operate. This may be partially compensated for by wiping the selected area very carefully with a fine brush dipped in dilute HCl prior to application of the acidic paste.

To isolate that portion of the test to be dissolved, it is possible to cover the test with a thin film of gum arabic. Let the test dry before applying the acid solution. It is also advisable to wet the specimen with a weak alkaline solution when the etching process is complete.

In the case of agglutinated tests, chemical etching is of little use, but a fine steel needle (for example, a dental tool), or glass needle can be used to rupture the test. Of course, rupturing the test in this manner works equally well on calcareous tests.

The fractured pieces of a calcareous test can be examined in plane transmitted and in polarized light to determine the optical characters of the crystals in the calcite walls (Table 16).

C. Impregnation

Making an internal mold of a complex canaliculate calcareous test may often be the only way to reveal the internal structure. The specimens must be free of internal

Table 16. Optical characters in calcareous walls (after Lewis, 1970).

Transmitted light	Polarized light	Wall type	
Amber or red-brown without pores	No optic figure, possible specks of colour	Porcellaneous	
With pores and colourless, milky, or pale blue	black cross with rings of colour mimicking a uniaxial figure	Radial	Hyaline
	mottled with specks of colour	Granular	

debris. Complete filling of the test can be aided by placing the clean test in a volatile solvent such as acetone and quickly transferring the test to a hot penetrant; e.g. Canada balsam. The solvent boils off and the liquified balsam is sucked into the test, filling all the cavities. After cooling, the test is dissolved in weak HCl and the internal mold can be studied. If the balsam is tinted with some colorant prior to use, thin-walled tests can be examined without being dissolved. Arnold (1965b) designed a vacuum oven which aids in the impregnation of small specimens.

D. Radiography

Schmidt (1952) was the first who proposed the use of contact X-ray photography to elucidate internal structure. The advantage of this method is that the test is preserved untouched.

Hedley (1957), Glaçon (1964), Caralp, Rechiniac & Vigneaux (1965) and Bé, Jongebloed & McIntyre (1969) also used X-ray micrography of foraminifera to describe internal structures. The proloculus diameter can be measured much more precisely than in thin section using this technique. The spacing and diameter of the pores are also easily determined in this manner.

E. Thin-sectioning

If none of the above methods are available or they do not yield satisfactory results, one must resort to a time-consuming but useful method – thin-sectioning.

Above all, one must note that this method, although it is not very complicated, requires a great deal of experience and patience. Many laboratories employ a specialist whose only task is the preparation of thin-sections.

The following is a description of one method for preparing a foraminiferal thin-section. Place a bit of Canada balsam, Lakeside 70 thermoplastic, epoxy resin or Rosenbusch cement on a clean, dry glass slide. Heat the slide but do not allow the cement to boil. This can be done by utilizing a burner, a hot plate or even an electric wire loop which can be used to heat the cement directly under the objective on the microscope stage. Wash the test in a volatile solvent, and then quickly drop it into the hot cement. This process allows a complete filling of the hollow portion of the test and the filled test is less likely to shatter once grinding begins. The test can be rotated to the desired orientation while the cement is still warm. Once cool, the cutting procedure can begin.

We recommend the thin-section grinding method first described by Morkhoven (1958) in which the grinding is done directly on the microscope stage in full view of the operator. The slide is placed firmly on the stage (it may be helpful to prepare a slide holder which fits securely to the stage) with the test positioned directly beneath the objective. Grinding is done with another glass slide which has been frosted on one side. Wet the specimen and the grinding slide and pass the grinding slide back and forth over the test using a rotary motion (Fig. 105). More rapid grinding can be

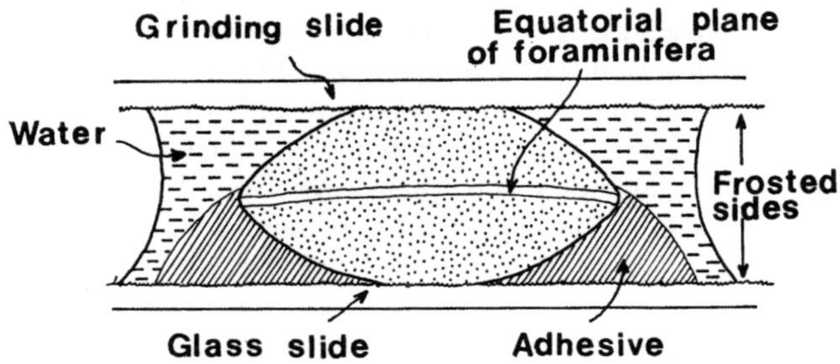

Fig. 105. Direct-visual method of thin-section preparation (after Morkhoven, 1958).

accomplished utilizing a small amount of fine aluminium oxide powder. The actual removal of test material and cement can be observed under the microscope with the use of transmitted and/or reflected light. Because the contact point of slide to specimen is small compared to the slide area, there is a danger of grinding a surface which will not be parallel to the bottom slide. To partially solve this problem, 4 drops of cement the same height as the test can be placed at the corners of the bottom slide. This produces a 5-point contact and insures a more parallel surface although it increases the grinding time. During the grinding process, accumulated debris should be cleaned off from time to time. Occasional reheating may be necessary to insure that the interior chambers are filled with cement. When the desired plane in the specimen is reached, the cement is heated and the specimen gently turned over. Grinding then can be resumed until the desired thickness is reached. The section is then preserved by putting a drop of Cedax or Canada balsam–xylene mixture on the specimen and placing a cover slip over the section. Heat the slide at 90–100 °C for a short time. Great care must be exercised in order to insure the absence of air bubbles beneath the cover slip. Aside from aesthetical considerations, the presence of bubbles can lead to erroneous interpretations.

The method described above is rather easily done with large specimens but great care and patience are necessary when sectioning small tests. Miklukho-Maclay

(1960) was able to obtain excellent sections of specimens whose diameter was less than 100 μ. Pohl & Browne (1973) described a technique for the precise orientation of small (100–300 μ) specimens in thin-section mounts. The specimens are mounted in epoxy resin on a small square of plastic which can be cut or ground in any orientation. Haake (1964) and Dupeuble (1965) proposed methods of preparing thin-sections involving the construction of a microscalpel which is used to cut a thin slice from the test. Arnold (1965a) also designed a sectioning device.

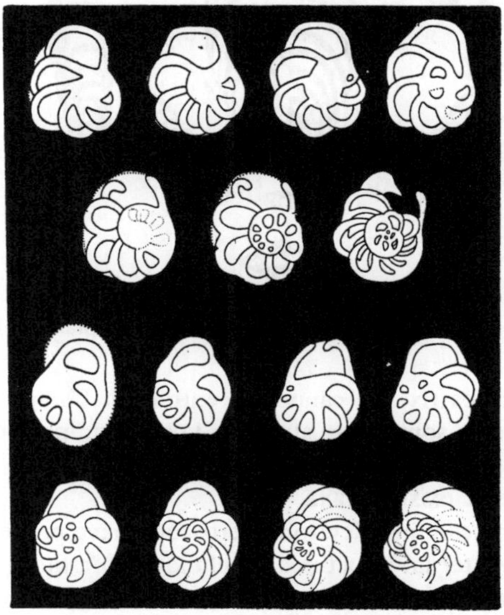

Fig. 106. Succesive (serial) thin-sections (after Chernysheva, 1940).

Hofker (1933) suggested a method for use with delicate tests. Instead of balsam, etc., paraffin is used as the mounting medium. The paraffin is heated to boiling in an aluminium container and the test dropped in. The test is then placed on a slide and sectioned with a razor blade using castor oil as a lubricant. Once cut, the paraffin is removed with xylene and the section is preserved using a cover slip.

Successive thin-sections can be prepared using any of these methods. To do this, a small portion of the test is ground or sliced away and a drawing, photograph, or acetate or carbon replica is made of the section. More material is removed and successive reproductions of the exposed surface are made. In this way the interior of the test can be reconstructed. This method was used by Chernysheva (1940) in a detailed taxonomic–biostratigraphic study of *Glyphostomella* resulting in the establishment of the new genus *Biseriammina* (Fig. 106). The use of a microtome can aid in the preparation of serial sections (Akpati, 1969).

Sections taken in different directions through the test can prove of great value in establishing the internal test relationships (Fig. 107). This method works well when

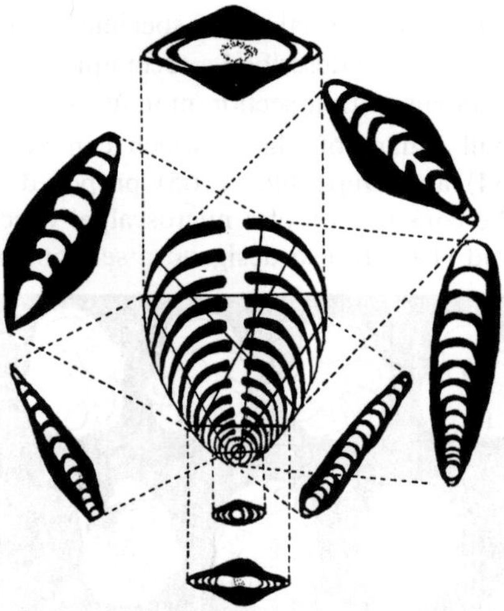

Fig. 107. Thin-sections of the same species taken in different planes (after Sosnina, 1960).

the test is encased in a matrix and cannot be extracted. Using this method Sosnina (1960) was able to clarify some long standing taxonomic problems with the genus *Pachyphloia*. Reichel (1946) showed how the use of multiple thin-sections could amplify the knowledge of existing taxa.

The transmission and scanning electron microscope have proven to be useful tools for the observation of internal structures. Techniques for the preparation of specimen replicas for electron microscope examination were given by Hyde & Krinsley (1964), Krinsley & Bé (1965) and Hansen (1969). Hansen (1967) described a photomosaic technique for compiling numerous electronmicrographs of the same specimen.

7. Wall mineralogy

The following method is suggested by Lewis (1970) to distinguish between calcite and aragonite tests: (a) Prepare Feigl's solution dissolving 11.8 g of $MnSO_4$ in 100 ml water; adding 1 g $AgSO_4$; bringing to a boil; cooling; filtering off the suspension; adding 1–2 drops of dilute NaOH solution to filtrate; filtering after 2 hours; and storing the solution in a dark airtight bottle away from sunlight where it will keep for years; (b) place the specimen on a slide; (c) etch the specimen surface by painting it with very dilute H_2SO_4; let dry for about one minute; (d) add several drops of alcohol to wash away excess acid; air dry; (e) paint on Feigl's solution; wait 5–10 minutes. If the specimen is unaffected by Feigl's solution, it is calcite; if it becomes black it is aragonite.

8. Wall lamellae

Lewis (1970) described a procedure to determine the lamellar nature of calcareous foraminifera. A thin-section is prepared as described in section 6. The grinding is done under the microscope using reflected and transmitted light. Layers (lamellae) making up the walls and septa can be seen under high magnification on a petrographic microscope (Table 17).

Table 17. Lamellar nature of calcareous walls observed in thin-section (after Lewis, 1970)

Lamella from each chamber continuing over previously formed exterior of test	Septa	Final chamber	Type of wall
Absent; test typically of single layer throughout	Of single lamella	Single lamella	Monolamellar
Present	Of single lamella	Single lamella	Monolamellar
Present	Double or partly double with canal system between lamellae	Single lamella	Rotalid wall
Present	Double, without canal system between lamellae	Double lamella	Bilamellar

9. Supplementary taxonomic traits

The classification of the foraminifera, both benthonic and planktonic, is based almost exclusively on studies of the external test morphology and the character of the wall. The internal structure of the test has been considered to a lesser extent in classification. As we have mentioned (Chapter VIII,3), these observations and data are insufficient to establish a natural classification. Because of this it is critically important to use other traits that might help establish the relationships that exist between various taxa.

Cytologic data will undoubtedly be of great help in the creation of a naturalistic classification of foraminifera (Lena, 1972). This is a fairly new approach and unfortunately one which has not yet been widely used for taxonomic purposes.

Another promising route was explored in the study of amino acid composition of the organic matrix of calcareous foraminifera. King & Hare (1972) studied 16 Recent planktonic species and determined that each is characterized by its original component of amino acids. Species which are similar morphologically; e.g. *Globorotalia menardii* and *G. tumida* contained very similar amino acid compositions. A study of amino acid composition of Recent and Early Miocene species belonging to the same genus (*Globoquadrina dutertrei*, *G. altispira* and *G. dehiscens*) showed that although their amino acid compositions were somewhat different, they had much in common. Quantitative analysis of the 16 Recent species revealed a strong correlation between amino acid composition and morphology. King & Hare proposed that amino acid composition of both Recent and fossil planktonic foraminifera can be used as a taxonomic character. The amino acids of the organic cement in agglutinated tests have also been examined (Hedley, 1963; Ross, 1963).

XIV. FAUNAL STUDIES AND THEIR APPLICATION

The first step in the study of a foraminiferal fauna is the qualitative taxonomic determination of the specimens. If the sampling methods permit, a subsequent quantitative study can be made. The data from these observations can be utilized in analysis of present ecological and oceanographic conditions. These data may also provide tools for the understanding of paleoecologic and paleooceanographic conditions. In this chapter several types of faunal studies and their interpretations will be described and discussed.

1. Qualitative and quantitative data and their display

A qualitative study is conducted without any consideration for the relative frequency of different taxa in the samples. This type of analysis yields the minimum amount of information because an abundant species is given the same weight as a rare species when presented in a list of taxa.

Even the most elementary of semiquantitative analysis is preferable to a faunal list. The simplest such analysis consists of erecting class sizes (very rare, rare, scarce, frequent, abundant, etc.) or establishing numerical bounds on classes (very rare = one specimen per sample; rare = 2–5, etc.). Various signs and symbols can be used to display these semi-quantitative data on tables and maps. Because the absolute number of specimens in a sample is of little value unless the volume, weight or surface area of the sample is known, this data should also be given.

To facilitate comparisons between data collected in different areas by different investigators, it is desirable to enumerate the exact quantity of each species in a sample. This data can be presented raw (number of specimens for each taxon) or converted to percentage form. The latter method gives the relative importance of each taxon in the sample fauna. The data can be presented most simply by establishing class limits and assigning appropriate symbols for each class. The total quantity of specimens in a sample should be included to give some ideas of the standing crop at the sample site. The Foraminiferal Number is probably the best indicator of quantity if living populations have not been identified (Chapter IV,8 and VII,3).

The number of specimens per sample that should be examined to insure an accurate prortrayal of the sample is not easily determined. There are two aspects of this sampling problem, each of which can be posed as a question: (a) how many specimens must be counted so that all or most of the species present will be encountered and (b) how many specimens must be counted to insure a precise understanding of the relative proportions of the species within the sample? The primary problem stems from our understanding of the words *most* and *precise*. It is reasonable to expect that as the number of specimens counted increases so will the number of species and the precision with which we can express the proportions.

In samples where species equitability is low; i.e. where there are a few dominant species and many minor ones, a greater number of specimens must be examined than in high diversity samples (Hulbert, 1963). Most distributions show a logarithmic relationship between the number of species and the number of individuals (Fisher, Corbett & Williams, 1943). An understanding of the peculiar distribution characteristics of a given sample will aid in deciding whether to expend the extra effort required to find an additional species. Dennison & Hay (1967) prepared a graph based on binomial sampling theory which showed that to be 95% certain of detecting the presence of a species whose contribution to the population was 1% would require the examination of about 300 specimens. As the degree of confidence desired increases, the sample size required also increases. As the proportional contribution of the species to the population increases, the sample size required to detect its presence or absence decreases.

An illustrative example of the effort required to find rare species was provided by Boltovskoy & Lena (1970b). They studied the benthonic fauna of Puerto Deseado Creek, Argentina for several years in repeated samplings and discovered 76 species. Continuation of the sampling program for several more years and the examination of hundreds of samples increased the number of species to 120. The additional species lived in small isolated biotopes and were very rare.

The precision with which the percentage contribution of a given species to a sample can be stated is dependent on (a) its actual proportion in the sample (the maximum error occurs when the proportion of a species in a sample is 0.50), and (b) the number of specimens that are counted. Ujiié (1962), Chang (1967) and Wright & Hay (1971) described these relationships and suggested various ways of selecting the optimum number of specimens to count. Many authors have suggested that between 100–300 specimens are adequate to portray the distribution of specimens in a sample. However, Wright & Hay (1971) showed that at the 95% confidence limits for a species whose actual contribution to a sample is 5%, it is necessary to count 1824 specimens if one wishes to express this contribution to the nearest percent. If the actual contribution of a species is greater than 5% even more specimens must be counted.

Once the optimum number of specimens has been determined, it is often necessary to divide the sample into fractions to facilitate examination. There are several devices available which can accomplish this splitting. Most of these utilize partitions which divide or split a sample into equal subsamples as the specimens pass over them (Otto, 1933; Grimsdale & Morkhoven, 1955; Kennard & Smith, 1961). A simple, easily constructed splitter was described by Parker (1948) (Fig. 108). If the split fraction is too large, it can be resplit until the desired size is obtained. The number of

Fig. 108. Simplified sample splitter.

specimens in the whole sample can be extrapolated from the number counted in the fractional part. An ingenious splitter whose use removes some of the bias inherent in linear splitters was described by Daniels (1970a). It consists of a rotating conical splitter divided into four parts which is fed by specimens falling off a vibrating chute (Fig. 109).

One of the problems in accurate counting is the bias that may be introduced when the sample is poured into the counting tray. The foraminiferal specimens are of different size and shape, and will therefore segregate when shaken out into the tray. Because of this spatial segregation on the tray, one should examine all specimens on a tray or only specimens found in randomly selected squares on the tray rather than beginning at one edge and counting until some predetermined, statistically significant number of specimens is reached. One method of selecting squares is by use of a random number table; another is used by European investigators who, using the standard German tray which has 45 squares (5×9), usually start in the lower left corner square and proceed diagonally up to the middle square on the top row and then drop diagonally to the lower right corner square.

It is advisable, however, after picking out the necessary number of specimens following this diagonal pattern, to quickly check the whole tray looking for the rare and/or interesting specimens which were not found in the predetermined squares.

An interesting technique for randomly distributing small objects on a tray was described by Moore (1973). The specimens are allowed to fall through a column of water to the slide surface. Random mixing in the settling tube insures randomness on the final slide.

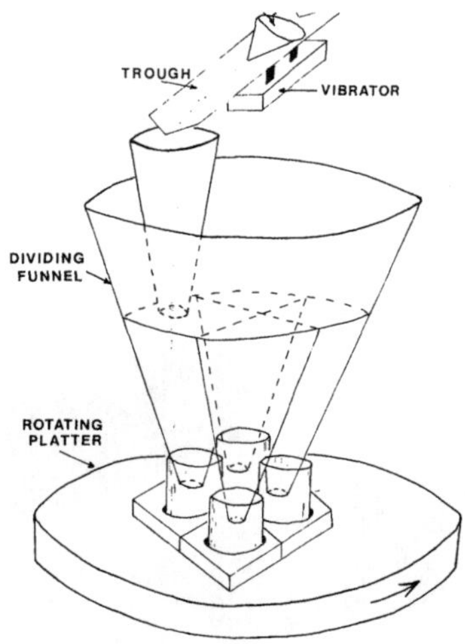

Fig. *109*. Automatic microsplitter (after Daniels, 1970a).

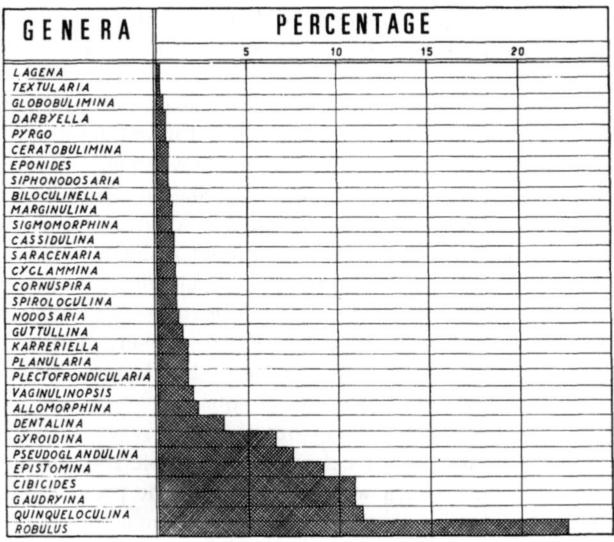

Fig. *110*. Relative quantity of foraminiferal genera arranged in rank order (after Rau, 1948).

There are various types of graphical techniques by which one can indicate the nature and distribution of the specimens within the sample or the sampling area.

Rau (1948) used a histogram with the genera listed in order of abundance to indicate the generic composition of a fauna (Fig. 110). The relative contribution of a genus with respect to any other is easily determined from this type of display.

Lowman (1949) introduced a linear type of diagram in which not only the relative frequency of taxa but ecological data are displayed. This type of representation is quite useful as it allows a direct visualization of all the data collected at a site, along a traverse, or in an area. This kind of diagram was made popular by Bandy (Fig. 111). There are two disadvantages to this sort of display. With an abundance of data (ecologic and/or number of taxa) the graph becomes cluttered and difficult to read. One must take care to remember that the *width* of the area representing a given taxon in the faunal part of the plot is indicative of its proportional contribution to the

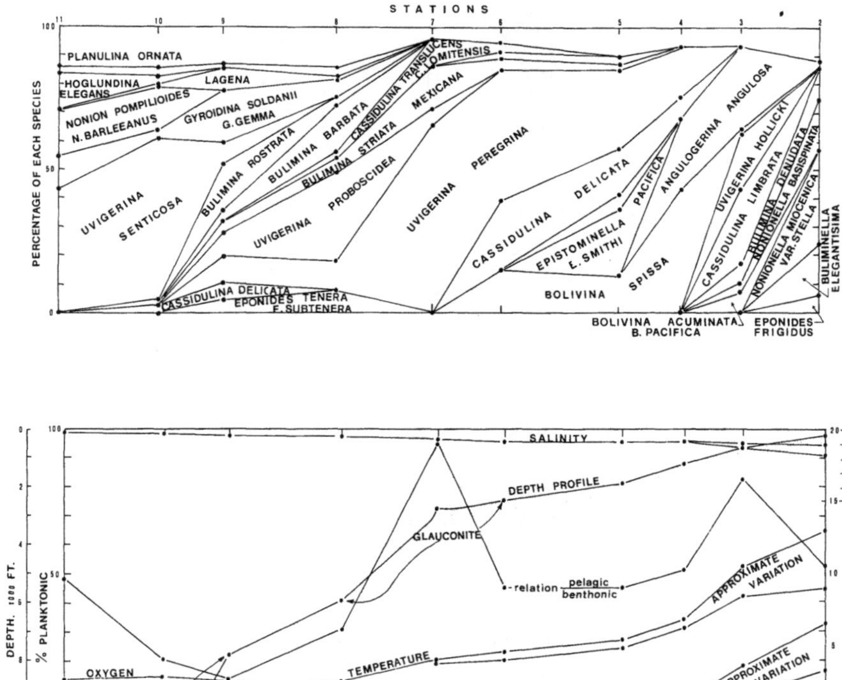

Fig. 111. Graph displaying geographic relationships and ecological parameters, Point Arguello, California (after Bandy, 1953).

fauna but that its vertical position on the chart means nothing. For example, the *Uvigerina peregrina* portion of Fig. 111 executes a 45° bend between Stations 5 and 7, but the contribution of the species to the fauna barely changes.

Pie diagrams are another traditional way of showing the contribution of various taxa. Nested pie diagrams can be used to yield even more information. An example is shown in Fig. 112 in which habitat, familial and generic affiliation are all shown without undue clutter. Very complex pie diagrams can be used without confusion (*see* Levy, 1967).

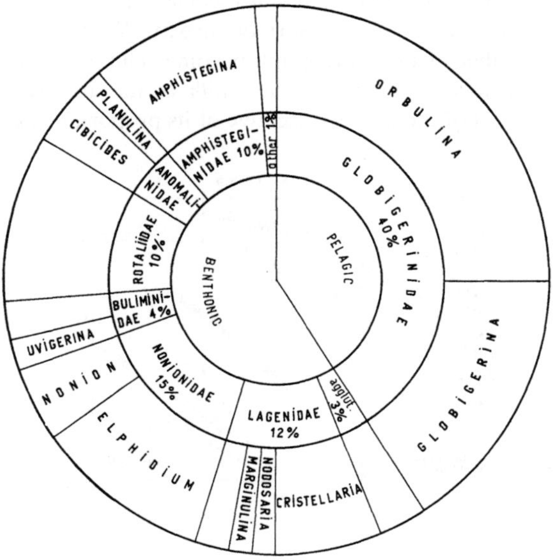

Fig. 112. Nested pie diagram (after Ciry, Tintant & Chareyron, 1948).

A vivid method of showing faunal distribution consists of making drawings of the species and placing them on a map in their appropriate geographic positions (Fig. 113).

It is probably best to treat separately the quantitative analysis of the benthonic and planktonic portion of a sample because these two groups occupy such different habitats and are influenced by different factors.

Since the advent of a simple and reasonably accurate method of distinguishing between living and dead foraminifera (Walton, 1952), it has been possible to examine both the living and the total (living plus dead) populations.

There are a variety of factors which can alter the distribution of empty tests after the death of the organism: (a) transport of the unattached tests by bottom water movement; (b) test dissolution, which is different for different species (occurring at great depth due to low temperature and high CO_2 partial pressure, and in some

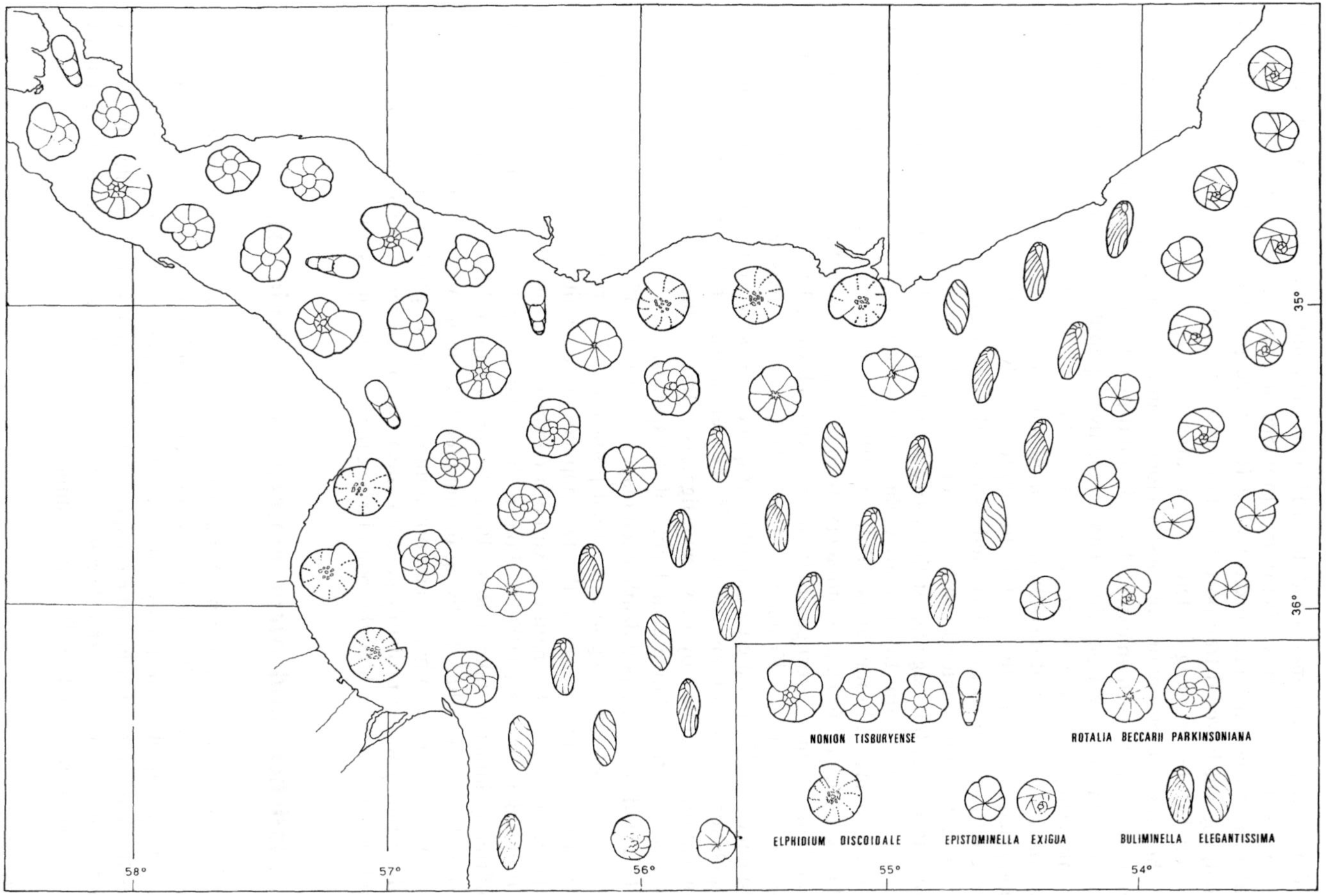

Fig. 113. Distribution of foraminifera in the Río de la Plata.

shallow water environments with low pH substrate conditions); (c) the presence of an empty test fauna which is relict from a previous time; e.g. an abundance of a shallow water glacial epoch fauna mixed with a Recent deeper water fauna; (d) submarine slumping which would transport a shallow water assemblage into deeper water (*see* section 13); (e) crushing of delicate tests after death.

The presence of any one of these factors in an area may alter the composition of the total fauna to such an extent that it is not truly reflective of the environment. It is obvious that whenever possible, data on the living population should be obtained. It is interesting and somewhat sad to note that of almost 350 investigations of benthonic foraminiferal distributions conducted between 1953–1973 only 42% distinguished living specimens. Of the approximately 24 000 samples these studies represent, only 9000 (37%) yielded data on the living populations.

It is not well understood just how accurately the total population reflects the living population; i.e. how significant the above factors are in altering the composition of the fauna after death. In some areas there seems to be a very high correspondence between living and total faunas. Phleger (1952a), Jell, Maxwell & McKellar (1965), Chiji & Lopez (1968) and Matoba (1970) indicated similar distributions of living and total fauna. Other investigators (Shifflett, 1961, Haynes, 1964; Perath, 1966; Murray, 1969; Ujiié & Kusukawa, 1969; Ikeya, 1970; Atkinson, 1971; Rosset-Moulinier, 1972) indicated a lack of correspondence between the living and the total population. It is not always clear if the relationship between living and total populations described by these investigators refers to the presence or absence of species or to the relative abundances of the species.

In a study of pelecypod death assemblages Johnson (1965) showed that the species composition of the living and the total populations is very similar but the relative proportions of the species in the two populations show little correspondence. Few detailed studies of this nature have been conducted on foraminifera.

Boltovskoy & Lena (1970a) showed that foraminiferal protoplasm may remain sufficiently intact to take a Rose Bengal stain (Chapter XI,3) up to 60 days after death. This means that the living foraminiferal fauna in a sample may include not only the standing crop but an additional month or two of production. Although such a discrepancy would probably have no effect on an ecological or spatial distribution study it may alter or blur the boundaries drawn in seasonal studies.

2. Statistical and other mathematical manipulations

Various statistical and other mathematical techniques have been used in the study of Recent foraminifera. These techniques have found use in the examination of taxonomy, distribution, ecology and population dynamics. It is not the purpose of this book to give a mathematical exposition of statistics and probability but we do wish to briefly describe and discuss some of the quantitative techniques that have been successfully used on Recent foraminifera.

Anyone considering the statistical treatment of Recent foraminiferal data should bear several things in mind. If data is to be subjected to a statistical analysis, it must have been collected with a particular analytical scheme in mind. A well conceived sampling design prepared *prior* to data collection is almost indispensable if one wishes to get the most from the data. Proper sample design includes such things as a sampling plan, size of sample and collection of environmental data at sampling sites. The specimens used in the analysis should generally be abundant and must always be selected randomly if any statistical probability statements are to made about the data. To avoid misleading probability statements it is necessary to have a clear understanding of the assumptions underlying a given statistical test. One must also remember that statistical treatment of data does not 'prove' any relationship but only allows the investigator to make well defined probability statements about the data.

The urge to statistically treat a set of data is a powerful one and must be resisted if such treatment substitutes for rather than complements a careful, thoughtful appraisal of the problem. Statistical analysis should never replace consideration of biological and ecological principles.

There are a variety of techniques which are quantitative but not probabilistic in nature. Care must be taken with these methods also to insure an understanding of the assumptions involved in their application.

There are a number of books and articles which deal with general statistical methods and techniques. We will mention a few because they have proven useful for us or for our colleagues (Lamotte, 1940; Swinnerton, 1940; Dixon & Massey, 1957; Miller & Kahn, 1963; Seal, 1964; Snedecor & Cochran, 1967; Sokal & Rohlf, 1969; Davies, 1971), but suggest that each investigator choose his own based on his own needs. We do recommend that any reference work contain examples, explain clearly each step of a process, list the assumptions and limitations of any given test or procedure, and briefly treat the probability theory on which the test or procedure is based.

The various uses to which statistical and other mathematical techniques have been put are discussed below.

A. Taxonomic determination

The variation exhibited by a set of specimens can be analysed by a variety of means for the purpose of constructing taxonomic categories. Probably the simplest and most widely used of these techniques is to plot measurements of two morphological variables on a rectangular graph. The resulting figure gives a visual expression of the type of variability possessed by the specimens and the presence of any irregularities in the data.

The statistical differentiation of species has been attempted rather often since the advent of widespread use of the digital computer. Now large sets of data and highly sophisticated techniques can be handled easily. Prior to describing these new investigations we wish to pay homage to the kind of study that can be accomplished with the aid only of imagination and diligent labour. We refer to the work of Schmid (1934) on the taxonomic problem of *Globorotalia menardii* and *G. tumida* and of P. Smith (1963) on bolivinids. The former two species were described by different authors as different species, synonyms, two sub-species and two *formae*. Schmid investigated a population of 600 specimens of *G. menardii/tumida* by graphically displaying the results of the measurement of several morphological traits. Schmid's results indicated that the two forms are part of a continuum and should be considered as one species (*G. menardii*, according to law of priority). This species, according to Schmid, is dimorphic and its macrospheric form corresponds approximately to the form described in the literature as *G. tumida* while the microspheric form corresponds to that of *G. menardii*. P. Smith examined approximately 4500 bolivinid specimens and discriminated 18 species by showing measurements of length, width, number of paired chambers and length:width ratio on rectangular plots and histograms. She also used triangular plots of morphologic parameters (Burma, 1948) to graphically separate members of different species.

A good clear description of univariate and bivariate statistical and graphical techniques that can be used for taxonomic discrimination was given by Imbrie (1956).

The treatment of a single morphological variable or an examination of the way two such traits covary is easily visualized because it can be displayed in two dimensions. The multiplicity of morphological characters that define a taxon and the possibility of interdependence among certain traits means that a multivariate study of form may prove more meaningful in terms of describing variation and determining taxa. Several books and articles deal with this problem (Burma, 1949; Blacklith & Reyment, 1971; Sneath & Sokal, 1973). To the uninitiated the wealth of techniques and procedures available may seem hopelessly confusing. To alleviate this turmoil, Reyment (1972) described the characteristics, capabilities and pitfalls of various multivariate techniques; e.g. factor analysis, principal components analysis, principal coordinate analysis, discriminant function analysis and canonical variate analysis. There have been several applications of these techniques on Recent and fossil foraminifera. Reyment (1961) made taxonomic discriminations on an Eocene planktonic foraminifer, *Globigerina yeguaensis*, using principal components analysis. Using the same technique, Reyment (1969) later performed a morphologic analysis of *Textilina mexicana*. Buzas (1966) examined the morphology of four groups of *Elphidium* by means of canonical variate analysis and concluded that only three of these groups were distinct enough to be separate species. For the most part, the statistical treatments in the studies described above have been applied to populations which have been classified according to conventional qualitative methods rather than to fresh, unclassified specimens.

B. Distributional studies

The quantitative delineation of an area into biofacies has been accomplished in a wide variety of ways ranging from simple graphs to complex multivariate statistical methods. Murray (1968b) distinguished between various lagoonal and estuarine environments by using a triangular diagram (Fig. 57) on which were plotted the percentage contribution of rotalid, textularid and miliolid components of the fauna. He also demonstrated that the faunal diversity in these different environments varied in such a manner that distinctions could be made. Wright & Murray (1972) used this technique to delineate Eocene shallow water saline environments.

The idea that population diversity could be used to define biofacies was also expressed by Beerbower & Jordan (1969) who devised statistical measures to establish environmental boundaries on the basis of diversity values. Niitsuma (1968) and Matoba (1970) described methods for using the Motomura diversity of sample populations (subsection D) to discriminate between populations and thereby to divide an area into biofacies.

Because foraminiferal populations are multispecies assemblages living in an area, the division of that area into zones containing a similar fauna lends itself to multivariate analysis. Buzas (1969b) described several of the different multivariate techniques that are applicable to biofacies quantification. Perhaps the simplest of these is a multivariate analysis of variance in which the species variance in pre-defined areas is examined for differences. One-way analysis of variance has been used to define the clumped nature of foraminiferal distribution (Lynts, 1966a; Buzas, 1968). In its multivariate form, as canonical variate analysis, the method was used by Buzas (1967, 1972a) to verify the existence of pre-defined biofacies.

The above techniques are useful if the investigator has some idea of the underlying biofacies pattern in an area. If no such *a priori* knowledge is available or the investigator is hesitant to introduce a possible bias into his analysis, there are a number of multivariate techniques which may indicate the faunal patterns or 'structure' underlying the area.

1. Cluster analysis

The degree of similarity between all pairs of samples in a study area can be calculated and these similarity values are then arranged together in a heirarchial order until clusters of similar samples are lumped together. These clusters constitute a biofacies. Presence/absence data or quantitative measures of species abundance have been used to calculate the measures of similarity. A large number of such measures have been devised, each having particular properties suitable to particular kinds of data or particular assumptions about the data. This technique has been used successfully by a number of investigators working with both benthonic (Kaesler, 1966; Mello & Buzas, 1968; Howarth & Murray, 1969; Gevirtz, Park & Friedman, 1971; Brooks, 1973; Johnson & Albani, 1973) and planktonic foraminifera (Parker & Berger, 1971; Ujiié & Negase, 1971; Ujiié, 1973).

2. Association analysis

A variation of cluster analysis uses the statistic chi-square as a measure of association between samples. The samples are lumped in a heirarchial fashion as in cluster analysis. Vilks (1967) and Vilks, Anthony & Williams (1970) applied this technique to foraminiferal populations with some success.

3. Factor analysis

This method of analysis attempts to account for the variance between objects (samples) with a limited number of 'factors'. Clusters of samples that respond in the same way are determined and presumably have ecological meaning when plotted on a map. There is an underlying assumption that the original correlated variables can be reduced into a smaller number of uncorrelated factors. Streeter (1964, 1972) and Howarth & Murray (1969) used factor analysis to discriminate different foraminiferal environments which had been previously analyzed by non-multivariate methods. In all cases there was a good correspondence between the factor analysis pattern and the prior interpretation.

4. Principal components analysis

The assumption underlying factor analysis that the data can be reduced as described above is not held in principal components analysis. The two methods are quite similar and are often mistaken for each other. Imbrie & Phleger (1963) analysed the benthonic foraminiferal fauna from San Diego, California using this technique. Malmgren & Kennett (1972) were able to determine the geographical distribution of various morphological variants of *Globigerina pachyderma* based in part on a principal components analysis.

5. Other methods of analysis

Trend surface analysis, a technique designed to reveal the anomalies in the spatial distribution of a variable, was utilized by Stehli (1965) and Rodionov (1968) to study the distribution of foraminiferal populations. Rank correlation, a non-parametric means of examining the statistical relationship between variables based on position in a sequence (rank), was used to clarify *Globigerina pachyderma* distribution in the South Pacific (Malmgren & Kennett, 1972).

C. Ecological studies

The quantitative relationships between foraminifera and various environmental parameters has been extensively examined (Chapter III). Many of these studies deal with the relationships between two variables, one foraminiferal, the other environmental; e.g. foraminiferal density as a function of depth, change in morphology of a particular species as a function of salinity, etc. However, an understanding of the complex interactions between many species and several environmental factors requires a multivariate analysis. The techniques used in these analyses are the same ones described above in the sections on taxonomic and distributional analyses. We will mention examples of particular problems on which these techniques were used.

Using the observations of Myers (1943c) on *Elphidium* (Chapter IX,4) as a model,

Lamy (1969) attempted to describe the life cycle and paleoecology of Miocene specimens of *Nonion dollfusi.* She measured test sizes of numerous specimens to establish a growth curve and suggested that different curves would reflect different environmental conditions.

Buzas (1969a) used a multiple analysis of variance to analyze the relationships between several species and environmental parameters in the estuary of the Choptank River, Maryland. A similar technique, a three level analysis of variance, was used by Smith & Kaesler (1970) to analyse the nature of detrital grains used in the construction of agglutinated tests (Chapter III,4).

Blanc, Blanc-Vernet & LeCampion (1972) analysed 14 species from four Mediterranean cores using a principal components analysis. They concluded that water temperature was a significant factor in the control of the species distribution. Factor analysis, a brother of principal components analysis, has been used by several investigators to establish ecological relationships in multivariate systems (Allen & Pujos-Lamy, 1972). Lynts (1972) described the method and its utility in ecological and paleoecological studies. This technique has the advantage of reducing the data to a simpler form. The investigator does not have to think in terms of the large masses of data that are the usual result of an ecological study. Although the data are reduced to a manageable level there is no guarantee that easily understandable patterns will result from the analysis. This technique was used by Hooper (1968, 1969) and Niitsuma (1971) to establish depth zonations of benthonic forms.

The relationship between various environmental parameters and certain morphological traits of *Ammonia beccarii* was determined by Chang & Kaesler (1974) using canonical variate analysis.

D. Population dynamics

The quantitative relationship between individuals and between species has been examined to try to assess such things as the nature of the species variation that occurs between samples, the causes of variations in species diversity in different environments, and the nature of an equilibrium community. An excellent review of species diversity was prepared by Buzas (1972b).

Vilks (1967) noted the variation between closely spaced samples and between samples taken at the same place at different times in Bras d'Or Lakes, Canada. He conducted an analysis of variance and concluded that about 50% of the sample to sample variation could be ascribed to a temporal component; i.e. short term and seasonal changes. An additional 25% of the variation was attributed to a geographical component.

The diversity of foraminiferal faunas appears to vary under different environmental conditions. Walton (1960) noted that the number of species around barrier islands in the Gulf of Mexico changes depending on the environmental variability. As the environment becomes more variable, the number of species decreases. He noted also that the degree of dominance, the proportional contribution of the most abundant species, increases near the barrier islands, zones of increased environmental variability.

One of the problems in examining biological diversity is disagreement over a quantitative measure of the concept. Qualitatively, diversity (or equitability) is a measure of the dispersal of individuals among the species in a population. The more uniformly distributed the individuals are among the species, and the greater the number of species, the greater the diversity. The confusion arises in trying to establish a parameter which measures these factors. Many diversity measures have been proposed and some of these have been utilized in foraminiferal population analyses.

One of the problems encountered in using the number of species as a measure of diversity is that sample sizes and methods vary from investigator to investigator. This means that the same area might yield different numbers of species if sampled by two different researchers. To eliminate this factor some parameter which is independent of the number of species is desirable. Simpson (1949) modified an equation of Yule (1944) which gives a measure of the probability that two individuals picked at random from a sample will be the same species.

$$\text{Diversity} = \frac{\sum n_i(n_i - 1)}{N(N-1)}$$

where n_i = number of individuals of the ith species; N = number of individuals in the sample. This measure varies from $0-1$ as defined by Simpson.

McGowan & Fraundorf (1966) suggested that the measure be revised by using the value of 1 minus the Simpson index so that the index would increase as diversity increases. The Simpson index or variants of it have been used by Gibson (1966) and Schnitker (1971a) in studies on benthonic foraminiferal populations. Gibson examined the same barrier islands that Walton (1960, see above) had described and reached the same conclusions regarding species diversity; low diversity values are associated with shallow water unstable or highly variable environments. Diversity increased in deeper more stable areas.

One of the characteristics of the Simpson index of diversity is that the more abundant species are weighted more heavily. This characteristic is not necessarily undesirable but has produced some criticism. Two similar measures of population diversity were devised which treat abundant and rare species alike. The earliest of these measures was described by Motomura (1932) who noted that the number of individuals per species exhibits a logarithmic progression which could be described:

$$\log y = b - ax$$

where x = the species rank by abundance (most abundant to least); y = frequency of occurrence of species x; a = slope of this relationship; b = the intercept.

If x and y are plotted against each other the data usually follow a straight line, the slope of which is a measure of diversity. Ikeya (1970) described the results of culture experiments which indicated that this geometrical progression has an experimental as well as theoretical basis for benthonic foraminifera. He concluded that under conditions of population equilibrium the rank-abundance data would follow a straight line and deviation from this pattern is the mark of a transitional fauna or, in fossil assemblages, of a mixed death assemblage. Niitsuma, Sakai & Hattori (1971) described various modifications of the Motomura progression which reveal additional changes in population structure.

Fisher, Corbett & Williams (1943) described a logarithmic relationship between the number of individuals and the number of species in a sample. Their relationship is similar in concept to that of Motomura. They devised an index of diversity, α, which is a function of the number of individuals and the number of species. Values for α can be read directly from a graph (Murray, 1968b). This index has been used extensively by Murray (1973b) to discriminate between different environments (subsection B).

Although the Fisher index treats rare species on an equal footing with abundant ones, it is independent of the distribution of individuals among the species; i.e. a sample with 45 individuals and 9 species will have the same α value regardless of how the 45 individuals are concentrated in these 9 species:

	Species								
	A	B	C	D	E	F	G	H	I
Case I	9	8	7	6	5	4	3	2	1
Case II	5	5	5	5	5	5	5	5	5

The α values for Case I and Case II are the same ($\alpha \approx 4$) whereas the modified Simpson index will indicate a higher diversity for Case II because of the more uniform distribution (0.909 versus 0.879).

The Shannon-Weaver diversity index based on the information function (Shannon & Weaver, 1963) is a measure of entropy or disorder in a population. The index is defined as:

$$H = -\sum p_i \ln p_i$$

where H = diversity index; p_i = proportional abundance of the ith species.

The index was used by Buzas & Gibson (1969) on assemblages along the East Coast of North America, by Beerbower & Jordan (1969) on the fauna in a brackish lake on the Texas Coast, and by Nichols & Norton (1969) in the James River Estuary, Virginia. These investigators also used as a measure of species equitability, the ratio between the number of taxa required to yield the observed diversity if distributed randomly among the species (e^H) and the actual number of taxa. Theyer (1971b) noted that benthonic foraminiferal diversity in the Pacific–Antarctic Basin as defined by this index is only half as great as that noted in comparable zones of the Atlantic by Buzas & Gibson. He suggested that this may be due to a differing community structure or contrasting environmental conditions.

Because some diversity indicies are weighted more heavily toward species richness and others more heavily toward species equitability or evenness, no single measure of diversity or community structure exists. Recognizing this impasse, Hurlbert (1971) described various parameters which would estimate different aspects of community structure. None of these measures has been used in describing foraminiferal populations.

3. Paleoclimatic interpretations based on foraminiferal species distribution

After studying the planktonic foraminifera from cores obtained during the German South Polar Gauss Expedition, Philippi (1910) mentioned that there was an inverse relationship between the abundance of *Globorotalia menardii* and *Globigerina bulloides* plus *G. inflata*. Schott (1935) confirmed these observations based on short (1 m) cores from the Meteor Expedition. He concluded that one could correlate the sub-Recent sediments of the Atlantic based on examination of the whole foraminiferal fauna. If the foraminifera were abundant, the cores could be correlated even though vertical changes in the lithology were lacking. Schott also noted the absence of *Globorotalia menardii* in that part of the core corresponding to the most recent glacial epoch, and its reappearance in a lower part of the core (representing an interglacial epoch).

A study of the foraminiferal fauna (primarily planktonic forms) in oceanic cores has become widespread. Many investigators have contributed greatly to the knowledge of planktonic faunas in cores and to the interpretation of this fauna in terms of water temperatures.[10]

The primary assumption on which most of these studies were conducted is that the planktonic foraminiferal faunas found in fairly recent sediments responded in the

[10] Bramlette & Bradley (1940), Cushman & Henbest (1940), Cushman (1941a), Phleger (1939, 1942, 1947, 1948, 1951b, 1955, 1960b), Phleger & Hamilton (1946), Ovey (1950), Arrhenius (1952), Schott (1952, 1954), Ericson, Ewing & Heezen (1952), Phleger, Parker & Peirson (1953), Boltovskoy (1954b, 1973b,c), Ericson, Wollin & Wollin (1954), Ericson *et al.* (1955), Ericson & Wollin (1956b, 1968), Hamilton (1957), Ewing, Ericson & Heezen (1958), Todd (1958, 1964), Parker (1958), Stschedrina (1958d), Saidova (1960b, 1961a), Brotzen & Dinesen (1960), Ericson *et al*, (1961), Olausson (1961), Ericson, Ewing & Wollin (1964), Herman (1964, 1965, 1968, 1969, 1970, 1972a), Lidz (1966a), Oba (1967, 1969), Caralp (1968, 1970), Caralp, Moyes & Vigneaux (1968), Frerichs (1968), Berggren & Boersma (1969), Blanc-Vernet, Chalmay & Froget (1969), Devereaux, Hendy & Vella (1970), Griggs *et al.* (1970), Kennett (1970b), Morin, Theyer & Vincent (1970), Napoli Alliata & Ornesi (1970), Napoli Alliata, Stefanini & Fioramont (1970), Kudrass & Thiede (1970), Ruddiman, Tolderlund & Bé (1970), Barash (1971), Clark (1971), Imbrie & Kipp (1971), Lynts & Judd (1971), Morin (1971d), Niitsuma (1971), Olausson *et al.* (1971), Olsson (1971b), Thiede (1971a,b), Blanc-Vernet (1972), Keany & Kennett (1972), Kennett & Huddlestun (1972a,b), Leclaire & Vergnaud-Grazzini (1972), McIntyre, Ruddiman & Jantzen (1972), Poag (1972), Rotschy *et al.* (1972), Sancetta, Imbrie & Kipp (1972), Steurwald & Clark (1972), Vincent (1972), Imbrie, van Donk & Kipp (1973), Ubaldo (1973), Lynts, Judd & Stehman (1973), Zobel (1973), Poore & Berggren (1974) and the voluminous works appearing in the Initial Reports of the Deep Sea Drilling Project, published by the National Science Foundation, Washington, D.C.

same way to climatic conditions as foraminifera do today. If this is true then the faunal variations observed in the Recent species distribution can be used to interpret the faunas from core samples.

Many of these authors considered not only qualitative and quantitative characteristics of the fauna but also other details which yielded information about changing conditions: $CaCO_3$ content of the core, coiling direction of specimens, appearance of new morphological forms, etc. It became clear soon after the onset of these investigations that there are layers of sediment that contain a fauna typical of water temperature cooler than that of the present surface layer. It was only logical to conclude that this fauna corresponded to the glacial epochs of the Pleistocene. Detailed studies revealed that it was possible to distinguish between the various episodes of glaciation and effect correlations from one core to another.

There are, however, pitfalls in the interpretation of these faunas. One must consider that the vertical change in a fauna may not always represent a world-wide climatic change but may represent a local change such as the movement of a warm or cold water current or a change in the position of a zone of convergence. Any stratification in the habitat of living planktonic species (Chapter VII,4) will affect the regularity of changes induced by climatic changes. The discovery that planktonic foraminifera are differentially dissolved in deep water also creates complications in the interpretation of foraminifera from cores.

Despite these obstacles, a great deal of interest has been shown in the paleoclimatic analysis of the taxa found in deep ocean cores.

Not all cores taken from the ocean floor serve equally. To be of greatest value they should be long ones, taken away from the continental shelf in moderately deep waters. On the shelf the velocity of sedimentation is too great and short cores represent only a short time span.

The problems with short cores taken in low latitudes was first illustrated by Hamilton (1957) who studied a core taken far from the coast in the Equatorial South Pacific. He encountered no faunal changes, although physiochemical analysis of the sediments indicated that the core was Pleistocene. Hamilton explained this observation by suggesting that the warm water faunas persisted in the equatorial regions during the glacial epochs.

Cores taken for the purpose of paleoclimatic interpretation should be treated in the following manner. As soon as possible the core should be unwrapped, cleaned, photographed and labelled. Cleaning is very important since the smallest amount of contamination can result in faunal mixing leading to erroneous conclusions. At least 2 mm of the outer portion of the core should be removed to prevent this mixing. While the core is still fresh it is wise to photograph it and describe its appearnace.

The core is divided into four longitudinal sections: one part put in sealed containers and preserved, one part set aside for foraminiferal and other microfossil studies, and the remaining half used for sedimentological and chemical analysis. Experience has shown that if a sample shows obvious mixing, displacement, or the effects of strong current action (coarse grains, no fines) it should be eliminated from the study.

The upper centimeter is removed for study of the Recent fauna and is treated in the manner described in Chapter XI. The remainder of the core can be sampled in a variety of ways (regular interval sampling, random sampling, samples every facies change, etc.). The method chosen depends on the ends of the investigator. As always, the greater number of samples taken the more detailed the results will be. The samples are washed and the specimens are separated from the residue (Chapter XII).

Different methods have been used for determining paleoclimates using the distribution and ecologic responses of Recent foraminiferal species as models. These can be divided into three categories: (a) whole fauna, (b) species ratios and (c) single species. The selection of method depends on the material, time available, acquaintance of the investigator with the fauna, and even on the taste of the worker.

Many investigators utilize more than one of these methods, but for the sake of simplicity we will discuss each one separately.

A. Studies of the whole planktonic assemblage

This method is perhaps the most common and has been utilized, with some modifications, by many investigators.

The method generally involves dividing the fauna into three groups; (a) species characteristic of cold and cold–temperate waters, (b) species characteristic of warm and warm–temperate waters and (c) cosmopolitan species. The latter group is not considered in the subsequent calculations. The ratio between the abundance of the first two groups is calculated for each sample. This ratio, which represents the cold:warm ratio, is then plotted against core depth to give a paleoclimatic curve.

The advantage of this method is that the maximum amount of information is used. Each species is contributing to the final interpretation. Errors in identification of a species will be partially cancelled in the final ratio. Unusual local conditions which may abnormally increase or decrease the abundance of a given species will be corrected by the remaining species. This method is probably the most reliable of the semi-quantitative methods described.

This method has, however, the following disadvantages: (a) It is time consuming. Even if only a pre-determined number (often 200–300) of specimens are picked from a sample aliquot, it often requires a lot of time to examine them taxonomically, to count them, and to determine their percentage contribution to the assemblage. (b) Different species within the warm and cold species groups are indicative themselves of different temperatures. Warm water species, such as *Pulleniatina obliquiloculata* or *Globorotalia menardii/tumida* are considered indicators of warmer water than *Globigerinoides ruber*. How to numerically treat or utilize effectively these within-group differences is a problem. (c) There is confusion with respect to the interpretation of some species. This is especially true in the case of *Globorotalia truncatulinoides* (*see* subsection C below).

Although this technique is best suited for use in temperate zones where both cold and warm water groups of foraminifera are present, it can be modified to work almost anywhere, even in high latitude areas where no real warm water species are present. In these areas, the whole fauna (or most of it) can be utilized by ranking the species according to their tolerance for low temperatures. The species at the extreme ends of the list can be considered as cold and warm indicators and those in the middle as neutral ones. Using the varying relative abundances of the extreme groups it is possible to construct a paleoclimatic curve (Barash, 1971). The most cold-tolerant species in these zones is always *Globigerina pachyderma*. Different species can fulfill the role of warm-water indicators; e.g. in an assemblage of *G. pachyderma*, *G. quinqueloba*, *G. bulloides*, *Globorotalia inflata* and *G. truncatulinoides*, the last two species would fill this role.

The changes in foraminiferal assemblages noted in core samples may be affected by factors other than climatic ones. The effects of shifting current systems and differential solution of tests may lead to misleading interpretations unless recognized. Ruddiman (1971) pointed out that a given area in the past may have been subjected to warmer or cooler than average temperatures due to shifts in current systems, systems which were intensified during the Pleistocene. The faunas in that area would alter although climatic conditions may not have changed. Olausson *et al.* (1971) interpreted changes in Indian Ocean cores as indicative of shifts in current patterns during the Pleistocene.

The differential susceptibility of planktonic foraminiferal species to solution (Chapter VII,3; IX,3) means that there will be a preferential enrichment in deep water sediments of solution resistant species. If unrecognized, this factor may alter paleoclimatic interpretations considerably. Thiede (1973) suggested that the degree of solution in a sample might be estimated by the degree of test fragmentation that was present. Because calcareous benthonic foraminifera are about three times less soluble than planktonic forms, Berger (1973) suggested that if one assumes the depth to have been constant and the relative productivity of benthonic and planktonic foraminifera to have remained constant, the planktonic:benthonic ratio may be used to predict the degree of dissolution that has occurred in an assemblage. Under conditions of intense dissolution where calcareous benthonic forms have begun to dissolve, the ratio between agglutinated and calcareous benthonic specimens may prove a useful index to dissolution (Theyer, 1971a).

Since the temperature preferences are known for most planktonic foraminifera, Berger (1973) also suggested that the change in abundance between species with the same temperature habitats but dissimilar resistances to solution would provide a dissolution index. He suggested the ratio between the abundances of the *Globorotalia menardii*, *G. tumida*, *Pulleniatina obliquiloculata* group and the *Globigerinoides ruber*, *G. sacculider*, *G. conglobatus* group. All of these species are warm and warm–temperate species but the first group is resistant to solution whereas the latter group is susceptible. For higher latitude cores he proposed the ratio between the *Globorotalia inflata*, *G. truncatulinoides*, *G. punctulata* group and *Globigerina bulloides*; all are cold and cold–temperate species but differ in their resistance to solution.

The results of a faunal examination are shown graphically (Fig. 114). A vertical column representing each core is divided longitudinally by a line. The left half is considered warm and the right half cold. The column is marked off to indicate the depth in the core. Suppose that eight samples have been studied (spaced 10 cm apart) and that the study indicates a greater proportion of warm water species near the top and a greater proportion of cool water species near the bottom. The results can be expressed in two ways: (a) comparing the changes to the first sample studied (surface)

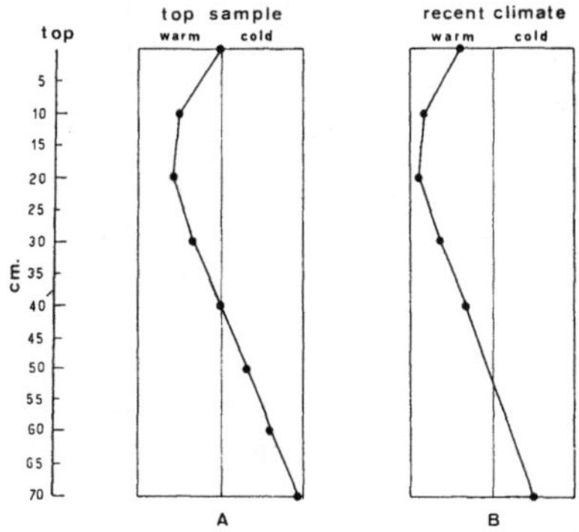

Fig. 114. Two methods of paleoclimatic data presentation.

and indicating that a given sample represents a time which was warmer or cooler than the present (Fig. 114A) or (b) locating a spot for the first sample which is indicative of the actual temperature of that sample; e.g. if the core was obtained in low latitudes the first point would appear on the warm side of the diagram (Fig. 114B). The two curves will have the same form but different lateral positions.

Given two cores the data from the paleoclimatic interpretations can be correlated (Fig. 115).

The method and its variations described above yield information on relative water temperatures. There have been several attempts to calculate absolute paleotemperatures from foraminiferal assemblages, a process which is much more complicated than estimating relative temperatures.

One of the first attempts to make such a determination was that of Kane (1956) who separated planktonic foraminifera into different groups whose water temperature was greater or less than 20 °C. Kane's conclusions were very tentative but do mark the first step toward absolute paleotemperature analysis.

It is possible to estimate absolute paleotemperatures by comparing the fossil foraminifera thanatocoenose with Recent assemblages found in waters whose temperatures are known. Morin (1971d) examined the relationships which exist between the surface water temperature and the faunal content of the California

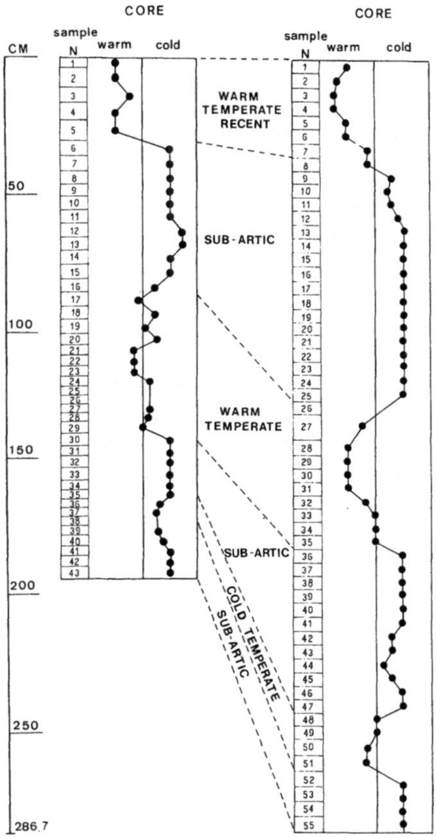

Fig. 115. Approximate relative water temperatures based on foraminifera in cores (after Phleger & Hamilton, 1946).

Current. By knowing the temperature limits of one or more given species in the area and making observations on their appearance in a core, he obtained some idea about the absolute temperature changes recorded in the core. However, this method requires a very good knowledge of the Recent fauna and is rather time consuming. Conclusions drawn from this method are very approximate.

Lynts & Judd (1971) utilized quantitative environmental data on living planktonic foraminifera in the Florida Straits to determine the optimum values of various hydrological factors for particular species. Having established these values, they then were able to estimate the value of a given hydrological parameter on a pre-Recent sample taken from the same area using an ecological response model which is weighted by the relative abundance of species in a core sample. They calculate the estimate for a given environmental parameter (H) using the following formula:

$$H = \frac{\sum p_i e_i}{\sum p_i}$$

where p_i = relative abundance of ith species;
e_i = optimum value of the hydrologic factor (H) being estimated.

Lynts, Judd & Stehman (1973) utilized this technique along with factor analysis to describe the Late Pleistocene history of an area in the Bahama Islands. This analysis was based on the assumption that the Pleistocene species had ecological responses analogous to their Recent counterparts.

A similar model using the entire foraminiferal assemblage was devised by Imbrie & Kipp (1971). It differs in concept from that of Lynts & Judd in that it is more sophisticated mathematically and uses the distribution of empty tests in the sediment to establish the response model. The Lynts & Judd model was based on living specimens. Imbrie & Kipp argue that the use of dead specimens on the bottom is more accurate because the fossil record is composed of death assemblages and so one must examine the analogous situation. They argue that dead faunas (a) are different than living ones, (b) have already undergone solution, (c) may have already undergone bioturbation mixing older and younger assemblages together and (d) can filter out seasonal variations in the living fauna. However, the utilization of the Lynts & Judd model on the Recent dead fauna in the Florida Straits allowed the estimation of the distribution of the living fauna in that area.

Imbrie & Kipp, whose model uses the percentage contribution of each species as an indicator of temperature, performed a factor analysis to compare faunal abundance data with environmental paramenters of different samples. They thus obtained a set of paleoecological equations. These equations allow a hydrologic factor to be predicted in terms of the factor of faunal assemblage. Data from core samples are fed into these equations and paleoenvironmental estimates are made.

The advantage of this technique over the use of tropical, warm–temperate, cold–temperate, cold water assemblage techniques described above is that this technique utilizes the fact that each species has a temperature optimum. The use of assemblages tends to gloss over the differences exhibited by each species. The use of this factor analysis technique requires that assumptions not only be made about the constancy of the ecosystem in the recent past, but also on the reducibility of the raw data into a limited number of factors.

Imbrie & Kipp used their model to calculate paleotemperatures but the model is designed so that it can be used for any hydrological factor.

Hecht (1973) devised a technique conceptually similar to that of Imbrie & Kipp. He established a temperature response model by: (a) calculating the similarity between pairs of Recent core top samples, (b) choosing one sample as a standard and plotting the similarity coefficients of all other samples with this standard sample against the surface water temperature for these samples (he also plotted the data for various subsurface water temperatures), (c) calculating the equation which best fits the data. This equation can then be used to estimate the paleotemperature for any core sample using the similarity coefficient between the core sample assemblage and the standard assemblage.

Hecht used this method on a core from the Caribbean and found that his results compared favourably with those of Imbrie & Kipp on a nearby core and with the oxygen isotope paleotemperatures (section 5) established in the same core.

The observations of Cifelli (1971) that there is no fixed relationship between cold-water species frequency and temperature among living planktonic foraminifera has some interesting implications for the quantitative techniques of estimating paleotemperatures. If the relationship between species abundance and water temperature is as imprecise as Cifelli indicated, the range of error in these estimates may be rather large.

Another factor which those who estimate absolute paleoclimatic parameters must contend with is the effect of selective dissolution on the frequency of species in an assemblage. Berger (1971b) demonstrated that because of differential solution the most abundant living species contribute relatively less to the sediment than do the less abundant living species. He also indicated that the species which live near the surface tend to contribute fewer tests to the bottom sediment than do subsurface species. This observation may imply that methods using bottom assemblages to estimate absolute paleotemperatures may be more indicative of the temperature of subsurface than surface waters.

Berger also demonstrated that kummerforms (Chapter III,11) are more common in subsurface waters and perhaps should be eliminated from an assemblage if an estimate of surface water temperature is desired.

The increase in faunal diversity in low latitudes (warmer waters) has been used as a paleoclimatic indicator in core samples (R. Olsson, 1974; Poore & Berggren, 1974) although Cifelli & Smith (1974) argued that there is no good evidence for these diversity gradients among planktonic foraminifera.

The composition of benthonic assemblages has also been used to describe paleoclimatic and paleocurrent systems (Caralp, Moyes & Vigneaux, 1968; Schnitker, 1973).

B. Studies of species abundance ratios

Various investigators have used the variations in the abundance ratios between two species or two groups of closely related species to note paleoclimatic changes. In a sense the qualitative whole fauna method described in section A above is based on the comparison of two groups of species, one 'cold' and the other 'warm'. However, the emphasis in the above methods is on the contribution of each species to the assemblage, whereas the following methods tend to concentrate on the ratios between two species.

One of the earliest uses of the ratio technique was that of Ewing, Ericson & Heezen (1958) who calculated the ratio between the *Globorotalia menardii* group (*G. menardii tumida, G. menardii flexuosa* and *G. menardii fimbriata*) and *G. inflata*. The use of this ratio was an outgrowth of the earlier use of increases in the abundance of the *G. menardii* group as an indicator of warm water (section C below). This ratio has been used a great deal in low latitude areas (Ericson, Ewing & Wollin, 1964; Sidner & Poag, 1972; Beard, 1973).

Ruddiman (1971) strongly critized the use of the *G. menardii: G. inflata* ratio because a great deal of the information in the samples is not used.

Lidz (1966a) suggested two rather complex ratios for use in low latitude cores. He calculated the ratio between the warm water species *Globorotalia menardii menardii* + *Sphaeroidinella dehiscens* and the cool water *G. inflata* + *Globigerinoides ruber*. He also suggested that the same ratio with the addition of *Hastigerina* spp. to the cool group was also useful. Both of these ratios correlated well with the paleotemperature variations obtained from oxygen isotope analyses (section 5).

The ratio of *Globigerinoides sacculifer* to *G. ruber* was used by Herman (1965, 1968). Although these two species have similar habitats (Chapter VII,2) the former is a stenothermal species whereas the latter is eurythermal. Herman interpreted an increase in the ratio between these species as an indication of warming conditions. Berggren & Boersma (1969) used an decrease in this same ratio as an indication of an increase in salinity in the Red Sea during times of Pleistocene sea level drop. The use of this ratio as a salinity index may only be valid over a small salinity range. *G. ruber* increases in abundance as the salinity increases above 36.0%; however, the abundance of this species also increases with respect to *G. sacculifer* when the salinity is less than 34.5% (Bé & Tolderlund, 1971).

In high latitude waters Herman (1970) used the increase in abundance of *Globigerina quinqueloba* with respect to *G. pachyderma* as an indication of warmer water.

Berger (1973) suggested several abundance ratios which would serve as temperature indicies but would be relatively insensitive to changes caused by test dissolution. He chose species which have similar resistences to solution but have different optimum temperatures. He suggested the following ratios, each of which is inversely proportional to temperature: (a) *Globigerina pachyderma: Globorotalia inflata* + *G. punctulata* + *G. truncatulinoides*, (b) *G. inflata* + *G. punctulata* + *G. truncatulinoides: G. menardii* + *G. tumida* + *Pulleniatina obliquiloculata* + *Sphaeroidinella dehiscens* and (c) *Globigerina bulloides: Globigerinoides ruber* + *G. sacculifer* + *G. conglobatus*. This last ratio is similar to the *Globigerina bulloides: Globigerinoides ruber* ratio suggested by Tolderlund & Bé (1971) for mapping thermal gradients.

Utilizing the ratio between *Globigerinoides ruber* and *Pulleniatina obliquiloculata*, Thompson & Saito (1974) recognized cycles of dissolution in equatorial Pacific Ocean cores (Fig. 116) and suggested that these correlated with glacial-interglacial events.

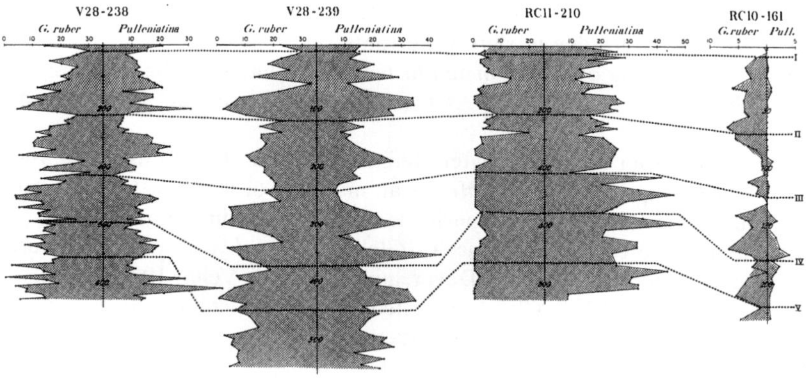

Fig. 116. Relative proportions of *Globigerinoides ruber* and *Pulleniatina obliquiloculata* in 4 Pacific cores showing dissolution cycles (after Thompson & Saito, 1974).

C. Studies of single species

This method was first used by Schott (1935) who, although he studied the whole fauna, drew his paleotemperature conclusions chiefly on the abundance of the warm-water species *Globorotalia menardii*. Ericson and his associates used this method widely and became its great advocates. They preferred to use not only *G. menardii, s.l.*, but the *G. menardii* complex adding *G. menardii tumida*, *G. menardii flexuosa* and *G. menardii fimbriata*. In addition, to simplify their calculations, they determined the ratio of the quantity of these species to the weight of all the planktonic foraminifera in the sample rather than to the numbers of specimens. Those who use this method emphasize that the main advantage is that it takes considerably less time than any other method.

This technique has drawn considerable criticism (Emiliani, 1964; Morin, Theyer & Vincent, 1970) on the grounds that it is incorrect to draw paleoclimatic conclusions from the study of only one (or even two) species which could easily be absent from the fauna.

Specific disadvantages of using the *Globorotalia menardii* complex have been cited. Lidz (1966a) compared the paleoclimatic curves prepared on the basis of the oxygen isotope method (section 5), the vertical abundance distribution of *G. menardii*, s.s., and the abundance distribution of the *G. menardii* complex. He found that the first two methods yield curves that coincide very well, but that the curve utilizing the third method did not correlate with the other two. Lidz concluded that *G. menardii* is a good temperature index, but the whole group cannot be used because different members have different temperature ranges during their lifetime. Berggren, Briskin & Gamper-Bravo (1972) suggested that an increase in the intensity of dissolution without an attendant climatic change would increase the proportion of *G. menardii* (solution resistant) in an assemblage. This increase would be interpreted as an indication of a warming trend but might only indicate an influx of cold Arctic or Antarctic Bottom Water due to shifting currents.

Another shortcoming of the *G. menardii* method is that it can be used only in low latitudes where the species occurs. As the latitude increases the proportion of this species declines.

In cores taken from the warm-water Red Sea, Herman (1968) used the absolute abundance of *Globigerinoides sacculifer* as an indicator of warm water. This method has not gained wide usage as this species is not abundant in many areas.

In high latitude cores the abundance of *Globigerina pachyderma* has been used as an indicator of water temperature (Devereaux, Hendy & Vella, 1970; Clark, 1971; Steuerwald & Clark, 1972). Clark suggested that cooler conditions would restrict photosynthesis and therefore reduce productivity, resulting in reduced abundances.

Frerichs (1968) used the abundance of the *Globigerina rubescens* complex (*G. rubescens* + *Globigerinoides tenellus*) as an indicator of water temperature in the Indian Ocean. The paleotemperature curve based on this complex agreed well with the curves based on *Globorotalia menardii* abundances and with isotopic paleotemperature data. Although the two species comprising this complex are typical subtropical species, Frerichs found that the intervals of greater abundance of the complex were indicative of glacial times. He suggested that the change in abundance of this complex might have been due to a shift in the geographical distribution of the group towards the Equator during glacial episodes. This method will not attain wide utilization as both species are rather infrequent except in local areas.

The benthonic genus *Amphistegina* has been used as an indicator of warming trends during the Tertiary in New Zealand (Hornibrook, 1968) and in the Pleistocene of the Gulf of Mexico (Sidner & Poag, 1972).

4. Stratigraphic and climatic studies based on coiling direction and other morphological variations

In Chapter III we spoke about the phenomenon of preferential coiling of trochospiral specimens. In this section we wish to examine the applications of this phenomenon.

There are numerous observations which indicate that the coiling direction of certain species varies with the distribution of the species. Ericson *et al.* (1955), on the basis of rich material, showed that in different areas of the Atlantic Ocean bottom there was a predominance of either right or left coiled *Globorotalia truncatulinoides* tests. Near the edges of these areas there was a mixing of coiling types and equal proportions of the two forms were found.

Studies done by the same authors (1954) of core samples demonstrated that during the Late Pleistocene the relationship between dextral and sinistral coiling changed and could be used as a stratigraphic indicator. Other studies done on fossil foraminifera (Vašíček, 1953; Nagappa, 1957; Kaever, 1960) confirm the utility of coiling direction as a useful stratigraphic tool.

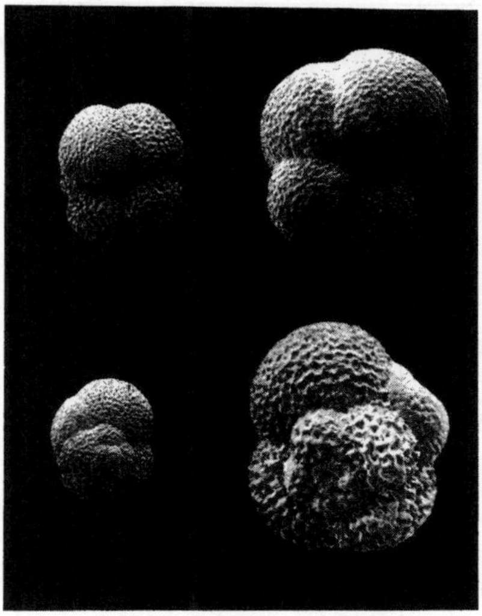

Fig. 117. Globigerina pachyderma, s.l.; Left: dextral; Right: sinistral.

The study of coiling direction not only yields data useful for correlation but also for paleoclimatic interpretation. Ericson (1959) showed that the coiling direction of *Globigerina pachyderma* (Fig. 117) was temperature controlled. Sinistral and dextral forms of this species are predominant on opposite sides of the 7.2 °C surface isotherm in the North Atlantic (Fig. 118) and the 9 °C surface isotherm in the South Atlantic (Boltovskoy, 1966b). This phenomenon is less well studied in other areas but some observations made in the South Pacific indicate that a comparable change occurs at a surface water temperature of 6.5–6.8 °C (Boltovskoy, 1969c). These observations coupled with the observation that populations composed of at least 95% left coiling *G. pachyderma* occur in waters colder than 3.5 °C (Boltovskoy, 1971b) indicate that it might be possible to estimate an upper or lower paleotemperature limit for samples of high latitude cores.

Studies of cores showed that those samples in which dextral specimens of *Globigerina pachyderma* were predominant were also characterized by a higher percentage of warm and temperate water planktonic species than were the core samples in which left coiling specimens predominate (Fig. 118 insert). It is quite clear that there is a very strong correlation between water temperature and the direction of coiling of this species. This phenomenon was also confirmed by many later studies. The change of coiling direction was so pronounced that Ericson *et al.* (1961) were able to distinguish between layers as thin as 4 cm using this trait as an indicator.

A similar temperature response was suggested for the fossil form *Globigerina daubjergensis* by Hofker (1962).

The change in proportions of right and left coiling *Globigerina pachyderma* has been used as a paleoclimatic indicator in several studies (Jenkins, 1967; Devereaux,

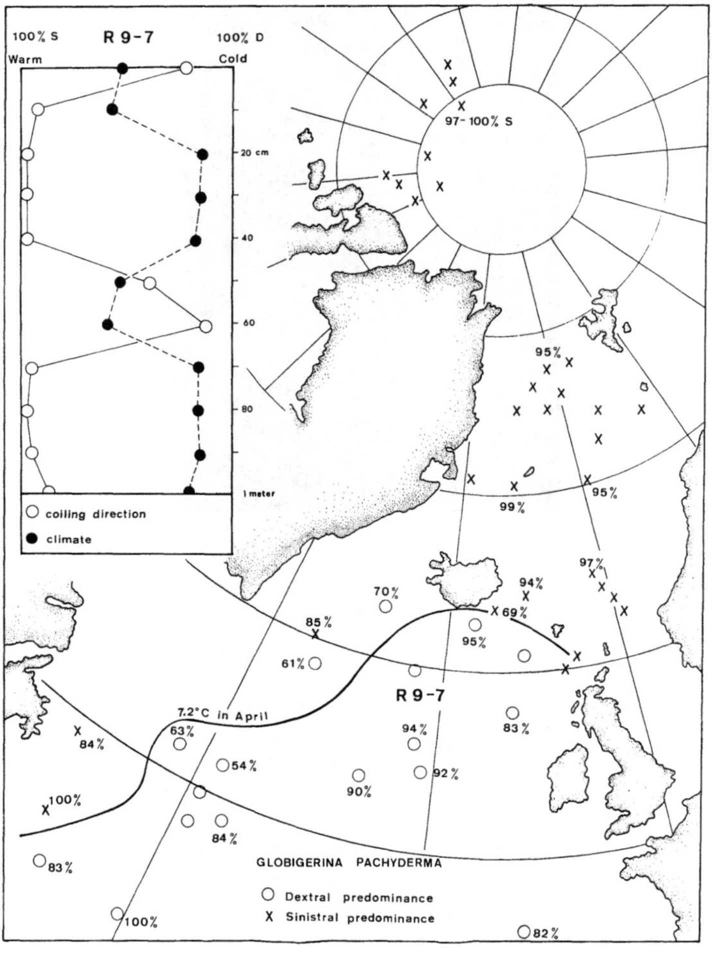

Fig. 118. Distribution of dextral and sinistral populations of *Globigerina pachyderma* (after Ericson, 1959).

Hendy & Vella, 1970; Kheradpir, 1970; Morin, 1971d; Keany & Kennett, 1972; Rotschy *et al.*, 1972; Williams, 1973; and numerous studies by Bandy). Bandy (1972a) summarized many of the Miocene–Pleistocene paleoclimatic interpretations that had been made on the basis of these coiling ratios.

There are two obvious disadvantages to the paleoclimatic method based on coiling direction of *Globigerina pachyderma:* (a) it can only be used in cold and cold–temperate zones, since only there is the species found in abundance and (b) the right coiled form often looks morphogically like *G. dutertrei.* This is especially true among fossil specimens and Recent specimens found along the geographic margins of the species. In addition, there are some authors, e.g. R. Olsson (1974), who have suggested that the ratio of sinistral to dextral specimens may not be as sensitive to temperature changes as has been thought. Vella (1974) indicated that only specimens of the same size fraction should be compared because the sinistral:dextral ratio varies with size in the same population.

A method utilizing the coiling direction of *Globorotalia truncatulinoides* has been used as an indication of paleoclimatic conditions. This method is based on observations of the Recent distribution of the dextral and sinistral *formae* of this species. The method is somewhat confusing because of disagreement concerning the temperature conditions the species represents. The species appears to be more tolerant of low temperatures in the Southern Hemisphere than in the Northern (Bé, 1969a; Boltovskoy 1969b, 1970b; Bé & Tolderlund, 1971). Ericson & Wollin (1968) stated that in the Early Pleistocene left coiled *G. truncatulinoides* were dominant during times of cold climate and right coiled specimens dominant in warmer periods. Ubaldo (1973) also stated that sinistral specimens predominated during Pleistocene glacial episodes.

Parker (1971) prepared a map showing the distributional patterns of sinistral and dextral specimens of *Globorotalia truncatulinoides* in the bottom sediments of the Recent oceans. The map does not give any clear indication as to what kind of water dextral and sinistral specimens prefer in the Northern Hemisphere. However, it is rather evident that in the Southern Hemisphere the sinistral form occupies high latitudes whereas the dextral one occupies medium and low latitudes. The same distribution was noted by Bé & Tolderlund (1971) who studied living planktonic species. Based on this data, it seems that in the South Atlantic an increase in sinistral population is related to a decrease in temperature. Whether or not this relationship existed during the Pleistocene is not well understood. In the Northern Hemisphere, particularly in the North Atlantic, the situation is still more confused. According to Takayanagi, Niitsuma & Sakai (1968) and Parker & Berger (1971), the dextral form is dominant in tropical waters and the sinistral one in the more temperate water masses. However, Wollin, Ericson & Ewing (1971) successfully utilized changes in coiling direction in the study of the paleoclimatological history of the North Atlantic during the last 75 000 years. They emphasized that the predominance of sinistral specimens was typical of warm water and dextral specimens of cold water. The results obtained by Thiede (1971b) seem to confirm this conclusion. Cores from the North Atlantic off the Portuguese Coast exhibited dextral specimens in the glacial samples and sinistral specimens in post-glacial samples.

Kennett & Huddlestun (1972a) reported that in cores from the Gulf of Mexico left coiled *Globorotalia truncatulinoides* was predominant in some cooler zones but that other cool zones were characterized by 100% right coiled forms.

Boltovskoy (1973b) compared several of the commonly used paleoclimatic methods listed above using material from cores collected in the temperate zone of the South Atlantic and concluded that the method based on a study of the whole fauna best represents the degree of relative climatic change and also yields more fine

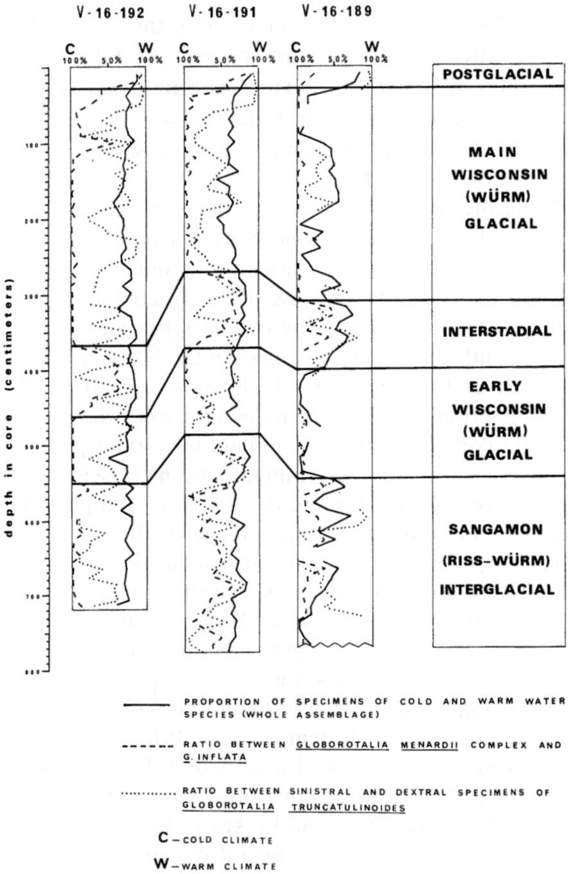

Fig. 119. Comparison of 3 paleoclimatic indicators, South Atlantic (after Boltovskoy, 1973b).

details, (e.g. insignificant and short climatic oscillations). The method based on the ratio between the *Globorotalia menardii* complex and *G. inflata* exaggerates the major temperature oscillations. However, because of this exaggeration both the separation of glacial, interglacial and interstadial epochs and their correlation in different cores is much easier. Paleoclimatic curves based on the sinistral:dextral ratio of *G. truncatulinoides* specimens did not coincide very well with those prepared from the two previous methods if the sinistral specimens were considered as cold-water indicators. The curves coincided even less if the sinistral specimens were interpreted as warm-water indicators (Fig. 119). Thus, the coiling direction of *G.*

truncatulinoides according to Boltovskoy proved of little use as a paleoclimatic indicator.

We conclude that there are too many contradictions and too much confusion in the coiling direction data of *Globorotalia truncatulinoides* for this method to be used to draw paleoclimatological conclusions.

The coiling direction of several other species has also been used for paleoclimatic purposes. Kudrass & Thiede (1970) and Thiede (1971a) used the coiling ratios of *Globorotalia hirsuta* and *G. crassaformis*. Herman (1972b) used these same species to determine changes in Pleistocene oceanic circulation patterns.

Herman (1965) also found that the use of left coiling *Globigerina bulloides* as a cold water indicator and left coiling *Globigerinoides sacculifer* as a warm water indicator was helpful in some Red Sea cores, but not in all. Boltovskoy (1973d) also suggested that the coiling of *Globigerina bulloides* might prove useful in estimating absolute paleotemperatures. This species is well known, easily recognized and rather abundant in cold and cold–temperate zones. In numerous studies of specimens in the Southwestern Atlantic and its Antarctic Sector the relationship between the sinistral:destral ratio of *G. bulloides* specimens and the surface water temperature was established. The relationship was discovered to be constant; with an increase in temperature the percentage of sinistral specimens always decreases. It was also determined that there are no abrupt changes in the ratio as in the case of *G. pachyderma*. Utilizing data collected during many voyages, this relationship is represented graphically (Fig. 120). After calculating.the dextral:sinistral ratio in a given core sample, it is possible to use this curve to estimate the surface water temperature at the time of deposition of the sample. The following should be noted, however, when using this technique: (a) The ratio should be calculated on the basis of as many specimens as possible, in no case less than 10; the use of too few specimens can yield misleading results. Some temperature variations were observed for different ratios. Nevertheless, even in low latitudes where the variations are most pronounced, the estimates are in error no more than 2.5 °C. (b) This particular curve may not be applicable to cold and cold–temperate zones in other geographic areas because the relationship between water temperature and coiling direction in other areas may be different. (c) The more ancient the samples the greater the risk that the relationship may have changed, or may not even have existed. Boltovskoy suggested that curves of this type be used to establish only Pleistocene and perhaps Pliocene paleotemperatures.

Longinelli & Tongiorgi (1960) suggested that the coiling direction in the benthonic species *Rotalia beccarii* might be temperature controlled. Their observations showed a greater proportion of left coiled specimens in cooler water. However, observations in other areas have not confirmed this relationship. The coiling preference in this species may be a function of reproductive mode rather than temperature or depth. If such is the case, its use as a paleoclimatic indicator would be misleading.

Morphological variations other than changes in coiling direction have also been suggested as paleoclimatic indicators. To use them it is necessary to recognize the

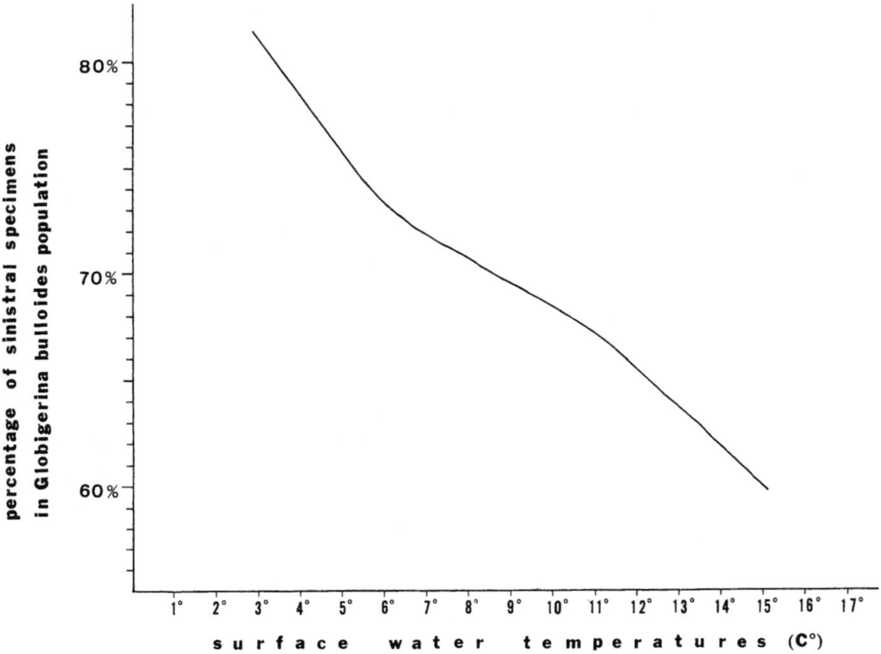

Fig. 120. Relationship between the percentage of sinistral forms of *Globigerina bulloides* and surface water temperature, SW Atlantic and its Antarctic sector (after Boltovskoy, 1973d).

slightest changes that depend of variations in some environmental parameter. Although we do not completely understand these changes, some of them may prove to be valuable paleoclimatic tools.

Boltovskoy (1959b) showed that specimens of *Globorotalia truncatulinoides* from different latitudes differ morphologically. Somewhat later the same observations were mentioned by Blair (1965), Boltovskoy (1966d) and Takayanagi, Niitsuma & Sakai (1968). Kennett (1968a), utilizing material from several areas, proved that an interdependence really does exist between the surface water temperature and the width-height ratio. This ratio is high in high latitudes where cold-water specimens have rather compressed tests. In temperate areas the tests are highly conical. This ratio is potentially useful in determining climatic change, but has not yet been completely worked out. It can be used only in the Southern Hemisphere as *G. truncatulinoides* does not reach high latitudes in the Northern Hemisphere.

Kennett (1968b) demonstrated morphological variations in *Globigerina pachyderma* specimens in the Southern Hemisphere which could be used as climate indicators. The dextral subantarctic and subtropical specimens exhibited a decrease in the number of chambers in the final whorl as the water temperature increased.

Subantarctic specimens have 4½–5 chambers per whorl whereas subtropical and northernmost subantarctic specimens possess 4.

Later studies of the same species by Srinivasan & Kennett (1974) revealed that the surface texture seemed to be a function of temperature. Specimens with a reticulate surface and round smooth microgranular edges were more common in core samples representative of warm intervals whereas specimens with a crystalline surface consisting of well developed rhombohedral crystals were abundant in samples representing cooler water.

Polar and subpolar specimens of *Globigerina bulloides* have 3½–5 chambers in the last whorl whereas temperate and warm–temperate specimens have 4 (Bandy, 1972b).

The number and size of chambers of some benthonic species has been shown to be a function of temperature (Lewis, 1968). The mean specimen size and number of chambers in the final whorl of the benthonic species *Nonionellina flemingi* was shown to be a function of temperature in Recent sediments near New Zealand (Lewis & Jenkins, 1969). This relationship was used to estimate paleotemperatures in the Upper Miocene strata of New Zealand. Lewis pointed out that the use of the foraminifera species as an indicator gave results contradictory to paleotemperature analysis obtained from molluscan data. He warned that the use of a single species for paleoclimatic reconstruction was risky unless a great deal is known about that species. He mentioned that the present distribution of a species may be restricted by factors other than temperature.

There are several other observations on morphological changes such as aperture size, test colour, etc. These changes have been described in some planktonic species (Boltovskoy, 1969c). The studies of these variations are not complete enough for these variations to be used as paleoclimatic indicators.

Hecht (1974) warned that statistical correlations between morphologic and oceanographic data may give workable results only in areas where the values of the ecological parameters are outside the ranges of the optimum values for the species. He suggested that it is only in these zones that the species responds in any statistically significant way to the environment.

5. Paleoclimatic studies based on oxygen isotope ratios

The work of Emiliani (1955b *et seq.*) with oxygen isotopes and foraminifera to determine paleoclimates is now well known but the background to it bears retelling as it is exemplary of the nature and process of discovery and of the multi-disciplinary techniques used to solve geologic and oceanographic problems.

The idea was originated in a lecture given by Urey at the Technische Hochschule in Zürich. Urey said that the isotopes of an element, even though they are chemically identical, may behave differently. He mentioned that if water in a glass is evaporated the three isotopes of oxygen (0^{16}, 0^{17}, 0^{18}) volatilize at different rates. During the period of evaporation 0^{16} will migrate to the surface with a slightly greater velocity than the others. Thus the water will be enriched in the heavier 0^{17} and 0^{18}. Urey concluded that since oceanic water has been undergoing evaporation for a greater period of time than terrestrial waters, it ought to have a greater proportion of 0^{17} and 0^{18}.

After the conference in Zürich, a coloquium was given in which the crystallographer Niggli deduced that if Urey were correct, the differences in the proportion of the different oxygen isotopes ought to be reflected in sediments of organic origin. He suggested that an analysis of calcareous shells might provide clues about the water in which they were precipitated.

This idea incited Urey to pursue and amplify his investigations into the problem. After many years of intense work he was able to describe a technique which proved of great interest and value (Urey, 1947). He showed that the relationship between the proportions of 0^{16} and 0^{18} (0^{17} not being considered because of its rarity in nature) in a calcareous shell depends on the temperature of the water in which the carbonate was secreted. Thus, an analysis of the ratio between the amounts of the two oxygen isotopes in calcite shells acts as a thermometer, yielding information about the temperature of the water in which the shell was made.

More detailed studies revealed that the problem was yet more complicated. Analysis of fossil calcareous material yields only relative temperature data; i.e. that the carbonate of a sample was secreted in water colder or warmer than some standard. To estimate absolute temperatures it was necessary to know the difference between the isotope ratio of the unknown sample and that of a standard.

There are several problems with this method. The $0^{16}/0^{18}$ ratio in sea water is not the same at all latitudes and distances from the coast. It is also very likely that the oxygen isotope ratio in sea water has changed with time. Because 0^{16} is removed preferentially from sea water, glacial ice would have had proportionally more 0^{16} and sea water less during Pleistocene glacial stages. It is necessary to estimate the degree to which the isotope ratio changed and apply this correction to values obtained from laboratory analyses. The extent of this change has been a source of heated debate.

Emiliani (1966a) and many of his students maintained that only about 30% of the isotopic paleotemperature variation between glacial and interglacial sea water is due to a change in the isotopic composition of sea water. Olausson (1965), Shackleton (1968) and Dansgaard & Tauber (1969) claimed that far more of the variation (up to 100%) could be explained in this way. These two assumptions would lead to different estimates of paleotemperature. A good summary of the methods used to estimate the temporal differences in isotopic composition of sea water was prepared by Emiliani & Shackleton (1974).

One of the assumptions necessary in the oxygen isotope method of estimating paleotemperatures is that the foraminifera secrete $CaCO_3$ in isotopic equilibrium with sea water. Most investigators believe this to be the case although Parker (1958) suggested that in planktonic foraminifera oxygen exchange may occur between the $CaCO_3$ and the oxygen being liberated by symbiotic algae in the test. Shackleton,

Wiseman & Buckley (1973) analysed the isotopic composition of 4 living planktonic foraminiferal species in the upper 50 m and discovered that each species had a different isotopic fractionation. This observation, if it holds true for other areas and at different depths, would invalidate the use of isotopic analysis to determine depth habitats of planktonic species (Chapter VII,4). An experiment conducted by Smith & Emiliani (1968) indicated that benthonic foraminifera deposit their $CaCO_3$ in equilibrium with sea water. Recent specimens from various depth and temperature ranges were analysed for their isotopic temperature. The results correlated well with the actual water temperature at the sample sites. However, when benthonic specimens from the same core sample were analysed isotopically (Duplessy, Lalou & Vinot, 1970) different species yielded different fractionation values. Acting on the assumption that these specimens all lived at the same bottom water temperature, they should have yielded similar isotopic ratios if calcite secretion was in equilibrium with the sea water. Duplessy & Vinot (1972) and Vinot-Bertouille & Duplessy (1973) claimed that these observations create serious difficulties in the use of benthonic species as paleotemperature indicators because many of them may yield values as much as 5 °C too high.

The problem of whether or not planktonic and/or benthonic species secrete $CaCO_3$ in equilibrium with sea water requires more study, as these recent observations indicate. Experiments on living specimens need to be conducted. Laboratory procedures for isotopic analysis are complicated and error may easily be introduced into the analysis. Results from different investigators are difficult to compare.

Because different planktonic species live at different depths in the ocean, an isotopic analysis of the whole fauna would include surface dwelling (warmer) and deeper dwelling (cooler) species. Such an analysis would give an average water temperature. Better results are obtained by utilizing a single species whose depth habitat is known. Shackleton (1968) pointed out that the interpretation of isotopic paleotemperatures based on a single species assumes that the depth distribution of that species has remained constant. It is possible that a species may move upward or downward as a result of changing water density during glacial episodes. He proposed that only surface water species be used in isotopic analysis because they, at least, cannot migrate upward.

It must be assumed that planktonic foraminiferal species secrete their tests over a narrow range of water depth or the isotopic temperatures will not be indicative of water temperature at a particular depth. Studies of living planktonic foraminifera (Chapter VII,6) indicate that this may not be a valid assumption. Any secondary calcification recognized on a test should be removed prior to performing an isotopic analysis.

Investigations still continue and there are reasons to think that the isotopes of other elements may also provide thermometers, some of which may be absolute temperature measures. Some of these elements are often associated with carbonates.

The techniques of oxygen isotope ratio calculation require laborious and careful chemical preparation techniques culminating in the use of the mass spectrometer.

Because of the small size of foraminiferal tests it is usually necessary to acquire large numbers (several thousands) of tests to obtain the 10–20 gm necessary for the analysis. The details of this technique fall beyond the scope of this book and will not be treated here.

In addition to Emiliani, several other investigators have used oxygen isotope fractionation as a means of estimating paleotemperatures (Lidz, 1966a; Donk & Mathieu, 1969; Oba, 1969; Devereaux, Hendy & Vella, 1970; Niitsuma, 1971; Herman, O'Neil & Drake, 1972; Rotschy *et al.*, 1972; Barash, Nikolaev & Bljum, 1973).

6. Relative sedimentation rates

If we encounter a site where a significant fraction of the sediments are foraminiferal tests, the sedimentation rate of inorganic and other non-foraminiferal organic materials is very low. This phenomenon occurs over large portions of the ocean basin where *Globigerina* ooze is found.

The discovery of only fossil foraminifera in bottom samples leads to the conclusion that erosion rates have exceeded sedimentation rates in the area. The tops of various sea mounts exhibit this characteristic.

To obtain an idea about the nature of the sedimentation rate it is possible to use the relationship between the quantities of living and dead tests. Of course, such a study requires the examination of Recent specimens which have been fixed and stained as discussed in Chapter XI. If the entire benthonic foraminiferal population, or most of it, consists of living specimens, the sedimentation rate of non-foraminiferal material must be rather high to cover all the empty tests. If, on the other hand, most of the population consists of empty tests, the sedimentation rate is low as there are few settling particles to cover the tests.

Phleger (1955) gave the following example of the calculation of a sedimentation rate. In a traverse in the Northeastern Gulf of Mexico 10 samples taken at depths of 40–84 m yielded 59 living benthonic foraminifera (an average of 6 per sample). In samples from deeper waters taken along the same traverse, less than one living benthonic specimen was found per sample. This indicates that the benthonic foraminiferal productivity in the shallow waters is about six times greater than that in deeper waters. The number of empty tests in the shallow samples averaged 67 per sample and in the deeper water 700 per sample; i.e. the deeper water samples contained about ten times as many specimens as the shallow water ones. Since the productivity in shallow water is six times greater and the number of tests is ten times less than in deep water, one can conclude that the low number of empty tests in the shallow water is due to the rapid accumulation of inorganic material in the area; i.e. due to a high sedimentation rate (about 60 times greater than in the deeper waters).

Phleger also calculated the living to dead ratio for many sites around the mouth of the Mississippi River. He was able to demonstrate convincingly how this ratio correlated well with the known sedimentation rates for this area.

Walton (1955) proposed the following calculation to determine the relative sedimentation rate:

$$R = \frac{\text{living benthonic population}}{\text{dead benthonic population}} \times 100$$

McGlasson (1959) estimated sedimentation rates based on this ratio in the area of Santa Catalina Island, California and described a reduction in the rate away from shore toward the open ocean. This conclusion corresponds to the findings of Phleger in the Gulf of Mexico and with independent observations on regional sedimentation patterns.

Lankford (1959) used the ratio of living:total population (living and dead) as a measure of relative sedimentation rate. His index is, in effect, the proportion of living tests in the entire population. In a study of the Mississippi River delta area he found rates that varied by a factor of 90. Phleger (1960a) made similar observations in Laguna Madre, Texas and noted widely different sedimentation rates in different parts of the lagoonal area. Ellison, Nichols & Hughes (1965) used the living:dead ratio in the Rapphannock Estuary, Virginia to conclude that the most rapid sedimentation was occurring in the mouths of tributaries to the estuary. Leslie (1965) calculated that sedimentation rates in the central parts of Hudson Bay were less than in the marginal areas.

There are certain assumptions that must be made when utilizing this living/dead or living/total relationship. One of these is that the productivity rate of the foraminifera is constant during the period of time in which the sediments accumulated. Nichols & Norton (1969) claimed that this assumption is invalid and that productivity rather than sedimentation rates is measured by this index. They found little correlation between the living:total ratios and 70 years of records of the sedimentation rate in the James River Estuary, Virginia.

This method fails to work in areas where there are no living specimens in a sample. This is a fairly common situation. In the work of McGlasson (1959) cited above, only 49 of the 71 samples studied contained living specimens. In general, the number of living specimens is small compared to the number of empty tests. The ratio of these two parameters, of course, depends on the thickness of sediment examined because benthonic foraminifera live both on the surface or buried within the sediments. The thickness of sediment studied must be constant for any comparison between sampling locations to be made.

Another problem is that the number of empty tests depends not only on the sedimentation rate but also on the degree of test preservation in the substrate. Some calcareous tests are dissolved in the substrate because of low pH conditions (Chapter IX,6). At depths below the calcite compensation depth in the open ocean (Chapter IX,3) empty tests also are subject to intense solution. In this respect, the observations of Parker (1954) are interesting. She showed that living examples of *Reophax hispidulus* in the Northeast Gulf of Mexico are much more abundant than empty tests in the sediment. She suggested that the empty tests of this species, because the agglutinated grains are weakly cemented to the test, are easily destroyed once the organism dies.

The patchy distribution of benthonic foraminifera (Chapter IV,6) may affect living:total ratios. Schafer (1971) found that these ratios varied as much as 58% for replicate samples taken from the same shallow-water sites. There was less variation in deeper water.

7. Absolute sedimentation rates

The problem of calculating absolute rates of sedimentation has been approached using both planktonic and benthonic foraminifera.

A. Planktonic specimens

The technique utilizing planktonic species is the following. A middle latitude, oceanic core is examined and the highest occurrence of cold-water planktonic species is noted. The distance between this point and the top of the core represents the amount of sediment that has accumulated since the last glacial epoch. Dividing this thickness by the time represented yields an estimate of the absolute sedimentation rate. This method is based on the assumption that there has been no erosion and that the highest occurrence of a cold-water fauna represents the most recent glacial epoch rather than an earlier one. The accuracy of this method depends directly on the accuracy of the time estimates of the termination of the most recent glaciation. Schott (1935, 1938, 1939) was the first to use this method. He calculated Atlantic Ocean sedimentation rates of 0.5–3.3 cm/1000 yr. This method was utilized by many later investigators working in different areas. The results of their studies were similar to those of Schott.

Phleger (1954a) also stated that sedimentation rates in the open ocean could be calculated using planktonic foraminifera. To do this requires a knowledge of the productivity of the organisms. He stated that data from repetitive samples collected in a site over a long period of time as well as that from culture studies would yield the productivity data necessary for sedimentation rate calculations. Planktonic foraminifera have proven unusually difficult to cultivate and it is only recently that *in situ* productivity studies of planktonic foraminifera have begun to be conducted (Berger, 1969b, 1970a,b, 1971b; Berger & Soutar, 1967). Sedimentation rates in water less than 3500 m deep based on productivity data are comparable to those obtained from isotope data and those calculated earlier by Schott. Because of carbonate solution the sedimentation rate drops rapidly in water deeper than 3500–4000 m.

B. Benthonic specimens

Uchio (1960a) suggested a method by which the sedimentation rate could be calculated using benthonic foraminifera. He calculated that an average of 225 living specimens per sample are found in waters 600–650 m deep in the San Diego Basin, California. The average total number of tests (living and dead) per sample at this depth is 12 400. Each sample was 10 cm^2 and was one centimeter thick. Uchio estimated that the reproductive cycle for benthonic foraminifera in the cold water at this depth was two years. Thus, the annual accumulation of tests per sample in that area is 225/2 = 128 tests. To accumulate 12 400 tests in one centimeter would require 12 400/128 = 97 years. He thus estimated a sedimentation rate of one centimenter per 100 years or 0.01 cm/yr.

There are several assumptions that must be made in order to utilize this method. For example one must assume that the reproductive rate of each species is the same. However, at the present time only a small fraction of the known living species have been cultured, but it is clear that different species have different reproductive periods.

Bartlett (1964b) used this method and calculated low sedimentation rates (0.005 – 0.009 cm/yr) in the center of St. Margaret Bay, Nova Scotia and higher rates in the intertidal zone (0.338 – 0.792 cm/yr). However, he criticized the method because of our inadequate knowledge of species reproductive cycles. He proposed that use of data from only one species whose reproductive history is well known would yield more reliable results. The presence of displaced specimens would artificially raise the calculated sedimentation rate. The destruction of empty tests in the substrate would lower the estimated rate. Despite these drawbacks, the method, when used carefully, may give some interesting data.

8. Changes in sea level

Uchio (1960a) described three methods for determining sea level changes utilizing benthonic foraminifera: (a) core studies, (b) bottom samples in traverses perpendicular to the coastline and (c) abundance relationships. The successful utilization of any of these three methods requires that the faunas be unaffected by mixing, displacement, slumping, current action or other movements.

A. Short core studies

If the depth distribution of each species in an area is known, it is possible to compare the biocoenose at the surface of the core with the thanatocoenose in selected portions of the core. If these contain assemblages typical of different depths, a measure of the sea level change is provided.

An example of this method was given by Phleger (1952a) who described different depth assemblages from the upper (Recent) and lower (glacial epoch) part of short cores from the North Atlantic near Portsmouth, Massachusetts. These assemblages indicated that the area had undergone a rise in sea level of 60–70 m. This conclusion, based wholly on the nature of the benthonic foraminifera found in the cores, concurs with independent geological data indicating a 75–120 m drop in sea level during the Pleistocene glacial epochs because of the great volume of water stored on the continents as ice.

Walton (1955) discovered the presence of foraminifera typical of the turbulent nearshore zone in cores taken on the shelf edge near Baja California, Mexico. He concluded that there had been a sea level change of 46–55 m.

The data of Saidova (1958a, 1960a,b, 1965, 1967a) with respect to these kinds of interpretations are very detailed. She studied a great number of bottom samples and cores from the Northeast Pacific Ocean (1960a) and was able to show that the faunas indicated a great deal of vertical variability. Since all the core species are also now living in the Pacific it was easy to use this vertical variability as a means of establishing sea level changes during the Pleistocene.

In a similar study in the Okhotsk Sea, Saidova (1960a) was able to establish the limiting ecological factors for the benthonic foraminifera. She then applied these data to the assemblages found in 42 cores (up to 33 m long) from the same area. As the fauna of the cores was well preserved and consisted of the same species, she could estimate the nature of the changes which had occurred during the deposition of the sediments in this area. The data indicated that there had been two episodes of sea level change.

This method can only be used successfully if the species utilized have not changed their depth habitat with time. Some species have undergone such changes (Chapter IX,13).

Several investigators have used this technique to estimate Pleistocene changes in sea level (Leslie, 1965; Golick, 1968; Bartlett, Molinsky & Slessor, 1972; Boltovskoy, 1973c: Weiss, 1974). Bock (1972) recorded Late Pliocene and pre-Miocene subsidence in Caribbean cores based on an interpretation of their benthonic faunas.

Seiglie (1968a) used the distribution of *Amphistegina*, a shallow warm water reef dwelling benthonic genus in Pleistocene sediments of the Gulf of Mexico and the Caribbean Sea to plot the location of fossil reefs and to note the times of lower water levels. Sidner & Poag (1972) also used the presence of this genus as an indicator of changing water depth.

B. Traverses

If the living benthonic fauna exhibits the same kinds of assemblage changes as does the dead fauna in samples taken from traverses perpendicular to the shoreline, it is possible that the depth difference between the location of these changes in the living and dead fauna is indicative of changes in water depth. For example, if an assemblage change in the living fauna occurs between 50–100 m, but the same faunal change in the dead fauna occurs between 70–120 m, a 20 m change in sea level has taken place.

C. Abundance studies

Studies from many localities indicate that the maximum benthonic foraminiferal population usually occurs between 150–250 m (Chapter IV,7). If the living and dead fauna in an area exhibit their greatest abundance at different depths, the difference may be due to a change in sea level.

9. Paleobathymetry

Planktonic foraminifera live in the open ocean and occupy water from the surface layer to depths of 2000 m or more. The number of planktonic specimens is less in shallow water than in deep (Chapter VII,3,5). The influx of fresh water and detrital material from the continents reduces the specimen abundance in shallow water as does the reduction in distance between the surface of the water and the bottom. If there are fewer planktonic specimens in shallow water there will be fewer of their tests accumulating on the sea floor. On the other hand, the number of benthonic specimens is maximum in shallow water and decreases beyond the edge of the continental shelf (Chapter IV,7). Phleger (1951b) therefore suggested that 99% or more of the tests found in water 2000 m deep should be of planktonic species. This ratio between the number of planktonic specimens and the number of benthonic specimens in a sample is, therefore, a measure of water depth. The greater the ratio, the greater the depth.

Grimsdale & Morkhoven (1955) demonstrated the reliability of this method by plotting the planktonic:benthonic ratio in Recent sediments (Fig. 121). The potential of this technique has been demonstrated in various parts of the world. It was

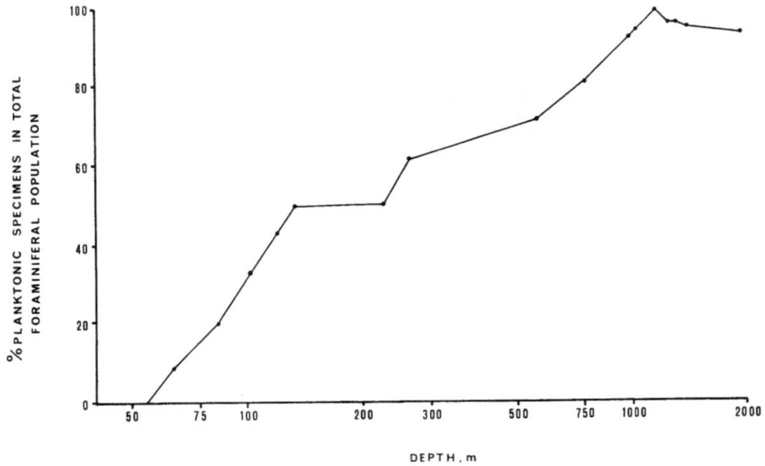

Fig. 121. Change in the relative proportion of planktonic and benthonic foraminifera with water depth, Gulf of Mexico (after Grimsdale & Morkhoven, 1955).

used by Stehli & Creath (1964) to estimate paleobathymetric conditions in the Cretaceous sediments of the Gulf Coast, by Poag (1972) in Late Quaternary sediments of the Gulf of Mexico, and by Eicher & Frush (1974) in Cretaceous sediments of France. The results consistently showed patterns similar to interpretations obtained from other sources.

One of the principal advantages of this method is that it requires no detailed taxonomic study of the fauna. It is sufficient to separate the fauna into only two groups, planktonic and benthonic. At depths greater than the lysocline, the depth at which calcareous tests begin to dissolve selectively, the method loses its value. In Polar areas the technique should be used with care as it is possible for non-indigenous benthonic specimens to be carried to a site by ice-rafting.

A thorough knowledge of the depth distribution of Recent benthonic taxa may provide analogies useful for interpreting the paleobathymetry of ancient sediments. Vella (1962, 1964) and Eade (1970) used the known depth distribution of benthonic species occurring in a Late Miocene coral thicket to ascribe a 1500–2500 m depth to the sea floor at that time. Fossil to Recent depth analogs have been used in a variety of areas and on rocks of different ages, including Quaternary sediments of Plymouth, England (Murray, 1965b), the Bay of Biscay (Caralp, Moyes & Vigneaux, 1970) and the Gulf of Mexico (Poag, 1972); Mio-Pleistocene sediments of the Gulf of Mexico (Poag & Sweet, 1972); and Cretaceous sediments of California and the Northeast Pacific (Sliter, 1972a; Sliter & Baker, 1972).

Blanc-Vernet (1969) demonstrated that the ratio between the abundance of two genera might be a useful bathymetric indicator. In the Mediterranean Sea the *Bolivina*:*Bulimina* ratio is high at depths less than 40–50 m but decreases rapidly below that depth. The *Textularia*:*Bigenerina* ratio behaves the same way at the 100–150 m level.

Stehli (1966) proposed a method to estimate paleobathymetry based on the increase in the number of planktonic foraminiferal taxa in deeper water sediments. Since there is a depth stratification of planktonic foraminiferal species (Chapter VII,5), shallow-water areas will not contain deep dwelling species whereas deep-water areas will contain shallow species. Kafescioglu (1971) used changes in species abundance to plot the Late Cretaceous bathymetry in Texas, Louisiana and Arkansas. The results were similar to those obtained from the use of the planktonic:benthonic ratio described above.

Barchuk *et al.* (1970) indicated that the trace element composition of foraminiferal tests might be related to environmental conditions. He suggested that the Mn concentration is a function of depth and could thus be used as a paleobathymetric tool.

10. Foraminiferal number and bottom topography

Use of the Foraminiferal Number (Chapter IV,8) may occasionally give interesting data which can be used to determine paleobathymetry. Saidova (1961a) calculated the FN for a large number of samples taken from the Northwest Pacific and adjacent seas. These were compared with hydrologic and bathymetric data for the same regions. These comparisons showed a marked correlation between FN and bottom topography. Saidova stated that the number of specimens in the lowest areas is always less than that in the elevated areas of the floor. She suggested that by using maps of the FN one could determine bottom topography, and conversely, one could determine foraminiferal abundance distribution from bottom contour maps.

For these observations to have value it is necessary to bear in mind that the maximum FN occurs at the edge of the continental shelf and adjacent regions of the upper slope.

As far as we know, there have been no studies of FN distribution in oceanic areas whose purpose has been to determine paleotopography.

11. Paleoclimates and planktonic test porosity

The first suggestion that the pores of planktonic foraminiferal species may reflect climatic conditions was made by Wiles (1967) who observed that the number of pores on specimens of *Globigerina eggeri* increased in interglacial core samples and decreased in those samples representing glacial epochs. Bé (1968) demonstrated that for Recent specimens pore size and pore concentration when taken individually show little correlation with latitude or temperature, but that when considered together, as test porosity, a significant relationship is revealed. Specimens in tropical and subtropical latitudinal belts exhibited greater than 10% porosity, specimens in temperate zone 5–10%, and those in Polar or sub Polar areas less than 5%. Bé suggested that these changes could be used in fossil specimens as a paleoclimatic indicator. Additional studies on the whole fauna and on individual species have confirmed this relationship between porosity and latitude (Frerichs *et al.*, 1972; Bé, Harrison & Lott, 1973). This technique has not received much use in deep-sea cores and Pflaumann (1971) expressed skepticism that it would prove useful because of post-depositional diagenesis which would alter the pore diameter. This drawback can be overcome if the porosity is determined by examining only the interior of the chamber wall as the pore diameters there are reflective of the size prior to secondary calcification and/or diagenesis. Unfortunately, the calculation of porosity in this way requires the destruction of the specimen.

12. Trace elements and paleosalinity

Knauff (1963) conducted experiments on specimens of *Patellina corrugata* that may prove useful in the determination of paleosalinities. He noted that specimens grown at 32.5% contained perceptible (0.03‰) amounts of Sr in their tests whereas those cultured at lower salinities (18‰ and 26‰) failed to take up any Sr. Knauff concluded that it might be possible to utilize trace element concentration in fossil forms to determine paleosalinities. To date no studies have been conducted to verify or utilize this method. Further discussion of the ecological effects of trace elements is found in Chapter IX,8.

13. Displaced faunas

The *Gazelle* expedition of the South Atlantic (1874–1876) encountered several areas of sand sized sediment quite distant from the coast whose source probably lay in water much shallower than that in which they were found. At first these observations did not seem unusual and no explanation was offered. However, as oceanographic observations increased and this phenomenon appeared in other areas, explanations were sought. The biggest boost to a rational explanation came when tests of shallow water benthonic foraminifera such as *Elphidium* were found in these abnormally coarse sediments. The concept of a physically displaced fauna and sediments was developed.

Although Norton (1930), Myers (1934) and Parker (1948) mentioned that currents could have a significant influence on the bathymetric distribution of benthonic foraminiferal tests, it remained for Ericson, Ewing & Heezen (1951) to provide the most detailed studies of this phenomenon. They studied cores from the Hudson River submarine canyon and concluded that the principal cause of the displacements is the action of turbidity currents which are capable of moving large quantities of sediment over considerable distances. Some benthonic foraminiferal tests have been displaced downslope for distances up to 500 km from where the organisms died.

When studying Recent foraminifera the distinction between living and dead tests aids greatly in the recognition of displaced faunas. When dealing with tests from cores the problem is more difficult because all the tests are empty; there is no biocoenose or thanatocoenose, only taphocoenose. The tests represent an accumulation of benthos and plankton, both of which are subject to post-mortem movement and diagenesis. There have, no doubt, been several studies of fossil foraminiferal forms in which the displaced nature of the assemblage was not recognized. A thorough understanding of assemblages and ecologic factors controlling the distribution of foraminifera should help in the discovery of such faunas.

One clue to the recognition of a displaced fauna is its state of preservation. After being transported over long distances many tests are broken and/or destroyed. However, the degree of destruction is not always a good indicator of a displaced fauna because even living tests can be broken or damaged.

Many investigators have recognized displaced faunas in both Recent and core studies.[11] As a general rule, the displaced fauna was found in an area of pronounced

[11] Phleger, 1951a; Natland & Kuenen, 1951; Hamilton, 1953; Bandy, 1953, 1964c; Phleger, Parker & Peirson, 1953; Parker, 1958; Resig, 1958; Christiansen, 1958; Zalesny, 1959; Green, 1960; Uchio, 1960a; Nesteroff *et al.*, 1968; Pujos, 1973; Ubaldo, 1973: and in numerous works in the Initial Reports of the Deep Sea Drilling Project, National Science Foundation, Washington, D.C.

bottom relief and often with fossil faunas both above and below it which were stratigraphically in place. Ericson & Wollin (1956a), after studying a large number of cores taken from the Atlantic, concluded that the processes which disturb the regular bottom stratification (slumping, turbidity currents, etc.) are much more common than previously believed. Thick continuous core sections representing undisturbed sedimentation are exceptional and most sections exhibit breaks in continuity.

The transportive agents of the displaced fauna ought to effect a granulometric sorting and the foraminiferal tests ought to be sedimented with detrital grains of similar hydraulic characteristics.

It is interesting to note the work of Grabert (1971) who was able to use benthonic forms as indicators of sand movement. She found that some foraminifera, e.g. *Trochammina*, were better indicators than others.

14. Biologic and hydrologic indicators

The oceans can be divided into regions, each of which is characterized by a different mass of water possessing its own physical, chemical and biological characteristics. Each species or assemblage is associated with a given water mass which represents its biotope. Each water mass contains its own flora and fauna (biocoenose).

The frequency of occurrence of the different species in a given water mass varies considerably. Because of this, failure to find a given species may not be an indication of its absence but rather of its low frequency. Planktonic distribution within a given water mass is uninterrupted even though this distribution may exhibit patches of high frequency adjacent to areas of low frequency. The phenomenon of uninterrupted distribution of plankton is especially true for the species of the open ocean, including the planktonic foraminifera.

As well as having horizontal limitations, each species also has vertical distribution limits which are controlled by the tolerance limits of the species and by the vertical distribution of the water mass. The organisms which characterize the different water masses and are utilized to differentiate them are called biologic indicators.

The term biologic indicator, in its oceanographic sense, has two meanings, one strict and the other broad. Under the broad sense of the term are included all those organisms (or their remains) that help to delimit or define any oceanographic phenomenon. Its use in this sense is very broad and diverse. Below is a brief description of some of the more important ways this term is used.

a. Ecologic indicators

Since each species lives within certain environmental conditions (defined by temperature, salinity, pH, oxygen, etc.), the presence of that species in an area is an indication of conditions within that area. The study of an assemblage of various species provides even better information than does a single species. Those species which are stenothermic, stenohaline, etc.; i.e. those with narrow ecologic limits, are the best ecologic indicators.

b. Geologic indicators

Organisms grouped under this heading yield information about the geology of an area. Such factors as sedimentation rate, location of turbidity currents and vertical motions of the sea floor are determined by the use of these organisms.

c. Icthyologic indicators

Certain organisms are associated with various kinds of fish and can be used to locate fishing grounds.

d. Hydrologic indicators

Under the strict sense of the concept of biologic indicators are those organisms which facilitate the definition of hydrologic characters in the ocean. It is probably more precise to refer to these organisms as hydrologic indicators. These indicators help in the identification of water masses and the determination of their movements.

A good hydrologic indicator ought to possess the following traits: (a) It should be a floating or drifting organism; i.e. it should not be capable of self-locomotion. If it does possess self-locomotion, the locomotion should be very weak; weaker than the currents which carry the organism; (b) It should be very abundant; (c) It should be easily recognized and identified; (d) It should lack daily vertical migration; (e) Its ecological tolerance limits and its biology should be well known; (f) It should be sensitive to environmental factors.

This final criterion requires some elaboration. To define and separate different water masses it is necessary to have indicators of great sensitivity which possess narrow ecologic tolerance limits. Those species which have broad tolerance limits and live in all or many types of water cannot serve as good indicators.

However, when using an organism to detect a current, it is preferable to use one capable of surviving rather large variations in some environmental parameters. An example of such a situation may be useful. Species typical of Tropical and Subtropical waters are carried by the Brazil Current from north to south. In tropical and subtropical zones these species live in waters of 20–24 °C. As the waters of the Current move south the temperature gradually decreases and then drops rapidly when these waters begin to mix with sub-Antarctic water. At the southernmost limits of the zone of mixture (48–49°S) the water temperature is 10–11 °C. If all the subtropical species had a low tolerance to temperature changes, no Brazilian forms would reach the southern limit. Many species do fail to pass the 16–18 °C isotherm; however, there are species which can resist the low temperature occurring at 48–49°S and their presence in the water at this latitude is an indication of the existence of the last remnants of the Brazil Current in this area. The tests of these tolerant indicator species are not common and are somewhat modified morphologically in the cooler water but are nevertheless present and useful. Thus, the best hydrologic indicators in this case are those species with wide tolerance limits.

The planktonic foraminiferal species used as hydrologic indicators are simple floaters. They can be compared to drift bottles tagged with an indication of their origin. As is well known, drift bottles were used in the past to determine the direction of surface currents. In more recent times they have been replaced by drift envelopes or drift cards made of polyethylene. A drift bottle, card or envelope would not be very useful if it floated only in waters of a limited temperature range.

It is understandable that a good hydrologic indicator cannot be cosmopolitan. Drawing on the analogy with drift bottles, a cosmopolitan species is like a bottle without a note inside. At the same time, as we have seen above, it is sometimes important that the indicator possess rather wide tolerance limits to some ecological factors.

Plankton are able to give a true indication of the origin of the water even when the water is mixed and its original physio-chemical characteristics have been altered. Fager & McGowan (1963) conducted a detailed water mass study on the plankton and the distribution of several physio-chemical traits (temperature, salinity, oxygen, etc.). They concluded that the geographic origin of the water is the principal factor controlling faunal characteristics; the physio-chemical factors are secondary.

The best hydrologic indicators are found among the microplankton, especially among the foraminifera. Planktonic foraminifera are very numerous, especially in middle and low latitudes and appear to lack vertical migration (Chapter VII,6). They can be easily identified with a low power stereomicroscope. Moreover, since they are identified by their hard parts they may be examined dry, a significant advantage when one is working aboard a moving ship at sea. Consequently it is possible to make water mass determinations on the spot where the plankton sample is taken. Because of these characteristics, the planktonic foraminifera serve as excellent hydrologic indicators. They are rather tolerant of temperature changes but somewhat less so to changes in salinity. Morphological changes often accompany environmental changes. Tests of *Globorotalia truncatulinoides*, as mentioned in section 4, when found in pure sub-Antarctic waters with a temperature of 6–10 °C are more flattened than the tests of this species found in zones of mixed waters.

Because of the subtle morphological changes produced by environmental shifts, it is sometimes necessary to use taxonomic categories lower than species level to express these differences. Sometimes the same species is found in two distinct water masses but the specimens of each water mass are morphologically different and can be described as such using the category of subspecies or the informal category of *forma*.

Thanks to the great abundance of foraminifera they can be used as drift bottles. The quantity of foraminifera is enormously greater than the number of drift bottles that have been or will be launched to determine current movement. This is one of the advantages of foraminifera over drift bottles as indicators. A disadvantage is that there is no way to determine the velocity of current motion as is true with drift bottles or cards where the date of launch and capture are known. Nevertheless, the foraminifera possess another advantage which can be illustrated using the following example. If we launch a drift bottle in a current, e.g. the Brazil Current, we can capture that bottle at any point of its path and find that the bottle will always appear the same. We can make deductions about its trajectory but we know nothing about the type of water it has passed through. But this kind of information can be obtained from the use of foraminiferal species as hydrologic indicators. The Brazil Current near its beginning, where it is composed of pure Subtropical waters, contains a qualitatively rich fauna. As it moves toward the south and begins to incorporate sub-Antarctic water, various species disappear and the remaining ones demonstrate various morphological changes. Specimens of *Globigerinoides ruber* become very small, lose their spines, exhibit a small principal aperture and almost lose their supplementary apertures. The species is the same, but the specimens are different morphologically. An analysis of the tests thus allows conclusions to be drawn about the nature of the water and position of the sample along the path of the current.

Although planktonic foraminifera are excellent hydrologic indicators, it must be noted that there are cases in which distinct ecological changes can provoke identical reactions in the specimens. In the South Atlantic, *Globigerina pachyderma,* forma sinistralis begins to predominate over *G. pachyderma,* f. dextralis in those areas where the temperature of the surface water is 9 °C or less. However, in the North Atlantic the same phenomenon occurs along the 7.2 °C isotherm. Bé (1969a) observed that *Globigerina inflata* is more tolerant of warm water in the North Atlantic than in the South Atlantic. *Globorotalia truncatulinoides* has a greater thermal tolerance in the Southern Hemisphere than in the Northern (Bé, 1969a; Boltovskoy, 1969b; Cifelli, 1971). Because of this the planktonic foraminifera are often more valuable as hydrologic indicators in localized areas than for global use. It is not possible to ascribe the same criteria to the same species in all oceanic bodies. It is not known whether or not this localization phenomenon also occurs with other planktonic organisms used as hydrologic indicators, but it would seem logical to suppose that it does.

Another problem is that the same species occasionally reacts in different ways in different water masses. Generally when the environmental factors affecting the life of planktonic foraminifera are studied, only factors such as temperature, salinity, depth, etc., are considered. This is understandable since these factors are apparently the most important and their determination is easily accomplished. However, they are not the only factors. There are others, which we can divide into three groups.

(a) Those factors whose values are easily determined, including such things as pH, density, oxygen content, phosphate and nitrogen concentration.

(b) Those factors whose values are difficult to determine including such things as trace elements, enzymes and biological interactions.

(c) Those factors about which we know nothing except that they probably exist.

The studies of Wilson (1951) and Wilson & Armstrong (1952, 1954, 1958) are of interest with respect to this last factor. They concluded that in addition to physio-chemical differences between two water masses there are also 'biological differences'. They suggested that some unknown property or properties exist in the water which vary from place to place, from time to time and with depth. Rae (1960) also noted the undeniable presence and influence of some unknown ecological factor which helps determine the variations in populations in space and time.

The conclusions of these authors were substantiated by the extensive plankton studies of Bary (1963) in the North Atlantic. He stated that the plankton distribution depended as much on an unknown oceanographic factor as on temperature and salinity. Fager & McGowan (1963) claimed that the commonly measured and studied hydrologic properties of sea water are not the ones that determine plankton distribution.

It is possible that it is this 'biological difference' that explains the different behaviour of *Globigerina pachyderma*, *Globorotalia inflata* and *G. truncatulinoides* in different areas. It may explain other phenomena which are yet mysteries; e.g. the absence of *G. truncatulinoides* in the eastern part of the Mediterranean Sea and its presence in the western part.

It is interesting to compare the uses and results of hydrologic indicators used for the delineation of water masses with those of the classic method of water mass delineation. The classic method to separate and identify the different water masses is based on the measurement of their temperature and salinity. Since these measurements are quantitative, they are much more easily manipulated than data derived from the study of hydrologic indicators. Numbers have a more convincing and impartial appearance than do the lengthy verbal descriptions in which the qualitative biological observations are presented. The primary question, however, is whether or not the classic method yields better results than the study of hydrologic indicators.

Using the classic method, physical oceanographers fail to take into consideration all those factors which cannot be quantitatively evaluated, including the 'biological difference' mentioned above. Nevertheless, these factors exist, are characteristic of the water mass, give it uniqueness and are recognizable by the effects they have on the organisms contained within the water mass.

The physical oceanographic technique also fails to take into account the fact that marine organisms are much more 'conservative' than physical hydrologic properties. Conservatism in this case refers to the retention of some indication of the origin of the water mass. The previously cited example of the Brazil Current illustrated this conservation. A further example should clarify it even more.

Waters of the Bering Sea pass through the Bering Straits and mix with those of the Arctic Ocean. As the Bering Sea waters are much warmer than those of the Arctic, they are cooled rapidly. Within a short distance north of the Straits, water temperature no longer serves as a means of identifying the entering water mass. The other physical properties are also altered and the water mass can only be recognized by a minute and detailed examination. However, the assemblage of planktonic organisms characteristic of the Bering Sea maintains its identity in the moving water mass for a long distance and provides clear evidence of the origin of the water.

The study of biological reactions or the detailed analysis of plankton and the comparisons between neighbouring areas is a job more difficult and complicated than the evaluation of temperature and salinity data. Because of these complications the biological methods are less widely used than the physical ones, and most hydrologic problems are resolved on the basis of physio-chemical data. Biological data and the interpretations made from them are commonly placed in a larger scheme which is based on the classic method. Nevertheless, it is quite probable that the future will see an increase in the use of hydrologic indicators, including foraminifera, as keys to understanding hydrological systems.

Beginning with the work of Murray & Renard (1891), the planktonic foraminifera have been considered as good hydrologic indicators. An increase in oceanographic activity in the second quarter of this century led to various bottom investigations which utilized planktonic foraminifera to study relationships between surface currents and bottom sediments (Schott, 1935; Correns, 1939; Revelle, 1944; Phleger & Hamilton, 1946; Arrhenius, 1952; Ericson, 1953; Phleger, Parker & Peirson, 1953; and others).

These studies established the fact that certain species always characterize the same thermal conditions as if they were bound to a specific mass of water. The regions of water mass convergence proved to be interesting areas for study in this regard. The sediments accumulating on the bottom in these zones are characterized by a mixed fauna containing elements of different water masses.

Until recently it was not possible to determine the configuration of currents solely by means of foraminifera. The first study of such a nature was conducted by Boltovskoy (1959b,c) who used both benthonic and planktonic species. In subsequent investigations this author expanded on the idea of foraminifera as hydrologic indicators. His work was conducted primarily in the eastern part of the South Atlantic. In this area the Malvin Current moves in a north–northeast direction. This current is composed of cold waters of sub-Antarctic origin. Boltovskoy noted that the benthonic foraminifera fauna underlying the Current was notably different from that found to the west on the broad continental shelf. The boundary between the two faunas is rather distinct. Boltovskoy was thus able to establish a sharp western limit of the Malvin Current wholly on the basis of foraminifera. By studying the fauna found beneath the current at its northern limits this investigator was able to show that the last remnants of this water mass continued as far north as Cabo Frio, Brazil, where it completely lost its identity.

This same investigator (Boltovskoy, 1961, 1962a), using planktonic foraminifera, determined the location of the zone of convergence between sub-Antarctic and sub-tropical waters in the Southwest Atlantic. This boundary has been determined several times by means of its physio-chemical traits, but the location of the zone as determined by different authors varies through 14° of latitude. This wide variation is due not only to the hydrologic complexities of the area but can also be ascribed to the fact that there is no unanimity among physical oceanographers regarding the specific criterion which should be used to separate water masses. Boltovskoy (1970b) pointed out that salinity values as disparate as 34.9 to 34.15% were used to separate sub-Antarctic from sub-tropical waters. This problem does not exist for an investigator working with planktonic foraminifera as hydrologic indicators. Pure

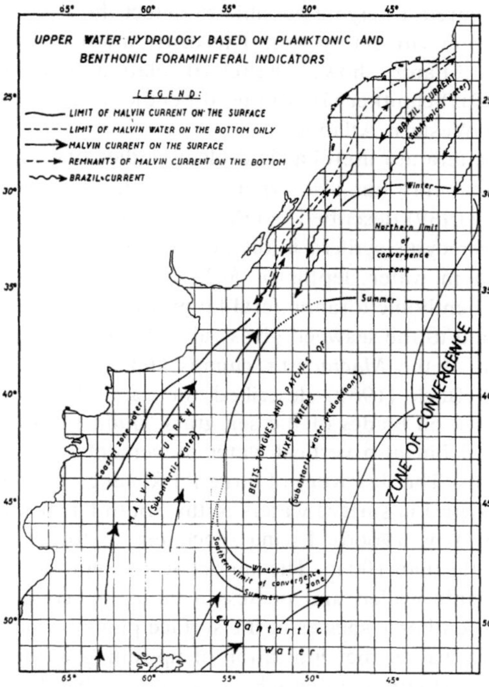

Fig. 122. Upper water hydrology, SW Atlantic (after Boltovskoy, 1969c).

sub-tropical, pure sub-Antarctic, Arctic and mixed water can be easily detected in the Southwest Atlantic on the basis of the planktonic foraminiferal taxa.

Only the exact character of the mixed water (predominance of sub-tropical or sub-Antartic species) presented any difficulties. As a result of the study of more than 3500 samples taken over a period of more than 10 tears, Boltovskoy (1969c) was able to prepare a summer and winter surface water hydrologic map based entirely on foraminiferal data (Fig. 122).

Other investigations of the relationships between planktonic foraminifera and current systems and the use of certain species as water mass indicators have been conducted by Bé (1960a, 1969a), Parker (1960), Uchio (1960a), Tsi-chung & Sau-yee (1962, 1963, 1964), Cifelli (1961, 1962a, 1965, 1967), Belyaeva (1966), Jones (1966 MS, 1967), Bé & Hamlin (1967), Herb (1968), Ujiié (1968), Cifelli & Smith (1969), Fossati Dutra Pereira (1969), Roettger (1970), Vilks (1970b, 1974), Bé & Tolderlund (1971), Bé, Vilks & Lott (1971) and Mateu (1971).

The distribution of empty benthonic tests may prove useful as a hydrologic indicator. If post-mortem transport by currents has occurred in an area, the concentration of the empty tests of a species will not correspond to the concentration of the living population of that species. If the areal distribution of the dead and living populations are compared it may be possible to detect the test transport pattern and consequently, the water current system. The distribution of deep water benthonic species in the North Atlantic shows a better correlation with water mass than with depth (Streeter, 1973). Schnitker (1974b) used benthonic assemblages to delineate 3 water masses in the Western Atlantic.

Benthonic foraminifera are useful as hydrologic indicators not only when encountered in bottom sediments but also when found attached to floating organisms. Boltovskoy & Lena (1969) described benthonic foraminifera found attached to the floating seaweed *Macrocystis pyrifera* in the Southwest Atlantic. This alga lives along the Argentine Coast in water depths of 5–25 m from 42°S to Tierra del Fuego. Because the coastal benthonic foraminiferal fauna varies along this zone, an analysis of the species attached to the seaweed gives valuable information about the location of the starting point of the *Macrocystis* clumps and therefore, data about surface current movements. Bock (1969) described the same kind of transportation of benthonic foraminifera on blades of the sea grass *Thalassia testudinum*.

Not only horizontal but also vertical water movements can be detected utilizing hydrological indicators. The discovery of large thick walled specimens of *Globigerina pachyderma*, forma typica in surface waters of the South Pacific indicate the presence of upwelling (Boltovskoy, 1966d). Living specimens of forma typica are generally found at depths greater than 100 m. The surface water representatives of this species (f. superficialis) are thin-walled and possess a larger aperture than f. typica. The discovery of several benthonic species in the surface plankton sample that yielded *G. pachyderma*, f. typica was further indication of upwelling water.

Pre-Recent ocean circulation patterns have been described by several investigators who used a variety of techniques, including that utilizing planktonic foraminifera as hydrologic indicators.

Stehli (1965) discussed a technique to describe paleocurrents which used the planktonic foraminiferal species diversity gradient; a gradient which is nearly parallel to the planetary temperature gradient. He used trend surface analysis to fit a smooth mathematically defined surface to the observed data on diversity. The difference between the smooth surface and the actual observations was interpreted by Stehli as a measure of ocean current movement of tests. These deviations from the fitted surface show the influence of warm- and cold-water movements.

Sliter (1972b) used the Recent distribution of planktonic foraminifera and their diversity gradients to calculate Cretaceous oceanic circulation patterns for the Eastern Pacific margins.

Several of the techniques described in this chapter (planktonic:benthonic ratio, distribution of benthonic species, planktonic species test porosity and isotopic data) were used by Bartlett & Hamdan (1972) to construct a Cretaceous to Recent history of oceanic circulation along the Canadian Atlantic Ocean margin.

Savin & Douglas (1973) devised a technique using oxygen isotope data from geographically diverse samples of the same age by which a paleotemperature map could be constructed and from which paleocirculation patterns could be interpreted.

15. Temperature–salinity–plankton diagrams

In order to utilize planktonic organisms as hydrologic indicators it is necessary to know which organisms are best suited for such purposes. A device known as the T–S–P diagram (temperature–salinity–plankton) is a convenient mechanism for learning more about some of the characteristics of the organisms. This diagram is an amplification of the well known T–S (temperature–salinity) diagram introduced by Helland-Hansen (1916) and widely used in physical oceanographic studies.

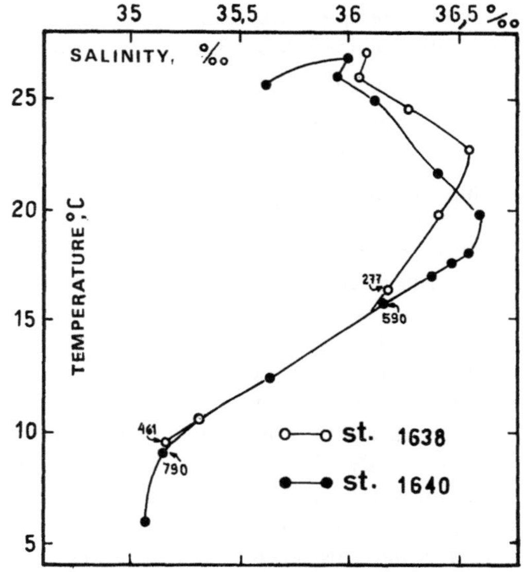

Fig. 123. Temperature–salinity diagram (after Sverdrup, Johnson & Fleming, 1942).

In the field of physical oceanography, water masses are generally characterized by a certain range of temperature and salinity. In certain special cases the limits set by these ranges can be altered. In very homogeneous water, a single data point (temperature and salinity) sufficies to characterize the mass. Helland-Hansen wanted to prepare a diagram which would graphically characterize water masses using just these two properties. Temperature data is plotted along the ordinate and salinity data along the abcissa. Data points representing a sequence of samples at different depths at the same station are connected to produce the T–S curve. This curve shows the temperature–salinity relationship at different depths, and any hydrographic anomalies can be easily observed. With curves from various stations it is possible to reconstruct the hydrology of an area. A typical T–S curve taken from two of the Atlantis stations is shown in Fig. 123. The two curves join at a point representing 277 m for one station and 590 m for the other. This indicates that the same water mass occupies different levels (313 m apart vertically) at different localities. The two curves are superimposed to a depth of 461 and 790 m respectively indicating a water mass about 200 m thick.

The T–S diagram is best suited to study subsurface waters because surface waters are subject to seasonal and local variations. Nevertheless, T–S diagrams for surface waters can be used to separate water masses. Data from surface samples of two different water masses should plot in distinct parts of the diagram with samples from transitional waters occupying intermediate sites. Data taken in the area of the Zone of Convergence, South Atlantic in 1961 shows that on March 10 two surface samples gave the following data:

Station 69: temperature, 24.3 °C; salinity 36.6%
Station 70: temperature, 23.5 °C; salinity 36.2%

Two days later the following data were obtained from two other stations:

Station 79: temperature 16.6 °C; salinity 33.5%
Station 80: temperature 13.0 °C; salinity 33.7%

On the first date the ship was in sub-tropical water; on the second in sub-Antarctic. The T–S diagram (Fig. 124) shows the physio-chemical differences between the two water masses. The data for the intermediate stations, obtained while the ship was in transitional waters, plotted in intermediate positions between the two groups.

The T–S diagram technique is a traditional and often used method in physical oceanography. Recently plankton have become more widely used as water mass indicators. Pickford (1946) initiated the use of planktonic organisms by superimposing faunal data on a T–S diagram to show that a particular planktonic cephalopod was intimately associated with a well defined water mass. Bary (1959) successfully used a T–S–P diagram in a study of biological indicators in superficial water near New Zealand. The procedures are simple: faunal (or floral) data are overlaid on T–S data obtained from the same samples. The distribution of certain species will correlate well with the T–S patterns whereas that of others will not. Those whose

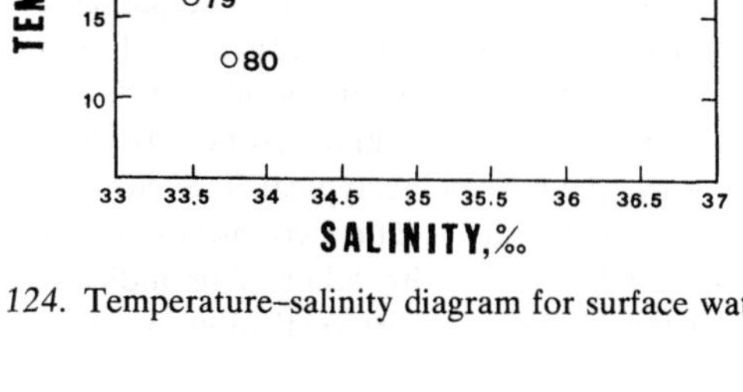

Fig. 124. Temperature–salinity diagram for surface waters.

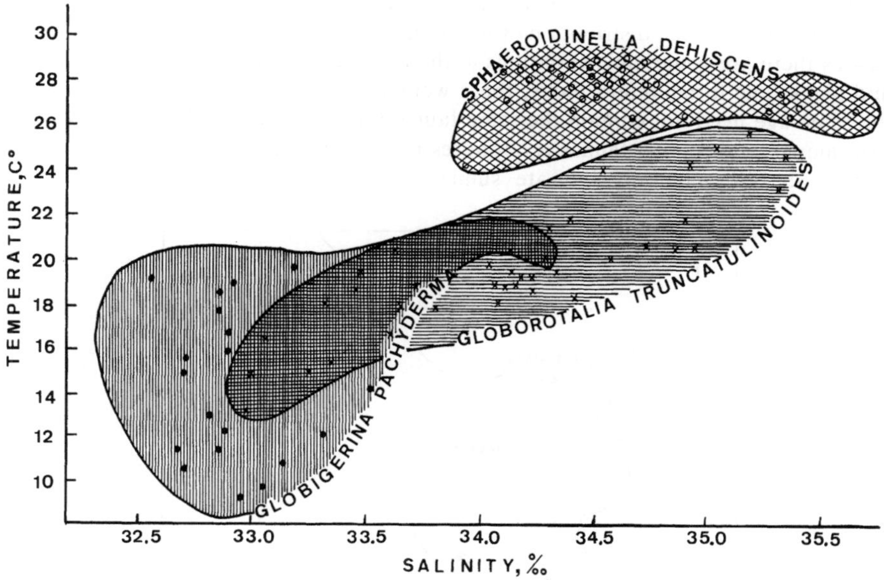

Fig. 125. Temperature–salinity–plankton diagram for 3 planktonic foraminifera, Pacific Ocean (after Bradshaw, 1959).

distribution does correlate can be considered as indicators. The first work of this kind on foraminifera was done by Bradshaw (1959) who dealt with three planktonic species in the Pacific (Fig. 125). The zones on the diagram occupied by the three species correspond to distinct water masses.

There is no doubt that the melding of biological indicator species with physio-chemical indicies will aid in the solution of oceanographic problems. Planktonic foraminifera should play a major role in this advancement.

16. Stratigraphy and correlation

Although this book deals with Recent foraminifera, it is often necessary to utilize data taken from fossil forms to clarify or add to information obtained from the Recent ones. In many cases the methods of examination, treatment and identification are the same for both groups. Moreover, fossil forms are sometimes used to resolve actual oceanographic problems. Examples of this sort of utilization are given in several sections in this chapter.

Even though the fields of micropaleontology and biostratigraphy are beyond the scope of this book, we wish to present some general ideas pertaining to stratigraphy and correlation in the light of what is known about Recent foraminifera.

Micropaleontologists studying the fauna in a stratigraphic section must keep in mind the nature of the distribution of the organisms at the surface of the earth today. Just as there are geographic variations in the distribution patterns of both benthonic and planktonic foraminifera today, there were such variations in the past. Rocks of the same age need not contain the same fauna. On the other hand, the discovery of two samples with similar faunas is not necessarily an indication that the rocks are the same age, but may only indicate similar environmental conditions, although a

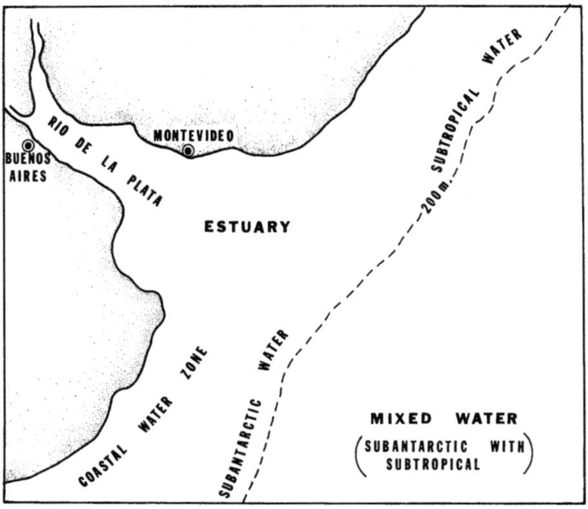

Fig. 126. Continental shelf area on which 3 different thanatocoenoses would occur.

detailed study of diachronous but similar faunas may show that there are slight differences due to evolutionary changes.

Local conditions have a great deal of influence on faunas. Marine rocks of the same age formed in three different environments, near shore, on the shelf and at depth will have completely different benthonic faunas. An examination of the lithology may help recognize this sort of situation and analogies made with modern faunas can serve as useful guides to environmental interpretation.

As an example of the effect that a relatively short geographic distance may have on faunal components we cite the situation in the northern part of the Argentine Continental Shelf (Fig. 126). Three distinct faunas occur there in waters of very similar depth. The north flowing, cold-water Malvin Current passes along the east margin of this area. The fauna of this current is sub-Antarctic. To the west is a coastal zone containing a typical near-shore fauna mixed with some Brazilian species. And finally there is the region influenced by the mouth of the Rio de la Plata where a qualitatively poor, brackish water fauna exists. These three distinct faunal types comprise the layer of sediments that are now being deposited in this relatively small area. If faunas arising from such a situation were encountered in fossil sediments, correct correlation could be effected if the investigator is careful to keep in mind all the different faunal changes that can be caused by environmental differences. An understanding of the ecology of Recent foraminifera is the key point in this process.

The stratigraphic value of a species depends in part on our knowledge of its distribution and environmental restraints. As more data about modern biotopes becomes available, a given species will be more valuable as an index in the past. Negative data, i.e. the absence of a given species, can be used also. However, this information should be used with caution since the absence of a species from a sample does not mean that it did not live in that environment, but only that it was not preserved there. It may be absent because of post-mortem transport, or destruction or because of inadequate sampling.

In preparing inter-regional correlations it may not be wise to draw conclusions only on the basis of the commonality of species. The principal criteria of inter-regional correlation are the stratigraphic limits of species. It is necessary to bear in mind that the same species may have different limits in different regions.

It should be evident that the process of correlation is very complicated and the investigator must have a broad knowledge of and ample experience with the group of organisms on which the correlation is based. It is usually wise not to base correlations only on one group of organisms nor even wholly on organic remains but to utilize all available data, both organic and inorganic.

A supplementary aid to correlation has been developed in the last decade. Cox, Doell & Dalrymple (1963) and MacDougall & Tarling (1964) discovered that certain extrusive volcanic rocks and deep-water marine sediments are magnetized. They also discovered that although the polarity of some samples was the same as now (normal), it was in the opposite direction (reversed) in others. Later studies (Cox, Doell & Dalrymple, 1965) revealed that this magnetization depends on the earth's magnetic field and not on any environmental factor. Thus, the changes from normal to reversed polarity occurred simultaneously around the earth and left paleomagnetic markers in the rocks which are of great help in correlation. The correlation of magnetic reversals has provided a framework in which foraminiferal correlations in deep-sea sediments can be accurately placed. The addition of radiometric dates to this framework provided even more accuracy.

Not only does the paleomagnetic reversal scheme provide a key to faunal correlations, but it has also been used to explain some faunal changes. Various investigators have suggested that an increase in genetic mutations (and subsequent faunal changes) were provoked by the abnormal environments during the period of time when the magnetic field reversed.

The last 4.5 million years has been divided into paleomagnetic epochs and events whose time of occurrence has been determined by radiometric dating (Fig. 127). In effect, the paleomagnetic events provide time lines; i.e. a temporal framework against which faunal changes can be interpreted. The studies that have used this

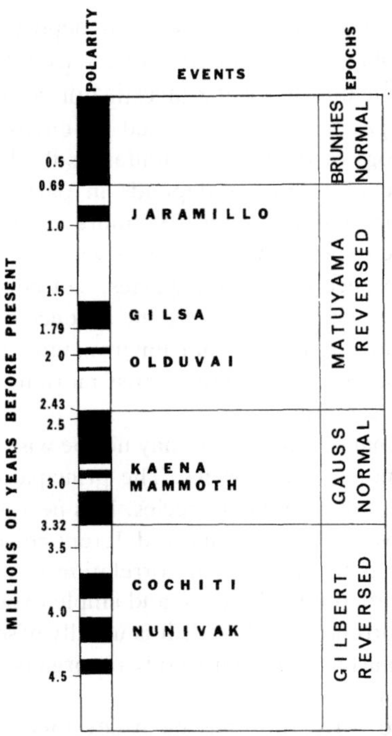

Fig. 127. Paleomagnetic epochs and events, their polarity and duration (after Bandy, Casey & Wright, 1971); black: normal polarity; white: reversed polarity.

framework to date give an idea of the kind of help it provides. Bandy, Casey & Wright (1971) defined the Pliocene–Pleistocene boundary on the basis of the appearance of *Globorotalia truncatulinoides*, an event occurring at or near the base of the Gilsa paleo-magnetic event. According to Kennett & Vella (1971), foraminiferal data indicating the first cold period of the Pliocene is found at the boundary between the Gauss and Matuyama epochs and the second cold period took place during the first half of the Olduvai event. They also place the Pliocene–Pleistocene boundary at the base of the Gilsa event based on the first appearance of *G. truncatulinoides*.

This paleomagnetic framework has just begun to be used in conjunction with paleoclimatic and stratigraphic studies using foraminiferal data. It should prove to be of great value in the future.

XV. FORAMINIFERAL CULTURES

The study of the soft parts of the living foraminifera has lagged considerably behind the study of the test and comparitively few studies of living organisms have been conducted. There appear to be two reasons for this phenomenon: (a) most students of foraminifera have been geologists whose background is in the study of the preservable remains of organisms (fossils), and (b) a generally held notion that maintainance of living foraminiferal cultures in the laboratory is difficult. The latter conclusion is not true and it is necessary to promote more widespread use of cultures because their study will solve many biological problems, allow the testing of various ecological hypotheses and resolve various taxonomic difficulties.

Unfortunately the number of investigators who cultivate foraminifera were and are few. Of the approximately 1000 persons working directly with foraminifera today, only 2–3 dozen maintain cultures.

Perhaps the first to examine living foraminifera was Blainville who described several observations in 1825. D'Orbigny had the opportunity to study specimens of living foraminifera on the French coast. Dujardin was the first to keep cultures of foraminifera. These cultures were maintained in tubes in a rather primitive manner in his Paris laboratory. Nevertheless, his observations of these living specimens led him to conclude that the foraminifera were not Cephalopoda, as had been previously thought, but Protozoa. At about the same time Ehrenberg was maintaining cultures of foraminifera in Berlin.

Later investigators observed living foraminifera in coastal areas, in plankton samples, or in laboratory cultures. They published their observations on the vital activity of these organisms.[12]

[12] Gervais (1847), Carter (1849), Harting (1851), Williamson (1848, 1851), Schultze (1854, 1856, 1860), Schulze (1875, 1877), Hertwig (1876, 1877), Bütschli (1880–1894), Gruber (1884), Moebius (1888), Verworn (1888–1889), Schaudinn (1893–1911), Rhumbler (1894–1911), J. Lister (1894–1906), Murray (1895–1897). Winter (1907) Prandtl (1907), Swarczewsky (1909), Lücke (1910), Hirsch (1912), Heron-Allen (1915), Schultz (1915), Heron-Allen & Earland (1920), Cushman (1921, 1930), Lwoff (1925), Jepps (1926–1956), Hofker (1922–1930), W. Schmidt (1929, 1937), Saedeleer (1932), Sandon (1934), Doyle (1935), Myers (1933–1957), Føyn (1936–1937), J. Le Calvez (1935–1950), Arnold (1948–1974), Slama (1954), Bradshaw (1955–1961), Nyholm (1951–1973), Grell (1952–1973), Hedley (1958–1968), Hedley & Bertaud (1962), Jahn & Rinaldi (1959), Czihak & Grell (1960), Lee *et al.* (1961–1974), Richter (1961–1967), Dahlgren (1962–1967), Murray (1963), Freudenthal, Lee & Pierce (1963), Narchi (1963) Sliter (1965, 1970b), Stephenson & Rees (1965a,b), Adshead (1966, 1967), Angell (1967a), Hedley, Parry & Wakefield (1967, 1968), Schnitker (1967), Sellier de Civrieux (1968), Marszalek & Hay (1968), Marszalek (1969), Lengsfeld (1969a,b), Muller & Lee (1969), Dietz-Elbrächter (1969, 1971), Berthold (1971), Christiansen (1971), Lutze, Grabert & Seibold (1971), Röttger (1972–1973), Walker (1971), Ross (1972) and Röttger & Berger (1972).

The great majority of studies and observations have been on benthonic species living in coastal zones and belonging to the following genera: *Acervulina, Allogromia, Ammobaculites, Archaias, Astrorhiza, Boderia, Bulimina, Bolivina, Calcarina, Calcituba, Cibicides, Cornuspira, Discobotellina, Discorbis, Discorinopsis, Elphidium, Entosolenia, Glabratella, Globobulimia, Globigerina, Gromia, Gypsina, Haliphysema, Heterostegina, Hippocrepinella, Iridia, Lagena, Marenda, Marginopora, Metarotaliella, Micrometula, Myxotheca, Nemogullmia, Nonion, Nubecularia, Ovammina, Patellina, Peneroplis, Phainogullmia, Planorbulina, Polymorphina, Pyrgo, Quinqueloculina, Rhumblerinella, Robulus, Rosalina, Rotalia, Rotaliella, Rubratella, Saccammina, Shepheardella, Spirillina, Spiroloculina, Textularia, Tinogullmia, Triloculina, Trochammina* and *Verneuilina*.

Observations on the biology of the planktonic foraminifera is considerably less abundant. It has proven much more difficult to transport living planktonic foraminifera to the laboratory and maintain them there than it has been for the benthonic forms. The most successful culture of planktonic forms to date lasted for only 3 months (Adshead, 1967) and consisted of *Globigerina bulloides* and *Orbulina universa*.

The first step in the cultivation of a laboratory foraminiferal population is the preparation of a base solution with nutrients. Various unicellular algae serve this function well. The food preferences of foraminifera are diverse (Chapter II). Nevertheless, experience has shown that certain algae (*Chlamydomonas, Chlorella, Dunaliella*) and diatoms (*Nitzchia, Navicula, Neidium affine, Amphora proteus* and *Phaeodactylum tricornutum*) can be used successfully as food for a variety of foraminiferal species.

1. Algal cultivation

Detailed descriptions of algal cultivation are given in works by Bold (1942), Pringsheim (1946, 1951), Provasoli, McLaughlin & Dropp (1957), Provasoli (1958) and Kabanova (1961) and we will only give some general instructions on the preparations of the algal cultures.

Algal cultures can be conveniently kept in Ehrlenmeyer flasks of about one litre volume. These flasks should be kept in a dark cool spot (a refrigerator is ideal) to prevent the excessive growth of bacteria.

Filtered sea water or synthetic sea water can be used as a medium for growth. Although not all algae will grow in synthetic sea water it may prove more convenient to use it than natural sea water. There is no universally accepted method of preparing artificial sea water. The formulas proposed by Pringsheim (1946) and Pantin (1960) work well. Pringsheim suggested the following ingredients be added to a litre of distilled water:[13]

NaCl	30 g
$MgCl_2$	4 g
KCL	10 g
$MgSO_4 \cdot 7H_2O$	5 g
$CaSO_4$	1 g

A small pinch of $CaCO_3$ can be added as well.

It is necessary to add additional nutrient material, either 'Earth extract' or Arnon solution, to the sea water solution to induce rapid growth of the algae. Earth extract is prepared in the following manner:

To 0.5 litre of sea water 100 g of top soil is added. This mixture is heated for an hour and a half. The volume is restored to about 600 ml by adding distilled water. The liquid is filtered several times and kept at a low temperature.

Then the following mixture, called *Erdschreiber*, is prepared in the following manner. To 473 ml of filtered sea water or synthetic sea water add 25 ml of earth extract, 1 ml of $NaNO_3$ (10% solution), and 1 ml of $NaHPO_4$ (2% solution). This mixture is placed in an autoclave for 15–20 minutes to kill all the bacteria. Immediately, before using this solution, it is advisable to heat it to about 80° C to further remove bacterial contamination.

To a flask of this solution is added an inoculation of algae. The flask is placed by a window or near an artificial light source. After 2–3 weeks the flask will develop a rich internal coating of algae.

To prepare a new culture some algal material is removed from algal culture and centrifuged to remove most of the liquid. The supernate liquid is decanted and clean sea water added and agitated. The sample is again centrifuged. Several repititions of this operation will remove various algal metabolic excretions whose presence in certain concentrations is toxic.

A different algal cultivation medium is Arnon solution whose ingredients are:

H_3BO_2	2.86 g/l.
$MnCl_2 \cdot 4H_2O$	1.81 g/l.
$ZnSO_4 \cdot 7H_2O$	0.22 g/l.
$CuSO_4 \cdot 5H_2O$	0.08 g/l.

Accorinti (1960) suggested that the zinc and copper sulfates be replaced by chlorides:

$CuCl_2 - 0.05$ g/l.; $ZnCl_2 - 0.11$ g/l.

[13] It is evident that the total addition of salts to this solution is 50 g/l. which is considerably more than that in normal sea water. However, for the cultivation of algae, this elevated salinity works well. Provasoli, McLaughlin & Dropp (1957) pointed out that algal cultures grow well between 17–50%. This artificial sea water would not work for the cultivation of foraminifera whose salinity tolerance is much less.

An experiment was conducted in the cultivation of *Dunaliella* in the following media: (a) Erdschreiber; (b) sea water with Arnon solution; (c) pure synthetic sea water; (d) mixture of artificial and natural sea water; (e) filtered sea water.

The sea water plus Arnon solution produced the greatest development of algae. The Erdschreiber solution also produced a good algal crop.

An initial culture of algae is easily obtained from some colleague who cultures foraminifera or algae. These starter cultures can be sent by mail in plastic tubes and can be held in a liquid medium or a solid one. The preparation of a solid medium is achieved by boiling 100 ml of sea water with 1.5 g of agar–agar. This mixture is poured into the tube, tilted to a low angle and the algal culture placed on top of the

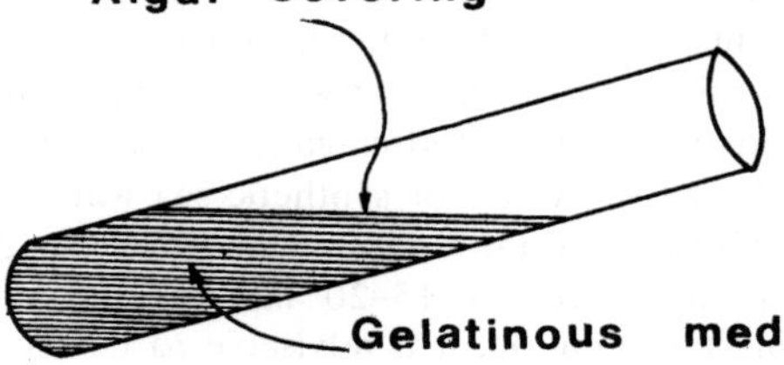

Fig. 128. Method of preparing algal cultures for shipment.

gelatinous medium (Fig. 128). The algae in this solution will withstand a voyage of several weeks.

2. Collection of living material

Arnold (1954c) and Lee (1974) made several suggestions for the collection of living foraminifera for culture purposes: (a) The collecting should be done at sites well known to the collector; (b) The littoral zone is the best zone for finding foraminifera for cultures; (c) Sites where there are very soft substrates or strong currents or tides should be avoided. Fine grained substrates and hard rocky bottoms covered with algae usually yield abundant specimens; (d) A good quantity of material is often obtained by gently washing marine algae as many specimens live attached to them. They may, however, be attached firmly to the algae and can be dislodged only with difficulty. To wash the algae they are placed in plastic buckets, agitated and allowed to stand for 15 minutes so that the foraminifera can settle to the bottom. The vegetation and water are replaced by fresh sea water. The samples should be kept at temperatures as close as possible to that of the natural habitat. This can be done by keeping them immersed in the water at the collecting site or by placing them in insulated containers. The sample containers should be filled with about 3 times as much sea water as sediment; (e) It may be possible to move the foraminiferal tests into suspension by agitation of the bottom. The tests are then collected by sweeping the water with a plankton net.

Our experience has been that the intertidal zone generally is not as productive as the zone just beneath low tide level.

Arnold (1974) described an extensive set of light weight, easily made equipment which increases the efficiency of collecting and examining living specimens in the field.

3. Separation and transport

After collection the living specimens must be separated from the detrital material. This is best done in the laboratory although some preliminary separation can be accomplished in the field. The sample and water are placed in a shallow wide dish. Gentle circular agitation of the dish will create a sorting of the grains so that the coarser non-foraminiferal fraction will concentrate in the center where it can be removed. Removal of the coarse and fine fraction can also be accomplished in the field by washing the material across two sieves without even removing the material from the water. The upper sieve should be coarse and and the lower one of fine mesh size.

If marine algae or sea grass is present in the collecting zone it can be harvested and rubbed briskly with the hands in a bucket of sea water. The agitation will loosen the specimens and they will be collected in the container.

The samples should be transported in thermos bottles to preserve the original temperature of the water at the site. Arnold suggested that the bottles should be opened and aerated every 4–6 hours if they are to be transported for a long time. This procedure may not be absolutely necessary as we know of cases in which living specimens arrived in perfect health after a voyage of 2 days completely sealed in a small cylindrical tube with algae. A $\frac{1}{2}$ litre flask containing bottom sediments and foraminifera contained healthly living specimens after being closed and in transit 6 days. The explanation for this may lie in the observation that foraminifera do not require a great deal of oxygen (Chapter IX,10).

Marszalek & Hay (1968) suggested techniques by which marine specimens, including foraminifera, can be transported great distances by automobile and kept aereated while in transit.

A fresh change of sea water may be given to the sample upon its arrival in the laboratory. The original water is removed as the new is added.

The foraminiferal specimens will migrate to the sediment surface in a few hours where they can be easily removed by a pipette (Fig. 129) and transferred to fresh culture containers. Lee (1974) suggested that the water be changed every day in a new culture for a week, then weekly for about 3 weeks and then less often.

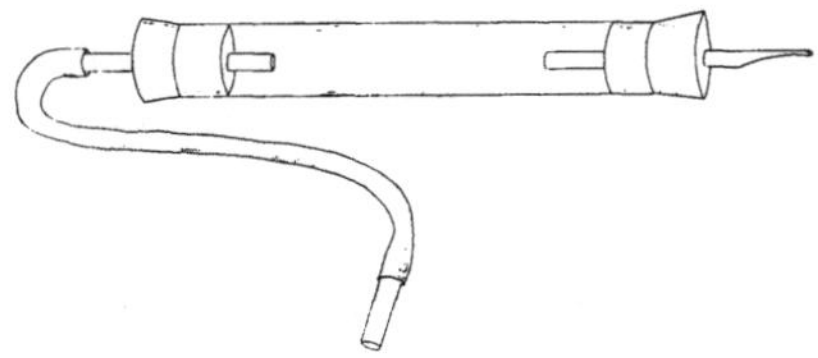

Fig. 129. Improvised mouth pipette.

Some investigators suggest the sterilization of the sea water used in the cultures. This technique, if performed by boiling, has the disadvantage of eliminating gasses from the water and altering the salinity equilibrium. The sterilized water will be altered chemically. The bacteria can be removed by filtering the water through an asbestos filter. Neither boiling nor filtering can remove all the bacteria from the culture. Some will be transported into the culture attached to the foraminiferal tests. Legarde & Arnold (1967a,b) suggested that the application of streptomycine (2 mg/l.) and chloramphenicol (3 mg/l.) to the cultures would control bacterial growth.

Which species to cultivate depends on the goals of the investigator and the fauna available. It is always more interesting to chose a species which has never before been cultivated in the laboratory. At the same time, it is necessary to consider that small species are generally easier to cultivate because their life cycle is usually shorter; e.g. *Allogromia laticollaris*, a unilocular species about 0.2 mm diameter, reproduces about every 2 weeks, whereas *Elphidium crispum*, a multilocular species 1 mm in diameter, reproduces about once a year.

Perhaps the easiest and most convenient mode of culture is to isolate a species in a culture and thereby try to observe the influence of various ecological factors on its development. This isolation presents a serious disadvantage, however. Culture conditions, at best, are only imitations of natural living conditions. Isolation of a particular species creates an even more artificial situation and consequently diminishes the value of the observations.

4. Foraminiferal cultures

Foraminiferal culture systems can range from simple arrangements of isolated specimens or specimens in small containers (Bradshaw, 1957; Lee & Pierce, 1963; Arnold, 1964, 1966b; Sliter, 1970b; Röttger, 1972a; Röttger & Berger, 1972) to moderately complex systems in which various cohabitants exist with the foraminifera (Myers, 1935a, 1937; Jepps, 1942; Arnold, 1954c, 1967a; Bradshaw, 1955; Marszalek & Hay, 1968; Marszalek, 1969) to complex multiorganism systems with tidal circulation systems (Lee *et al.*, 1961; Freudenthal, Lee & Pierce, 1963). These latter systems are beyond the reach of researchers in most laboratories and simpler arrangements usually must suffice.

The simplest culture system is set up in a crystallization or Petri dish 4–8 cm high, 10–15 cm in diameter which is covered with a watch glass. Containers should have a lot of surface area with respect to volume. The cover should be removed from time to time to allow for aeration. Better aeration can be obtained by covering the dish with a glass plate held above the rim of the dish by two thin glass rods. Under these conditions, however, salinity will gradually increase due to evaporation of the water and it will be necessary to add distilled water to maintain salinity. Ideally the dish should be completely enclosed and oxygen supplied through a pumping system, but this is difficult in small cultures. The small culture dish has the advantage of being able to be studied directly under the microscope. In this way the foraminiferal specimens do not have to be disturbed. The culture system is a poor representation of natural conditions as there is no wave energy, competition is reduced or unnatural, the food supply is different, and there are subtle differences in water chemistry. Any additional disturbance of the individuals in order to observe them further isolates them from their natural habitat and makes assessment of the observation difficult. When viewing the small culture dish under the microscope one must be careful that the microscope light does not overheat the water.

Cultures in which the non-foraminiferal organisms have been removed will not yield reliable results for studies of community and ecological behaviour. For purposes of studying the life cycle they may work well because the influences of the other organisms is minimal.

If the foraminiferal population increases too rapidly it may be necessary to transfer individuals to new containers. The transported specimens should be thoroughly washed in clean sea water before being placed in a new culture.

Foraminiferal cultures seem to thrive best in natural sea water which has been filtered and maintained in a refrigerator. These precautions are necessary to prevent the excessive development of bacteria in the culture. Attempts to establish bacteria free cultures (Lee & Pierce, 1963; Lee *et al.*, 1966) met with failure and Muller & Lee (1969) suggested that the presence of some bacteria might be indispensable to foraminiferal growth.

If necessary, synthetic sea water can be used for foraminiferal cultures. The formula of Pringsheim described in the section on algal cultivation results in salinity values which are too high for foraminifera. Lyman & Fleming (1940) suggested that the following ingredients be added to one litre of distilled water to yield artificial sea water of normal salinity:

$NaCl$	23.476 g	$NaHCO_3$	0.192 g
$MgCl_2$	4.981 g	KBr	0.096 g
Na_2SO_4	3.917 g	H_3BO_3	0.026 g
$CaCl_2$	1.102 g	$SrCl_2$	0.024 g
KCl	0.664 g	NaF	0.003 g

The use of synthetic sea water does not seem to result in the best cultures. According to Bradshaw (personal communication) species cultured in artificial sea water survived but did not grow or reproduce. There are minor constituents in sea water in addition to the 10 listed above, and some of these may play a significant or even critical role in the life cycle of foraminifera. Consequently it is necessary to add a bit of natural sea water to a culture growing in synthetic sea water. This natural sea water should probably come from the same locale as the foraminiferal specimens.

One method of preparing sea water is to evaporate natural sea water along the coast and save the salts. These can later be reconstituted by adding about 35 g of salt to a litre of distilled water. The water produced by this method is not exactly the same as the original because the proportions of salts vary somewhat and some organic compounds are lost. A better method would be to evaporate sea water until a concentrated solution remains (about $\frac{1}{7}$ of the original volume). Distilled water is added in the laboratory to restore the solution to its original volume. The distilled water used in foraminiferal cultures should be free of any metalic ions. Distillation in borosilicate glass is the best method for obtaining clean water.

Warm-water foraminifera are easiest to cultivate as they can be maintained in most laboratories under room temperature conditions. Cold water species require more elaborate conditions, usually a temperature controlled water bath, to maintain them at the proper temperature.

The food supply for the foraminifera can be cultured with the foraminifera themselves or separately. In the case of simple cultures a separate food culture is probably more convenient. Nutrient material is transferred to the culture on a regular basis, usually about twice a week. The transfer should be accomplished using graduated pipettes so that a constant and controlled supply of food is added each time. Some investigators have successfully cultured foraminifera using a single nutrient species whereas others suggest that a mixture of species works better.

If the foraminifera and the algae are cultivated together it is advisable to change the liquid periodically. This is done by removing by pipette about 30% of the liquid and replacing it with new fluid. This removal must be done carefully to prevent removal or disturbance of the foraminifera.

For the production of cultures in which all the organisms present are known (gnotobiotic) it is necessary to free the foraminiferal specimen from associated microorganisms and place it in a culture of known and controlled food sources. The foraminiferal specimen is cleaned by placing it in a sterile 9 hole plate whose cavities are filled with sterile sea water. The specimen is then carefully transferred from one spot to another to gradually remove the microorganisms. About 20 serial washings over a 4 hour period should be sufficient (Lee, 1974).

Schultze (1854) suggested that foraminifera are very tolerant of decomposing material. He observed that Rotaliidae and Miliolidae specimens maintained an active existence in a small glass filled with lagoonal mud. This observation would suggest that it is unnecessary to change the water in a culture. We are sceptical of this conclusion as the decomposition of organic material in a stagnant environment lowers the pH in the substrate to the point that calcareous tests begin to dissolve and specimens die. Our experiments demonstrated that sea water (pH 8) in a 10 cm diameter Petri dish with specimens of *Elphidium* and *Quinqueloculina* developed a pH of 7 in less than a month. Marszalek & Hay (1968) proposed covering the bottom of the culture receptacle with carbonate sand which acts as a buffer to pH changes. They were able to maintain a pH of 8.2 in enclosed aquaria for 2 years without the addition of new sea water.

It is difficult to determine how many specimens of foraminifera to cultivate in a single receptacle. The number depends on the particular species, its food demands, etc. An abundance of nutrient material will cause an increase in the bacterial population and a nutrient deficit will inhibit the foraminiferal development. An overabundance of filamentous algae in the container leads to reduced mobility and entrapment of the foraminifera. The foraminifera try to feed as much as possible and become completely covered by the algae and their pseudopodial activity is thus impeded (Arnold, 1954c). This becomes especially harmful to the organism during reproduction. Arnold described having to liberate juvenile forms by means of 'Cesarean section'. Even those algae which are not filamentous can be harmful to the foraminifera if overabundant. Various by-products of algal metabolism may be toxic to foraminifera.

The effect of overfeeding on the growth and development of foraminifera was demonstrated by Bradshaw (1955) who showed that initial population growth was retarded when the foraminifera were fed twice a day instead of once (Fig. 130).

For the above reasons it is suggested that a limited supply of nutritive material be present in the cultures. If inert algae are used they should cover the bottom of the container in a thin layer. If flagellated algae are used they should be present in quantities insufficient to give colour to the water. Trial and error are probably the best instructors for the establishment of an equilibrium nutritional condition.

Some foraminifera prefer dead algae as food. The algae are killed by heating the algal culture to 50–60 °C. The use of heat-killed food avoids the build up of the algal metabolites mentioned above.

It may be necessary to translate the cultured specimens to a new culture medium from time to time as toxic substances build up in the container. The discovery of numerous dead juvenile forms among the population is a good indicator that a fresh medium is needed. If the dead are adult forms, there is probably no problem as this is merely a consequence of reproduction.

Water temperature will have significant influence on the survival, growth, and reproductive rates of the specimens in culture (Chapter IX,1) Just as in the case of nutrition, only trial and error will determine the optimum temperature conditions for a particular culture situation.

Laboratory cultures pass through 3 stages: (a) the juvenile stage, during which the reproductive and growth rates increase, (b) the adult stage, during which reproductive and growth rates are maintained at a constant level and (c) the senile stage, during which these rates decline. If the temperature is lowered during this final stage, the organisms may be kept alive for an extended period of time.

The success of many investigators, operating with both simple and complex culture

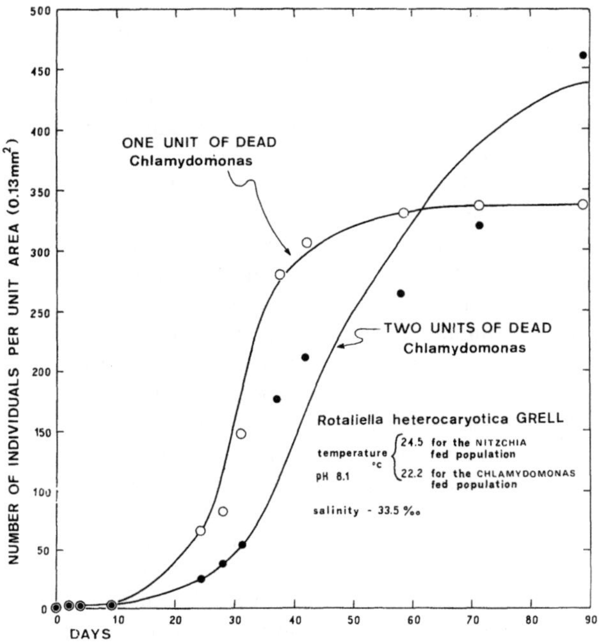

Fig. 130. Unfavorable influence of excess food (after Bradshaw, 1955).

systems, has shown that the cultivation of benthonic foraminifera does not present great difficulties. It is always necessary to remember that laboratory conditions, no matter how elaborate, do not exactly duplicate natural conditions. The reciprocal action of all the various organisms in the environment, and the constant influx of new water are but two of the dynamic equilibrium conditions difficult to duplicate in the laboratory.

The absence of perfectly natural conditions in laboratory cultures may explain the abnormally high proportion of monstrous, degenerate and deformed specimens that so often occur there. One of the primary tasks that face those who culture foraminifera is the construction of systems which better duplicate nature.

XVI. STORAGE OF SAMPLES AND SPECIMENS

Laboratory collections are of importance because they constitute material from which past investigations can be verified and new studies initiated. Laboratory collections allow effective comparisons based on real specimens rather than on drawings or written descriptions. Taking all this into account it is not an exaggeration to say that a well maintained foraminiferal collection is of international value. Consequently, it is necessary to devote serious attention to its organization.

There have been numerous publications dealing with the installation of foraminiferal laboratories and with different laboratory techniques for use on foraminifera, but little has been written about the organization of a collection. The present chapter is based on an article by Boltovskoy (1958c) devoted especially to this problem.

1. Sample storage

All samples should be catalogued immediately upon their arrival in the laboratory. One method is to utilize different catalogues based on (a) geologic age, (b) sampling method, (c) geographic location, etc. It is much more convenient to use only a single catalogue, however. The samples should be numbered sequentially, and stored in the order of their entry into the collection. This method will save a great deal of both space and time. The various categories of samples can be best handled by making a separate card for each sample on which is recorded (a) the catalogue number, (b) stratigraphic unit, (c) age and (d) location. (Fig. 131). By storing this data on separate cards and by keeping it to a minimum it is possible to effect quick searches through the cards for samples of interest.

OCEAN BOTTOM

RECENT

BAHIA CONCEPCION, CHILE

Fig. 131. Quick reference card for samples.

The grouping of the data cards into categories depends upon the interests of the investigator. It may be convenient to separate the fossil from the Recent material and to subdivide the latter by method of collection (dredge, core; qualitative, quantitative). A further subdivision by geographic location (ocean basin, latitude and longitude sectors) or water depth (littoral, shelf, etc.) may prove helpful.

The cards representing fossil samples should be arranged chronologically and then geographically. If cards of different color are used for each chronologic division, sorting becomes much easier. The international geologic map colours representing each age are the most logical colours to use.

SAMPLE CATALOG

CATALOG NUMBER	STRATIGRAPHIC DESIGNATION	AGE	LOCALITY	COLLECTOR	ORIGINAL SAMPLE NUMBER	COLLECTION DATE	SAMPLE DESCRIPTION	WEIGHT	DATE RECEIVED IN LABORATORY	OBSERVATIONS

SPECIMEN CATALOG

CATALOG NUMBER	DATE	NAME	LOCALITY	AGE	COLLECTOR	IDENTIFIED BY	DONOR	NUMBER OF SPECIMENS	OBSERVATIONS

Fig. 132. Catalog layout for samples and specimen slides.

If the samples are from well cuttings it may be helpful to arrange the cards by company and by the company well name or number. A geographic or stratigraphic division can be made with these cards.

It is very convenient to have a large scale map on which the location and number of all the samples is plotted. This sort of plot provides a very good representation of the degree of sample coverage. The map plot is also useful if a sample from a particular geographic locality is desired. A quick glance at the map indicates if the collection contains any material from that area and which sample numbers are pertinent. The catalogue entry will indicate the exact location of the samples in question as well as other data which will indicate whether or not a given sample may be of interest.

The catalogue entry for a sample should be as complete as possible (Fig. 132). There should be correlation between the sample number in the catalogue, on the file card and on the sample itself.

2. Specimen storage

Identified foraminiferal specimens are housed on slides described in Chapter XII. For the organization of the collection it is advisable to have at least two types of these slides, (a) those with a central circular or square cavity for housing specimens of a single species and (b) those with a rectangular cavity divided into numbered squares (Fig. 102); these latter slides are useful for carrying an entire faunal association. The depth of the cavity should vary from 1–4 mm.

Specimens may be strewn loosely in the slide cavity or glued to the base, usually with water soluble gum tragacanth. This glue holds the specimen firmly in place once dry but can easily be dissolved again with a drop of water. If an entire faunal assemblage is carried on a slide the specimen must be glued to keep each specimen in its appropriate spot on the slide. The various type specimens should also be glued firmly to the slide.

The use of foraminiferal slides and gum tragacanth as described above is very convenient but does have one disadvantage. Various investigators, among them, Heron-Allen & Earland and Triebel, noted that among old collections (19th Century collections of d'Orbigny and Brady) some foraminiferal specimens decompose completely. Boltovskoy (1953) suggested that this decomposition was due to long term bacterial activity in the glue on the slide. Consequently, for long term storage, the glueing of a specimen to a slide may not be desirable. Boltovskoy suggested that long term safe storage can be accomplished by storing the specimens in glass tubes. A 8–10 mm diameter glass tube is sealed at one end by a flame. The specimen is placed inside with a slip of heat resistant paper containing the pertinent data. The open end of the tube is sealed by flame and the tube sterilized at 100 °C for an hour or more. Cursory investigation of the specimen can be accomplished through the tube but detailed study will require breaking the sealed tube. After examination the specimen should be again sealed off from the environment.

Although this method is much more secure than the use of slides and glue, it is much more complicated and time consuming. It is probably most convenient to continue to use the slide and glue method except for specimens which are very valuable; i.e. holotypes or members of old irreplaceable collections.

The foraminiferal slide should bear the following data: (a) name of the collection, (b) catalogue number (see below), (c) age, (d) locality, (e) taxonomic designation, (f) name of identifier, (g) indication if the specimen is a type and (h) publication in which it is figured (Fig. 133).

The locality is written in the middle part of the left side using the mimimum amount of information: geographic region (bay, gulf, strait, etc.), longitude, latitude and depth. This information is sufficient.

The age is written above the location. For fossil specimens we suggest that the age be written in the colour which corresponds to that period of time in the international geological map colour scheme (Cretaceous-green; Jurassic-blue, etc.). This colouration of the slides makes rapid searches through the collection possible.

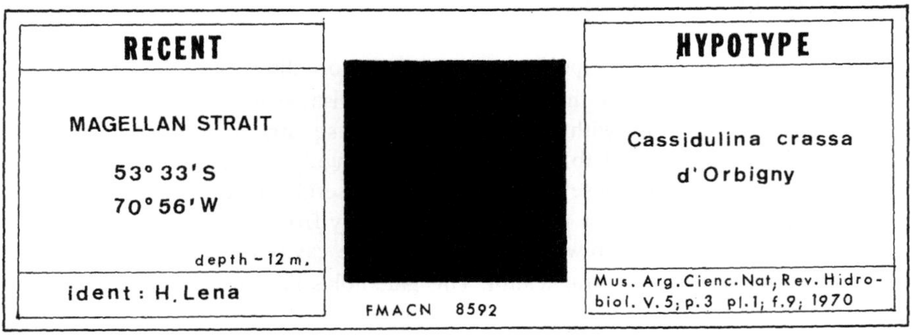

Fig. 133. Layout of typical foraminiferal slide showing pertinent data.

The name of the person who identified the specimen is placed in the lower left corner. This information is useful because it permits requests for supplementary material or additional information about the specimens. Knowledge of the identifier allows us to get a feel for that individual's taxonomic 'taste'.

The upper right hand corner is left free if the specimens are not types. If they are types, the particular kind of type is written here. We have found that using different colours for holotypes and paratypes (red), topotypes (green) and hypotypes (light blue) proves very helpful in rapid searches through the collection.

The taxonomic designation is placed in the middle of the right hand side. If the specimen is a figured type, the bibliographic reference to the figure is given below in the lower right hand corner.

Each foraminiferal slide should have an entry in a specimen catalogue laid out as in Fig. 132. The specimen number is entered in the catalogue and in the lower middle part of the slide.

Some planktonic foraminifera contain long fragile spines and should not be mounted on common foraminiferal slides and in a dry state or the spines will break. Banner & Blow (1960) suggested placing these kinds of specimens in a glycerine jel in a small plastic container fixed to a slide. The necessary specimen information can be affixed to the slide.

The foraminiferal slides should be housed in an easily accessible location. They may be arranged according to (a) publications or studies in which they occur, (b) systematic position or (c) alphabetical order.

(a) Arrangement by study has the single advantage that the fauna from a single region is housed together, but it is difficult to find all the various specimens of a particular species if this arrangement is followed. Persons unfamiliar with the collection or the studies represented will be particularly confused.

(b) A systematic arrangement, although ideal in theory, is not particularly convenient in practice because there is no universally accepted classification. Visitors using the collection will probably have difficulty finding particular specimens. Each change in systematics requires a reorganization of the collection, and the time spent on this could be better spent in examining the specimens.

(c) Alphabetic arrangement is the most convenient method. Any visitor may locate a particular taxon easily, without help. If species are reclassified to other genera it is advisable to leave a mock slide or piece of cardboard in the collection at the old site with a notation of the transfer. Planktonic specimens should probably be housed separately from benthonic specimens.

Even if an alphabetical arrangement is utilized, some of the advantages to the first method of arrangement can be maintained by preparing slides containing faunal suites from the different regions and areas that have been studied. Each small square on the slide can be filled with a different species from the area. These can even be proportioned on the slide to represent the actual proportions in the area. This proportional method is more difficult to achieve because of the tremendous variations which often occur in species abundances in a given region. Nevertheless, it is a very revealing method. If the differences in abundance between the dominant species and the rare are too great, the logarithm of the abundance can be used or only the most abundant specimen mounted.

It is also convenient to prepare a card file listing all the species that have been studied. This arrangement permits a convenient and rapid means of sorting and comparing new specimens with those already examined. The following information should prove useful on the cards: (a) taxonomic determination, (b) a photograph or drawing of the species, (c) a brief diagnosis, (d) a notation of the typical morphology and its variations in the samples studied, (e) geographic or stratigraphic distribution of the studied specimens and, if possible, (f) data about the synonymy and geographic distribution in other areas. Preparation of a card file of this sort requires a great deal of time and effort but will prove extremely useful.

XVII. SYNONYMY AND ILLUSTRATION

The foundation on which all foraminiferal studies is based is the correct identification of the taxa in the fauna. The taxonomic interpretation of each investigator is contained in the systematic part of the paper, an area of great importance in each study because it allows the reader to form an opinion about the fauna and the author's interpretations. Of course the best interpretation should be based on actual examination of the specimens themselves, but this is usually impossible. The increasingly common practice, in recent years, of saving space by merely listing the taxa which have been studied is regretable because almost no information can be obtained from such lists. The chaotic nature of foraminiferal nomenclature precludes the use of such lists. Synonyms abound and some species have such widely different interpretations that the appearance of a species name in a list often conveys little information to the reader. Only by giving careful consideration to the systematic portion of the work and to the illustrations can the author meet his responsibilities to the reader.

The description of a new species should be done with great care and thoroughness (Chapter VIII,2). Because of space problems it is usually not convenient to prepare a detailed description of those pre-existing species encountered in a study. Most authors limit their comments on these established species to a synonymy and brief observations in which unique or unusual features are mentioned. The lack of a detailed description means that the synonymy and the illustrations take on even more importance than usual and should be done as well as possible.

1. Synonymy

The synonymy is an abbreviated resume of the nomenclatural history of a given species. It includes data about the original designation, subsequent examination and the various name changes the species has undergone.

It is not wise to include all possible references in the synonymy. Those references which lack illustrations or detailed descriptions should be avoided. In the case of a local or regional study it may be misleading to include references from studies of far distant regions or of rocks of distinctly different ages. A synonymy which has been made without consideration of geologic age, location of study or thoroughness of the study loses its value.

Even the incorporation of references from articles containing detailed descriptions and good illustrations may be misleading because each presentation of a species is subjective to a greater or lesser extent. Study of the literature may give a good idea of the species but is not a substitute for study of the original types of the investigator whose works are to be cited in the synonymy. A synonymy which includes references to type material has the greatest value because it is based on a direct visual comparison of specimens. From it the reader can obtain a good idea of the author's concept of that particular species.

A synonymy of this sort need not be very extensive and can follow the form used by the 'English school' (Williamson, Brady, Heron-Allen and Earland) and recommended by Richter (1948) in his fundamental work on the subject. According to Richter the entries in the synonymy should be arranged in the following manner. The first notation is the year of publication which is followed by the species name appearing exactly as it does in the original work. This is followed by the name of the author, an abbreviated title of the work and the appropiate page, plate and figure numbers of the illustrations. Richter suggested that space be saved by abbreviating the title of the article as much as possible and that the jouunal name be omitted because most of those who work with foraminefera are familar with the articles and monographs by title and author rather than journal; who would doubt that the notation 'd'Orbigny, Cuba' refers to d'Orbigny's work on the foraminifera of Cuba whose full notation would have to be 'd'Orbigny, in de la Sagra, Hist. Fis. Pol, Nat. Cuba'. As an aid to those who may not be familar with the works cited in the synonymy, the full reference should appear in the bibliography at the end of the article.

The first entry in the synonymy is the original description of the species.

A synonymy of this sort, based on the examination of type material, should be accompanied by some sort of indication of which and what kinds of types have been examined. Such an indication, a critical synonymy, is of much more value than the common synonymy which is no more than a list. Richter suggested several notations, the most significant of which will be mentioned here.

A reference preceeded by a 'V' (*vidi*) indicates that the author has seen type material. This sign appears in front of the year designation for an entry. Further symbols are needed to indicate the particular kind of type examined. An example will serve to illustrate the possibilities. Suppose that the author has investigated a foraminiferal fauna from the southern Brazilian coast and finds specimens which he identifies as *Quinqueloculina lamarckiana* d'Orbigny. He has topotype material from the original locales studied by d'Orbigny. The purpose of his study is systematic; he wants to present a list of the species which live in this area and compare the fauna to other studies in the same and adjacent areas. For this reason he will include in the synonymy only works on the Brazilian and Northern Argentine Coast. The author was able to compare his specimens with the actual specimens identified and illustrated in these works and discovered strong similarity between his specimens and particular ones in these other works. Consequently, he indicated this examination and similarity by including the references in the synonymy and placing the notation 'V.hy.' before the date of the appropriate entry. This signifies that he has seen (V) the hypotypes (hy) of the cited investigation.

Suppose also that he was unable to see the original material of all the investigations done in the area. The synonymy might appear as follows:

Quinqueloculina lamarckiana d'Orbigny

V.t. 1839 *Quinqueloculina lamarckiana* d'Orbigny–d'Orbigny, Cuba, p.189, pl. 11, figs. 14,15

V.hy. 1931 *Quinqueloculina lamarckiana* d'Orbigny–Cushman & Parker, Atlant. S. Amer., p. 4, pl. l, fig. 7.

V.hy. 1952 *Quinqueloculina lamarckiana* d'Orbigny–Carvalho & Chermont, Sào Paulo, p. 81, pl. l, fig. 2.

1954 *Quinqueloculina lamarckiana* d'Orbigny–Boltovskoy, San Blas, p. 259, pl. 20, fig. 4.

1955 *Quinqueloculina lamarckiana* d'Orbigny–Tinoco, Cabo Frio, p. 16, pl. 1, fig. 6.

V.hy. 1958 *Quinqueloculina lamarckiana* d'Orbigny–Boltovskoy, S. Brasil, p. 45, pl. 3, fig. 7.

This synonymy indicates that the author compared and verified his specimens with topotypes of the species and with hypotypes of the collections of Cushman & Parker (1931), Carvalho & Chermont (1952), and Boltovskoy (1958) and concluded that these specimens were identical. He did not see original material of Boltovskoy (1954) and Tinoco (1955). He has the option of either not including these references in the synonymy or including them without the 'V'. He selected the latter option because in one case the material comes from the same area, and because in the other the interpretation of the species by Boltovskoy is known from his 1958 work. Therefore the inclusion of both these works in the synonymy has a legitimate rationale.

The specimens included in the synonymy, although identical from a taxonomic viewpoint, are bound to exhibit some differences. These differences should be mentioned and discussed in the 'observations' section which appears directly below the synonymy. Above all, deviations from the holotypes, paratypes and in the case just cited, topotypes, should be discussed. It is important to remember that the type specimens are the standards of comparison.

For the systematic study of an area, the kind of synonymy just described has much more scientific value than does an enormous list composed only of collected bibliographic references representing specimens which have not been seen.

An abbreviated space-saving method of preparing a synonymy has been developed recently in which only the original bibliographic citation for a species is listed. This citation is given together with a list of all species at the end of an article rather than with the 'observations' section in the text. In the case of the above specimens, the designation *Quinqueloculina lamarckiana* d'Orbigny would be cited by the author and observations on his specimens would follow. At the end of the article, the original reference (d'Orbigny, 1839) would be given in a list along with the original citation of all other species identified by our author.

2. Illustration

The illustration of specimens is probably more important than the synonymy and the description from the point of view of correct interpretation of a species. With an illustration, or better yet several illustrations, it is possible to give a visual idea of the principal traits and range of variations within the taxon.

During the process of identification the investigator commonly makes schematic sketches of the specimens which will help in comparison with published illustrations. These sketches can be made by using the camera lucida or by visual recollection. The former method is better as it gives a more correct relationship between different parts of the test. These sketches, if retouched and shaded to give an indication of relief, may serve very well as final illustrations.

Even the best of drawings are subjective; that is, they represent not only what the drawer sees but to some extent what he wants to see. Because of the subjectivity introduced into drawings it is advisable to prepare illustrations using photographs. Aside from being impartial, photographs are generally cheaper and faster to prepare.

The appearance in recent years of the scanning electron microscope (Chapter III) has vastly improved the quality and ease of preparation of foraminiferal illustrations. Many of the problems which plagued the photographer when light microscopes were used (resolution, depth of field, etc.) are now overcome with the use of illumination by an electron beam.

The use of the scanning electron microscope to examine the microtopography of foraminiferal tests has yielded a great quantity of information. So revealing is the detail shown by this method that Hay (1971) suggested that a good SEM photomicrograph could provide more information of higher quality than a verbal description. We must qualify this conclusion with our observations that SEM photographs of specimens with little or no relief, or with transparent walls through which internal detail can be seen, are less revealing than examination with a light microscope. SEM viewing is most appropriate for sculptured specimens.

Methods and techniques for the use of the SEM were described by Hansen (1968a), Honjo & Okada (1968) and Sylvester-Brady (1969).

Unfortunately, not all institutions and few individuals can afford to buy or maintain one of these instruments and so we will discuss the photography of specimens using the conventional method of photomicrography with visible light. Those interested in the optical aspects of such photography are referred to Fournier (1956, 1957), Triebel (1958), McLean (1959), Pessagno (1960), Krivoborskij (1961), LeRoy & Crane (1964), Albani (1964) and Harland & Sutherland (1972). The preparation of specimens for photography will be treated below.

The goal of specimen preparation is to bring out all the morphological details of the test. Since these details are different in different species, a variety of methods must be utilized.

The specimen should be intact and clean. Cleaning can be accomplished by washing or by immersion in an ultrasonic bath, although this may damage delicate tests. Specimens of the same species may often be both translucent and opaque. The latter usually photograph better. If only translucent forms are present, they can be transformed into opaque specimens by heating over a burner for several seconds. Heating may also bring out morphological detail but it may damage fragile specimens.

The details of some tests become clearer when they are immersed in refractive liquids such as water or glycerine. The use of transmitted and/or polarized light sometimes yields better results than reflected light.

There are a variety of coloration techniques which can be used to accentuate external morphological detail (Chapter XIII,5). Simple colouring agents may be applied to the test; e.g. fuchsine or ink. The specimen is placed in the coloring agent for one or two minutes and then washed to remove excess colorant. Additional application of the agent may give the desired result. This technique may be used on aglutinated and calcareous specimens, although it does not generally work well on smooth calcareous specimens as the effectiveness of this method is based on the absorption of colouring agent by fine sediment trapped in the sutures and other depressions in the test.

If coloration does not work, the plating technique described by Triebel (1947) may help. The test is inmersed for several seconds in a 5–20% solution of silver nitrate ($AgNO_3$), removed, placed on filter paper and heated. Kugler (1957) suggested that this heating be done on platinium foil. If the specimen turns dark brown or almost black the plating is successful. If it has a brilliant metallic sheen, the nitrate solution was too concentrated.

Cummings (1956) described a similar silver plating treatment. He suggested placing the specimen in a 10% silver nitrate solution for 2–5 minutes and then putting the specimens in common photographic developer. The intensity of coloration is a function of the time left in the silver nitrate solution and in the developer. The silver plating can be removed by bathing the specimen in warm water to which a little potassium ferrocynide ($K_3Fe(CN)_6$) has been added.

Agglutinated species can be treated with ammonium chloride powder (NH_4Cl). A gram of this chemical is placed in the bulging part of a pipette and the open end is plugged with a gum ball to which is attached the specimen. The pipette is heated over an alcohol burner and the volitalized chemical produces a powder which covers the specimen with a white patina, nicely outlining the relief of the test.

After colouring the specimen in some way it is necessary to mount it against some background in the proper position for photography. The foraminiferal slide generally does not provide a good photographic background because the paper fibers are too dense and reflect light in undesirable ways. The walls of the cavity on the slide may impede good illumination. The best background is a clean smooth glass slide. For a black background the slide alone can be used; if a white background is desired a sheet of clean white paper is placed beneath the slide. To avoid reflections off the glass slide a polaroid filter can be used. The specimen can be oriented properly on the slide using plasticine, wax or paraffin. Unfortunately these substances often soften under the heat of the lamp; Wood metal (Chapter XIII,5) will not.

Although the treatments described above will usually result in a better illustrated specimen, they are generally irreversible processes and the specimen becomes permanently altered.

Untreated specimens can be photographed but the results generally show little of the sculpture and relief of the test. These photos, if prepared on matte paper, may be retouched using a fine pen.

Despite the distinct advantages of illustration by good photography, there is one distinct advantage of a drawing. The illustrator can emphasize details of the test that are visible only to the trained and experienced observer of foraminifera and can de-emphasize the less important traits. This subjectivity, which may be a detriment, can also result in an illustration showing the pertinent detail of a specimen better than a photograph. Wallis (1955) offered several suggestions for the production of good drawings using the microscope.

In each illustration of foraminifera, whether they be drawings or photographs, the light source should appear to come from the upper left corner at a vertical angle of 45°.

The magnification of the illustration should be indicated either by a bar scale on the plate or a written notation accompanying the plate description.

It is always better, if time and space permit, to illustrate several specimens of the same species in order to give an idea of the variability of the species.

XVIII. UNRESOLVED PROBLEMS

The study of foraminifera has produced a vast set of unanswered questions. The enumeration of all the problems that require solution would use too much space; however, an abbreviated list of some fundamental problems is in order.

The initiation of the study of foraminifera on a large scale occurred only with the discovery of their stratigraphic utility in mineral and fossil fuel exploration. Because of this, far more attention has been given to the test than to the protoplasmic constituents of the organism. Consequently, one of the more urgent tasks in the future study of foraminifera is the investigation of living specimens, their reproductive cycles, their physiology, ontogeny and ecology. These studies must be conducted in laboratory cultures and in the natural habitat. Laboratory culture systems must be devised which are good replicas of nature. Long term studies of the seasonal occurrences of particular benthonic species should be undertaken. Most seasonal studies to date have dwelt with an entire fauna rather than with particular species. The effects of certain environmental factors on distribution, abundance, size and reproductive health of the foraminifera have not been well studied. There have been numerous studies on the effects of temperature, salinity, depth and substrate, but few which have examined the effects of trace element concentration, illumination, tidal currents and various pollutants.

Another problem of great importance which requires solution in the near future is the chaotic classification system. The enormous burden of synonyms carried by the system makes work with the foraminifera increasingly difficult. The addition of useless names to the nomenclature continues and there is a danger that the foraminifera will lose their importance unless a rather drastic curtailment of this excessive naming occurs. The existing nomenclature needs to be better organized. This will require restudy of original type material using new methods. A step in this direction has been taken by the extensive restudy and reclassification done by Loeblich & Tappan (1964) and Rauser-Chernoussova & Furssenko (1959). Renewed study of the protoplasmic parts of the organism, particularly of the cytoplasm and pseudopodia, should aid in the discovery of new discriminatory traits at the higher taxonomic levels. Increased study of intraspecific variation and wall structure should also help in classification. Of great help in this search for new classificatory traits will be the widespread use of the transmission and scanning electron microscope.

There is no doubt that numerous families, genera and even species should be redescribed.

Another problem requiring further resolution, although less urgently than the above, concerns our biased knowledge of the geographic and stratigraphic distribution of foraminifera. Studies of the distribution of Recent foraminifera have been concentrated in certain geographic regions and types of environments at the expense of others; e.g., the well studied Gulf of Mexico versus the west coast of South

America and the Continental Shelf versus the abyssal zones. Studies of fossil foraminifera have been concentrated in certain areas and stratigraphic zones while others are relatively unstudied; e.g. Europe versus South America and Africa; the Cretaceous and Neogene versus the Cambrian, Ordovician, Silurian and Triassic. Each investigation, even if it be of a preliminary and qualitative character, of one of these lesser known areas, will constitute a significant contribution to our knowledge of foraminifera. We hope that these blank spots in our knowledge of the temporal and spatial distribution of foraminifera will be gradually filled.

In the first edition of this book one of the suggestions for future study was in the application of foraminiferal data to geologic and oceanographic problems. The intervening 12 years have seen an extraordinary use of foraminifera in the study of paleoclimates, water mass definition and other such problems (see Chapter XIV). This has been a significant advance in the study and use of foraminifera.

A large number of investigations have been devoted to the vertical and horizontal zonation of benthonic foraminifera (Chapter IV). Nevertheless, many problems remain. There is no general agreement among many investigators regarding zonation. This is an area which requires additional study because its understanding provides a key to the correct interpretation of faunal changes that are encountered in various other kinds of studies.

Apart from these problems of a general nature, there are a large number of specific questions that remain unanswered or are poorly stated. We will list but a few that seem significant to us: (a) What is the nature and function of the vesicular system? (b) Can the foraminifera feed osmotically? (c) Are there differences between the cytoplasm of planktonic and benthonic foraminifera? (d) Do microspheric and megalospheric generations have different environmental requirements? Do they have different coiling ratios? (e) What conditions facilitate reproduction? (f) What is the function of the fistulose appendages in *Polymorphina*? (g) How do unilocular tests grow? (h) What is the function of pores in those foraminifera which have an imperforate inner organic membrane? (i) What is the function of sieve plates? (j) What is the function of tooth plates? (k) What is the function of a canal system? (l) What is the productivity of benthonic and planktonic foraminifera? (m) How do foraminifera fit into the trophic structure of marine communities? What is their role in carbon and nutrient recycling? (n) What are the effects of sample preparation on data obtained from foraminiferal concentrations in plankton tows? (o) Does the secretion of calcite by foraminifera occur in oxygen isotopic equilibrium with sea water? (p) What are the ecological causes of monstrous or pathological specimens?

These are but a few of the broad and specific problems that face the foraminiferal investigator today. As with all investigations into natural phenomena, our solution of particular problems will always create new questions.

'I am thinking ... how nothing is ever solved finally. In every age, from every angle, we are facing the same set of natural phenomena .... We make idolatrous attempts to encase them in a conceptual frame. And all the time they change under our very nose'.

Prospero's Cell (Durrell)

BIBLIOGRAPHY

Abrard, R. 1928. Contribution a l'étude de l'evolution des Nummulites: Soc. Géol. France, Bull., (4), 28: 161–182.

Accorinti, J. 1960. Cultivo unialgal y masivo de '*Scendesmus obliquus*' Turp. Ktz. Técnicas de obtención: Mus. Argent. Cienc. Nat., Com., Bot. 1, 9: 21–29.

Adams, G. C. 1957. A study of the morphology and variation of some Upper Lias Foraminifera: Micropaleontol., 3, 3: 205–226.

——. 1962. *Alveolina* from the Eocene of England: *Ibid.*, 8, 1: 45–54.

Adams, T. D. & Frampton, J. 1965. A note on some Recent Foraminifera from northwest Iceland: Cushman Found. Foram. Res., Contr., 16, 2: 55–59.

Adegoke, O. S., Dessauvagie, T. F. J. & Kogbe, C. A. 1971. Planktonic foraminifera in Gulf of Guinea sediments: Micropaleontol., 17, 2: 197–213.

Adshead, P. C. 1966. Observations on living planktonic Foraminifera in cultures [abstr.]: Amer. Assoc. Petroleum Geol. Bull., 50, 3: 603–604.

——. 1967. Collection and laboratory maintenance of living planktonic foraminifera: Micropaleontol., 13, 1: 32–40.

Akers, W. H. 1971. Estuarine foraminiferal associations of the Beaufort area, North Carolina: Tulane Stud. Geol. Paleontol., 8, 3: 147–165.

Akpati, B. N. 1969. A new technique for serial sectioning of foraminifera: Jour. Paleontol., 43, 4: 1046–1048.

Albani, A. D. 1964. Photography of microfossils: Micropaleontol., 10, 3: 396–398.

——. 1965. The Foraminifera in a sample dredged from the vicinity of Salisbury Island, Durban Bay, South Africa: Cushman Found. Foram. Res., Contr., 16, 2: 60–66.

——. 1968a. Recent Foraminiferida from Port Hacking, New South Wales: *Ibid.*, 19, 3: 85–119.

——. 1968b. Recent foraminifera of the central coast of New South Wales: AMSA Handbook, 1: 1–37.

——. 1970. A foraminiferal fauna from the eastern continental shelf of Australia: Cushman Found. Foram. Res., Contr., 21, 2: 71–77.

Alldredge, A. L. & Jones, B. M. 1973. *Hastigerina pelagica:* Foraminiferal habitat for planktonic dinoflagellates: Mar. Biol., 22, 2: 131–135.

Allen, G. P. & Pujos-Lamy, A. 1970. Application de l'analyse factorielle (Mode-Q) a l'étude micropaléontologique d'une carotte marine: Soc. Géol. France, Bull., (7), 12, 5: 196–225.

Alpatov, V. V. 1957. Levizna-pravizna v stroenii rastitel'nykh i zhivotnykh organizmov: Moskov. O-vo Isp. Prir., Bjull., 62, 5: 19–27.

Amicis, G. 1895. Sopra alcune forme nuove di Foraminiferi del Pliocene Inferiore: Soc. Toscana Sci. Nat., Atti, Mem., 14: 18–30.

Anderson, G. J. 1963. Distribution patterns of Recent Foraminifera of the Bering Sea: Micropaleontol., 9, 3: 305–317.

Anderson, J. B. 1968. Ecology of foraminifera from Mississippi Sound and surrounding waters: Jour. Alabama Acad. Sci., 39, 4: 261–269.

Anderson, L. & Hoffman, E. 1963. Concentration of microfaunal samples by the use of a syntron: Micropaleontol., 9, 1: 105–106.

Angell, R. W. 1967a. The process of chamber formation in the foraminifer *Rosalina floridana* (Cushman): Jour. Protozool., 14, 4: 566–574.

Angell, R. W. 1967b. The test structure and composition of the foraminifer *Rosalina floridana:* Jour. Protozool., 14, 2: 299–307.

——. 1967c. Test recalcification in *Rosalina floridana* (Cushman): Cushman Found. Foram. Res., Contr., 18, 4: 176–177.

——. 1971. Observations on gametogenesis in the foraminifer *Myxotheca:* Jour. Foram. Res., 1, 1: 39–42.

Arnal, R. E. 1957. Some occurrences of abnormal Foraminifera: Compass, 32: 185–194.

——. 1958. Rhizopoda from the Salton Sea, California: Cushman Found. Foram. Res., Contr., 9, 2: 36–45.

——. 1961. Limnology, sedimentation, and microorganisms of the Salton Sea, California: Geol. Soc. Amer., Bull., 72: 427–478.

Arnold, Z. M. 1948. A new Foraminifera belonging to the genus *Allogromia:* Amer. Micr. Soc., Trans., 67, 3: 231–235.

——. 1953a. An introduction to the study of movement and dispersal in *Allogromia laticollaris* Arnold: Cushman Found. Foram. Res., Contr., 4, 1: 15–21.

——. 1953b. Paleontology and study of variation in living Foraminifera: *Ibid.*, 4, 1: 24–26.

——. 1954a. *Discorinopsis aguayoi* (Bermúdez) and *Discorinopsis vadescens* Cushman & Brönnimann: A study of variation in cultures of living Foraminifera: *Ibid.*, 5, 1: 4–13.

——. 1954b. Variation and isomorphism in *Allogromia laticollaris:* a clue to foraminiferal evolution: *Ibid.*, 5, 2: 78–87.

——. 1954c. Culture methods in the study of living Foraminifera: Jour. Paleontol., 28: 404–416.

——. 1954d. A note on Foraminiferan sieve-plate: Cushman Found. Foram. Res., Contr., 5, 2: 77.

——. 1955a. Life history and cytology of the Foraminifera *Allogromia laticollaris:* Univ. Calif. Publ. Zool., 61, 4: 167–252.

——. 1955b. An unusual feature of miliolid reproduction: Cushman Found. Foram. Res., Contr., 6, 3: 94–96.

——. 1956. The contribution of Jean Le Calvez to the study of the Foraminifera: *Ibid.*, 7, 1: 1–14.

——. 1962. A high-speed plankton sampler for manual operation: Micropaleontol., 8, 4: 515–518.

——. 1964. Biological observations on the foraminifer *Spiroloculina hyalina* Schulze: Univ. Calif. Publ. Zool., 72: 1–93.

——. 1965a. An improved instrument for sectioning microfossils: *in*, Handbook of Paleontological Techniques: W. H. Freeman, San Francisco,: 232–238.

——. 1965b. A vacuum oven for impregnating microfossils: *Ibid.*,: 262–263.

·. 1966a. Observations on the sexual generation of *Gromia oviformis* Dujardin: Jour. Protozool., 13, 1: 23–27.

——. 1966b. A laboratory system for maintaining small-volume cultures of foraminifera and other organisms: Micropaleontol., 12, 1: 109–118.

——. 1967a. Biological observations on the foraminifer *Calcituba polymorpha* Roboz: Arch. Protist., 110: 280–304.

——. 1967b. Biological clues in interpreting the biogeography of the foraminifer *Nubecularia lucifuga* Defrance: Inter. Conf. Trop. Oceanogr., Miami, Proc.

——. 1968. The uniparental species concept in the Foraminifera: Trans. Amer. Micr. Soc., 87, 4: 431–442.

——. 1974. Field and laboratory techniques for the study of living foraminifera: *in*, R. H. Hedley & C. G. Adams (eds.), Foraminifera, 1, Academic Press, London: 153–206.

Arrhenius, G. 1950. Foraminifera and deep sea stratigraphy: Science, 3: 288.

——. 1952. Sediment cores from the East Pacific: Swedish Deep-Sea Expd., Repts., 5, 1: 6–227.

Atkinson, K. 1969. The association of living Foraminifera with algae from the littoral zone, south Cardigan Bay, Wales: Jour. Nat. Hist., 3, 4: 517–542.
——. 1971. The relationship of Recent foraminifera to the sedimentary facies in the turbulent zone, Cardigan Bay: *Ibid.*, 5, 4: 385–439.
Aves, C. 1958. Foraminiferal fixatives and preservatives: Gulf Coast Assoc. Geol. Soc., Field Trip Guidebook: 48–56.
Avnimelech, M. 1952. Revision of the tubular Monothalamia: Cushman Found. Foram. Res., Contr., 3, 2: 61–68.
Awerinzew, S. 1901. O strukture izvesti v rakovinakh kornenozhek: Sankt-Petersb. O-vo Estestv., Trudy: 32.
——. 1903. Über die Struktur der Kalkschalen mariner Rhizopoden: Z. Wiss. Zool., 74: 478–490.
——. 1908. Otchet zavedujushchego Murmanskoj biologichiskoj stanthiej za 1906 g: Sankt-Petersb. O-vo Estestv., Trudy: 38, 2.
——. 1911. Zur Foraminiferen-Fauna des Sibirischen Eismeeres: Acad. Imp. Sci. Sankt-Petersb. Mem., 8, 29: 1–28.
Ayala-Castañares, A. 1963. Sistemática y distribución de los Foraminíferos recientes de la Laguna de Terminos, Campeche, México: Méx. Univ. Nac., Inst. Geol., Bol. 67, 3: 1–130.
Ayala-Castañares, A. & Segura, L. R. 1968. Ecología y distribución de los Foraminíferos recientes de la Laguna Madre, Tamaulipas, México: *Ibid.*, Bol. 87: 1–89.
Azor, W. 1958. The use of a large capacity portable pump for plankton sampling with notes on plankton patchiness: Jour. Mar. Res., 16, 2: 158–173.
Bandy, O. L. 1953. Ecology and paleoecology of some California Foraminifera; Part I. The frequency distribution of Recent Foraminifera off California: Jour. Paleontol., 27, 1: 161–182.
——. 1954. Distribution of some shallow water Foraminifera in the Gulf of Mexico: U.S. Geol. Surv. Prof. Pap. 254-F: 125–141.
——. 1955. Evidence of displaced Foraminifera in the Pursima Formation of the Halfmoon Bay area, California: Cushman Found. Foram. Res., Contr., 6, 2: 77–79.
——. 1956. Ecology of Foraminifera in north-eastern Gulf of Mexico: U.S. Geol. Surv. Prof. Pap. 724-G: 179–204.
——. 1960a. General correlation of foraminiferal structure with environment: Inter. Geol. Congr., 21st, Copenhagen, 22: 7–19.
——. 1960b. The geologic significance of coiling ratios in the Foraminifer *Globigerina pachyderma* (Ehrenberg): Jour. Paleontol., 34, 4: 671–681.
——. 1960c. Planktonic foraminiferal criteria for paleoclimatic zonation: Tohoku Univ. Sci. Rep., Geol., Spec. Vol. 4: 1–8.
——. 1961. Distribution of Foraminifera, Radiolaria and Diatoms in sediments of the Gulf of California: Micropaleontol., 7, 1: 1–26.
——. 1963a. Larger living Foraminifera of the continental borderland of southern California: Cushman Found. Foram. Res., Contr., 14, 4: 121–126.
——. 1963b. Dominant paralic Foraminifera of southern California and the Gulf of California: *Ibid.*, 14, 4: 127–134.
——. 1964. Foraminiferal trends associated with deep-water sands, San Pedro and Santa Monica basins, California: Jour. Paleontol., 38: 138–148.
——. 1972a. Origin and development of *Globorotalia* (*Turborotalia*) *pachyderma* (Ehrenberg): Micropaleontol., 18, 3: 294–318.
——. 1972b. Environmental significance of variations in *Globigerina bulloides* d'Orbigny: Inter. Geol. Congr., 24th, Montreal, sec. 7: 3.
Bandy, O. L. & Arnal, R. E. 1957. Distribution of Recent Foraminifera off the west coast of Central America: Amer. Assoc. Petrol. Geol., Bull., 41, 9: 2037–2053.
——. 1960. Concepts in foraminiferal paleoecology: *Ibid.*, 44, 12: 1921–1932.

Bandy, O. L., Casey, R. E. & Wright, R. C. 1971. Late Neogene planktonic zonation, magnetic reversals, and radiometric dates, Antarctic to the tropics: Biol. Antarctic Seas IV, Antarctic Res. Ser., 15: 1–26.
Bandy, O. L. & Chierici, M. A. 1966. Depth-temperature evaluation of selected California and Mediterranean bathyal Foraminifera: Mar. Geol., 4, 4: 254–271.
Bandy, O. L. & Echols, R. J. 1964. Antarctic foraminiferal zonation: Biol. Antarctic Seas, Antarctic Res. Ser., 1: 73–91.
Bandy, O. L., Ingle, J. C. & Resig, J. M. 1964a. Foraminiferal trends, Laguna Beach outfall area, California: Limnol. Oceanog., 9, 1: 112–123.
——. 1964b. Foraminifera, Los Angeles County outfall area, California: *Ibid.*, 9, 1: 124–137.
——. 1964c. Facies trends, San Pedro Bay, California: Soc. Amer., Bull., 75, 5: 403–423.
——. 1965a. Foraminiferal trends, Hyperion outfall, California: Limnol. Oceanog., 10, 3: 314–332.
——. 1965b. Modification of foraminiferal distribution by the Orange County outfall, California: Ocean Sci. and Ocean Eng., June, 1965: 55–76.
Bandy, O. L. & Rodolfo, K. S. 1964. Distribution of foraminifera and sediments, Peru-Chile trench area: Deep-Sea Res., 11, 5: 817–837.
Bandy, O. L. & Theyer, F. 1971. Growth variation in *Globorotalia pachyderma* (Ehrenberg): Antarctic Jour., 6, 5: 172–174.
Banerji, R. K., Schafer, C. T. & Vine, R. 1971. Environmental relationships and distribution of planktonic foraminifera in the Equatorial and Northern Pacific waters: Bedford Inst., Atlant. Oceanog. Lab., Rept 71-7: 1–65.
Banner, F. T. 1970. A synopsis of the Spirocyclinidae: Rev. Españ. Micropaleontol., 2: 243–290.
——. 1971. A new genus of the Planorbulinidae: an endoparasite of another foraminifer: *Ibid.*, 3, 2: 113–128.
Banner, F. T. & Blow, W. H. 1960. Some primary types of species belonging to the superfamily Globigerinaceae: Cushman Found. Foram. Res., Contr., 11, 1: 1–41.
Banner, F. T., Sheehan, R. & Williams, E. 1973. The organic skeletons of rotaline foraminifera: a review: Jour. Foram. Res., 3, 1: 30–42.
Banner, F. T. & Williams, E. 1973. Test structure, organic skeleton and extrathalamous cytoplasm of *Ammonia* Brünnich: *Ibid.*, 3, 2: 49–69.
Barash, M. S. 1970. Planktonnye foraminifery v osadkakh severnoj Atlantiki: Nauka, Moscow, 94pp.
——. 1971. Paleoklimaticheskie rekonstrukthii po planktonnym foraminiferam: Okeanol., 11, 6: 1049–1056.
Barash, M. S., Nikolaev, S. D. & Bljum, N. S. 1973. Paleotemperaturnyj analiz trekh kolonok osadkov severnoj Atlantiki: *Ibid.*, 13, 6: 1052–1058.
Barbieri, F. & Medioli, F. 1969. Distribution of foraminifera on the Scotian Shelf, Canada: Riv. Ital. Paleontol. Stratigr., 75, 4: 849–865.
Barchuk, I. F., Venhlins'kyi, I. V., Bazarov, D. O., Holyshkin, V. I. & Ohorodnyk, A. K. 1970. Microelementy v cherepashkakh foraminifer ta ostrakod: Dupov. Akad. Nauk, URSR, 32, 2: 99–101.
Barker, R. W. 1960. Taxonimic notes on the species figured by H. B. Brady in his "Report on the Foraminifera dredged by H.M.S. *Challenger* during the years 1873–1876": Soc. Econ. Paleontol. Mineral., Spec. Publ. No. 9, Tulsa, Oklahoma, 238 pp.
Barnard, T. 1954. *Hantkenina alabamensis* Cushman and some related forms: Geol. Mag., 91, 5.
Barnard, T. & Banner, F. T. 1953. Arenaceous Foraminifera from the Upper Cretaceous of England: Geol. Soc. London, Quart. Jour., 109, 2: 173–216.
Bartenstein, H. 1938. Foraminiferen der meerishen und brackischen Bezirke des Jade-Gebietes: Senckenberg., 20: 386–412.

Bartenstein, H. 1939. Agglutinierende Brackwasser-Foraminiferen in Quell-Tumpeln Mitteldeutschlands: *Ibid.*, 21: 374–378.

Bartenstein, H. & Brand, E. 1938. *Jadammina polystoma* n.g., n.sp.: *Ibid.*, 20: 381–385.

Bartenstein, H. & Heinemann, W. 1954. Brackwasser-Foraminiferen im Oberen Aquitan des Mittelrhein-Gebietes: *Ibid.*, 35: 23–35.

Bartlett, G. A. 1964a. A preliminary study of Foraminifera distribution on the Atlantic Continental Shelf, southeastern Nova Scotia: Canada Geol. Surv. Pap. 64-5, 19 pp.

——. 1964b. Benthonic foraminiferal ecology in St. Margaret Bay and Mahone Bay, southeast Nova Scotia: Bedford. Inst. Oceanog., Rpt. 64-8, 162 pp.

——. 1966. Foraminifera distribution in Tracadie Bay, Prince Edward Island: Canada Geol. Surv. Pap. 66-20, 54 pp.

Bartlett, G. A. & Hamdan, A. R. A. 1972. The Canadian Atlantic continental margin-biostratigraphy, paleoecology and paleooceanography from Cretaceous to Recent: Inter. Geol. Congr., 24th, Montreal, (8): 3–15.

Bartlett, G. A., Molinsky, L. & Slessor, D. K. 1972. Quaternary history of the Gulf of St. Lawrence [abstr.]: *Ibid.*, Resumes: 365.

Bary, B. M. 1959. Species of zooplankton as a means of identifying different surface waters and demonstrating their movement and mixing: Pacific Sci., 13: 14–54.

——. 1963. Distribution of Atlantic pelagic organisms in relation to surface water bodies: Roy. Soc. Canada, Spec. Publ., 5: 516–567.

Basov, B. A. 1961. Sostav i raspredelenie foraminifer v donnykh osadkakh rajona zemli Frantha-Iosifa: N.I.I.G.A., Trudy, 124, 2: 61–65.

Basov, V. A. & Slobodin, V. Ja. 1965. Kompleksy foraminifer iz sovremennykh i pozdnekaj-nozojskikh otlozhenij zapadnoj chasti sovetskoj Arktiki: *Ibid.*, 143: 190–210.

Batsch, A. J. G. K. 1791. Sechs Kupfertafeln mit Conchylien des Seesandes, gezeichnet und gestochen: Jena.

Bé, A. W. H. 1959a. Ecology of Recent planktonic Foraminifera: Pt. 1 – Areal distribution in the western North Atlantic: Micropaleontol. 5, 1: 77–101.

——. 1959b. A method for rapid sorting of Foraminifera from marine plankton samples: Jour. Paleontol., 33: 846–848.

——. 1960a. Ecology of Recent planktonic Foraminifera: Pt. 2 – Bathymetric and seasonal distributions in the Sargasso Sea of Bermuda: Micropaleontol., 6, 4: 373–392.

——. 1960b. Some observations on Arctic planktonic Foraminifera: Cushman Found. Foram. Res., Contr., 11, 2: 64–68.

——. 1965. The influence of depth on shell growth in *Globigerinoides sacculifer* (Brady): Micropaleontol., 11, 1: 81–97.

——. 1968. Shell porosity of recent planktonic Foraminifera as a climatic index: Science, 161: 881–884.

——. 1969a. Planktonic Foraminifera, *in*, Distribution of selected groups of marine invertebrates in waters south of 35° S latitude: Antarctic Map Folio Ser., Amer. Geogr. Soc., Folio 11: 9–12.

——. 1969b. Microstructural evidence of the close affinity of *Globigerinella* Cushman to *Hastigerina* Thompson: Inter. Conf. Plank. Microfossils, 1st, Geneva, Proc., 1: 89–91.

——. 1972. *Orbulina universa* d'Orbigny – foraminiferal indicator species of latitudinal occurrences in the Indian Ocean: Inter. Geol. Congr., 24th, Montreal, (7): 4.

——. An ecological, zoogeographical and taxonomic review of Recent planktonic foraminifera: (in press).

Bé, A. W. H. & Anderson, O. R. 1973. On the preservation of planktonic foraminifera and other calcareous plankton: SCOR/UNESCO WG 23 Manual, "Laboratory Methods in Marine Zooplankton", Chap. 8, "Microplankton", 15 pp.

Bé, A. W. H., Croce, N. D., Boudillon, A., de Decker, A., Kimer, B., Hagmeier, E. & Bognov,

B. 1968. Smaller mesoplankton. Report of Working Party No. 2; *in*, Zooplankton Sampling: UNESCO, Geneva: 153–159.
Bé, A. W. H., Ewing, M. & Linton, L. W. 1959. A quantitative multiple opening-and-closing plankton sampler for vertical towing: Jour. Conseil Inter. Explor. Mer, 25, 1: 36–46.
Bé, A. W. H. & Hamlin, W. M. 1967. Ecology of Recent planktonic Foraminifera. Part 3 - Distribution in the North Atlantic during the summer of 1962: Micropaleontol., 13, 1: 87–106.
Bé, A. W. H., Harrison, S. M. & Lott, L. 1973. *Orbulina universa* d'Orbigny in the Indian Ocean: *Ibid.*, 19, 2: 150–192.
Bé, A. W. H. & Hemleben, C. 1970. Calcification in a living planktonic foraminifer, *Globigerinoides sacculifer* (Brady): Neues Jahrb. Geol. Palaeontol., Abh., 134, 3: 221–234.
Bé, A. W. H., Jongebloed, W. L. & McIntyre, A. 1969. X-ray microscopy of recent planktonic foraminifera: Jour. Paleontol., 43, 6: 1384–1396.
Bé, A. W. H. & Tolderlund, D. S. 1971. Distribution and ecology of living planktonic foraminifera in surface waters of the Atlantic and Indian Oceans, *in*, The Micropaleontology of the Oceans, Cambridge Univ. Press: 105–149.
Bé, A. W. H., Vilks, G. & Lott, L. 1971. Winter distribution of planktonic foraminifera between the Grand Banks and the Caribbean: Micropaleontol., 17, 1: 31–42.
Beard, J. H. 1973. The Pleistocene-Holocene boundary and Wisconsinan climatic substages, *in*, The Wisconsinan Stage: Geol. Soc. Amer., Mem. 136: 277–316.
Beccarius, J. B. 1731. De Bononiensi arena quadam: Com. Bonon. Sci. Art. Inst., 1.
Beerbower, J. R. & Jordan, D. 1969. Application of information theory to paleontologic problems: Jour. Paleontol., 43, 5: 1184–1198.
Beers, J. R. & Stewart, G. L. 1970. Numerical abundance and estimated biomass of microplankton, *in*, The ecology of the plankton off La Jolla, California in the period April through September, 1967, pt. 6, Scripps Inst. Oceanogr., Bull., 17: 67–87.
Beers, J. R., Stewart, G. L. & Strickland, J. D. H. 1967. Pumping system for sampling small Plankton: Jour. Fish. Res. Bull. Canada, 24, 8: 1811-1818.
Behm, H. J. & Grekulinski, E. F. 1958. The ecology of Foraminifera of Main and Richmond Creeks, Staten Island, New York: Staten Island Inst. Art. Sci., Proc., 20: 53–66.
Beklemishev, K. V., Nejman, A. A., Parin, N. V. & Semina, G. I. 1973. Estestvennye uchastki morskoj sredy obitanija s biothenoticheskoj tochki zrenija: VNIOR. 84: 7–32.
Belyaev, G. M., Vinogradova, N. G. & Filatova, Z. A. 1960. Issledovanie donnoj fauny glubokovodnoj vpadiny juzhnoj chasti Tikhogo okeana: Akad. Nauk SSSR, Inst. Okeanol., Trudy, 41: 106–122.
Belyaeva, N. V. 1960. Raspredelenie foraminifer v zapadnoj chasti Beringova morja: *Ibid.*, 32: 158–170.
——. 1961. Ehkologija planktonnykh foraminifer Indijskogo okeana: Mosk. Obshch. Isp. Prir., Otd. Geol., 36, 6: 122–123.
——. 1962a. Raspredelenie planktonnykh foraminifer v tolshche vod Indijskogo okeana: *Ibid.*, 37, 3: 98–101.
——. 1962b. Raspredelenie planktonnykh foraminifer v vodakh i osadkakh Indijskogo okeana: Akad. Nauk SSSR, Geol. Inst.: 1–18.
——. 1963. Raspredelenie planktonnykh foraminifer na dne Indijskogo okeana: Voprosy Micropal., 7: 209–222.
——. 1964. Raspredelenie planktonnykh foraminifer v vodakh i na dne Indijskogo okeana: Inst. Okeanol., Trudy, 68: 12–83.
——. 1966. Climatic and vertical zonality in the distribution of planktonic foraminifera in the sediments of the Pacific Ocean: 2nd Inter. Oceanol. Congr., Abstr.: 35–36.
——. 1968. Kolichestvennoe raspredelenie rakovin planktonnykh foraminifer v sovremennykh osadkakh Tikhogo okeana: Okeanol., 8, 1: 111–115.

Belyaeva, N. V. 1969a. Planktonnye foraminifery v osadkakh Mirovogo okeana: Vopr. Mikropal., 12: 168–172.
——. 1969b. Zakonomernosti raspredelenija planktonnykh foraminifer v vodakh i osadkakh Juzhnogo okeana: Okeonol., 9, 6: 1063–1070.
——. 1973. Osobennosti khimicheskogo sostava rakovin planktonnykh foraminifer: *Ibid.*, 13, 2: 303–306.
Benda, W. K. & Puri, H. S. 1962. The distribution of Foraminifera and Ostracoda off the Gulf Coast of the Cape Romano area, Florida: Gulf Coast. Assoc. Geol. Soc., Trans., 12: 303–341.
Benson, R. H. & Kaesler, R. L. 1965. The Benson-Kaesler Bolapipe Dredge: Micropaleontol. 11, 3: 369–372.
Berger, W. H. 1966. Distinguishing juvenile Foraminifera in plankton samples by using biological dyes: Jour. Paleontol., 40, 6: 1678–1682.
——. 1967. Foraminiferal ooze: solution at depths: Science, 156: 383–385.
——. 1968. Planktonic Foraminifera: selective solution and paleoclimatic interpretation: Deep-Sea Res., 15, 1: 31–43.
——. 1969a. Planktonic Foraminifera: basic morphology and ecologic implications: Jour. Paleontol., 43, 5: 1369–1383.
——. 1969b. Ecologic patterns of living Planktonic Foraminifera: Deep-Sea Res., 16, 1: 1–24.
——. 1970a. Planktonic Foraminifera: selective solution and the lysocline: Mar. Geol., 8, 2: 111–138.
——. 1970b. Planktonic Foraminifera: differential production and expatriation off Baja California: Limnol. Oceanogr., 15, 2: 183–204.
——. 1971a. Sedimentation of planktonic Foraminifera: Mar. Geol., 11, 5: 325–358.
——. 1971b. Planktonic Foraminifera: sediment production in an oceanic front: Jour. Foram. Res., 1, 3: 95–118.
——. 1973. Deep-Sea carbonates: Pleistocene dissolution cycles: *Ibid.*, 3, 4: 187–195.
Berger, W. H. & Parker, F. L. 1970. Diversity of planktonic foraminifera in deep-sea sediments: Science, 168: 1345–1347.
Berger, W. H. & Piper, D. J. W. 1972. Planktonic Foraminifera: differential settling, dissolution and redeposition: Limnol. Oceanogr., 17, 2: 275–287.
Berger, W. H. & Soutar, A. 1967. Planktonic Foraminifera: field experiment in production rate: Science, 156: 1495–1497.
——. 1970. Preservation of plankton shells in an anaerobic basin off California: Geol. Soc. Amer., Bull., 81, 1: 275–282.
Berggren, W. A. & Boersma, A. 1969. Late Pleistocene and Holocene foraminifera from the Red Sea, *in*, Hot Brines and Recent Heavy Metal Deposits in the Red Sea: a Geochemical and Geophysical Account: Springer, New York: 282–298.
Berggren, W. A., Briskin, M. & Gamper-Bravo, M. 1972. Late Pliocene-Pleistocene quantitative paleoclimatology of the North and South Atlantic [abstr.]: Inter. Geol. Congr., 24th, Montreal, Resumes: 534.
Bergmann, C. 1847. Über die Verhältnisse der Wärmeekonomie der Thiere zu irher Grösse: Göttinger Studien, I: 595–708.
Bermúdez, P. J. 1952. Estudio sistemático de los Foraminíferos rotaliformes: Min. Hidrocarb., Dir. Geol., 2, 4: 1–230.
Bermúdez, P. J. & Seiglie, G. A. 1963. Estudio sistemático de los foraminíferos del Golfo de Cariaco: Univ. Oriente, Inst. Oceanog., (B), 2, 2: 1–267.
Berthois, L. & Le Calvez, Y. 1959. Deuxième contribution a l'étude de la sédimentation dans le golfe de Gascogne: Rev. Trav. Inst. Pêch. Marit., 23, 3: 323–376.
——. 1960. Étude de la vitesse de chute de Foraminifères plactoniques dans un fluide comparativement a celle des grains de quartz: *Ibid.*, 24: 293–301.
Berthois, L., Le Calvez, Y. & Crosnier, A. 1968. Répartition des foraminifères dans les

sédiments du plateau continental camerounais; éssai d'écologie: Acad. Sci., C.R., (D), 266, 7: 660–662.
Berthois, L., Le Calvez, Y. & Guilcher, M. A. 1964. Influence des facteurs dynamiques sur la répartition des foraminifères dans le lagon de l'ile Mayotte (Comores): *Ibid.*, 258: 1292, 1294.
Berthold, W. 1971. Untersuchungen über die sexuelle Differenzierung die Foraminifere *Patellina corrugata* Williamson mit einem Beitrag zum Entwicklungsgang und Schalenbau: Arch. Protist., 113, 3: 147–184.
Betjeman, K. J. 1969. Recent foraminifera from the western continental shelf of Western Australia: Cushman Found. Foram. Res., Contr., 20, 4: 119–138.
Bettenstaedt, F. 1962. Evolutionsvorgänge bei fossilen Foraminiferen: Geol. Staatsinst. Hamburg, Mitt., 31, 1962.
Bhalla, S. N. 1968. Recent Foraminifera from Visakapatnam beach sands and its relation to the known foramgeographical provinces in the Indian Ocean: Bull. Nat. Inst. Sci. India, 38: 376–392.
——. 1970. Foraminifera from Marina beach sands, Madras, and faunal provinces of the Indian Ocean: Cushman Found Foram. Res., Contr., 21, 4: 156–163.
Bhatia, S. B. 1956. The study of variation in some smaller Foraminifera: Jour. Paleontol. Soc. India, 1: 142–152.
——. 1956. Recent Foraminifera from shore sands of western India: Cushman Found. Foram. Res., Contr., 7, 1: 15–24.
Bignot, G., Depeche, E. & Neumann, M. 1971. Utilization du microscope électronique a balayage dans l'étude des Foraminifères. II. Contribution a l'étude du genre *Peneroplis* Montfort: Rev. Micropaléont., 14, 2: 135–145.
Blackith, R. E. & Reyment, R. A. 1971. Multivariate Morphometrics: Acad. Press, London, 412 pp.
Blackmon, P. D. & Todd, R. 1959. Mineralogy of some Foraminifera as related to their classification and ecology: Jour. Paleontol., 33, 1: 1–15.
Blainville, H. M. D. 1834. Manuel d'actinologie ou de zoophytologie: Levrault, Paris, 644 pp.
Blair, D. G. 1965. The distribution of planktonic foraminifera in deep-sea cores from the Southern Ocean, Antarctica: Fla. State Univ., Dept. Geol., Sed. Res. Lab., Contr., 10: 1–141.
Blanc, F., Blanc-Vernet, L. & Le Campion, J. 1972. Application paleoécologique de la méthode d'analyse factorielle en composantes principales: interprétation des microfaunes de foraminifères planctoniques Quaternaires en Méditerranée, 1. Edute des espèces de Méditerranée occidentale: Tethys, 4, 3: 761–778.
Blanc-Vernet, L. 1958. Les milieux sédimentaires littoraux de la Provence occidentale (côte rocheuse). Relations entre la microfaune et la granulométrie du sédiment: Inst. Océanogr. Monaco, Bull., 1112: 1–45.
——. 1965a. Note préliminaire sur quelques dragages effectués au large de Marseille (Canyon de Planier): Rec. Trav. St. Mar. End., Bull. 36, 52: 185–190.
——. 1965b. Note sur la répartition des foraminifères au voisinage des côtes de terre Adélie (Antarctique): *Ibid.*, p. 191–205.
——. 1969. Contribution a l'étude des foraminifères de Mediterranée: *Ibid.*, Bull. 64-48: 1–281.
——. 1972. Données micropaléontologiques et paléoclimatiques d'après des sédiments profonds de Méditerranée, *in*, The Mediterranean Sea: A natural sedimentation laboratory: Dowden, Hutchinson & Ross, Inc., Stroudsberg, Pa.: 115–127.
Blanc-Vernet, L., Chamley, H. & Froget, C. 1969. Analyse paléoclimatique d'une carrote de Méditerranée nord-occidentale. Comparaison entre résultats de trois études: Foraminifères, Pteropodes, fraction sédimentaire issue du continent: Palaeogeogr., Palaeoclimatol., Palaeoecol., 6: 215–235.

Blow, W. H. 1969. Late Middle Eocene to Recent planktonic foraminiferal biostratigraphy: Inter. Conf. Plank. Microfossils, 1st, Proc., Geneva, 1: 199–475.
Bock, W. D. 1968. Two new species of foraminifera from the Florida Keys: Cushman Found. Foram. Res., Contr., 19, 1: 27–29.
——. 1969. *Thalassia testudinum*, a habitat and means of dispersal for shallow water benthonic foraminifera: Gulf Coast Assoc. Geol. Soc., Trans., 19: 337–340.
——. 1972. The use of foraminifera as indicators of subsidence in the Caribbean: Carib. Geol. Conf., Trans., 6: 439–440.
Bock, W. D. & Moore, D. R. 1968. A commensual relationship between a foraminifer and a bivalve mollusk: Gulf Res. Rept., 2, 3: 273–279.
——. 1971. The Foraminifera and micromollusks of Hogsty Reef and Serrana Bank and their paleoecological significance: Trans. 5th Carib. Geol. Conf., Geol. Bull. 5: 143–146.
Boersma, A. & Berggren, W. A. 1972. Paleogene evolution and biogeography of the benthonic foramaniferal genus *Uvigerina* [abstr.]: Inter. Geol. Congr., 24th, Montreal, Resumes: 218.
Bogdanowicz, A. K. 1950a. Tarkhanskie otlozhenija Kubani v svete izuchenija microfauny: VNIGRI, Trudy, nov. ser., 51: 113–128.
——. 1950b. Chokrakskie Foraminifery zapadnogo Predkavkaz'ja: *Ibid.*, p. 129–176.
——. 1952. Miliolidy i peneroplidy: *Ibid.*, 64, Iskop. Foram. SSSR, 64: 1–138.
——. 1973. O regenerathii miliolid: Voprosy Micropal., 16: 111–117.
Bold, H. C. 1942. The cultivation of algae: Bot. Rev., 8: 69–138.
Bolli, H. M. 1950. The direction of coiling in the evolution of some Globorotaliidae: Cushman Found. Foram. Res., Contr., 1, 3–4: 82–89.
——. 1951. Notes on the direction of coiling of Rotalid Foraminifera: *Ibid.*, 2, 4: 139–143.
——. 1952. Note on the disintegration of indurated siliceous rocks: Micropaleontologist, 6, 1: 46–48.
——. 1971. The direction of coiling in planktonic Foraminifera, *in* The Micropaleontology of the Oceans: Cambridge Univ. Press: 639–648.
Boltovskoy, E. 1953. Uber Zersetsungerscheinungen bei mikropaläontologischem Sammlungsmaterial: Paläontol. Zeit., 27, 3–4: 237–240.
——. 1954a. Beobachtungen über den Einfluss der Ernährung auf die Foraminiferenschalen: *Ibid.*, 28, 3–4: 204–207.
——. 1954b. Foraminíferos del golfo San Jorge: Inst. Nac. Cienc. Nat., Rev., Geol., 3, 3: 79–246.
——. 1954c. Foraminíferos de la bahia San Blas: *Ibid.*, 3: 247–300.
——. 1955. Recent Foraminifera from shore sands at Quequén, province of Buenos Aires and changes in the foraminiferal fauna at the north and south: Cushman Found. Foram. Res., Contr., 6, 1: 39–42.
——. 1956a. Contribución al conocimiento de las tecamebas del Río de la Plata: Acta Geol. Lilloana, 1: 299–314.
——. 1956b. Applications of chemical ecology in the study of the Foraminifera: Micropaleontol., 2, 4: 321–325.
——. 1956c. On the cyclical occurrence of Foraminifera: Dusenia, 7, 4: 211–218.
——. 1957a. Las anormalidades en los caparazones de Foraminíferos y el "Indice de regeneramiento": Ameghiniana, 1: 80–84.
——. 1957b. Los Foraminíferos del estuario del Río de la Plata y su zone de influencia: Inst. Nac. Cienc. Nat., Rev., Geol., 6: 1–77.
——. 1958a. The Foraminiferal fauna of the Rio de la Plata and its relation to the Caribbean area: Cushman Found. Foram. Res., Contr., 9, 1: 17–21.
——. 1958b. Problems in taxonomy exemplified by *Nonion affine* (Reuss): Micropaleontol., 4, 2: 193–200.
——. 1958c. On the organization of Foraminiferal collections: Cushman Found. Foram. Res., Contr., 9, 4: 99–102.

Boltovskoy, E. 1959a. Foraminíferos recientes del sur de Brasil y sus relaciones con los de Argentina e India del oeste: Serv. Hidr. Nav., Argentina, H. 1005: 1–120.
——. 1959b. Foraminifera as biological indicators in the study of ocean currents: Micropaleontol., 5, 4: 473–481.
——. 1959c. La corriente de Malvinas (un estudio en base a la investigación de Foraminíferos): Serv. Hidr. Nav., Argentina, H. 1015: 1–96.
——. 1961. Línea de convergencia subantártica en al Atlántico Sur y su determinación usando los indicadores biológicos-Foraminíferos: *Ibid.*, H. 1018: 1–35.
——. 1962a. Planktonic Foraminifera as indicators of different water masses in the South Atlantic: Micropaleontol., 8, 3: 403–408.
——. 1962b. Foraminíferos de la plataforma continental entre el cabo Santo Tomé y la desembocadura del Río de la Plata: Mus. Arg. Cienc. Nat., Rev., Zool., 6, 6: 249–346.
——. 1963a. The littoral foraminiferal biocoenosis of Puerto Deseado (Patagonia, Argentina): Cushman Found. Foram. Res., Contr., 14, 2: 58–70.
——. 1963b. Sobre las relaciones enter Foraminíferos y Turbelarios: Neotrópica, 9, 29: 55–60.
——. 1963c. Foraminíferos y sus relaciones con el medio: Mus. Arg. Cienc. Nat., Rev., Hidrobiol., 1, 2: 21–107.
——. 1964a. Distribución de los Foraminíferos planctónicos vivos en el Atlántico Ecuatorial, parte Oeste: Serv. Hidrogr. Nav., Argentina, H. 639: 1–54.
——. 1964b. Seasonal occurrences of some living Foraminifera in Puerto Deseado (Patagonia, Argentina): Jour. Cons. Perm. Inter. Explor. Mer., 29, 2: 136–145.
——. 1964c. Provincias zoogeográficas de América de Sur y su sector antártico según los Foraminíferos bentónicos: Bol Inst. Biol Mar., Mar del Plata, 7: 93–99.
——. 1965a. Twilight of foraminiferology: Jour. Paleontol., 39, 3: 383–390.
——. 1965b. Recolección de Foraminíferos en las aguas someras y su preparación: Centro Inves., Biol. Marina, Contr. Tech., 1: 1–11.
——. 1965c. Datos nuevos con respecto a la ubicación de la zona de convergencia subtropical/subantártica en base al estudio de los Foraminíferos planctónicos: Anais Acad. Brazil. Cienc., 37: 146–155.
——. 1965d. Beitrag zur Kenntnis der Jahreszyklen der Foraminiferen: Int Rev. Ges. Hydrobiol., 50, 2: 293–296.
——. 1966a. Methods of sorting Foraminifera from plankton samples: Jour. Paleontol., 40, 5: 1244–1246.
——. 1966b. The subtropical/subantarctic zone of convergence (Atlantic Ocean, western part): Serv. Hidrogr. Nav., Argentina, H. 640: 1–69.
——. 1966c. Depth at which Foraminifera can survive in sediments: Cushman Found. Foram. Res., Contr., 17, 2: 43–45.
——. 1966d. Zonación en las latitudes altas del Pacífico Sur según los Foraminíferos planctónicos vivos: Mus. Arg. Cienc. Nat., Rev., Hidrobiol. 2, 1: 1–56.
——. 1966e. Resultados oceanográficos sobre la base del estudio del plancton recogido durante la compaña 'Cosetri II': Serv. Hidrogr. Nav., Argentina, Bol. 3, 2: 105–114.
——. 1967a. Indicadores biológicos en la oceanografía: Cienc. Invest., 23, 2: 66–75.
——. 1967b. Campaña oceanográfica 'Corrientes Drake VI' (distribución de masas de aguas superficiales según el plancton): Serv. Hidrogr. Nav., Argentina, Bol., 4, 1: 5–16.
——. 1968a. Living planktonic foraminifera of the eastern part of the tropical Atlantic: Rev. Micropaléontol., 11, 2: 85–98.
——. 1968b. Hidrología de las aguas superficiales en la parte occidental del Atlántico Sur: Mus. Arg. Cienc. Nat., Rev., Hidrobiol., 2, 6: 199–224.
——. 1969a. Distribution of planktonic Foraminifera as indicatores of water masses in the western part of the Tropical Atlantic: Proc. Symp. Oceanogr. Fish. Resour. Trop. Atlantic, Unesco, Paris, 1969: 45–55.

Boltovskoy, E. 1969b. Living planktonic foraminifera at the 90°E meridian from the equator to the Antarctic: Micropaleontol., 15, 2: 237–248.
——. 1969c. Foraminifera as hydrological indicators: Inter. Conf. Plank. Microfossils, 1st, Geneva, Proc., 2: 1–14.
——. 1970a. Distribution of the marine littoral Foraminifera in Argentina, Uruguay and southern Brazil: Mar. Biol, 6, 4: 335–344.
——. 1970b. Masas de agua (característica, distribución, movimientos) en la superficie del Atlántico sudoeste, según indicatores biológicas – Foraminíferos: Serv. Hidrogr. Nav., Argentina, H. 643: 1–99.
——. 1971a. Planktonic foraminiferal assemblages of the epipelagic zone and their thanatocoenoses; *in* The Micropaleontology of the Oceans, Cambridge Univ. Press: 277–288.
——. 1971b. Ecology of the planktonic foraminifera living in the surface layer of Drake Passage: Micropaleontol., 17, 1: 53–68.
——. 1971c. Patchiness in the distribution of planktonic foraminifera: Inter. Conf. Plank. Microfossils, 2d, Rome, Proc., 1: 107–115.
——. 1971d. Relationship between benthonic foraminiferal fauna and the substrate in the littoral zone: Jour. Mar. Geol., 7, 1: 26–30.
——. 1972. Nota sobre los valores mínimos de oxigenación que pueden soportar los foraminíferos bentónicos: Bol. Soc. Biol. Concepción, 45: 135–143.
——. 1973a. Daily vertical migration and absolute abundance of living planktonic foraminifera: Jour. Foram. Res., 3, 2: 89–94.
——. 1973b. Reconstruction of post-Pliocene climatic changes by means of planktonic Foraminifera: Boreas, 2, 2: 55–68.
——. 1973c. Estudio de testigos submarinos del Atlántico sudoccidental: Mus. Arg. Cienc. Nat., Rev., Geol., 7, 4: 215–339.
——. 1973d. Note on the determination of absolute surface water paleotemperatures by means of the foraminifer *Globigerina bulloides*: Paläont. Zeit., 47, 3–4: 152–155.
——. New observations on the solution of planktonic foraminiferal tests and spines: (in press).
Boltovskoy, E. & Boltovskoy, A. 1968. Foraminíferos y Tecamebas de la parte inferior del Río Quequén Grande, Prov. de Buenos Aires, Argentina: Mus. Arg. Cienc. Nat., Rev., Hidrobiol., 2, 4: 127–164.
Boltovskoy, E. & Boltovskoy, D. 1970. Foraminíferos planctónicos vivos del Mar de la Flota (Antártica): Rev. Españ. Micropaleontol., 2, 1: 27–44.
Boltovskoy, E. & Gualancañay, E. Foraminíferos bentónicos actuales de Ecuador. 1. Provincia de Esmeraldas (in press).
Boltovskoy, E. & Lena, H. 1966a. Foraminíferos Recientes de la zona litoral de Pernambuco (Brasil): Mus. Arg. Cienc. Nat., Rev., Hidrobiol., 1, 8: 269–367.
——. 1966b. Unrecorded foraminifera from the littoral of Puerto Deseado: Cushman Found. Foram. Res., Contr., 17, 4: 144–149.
——. 1969a. Microdistribution des foraminifères benthoniques vivants: Rev. Micropaléontol., 12, 3: 177–185.
——. 1969b. Seasonal occurrences, standing crop and production in benthic Foraminifera of Puerto Desado: Cushman Found. Foram. Res., Contr., 20, 3: 87–95.
——. 1969c. Los epibiontes de *Macrocystis* flotante como indicadores hidrológicos: Neotrópica, 15, 48: 135–137.
——. 1970a. On the decomposition of the protoplasm and the sinking velocity of the planktonic foraminifers: Int. Rev. Ges. Hydrobiol., 55, 5: 797–804.
——. 1970b. Additional note on unrecorded foraminifera from littoral of Puerto Deseado (Patagonia, Argentina): Cushman Found. Foram. Res., Contr., 21, 4: 148–155.
——. 1971. The foraminifera (except family Allogromiidae) which dwell in fresh water: Jour. Foram. Res., 1, 2: 71–76.

Boltovskoy, E. & Lena, H. 1974. Foraminiferos del Río de la Plata: Serv. Hidrogr. Nav., Argentina, H. 660: 1–30.
Boltovskoy, E., Lena, H. & Asensi, A. 1974 Ms. Algae as a substrate for Foraminifera in Puerto Deseado area (Patagonia).
Boltovskoy, E. & Theyer, F. 1965. Neuere Daten über rezente Foraminiferen Zentralchiles: Beitr. Neotrop. Fauna, 4, 3: 143–149.
——. 1970. Foraminíferos Recientes de Chile Central: Mus. Arg. Cienc. Nat., Rev., Hidrobiol., 2, 9: 279–379.
Bouysse, Ph. & Le Calvez, Y. 1967. Étude des fonds marins compris entre Penmarch et Groix (Sud-Finistère): Bull. Bur. Rech. Géol. Min., 2: 39–73.
Bowen, R. N. C. 1957. Aperture in Foraminifera: Cushman Found. Foram. Res., Contr., 8, 2: 66–69.
Brachet, J. 1953. The use of basic dyes and ribonuclease for the cytochemical detection of RNA: Quart. Jour. Microsc. Sci., 94: 1.
Bradshaw, J. S. 1955. Preliminary laboratory experiments on ecology of foraminiferal populations: Micropaleontol., 1, 4: 351–358.
——. 1957. Laboratory studies of the rate of growth of the Foraminifera *Streblus beccarii* (Linné), var. *tepida* Cushman: Jour. Paleontol., 31, 6: 1138–1147.
——. 1959. Ecology of living planktonic Foraminifera in the north and equatorial Pacific Ocean: Cushman Found. Foram. Res., Contr., 10, 2: 25–64.
——.1961. Laboratory experiments on the ecology of Foraminifera: *Ibid.*, 12, 3: 87–106.
——. 1968. Environmental parameters and marsh Foraminifera: Limnol. Oceanogr., 13, 1: 26–38.
Brady, H. B. 1870. The Ostracoda and Foraminifera of Tidal Rivers: Ann. Mag. Nat. Hist., (4) 6: 273–306.
——. 1878. On the Reticularian and Radiolarian Rhizopoda (Foraminifera and Polycystina) of the North Polar Expedition of 1875–1876: *Ibid.*, (5), 1: 425–440.
——. 1881a. On some Arctic Foraminifera from soundings obtained on the Austro-Hungarian North Polar Expedition of 1872–1874: *Ibid.*, (5), 8: 393–418.
——. 1881b. Notes on Rhizopoda obtained from Captain Markham's soundings on the shores of Novaya Zemlya, *in*, Markham, A. H., A Polar Reconnaissance, 1879: C. Paul & Co., London: 346.
——. 1882. Über eine Arktische Tiefsee Foraminiferen gesammelt während der österreichischungarischen Nordpol-Expedition in der Jahren 1872–1874: Akad. Wiss. Wien., Denkschr., Math-Nat. Kl., 43: 9–110.
——. 1884. Report of the Foraminifera dredged by H.M.S. 'Challenger' during the years 1873-1876: Rept. Voy. Challenger, Zool., 9: 1–814.
——. 1888. Note on the reproductive conditions of *Orbitolites complanata*, var. *lacinata*: Roy. Microsc. Soc. London, Jour., 2: 693–697.
Braga, J. M. 1960. Foraminíferos da costa de Mosambique: Junta Inves. Ultramar, Est. Ens. Doc., 67: 1–211.
Braga, J. M. & Galhano, M. H. 1965a. Foraminíferos do Arquipelago da Madeira, I: Ann. Fac. Cienc., Porto, 48, 1-2: 135–190.
——. 1965b. *Idem*, II.: *Ibid.*, 3-4: 205–264.
Bramlette, M. N. & Bradley, W. H. 1940. Geology and biology of North Atlantic deep-sea cores between Newfoundland and Ireland, Pt. 1, Lithology and geologic interpretations: U.S. Geol. Surv., Prof. Pap., 196-A: 1–34.
Brenner, G. J. 1962. A zoogeographic analysis of some shallow water Foraminifera in the Gulf of California: Amer. Mus. Nat His., Bull., 123, 5: 249–298.
Brodniewicz, I. 1965. Recent and certain Holocene Foraminifera of the Southern Baltic Sea: Acta Paleontol. Polon., 10, 2: 131–236.

Brodsky, A. L. 1928. Foraminifera (Polythalamia) v kolodthakh pustyni Kara-Kum: Sr.-Aziat. Gos. Univ., Trudy, ser. 8, Zool., 5: 1–16.

——. 1929. Fauna vodoemov pustyni Kara-Kum: Sr.-Aziat. Inst., Trudy, 5: 1–42.

Brönnimann, P. 1951. Internal structure of *Cyclammina cancellata:* Jour. Paleontol., 25: 756–761.

Bronte-Gatenby, J. & Beams, H. W. 1950. The microtomists vade-medum: A handbook of methods of animal and plant microscope techniques: J. and A. Churchill Ltd., London.

Brooks, A. L. 1967. Standing crop, vertical distribution and morphometrics of *Ammonia beccarii* (Linné): Limnol. Oceanogr., 12, 4: 667–684.

Brooks, W. W. 1973. Distribution of Recent foraminifera from the southern coast of Puerto Rico: Micropaleontol., 19, 4: 385–416.

Brotzen, F. 1950. Method and techniques in routine work: The Micropaleontologist, 4, 2: 18–20.

Brotzen, F. & Dinesen, A. 1960. On the stratigraphy of some bottom sections from the central Pacific: Swedish Deep-sea Exped., 10, Spec. Invest., 3: 45–55.

Buchanan, J. B. & Hedley, R. H. 1960. A contribution to the biology of *Astrorhiza limicola* (Foraminifera): Jour. Mar. Biol. Assoc. U.K., 39: 549–560.

Buchner, P. 1940. Die Lagenen des Golfes von Neapel und der marinen Ablagerungen auf Ishia: Nova Acta Leopoldina, n. F., 9, 62: 363–560.

Buckley, D. E., Owens, E. H., Schafer, C. T., Vilks, G., Cranston, R. E., Rashid, M. A., Wagner, F. J. E. & Walker, D. A. 1974. Canso Strait and Chedabocto Bay: A multidisciplinary study of the impact of man on the marine environment: Geol. Surv. Canada, Pap. 74-30, 1: 133–160.

Burdett, I. D. J., Hedley, R. H., Hornibrook, N. B. & Hurdle, C. M. 1963. *Gaudryina convexa* (Karrer) 1865, Upper Eocene to Recent; an example of variation and synonomy among Foraminifera: New. Zeal. Jour. Sci., 6, 4: 513–530.

Burma, B. H. 1948. Studies in quantitative paleontology; I, Some aspects of the theory and practice of quantitative invertebrate paleontology: Jour. Paleontol., 22: 725–761.

——. 1949. *Idem*; II, Multivariate analysis – a new analytic tool for paleontology and geology: *Ibid.*, 23: 95–103.

Butcher, W. S. 1951. Foraminifera, Coronado Bank and vicinity, California: Scripps Inst. Oceanogr., Submar., Geol. Rpt., 19.

Butler, E. A. 1963. Species concepts and Foraminifera: Gulf. Coast Assoc. Geol. Socs., Trans., 13: 121–125.

Bütschli, O. 1880. Protozoa: Bronn's Thierreich, 1, 1: 1–1097.

——. 1886. Kleine Beiträge zur Kenntnis einiger Rhizopoden: Morph. Jahr., 11: 78–101.

——. 1894. Protoplasm and Microscopic forms, London.

Buzas, M. A. 1965. The distribution and abundance of Foraminifera in Long Island Sound: Smithsonian Misc. Coll., 149, 1: 1–89.

——. 1966. The discrimination of morphological groups of *Elphidium* (Foraminifer) in Long Island Sound through canonical analysis and invariant characters: Jour. Paleontol. 40, 3: 585–594.

——. 1967. An application of canonical analysis as a method for comparing faunal areas: Jour. Animal Ecol., 36: 563–577.

——. 1968. On the spatial distribution of Foraminifera: Cushman Found. Foram. Res., Contr., 19, 1: 1–11.

——. 1969a. Foraminiferal species densities and environmental variables in an estuary: Limnol. Oceanogr., 14, 3: 411–422.

——. 1969b. On the quantification of biofacies: Proc. North Amer. Paleont., 1969, (B): 101–116.

——. 1970. Spatial homogeneity: statistical analysis of unispecies and multispecies populations of Foraminifera: Ecology, 51, 5: 874–879.

Buzas, M. A. 1972a. Biofacies analysis of presence or absence data through canonical variate analysis: Jour. Paleontol., 46, 1: 55–57.
——. 1972b. Patterns of species diversity and their explanation: Taxon, 21: 275–286.
——. 1974. Vertical distribution of *Ammobaculites* in the Rhode River, Maryland: Jour. Foram. Res., 4, 3: 144–147.
Buzas, M. A. & Gibson, T. G. 1969. Species diversity and benthonic Foraminifera in western North Atlantic: Science, 163: 72–75.
Camp, W. H. 1952. Phytophyletic patterns on land bordering the South Atlantic basin: Amer. Mus. Nat. Hist., Bull., 99, 3: 205–212.
Caralp, M. 1968. Variations climatiques Pléistocènes dans le Golfe de Gascogne d'après les Foraminifères planctoniques: Bull. Inst. Géol. Bass. Aquit., 5: 87–110.
——. 1970. Essai de stratigraphie du Pléistocène marin terminal d'après les paleoclimats observés dans des carottes du Golfe de Gascogne:Bull. Soc. Geol. Fran. (7), 12, 3: 403–412.
Caralp, M., Lamy, A. & Pujos, M. 1970. Contribution a la connaissance de la distribution bathymétrique des foraminifères dans le Golfe de Gascogne: Rev. Españ. Micropaleontol., 2, 1: 55–84.
Caralp, M., Moyes, J. & Vigneaux, M. 1968. La microfauna actuelle et subrecente d'une carotte atlantique (Golfe de Gascogne); observations écologiques et climatiques: Bull. Soc. Géol. France, (7), 9, 3: 418–424.
——. 1970. Essai d'utilisation des mélanges de microorganismes benthiques dans la reconstitution des environnements; application à un canyon du Golfe de Gascogne: Deep-Sea Res., 17, 3: 661–670.
Caralp, M., Rechiniac, A. & Vigneaux, M. 1965. Utilisation de microscope a rayons – X dans l'étude des structures internes des microfossiles: Jour. Microscopie, 4: 451–460.
Carcelles, A. R. 1950. Catálogo de moluscos marinos de la Patagonia: Mus. Mahuel Huapi, An., Extra, n.s., 8: 1–99.
Carcelles, A. R. & Williamson, S. T. 1951. Catálogo de los moluscos marinos de la provincia magallánica: Inst. Nac. Invest. Cienc. Nat., Rev. Zool., 2, 5: 225–383.
Carpenter, W. B. 1856. Researches on the Foraminifera, I: Roy. Soc. London, Phil. Trans., 146: 547–569.
——. 1875. Remarks on Prof. Wyville Thomson's preliminary notes on the nature of the sea bottom procured by soundings of H.M.S. 'Challenger': Roy. Soc. London, Proc., 23.
——. 1876. Foraminifera; *in* Jeffereys, J. G., Preliminary reports of a cruise in H.M.S. 'Valporous' to Davis Straits in 1875: Roy. Soc. London, Proc., 25: 223–227.
Carpenter, W. B., Parker, W. K. & Jones, T. R. 1862. Introduction to the study of Foraminifera: Roy. Soc. London, 319 pp.
Carsola, A. J. 1952 Ms. Marine geology of the Arctic Ocean off Alaska and Northwestern Canada: Univ. Calif, Los Angeles, Ph.D. thesis.
Carter, H. J. 1849. On Foraminifera, their organisation and their existence in a fossilised state in Arabia, Sindh, Kutch and Khattyawar: Jour. Bombay Branch Roy. Asiatic Soc., 3: 158–173.
Carvalho, J. & Chermont, E. M. L. 1952. Sobre alguns Foraminífera da costa do estado de São Paulo: Inst. Oceanogr., 3, 1-2: 77–97.
Cebulski, D. E. 1969. Foraminiferal populations and faunas in barrier-reef tract and lagoon, British Honduras: Amer. Assoc. Petrol. Geol., Mem., 11: 311–328.
Chamney, T. P. 1957. A micropaleontological disintegration method: Alberta Soc. Petrol. Geol., February, 1957.
Chang, Y. M. 1967. Accuracy of fossil percentage estimation: Jour. Paleontol., 41: 500–502.
Chang, Y. M. & Kaesler, R. L. 1974. Morphological variation of the foraminifer *Ammonia beccarii* (Linné) from the Atlantic Coast of the United States: Univ. Kansas Paleontol. Contr., Pap. 69, 23 pp.

Chang, Y. M., Kaesler, R. L. & Merrill, W. M. 1974. Simulation of growth of the Foraminiferida *Ammonia beccarii* by computer: Geol. Soc. Amer., Bull., 85, 5: 745–748.
Chapman, F. 1916. Report on the Foraminifera and Ostracoda out of marine muds from soundings in the Ross Sea: Brit. Antarc. Exped. 1907–1909, Geol., 2, 3: 53–80.
Chapman, F. & Parr, W. J. 1936. A classification of the Foraminifera: Roy. Soc. Victoria, Proc., 49, 1: 139–151.
——. 1937. Foraminifera: Australasian Antarc. Exped. 1911–1914, Sci. Rept., ser. C, Zool. Bot., 1, 2: 7–10.
Charlton de Rivero Palacio, F. & Bermúdez, P. 1963. Micropaleontología General: Univ. Centr. Venezuela, Caracas, 508 pp.
Chatterjee, B. P. & Gururaja, M. N. 1968. Foraminifera off Mangalore Coast, South India: Bull. Nat. Inst. Sci. India, 38: 393–397.
Chekhovskaja, M. P. 1973. Raspredelenie bentosnykh foraminifer v severo-vostochnoj chasti Beringova morja: Oceanol., 13, 4: 691–699.
Chen, C. 1966. Calcareous zooplankton in the Scotia Sea and Drake Passage: Nature, 212: 678–681.
Cherif, O. H. & Flick, H. 1974. On the taxonomic value of the wall structure of *Quinqueloculina:* Micropaleontol., 20, 2: 236–244.
Chernysheva, N. E. 1940. K stratigrafii nizhnego karbona Makarovskogo rajona Juzhnogo Urala po faune foraminifer: Mosk. O-vo Isp. Prir., Bull., Geol., 18, 5-6: 113–135.
Chierici, M. A., Busi, M. T. & Cita, M. B. 1962. Contribution a un étude écologique des Foraminifères dans le mer Adriatique: Rev. Micropaléontol., 5, 2: 123–142.
Chiji, M. & Lopez, S. M. 1968. Regional foraminiferal assemblages in Tanebe Bay, Kii Peninsula, central Japan: Publ. Seto Marine Biol. Lab., 16, 2: 85–125.
Chiji, M. & Konda, I. 1970. Depth distribution of foraminiferal assembleges in the bottom sediments around Okushiri Island, North Japan Sea: Bull. Osaka Nat. Hist., 23: 35–50.
Christiansen, B. 1958. The Foraminiferal fauna in the Drobak Sound in the Oslo Fjord (Norway): Nytt Mag. Zool., 6: 5–91.
——. 1964. *Spiculosiphon radiata*, a new Foraminifera from Northern Norway: Astarte, 25: 1–8.
——. 1965a. A bottom form of the planktonic foraminifer *Globigerinoides rubra* (d'Orbigny) 1839: Publ. Stazione Zool. Napoli, 34, 2: 197–202.
——. 1965b. Notes on the littoral fauna of Bear Island: Astarte, 26: 1–15.
——. 1971. Notes of the biology of foraminifera: Vie et Milieu, suppl. 22: 465–478.
——. 1974 Ms. Notes on the life habitats of the Foraminiferida.
Cifelli, R. 1960. Variation of English Bathonian Lagenidae and its phylogenetic significance: Jour. Paleontol., 34: 556–569.
——. 1961. *Globigerina incompta*, a new species of pelagic Foraminifera from the North Atlantic: Cushman Found. Foram. Res., Contr., 12, 3: 83–86.
——. 1962a. Some dynamic aspects of the distribution of planktonic Foraminifera in the western North Atlantic: Jour. Mar. Res., 20, 3: 201–213.
——. 1962b. The morphology and structure of *Ammonia beccarii* (Linné): Cushman Found. Foram. Res., Contr., 13, 4: 119–126.
——. 1965. Planktonic Foraminifera from the Western North Atlantic: Smithsonian Misc. Coll., 148, 4: 1–36.
——. 1967. Distributional analysis of North Atlantic Foraminifera collected in 1961 during cruises 17 and 21 of the R/V *Chain:* Cushman Found. Foram. Res., Contr., 18, 3: 118–128.
——. 1971. On the temperature relationships of planktonic foraminifera: Jour. Paleontol., 1, 4: 170–177.
Cifelli, R. & Sachs, K. N. 1966. Abundance relationships of planktonic Foraminifera and Radiolaria: Deep-Sea Res., 13: 751–753.
Cifelli, R. & Smith, R. K. 1969. Problems in the distribution of planktonic foraminifera and

their relationships with water mass boundaries in the North Atlantic: Inter. Conf. Plank. Microfossils, 1st, Geneva, Proc., 2: 112–125.
Cifelli, R. & Smith, R. K. 1974. Distribution patterns of planktonic foraminifera in the Western North Atlantic: Jour. Foram. Res., 4, 3: 112–125.
Ciry, R., Tintant, H. and Chareyron, H. 1948. Sur la microfaune du Pliocène de Freius: Bull. Scien. Bourgogne, vol. 12.
Clark, D. F. 1971. Effects of aquaculture outfall on benthonic Foraminifera in Clam Bay, Nova Scotia: Maritime Seds., 7, 2: 76–84.
Clark, D. L. 1971. The Arctic ice cover and its late Cenozoic history: Geol. Soc. Amer. Bull., 82: 3313-3324.
Clarke, G. L. & Bumpus, D. F. 1950. The plankton sampler, an instrument for quantitative plankton investigations: Limnol. Oceanogr. Sp. Publ. 5: 2–8.
Closs, D. 1962. Foraminíferos e Tecamebas da Lagoa dos Patos: Escol. Geol., Univ. Rio Grande do Sul, 11: 1–130.
Closs, D. & Barberena, M. C. 1960a. Foraminíferos recentes da praia do Cassino: *Ibid.*, 5: 1–29.
——. 1960b. Foraminíferos recentes do praia da Barra: *Ibid.*, 6: 1–50.
——. 1962a. Foraminíferos recentes das praias do Litoral Sul-Brasileiro: Inst. Cienc. Nat., Univ. Rio Grande do Sul, 16: 7–57.
——. 1962b. Faunal studies of recent Foraminifera from the shore sands of the state Rio Grande do Sul in Southern Brazil: Cushman Found. Foram. Res., Contr., 13: 74–78.
Closs, D. & Madeira, M. 1962. Tecamebas e foraminíferos do arroio Chui (Santa Vitoria do Palmar, R.G.S.): Iheringia, Zool., 19: 3–43.
——. 1967. Foraminíferos e tecamebas aglutinantes da Lagoa de Tramadai (R.G.S.): *Ibid.*, 35: 7–31.
——. 1968. Seasonal variations of brackish foraminifera in the Patos lagoon, southern Brazil: Escol. Geol., Univ. Rio Grande do Sul, 15: 1–41.
Closs, D. & Medeiros, V. M. F. 1965. New observations on the ecological subdivision of the Patos Lagoon in Southern Brazil: Inst. Cienc. Nat., Univ. Rio Grande do Sul, 24: 7–33.
——. 1967. Thecamoebina and Foraminifera from the Mirim Lagoon, Southern Brazil: Iheringia, Zool., 35: 75–88.
Cockbain, A. E. 1963. Distribution of Foraminifera in Juan de Fuca and Georgia straits, British Columbia, Canada: Cushman Found. Foram. Res., Contr., 14, 2: 37–57.
Cohen, A. D. & Guber, A. L. 1968. Production of pollen-sized 'microforaminifera' from 'normal' foraminifera: Micropaleontol., 14, 3: 361–362.
Collen, J. D. 1973. Morphology and development of the test surface in some species of *Notorotalia* (Foraminiferida): Rev. Españ. Micropaleontol., 5, 1: 113–132.
Collins, A. C. 1958. Foraminifera: Great Barrier Reef Exped. 1928–1929, Scient. Rept., 6, 1: 335–437.
Colom, G. 1946. Introducción al estudio de los Microforaminíferos fósiles: Consejo Sup. Inv. Cient., Madrid, 308 pp.
——. 1950. Estudio de los Foraminíferos de muestras de fondo recogidas entre los cabos Juby y Bojador: Inst. Españ. Oceanogr., 28: 1–45.
——. 1952. Foraminíferos de las costas de Galicia: Inst. Españ. Oceanogr., Bol., 51: 3–59.
Cooper, S. C. 1964. Benthonic Foraminifera of the Chukchi Sea: Cushman Found. Foram. Res., Contr., 15, 3: 79–104.
——. 1965. A new morphologic variation of the foraminifer *Cibicides lobatulus*: *Ibid.*, 16, 4: 137–139.
Cooper, W. C. 1961. Intertidal Foraminifera of the California and Oregon coast: *Ibid.*, 12, 2: 47–63.
Correns, C. W. 1939. Pelagic sediments of the North Atlantic Ocean, *in* Trask, P.D., ed., Recent Marine Sediments: Amer. Assoc. Petrol. Geol., Symposium, Tulsa: 373–395.

Cosijn, A. J. 1938. Statistical studies on the phylogeny of some Foraminifera: Techn. Hoogesch. Delft Proefschr., Leiden, 70 pp.
Cox, A., Doell, R. R. & Dalrymple, G. B. 1963. Geomagnetic polarity epochs and Pleistocene geochronometry: Nature, 198: 1049–1051.
——. 1965. Quaternary paleomagnetic stratigraphy; *in*, Wright & Frey, eds., Quaternary of the United States, Princeton Univ. Press: 817–830.
Craig, G. Y. & Hogg, J. 1962. A rapid sorting device for microfossils: Micropaleontol., 8, 1: 107–108.
Crespin, I. 1960. Some Recent Foraminifera from Westfold Hills, Antarctica: Tohoku Univ., Science Rept., Geol. Spec. Vol., 4: 19–34.
Crouch, R. W. 1952. Significance of temperature on foraminifers from deep basins off Southern California coast: Amer. Assoc. Petrol. Geol., Bull., 36: 807–843.
——. 1954. Paleontology and paleoecology of the San Pedro shelf and vicinity: Jour. Sed. Petrol., 24: 182–190.
Cummings, R. H. 1956. Preparation of microfossils for photography: Micropaleontol., 2, 4: 402.
Cushman, J. A. 1920a. Report of the Canadian Arctic expedition 1913–1918. Part M., Foraminifera: 1–13.
——. 1920b. Observations on living specimens of *Iridia diaphana*, a species of Foraminifera: U.S. Nat. Mus., Proc., 57: 153–158.
——. 1921a. Results of the Hudson Bay Exped. 1920. I. The Foraminifera: Canadian Biol., Contr., 9: 135–147.
——. 1921b. Foraminifera from the north coast of Jamaica: U.S. Nat. Mus., Proc., 59: 47–82.
——. 1927. An outline of a re-classification of the Foraminifera: Cushman Lab. Foram. Res. Contr., 3, 1: 1–105.
——. 1930. About the interrelations of foraminifera and algae: Jour. Wash. Acad. Sci., 20: 395–396.
——. 1933. New Arctic Foraminifera collected by Capt. R. A. Bartlett from Fox basin and off the northeast coast of Greenland: Smithsonian Misc. Coll., 89, 9: 1–8.
——. 1941a. A study of the Foraminifera contained in cores from the Bartlett Deep: Amer. Jour. Sci., 239: 128–143.
——. 1941b. Recent foraminifera from Old Providence Island collected on the Presidential cruise of 1938: Smithsonian Misc. Coll., 99, 9: 1–14.
——. 1945. Foraminifera of the United States Antarctic Service Expedition 1939–1941: Amer. Phil. Soc., Proc., 89: 285–288.
——. 1948a. Foraminifera, their classification and economic use: Harvard Univ. Press, 4th ed., 605 pp.
——. 1948b. Arctic Foraminifera: Cushman Lab. Foram. Res., Spec. Publ., 23: 1–95.
Cushman, J. A. & Henbest, L. G. 1940. Geology and biology of North Atlantic deep-sea cores between Newfoundland and Ireland, Pt. 2, Foraminifera: U.S. Geol. Surv., Prof. Pap., 196-A: 35–50.
Cushman, J. A. & Jarvis, P. W. 1930. Miocene Foraminifera from Buff Bay Jamaica: Jour. Paleontol., 4: 353–368.
Cushman, J. A. & Kellett, B. 1929. Recent Foraminifera from the west coast of South America: U.S. Nat. Mus., Proc., 75, 25: 1–17.
Cushman, J. A. & Parker, F. L. 1931. Recent Foraminifera from the Atlantic coast of South America: *Ibid.*, 80, 3: 1–24.
Cushman, J. A. & Todd, R. 1949. Species of the genera *Allomorphina* and *Quadrimorphina:* Cushman Lab. Foram. Res., Contr., 25, 3: 68.
Cushman, J. A. & Wickenden, R. T. S. 1929. Recent Foraminifera from off Juan Fernandez Islands: U.S. Nat. Mus. Proc., 75, 9: 1–16.

Czihak, G. & Grell, K. G. 1960. Zur determination der Zellkerne bei der Foraminifere *Rotaliella heterocaryotica:* Naturwiss, 47: 211–212.
Daday, E. 1884. Über einige Polythalamia der Kochsalztumpel bei Deva in Siebenbürgen: Z. Wiss. Zool., 40: 465–480.
Dahlgren, L. 1962. *Allogromia crystallifera*, n.sp., a monothalamous foraminifer: Zool. Bidr. Uppsala, 35: 451–455.
——. 1964. On nuclear cytology and reproduction in the monothalamous foraminifer *Ovammina opaca* Dahlgren: *Ibid.*, 36: 315–334.
——. 1967a. On the ultrastructure of the gamontic nucleus and the adjacent cytoplasm of the monothalamous foraminifer *Ovammina ovata* Dahlgren: *Ibid.*, 37: 77–112.
——. 1967b. On the nuclear distribution of RNA and DNA and on the ultrastructure of nuclei and adjacent cytoplasm of the foraminifers *Hippocrepinella alba* Heron-Allen & Earland and *Globobulimina turgida* (Bailey): *Ibid.*, 37: 113–138.
Daniels, C. H. 1970a. Quantitative ökologische Analyse der zeitlichen und räumlichen Verteilung rezenter Foraminiferen im Limski kanal bei Rovinj (nördliche Adria): Göttinger Arb. Geol. Paläontol., 8: 1–108.
——. 1970b. Jahreszeitliche ökologische Beobachtungen an Foraminiferen im Limski Kanal bei Rovinj/Jugoslawien (nördliche Adria): Geol. Rundsch., 60, 1: 192–204.
Dansgaard, W. & Tauber, H. 1969. O^{18} content and Pleistocene ocean temperatures: Science, 166: 499–502.
Dardenne, M. 1954. Paléontologie et écologie du Miocène Marocain Région du Zegotta: Ser. Géol. Maroc., Not., 9.
Darlington, C. D. & LaCour, L. F. 1962. The handling of chromosomes: G. Allen and Unwin, London, 263 pp.
Davies, R. G. 1971. Computer Programming in Quantitative Biology: Academic Press, London, 492 pp.
Davis, R. A. Jr. 1964. Foraminiferal assemblages of Alacran Reef, Campeche Bank, Mexico: Jour. Paleontol., 38, 2: 417–421.
Decloitre, L. 1966. Comment compter le nombre de thécamoebiens dans une récolte: Limnologica, 4, 3: 489–492.
Delaca, T. E. & Lipps, J. H. 1972. The mechanism and adaptive significance of attachment and substrate pitting in the foraminiferan *Rosalina globularis* D'Orbigny: Jour. Foram. Res., 2, 2: 68–72.
Dennison, J. M. & Hay, W. W. 1967. Estimating the needed sampling area for subaquatic ecologic studies: Jour. Paleontol., 41, 3: 706–708.
Derjugin, K. 1915. Fauna Kol'skogo zaliva i uslovija ee sushchestvovanija: Akad. Nauk., Zap., 24.
Derville, H. 1936. Manière d'être des algues dans les calcaires a Nubeculaires: Soc. Géol. France, Bull., (5) 6: 487–493.
Detling, M. R. 1958. Some littoral Foraminifera from Sunset Bay, Coos County, Oregon: Cushman Found. Foram. Res., Contr., 9, 2: 23–52.
Devereux, I., Hendy, C. H. & Vella, P. 1970. Pliocene and Early Pleistocene sea temperature fluctuations, Mangaopari Stream, New Zealand: Earth and Planet. Sci. Let., 8: 163–168.
Dhillon, D. S. 1969. An abnormal *Astacolus* from the Holocene, Gantheaume Bay, Western Australia: Cushman Found. Foram. Res., Contr., 20, 3: 99.
——. 1970. An abnormal *Ammotium* from the Holocene, West Malaysia: Micropaleontol., 16, 1: 59–60.
Diener, C. 1925. Grundzüge der Biostratigraphie: 304 pp., Deuticke, Wien and Leipzig.
Dietz-Elbrächter, G. 1969. Zur Valenz der Foraminifere *Ammonia beccarii* zur passiven Verbreitung durch Vögel: Progr. Protozool, 3d, Inter. Congr. Protozool,: 191.
——. 1971. Untersuchungen über die Zooxanthellen der Foraminifere *Heterostegina depressa* Orbigny 1826: 'Meteor' Forsch. Ergeb., Reihe C, 6: 41–47.

Dihdkov'skij, V. Ja. 1958. Pro faunu foramihnihfer Azovs'kogo morja: Akad. Nauk URSR, Dokl., 10.
——. 1959. Vikopnih Peneroplihdi pihvdenno-zakhidnoih chastinii Radjan'skogo Sojuzu: Akad. Nauk URSR, Inst. Geol., Trud., Strat. Paleontol., 28.
Dixon, W. J. & Massey, F. J. 1957. Introduction to Statistical Analysis: McGraw-Hill, New York, 488 pp.
Dmitrieva, R. G. 1952. O prirode tak nazyvaemykh 'pelethipodovykh' mergelej iz miothenovykh otlozenij Zapadnogo Predkavkaz'ja: Akad. Nauk USSR, Dokl., 82.
Dobson, M. & Haynes, J. 1973. Association of foraminifera with hydroids on the deep shelf: Micropaleontol., 19, 1: 78–90.
Dolgopol'skaja, M. A. & Pauli, L. V. 1932. Foraminifera Chernogo morja rajona Karadagskoj biologicheskoj stanthii: Tr. Karadagskoj Biol. Stant., 4: 23–47.
Dommasnes, A. 1969. On the fauna of *Corallina officinalis* in Western Norway: Sarsia, 38: 71–86.
Donk, J. van & Mathieu, G. 1969. Oxygen isotope composition of foraminifera and water samples from the Arctic Ocean: Jour. Geophys. Res., 74, 13: 3396–3407.
Douglass, R. C. 1960. Introducción al estudio de los Microfóssiles: Univ. Chile, Esc. Geol., Santiago, 86 pp.
Douville, H. 1906. Evolution et enchainements des Foraminifères: Soc. Géol. France, Bull., (4) 6, 7: 583–602.
Doyle, W. L. 1935. Distribution of mitochondria in the Foraminiferan *Iridia diaphana:* Science, 81: 387.
Drew, E. A. 1973. The biology and physiology of alga-invertebrate symbioses. III. *In situ* measurements of photosynthesis and calcification in some hermatypic corals: Jour. Exp. Mar. Biol. Ecol., 13: 165–179.
Dreyer, F. 1898. *Peneroplis* – eine Studie zur biologischen Morphologie und zur Speziesfrage: Leipzig, 119 pp.
Drooger, C. W. & Kaasschieter, J. P. G. 1958. Foraminifera of the Orinoco–Trinidad–Paria Shelf: Orinoco Shelf Exped., Rept. 4: 1–108.
Dujardin, F. 1835. Observations nouvelles sur les prétendus Céphalopodes microscopiques: Ann. Sci. Nat., (2) 3, Zool.: 36.
——. 1841. Histoire Naturelle des Zoophytes: Paris, 684 pp.
Dupeuble, P. A. 1962. Polymorphisme chez les Cibicidinae actuels de la région de Roscoff (Finistère): Rev. Micropaléontol., 4: 197–202.
——. 1963. Apercu sur la répartition des principales familles de foraminifères dans trois facies de la région de Roscoff (Finistère): *Ibid.*, 5, 4: 277–279.
——. 1964a. Contribution a l'étude des sédiments et de la microfaune de la région de Roscoff: Thèse de 3° Cycle, Fac. Sci. Paris, 212 pp.
——. 1964b. Sur une association microfaunique characteristique de la Manche et des régions voisines: Soc. Géol. Normandie, Bull., 54: 52–56.
——. 1965. Un procédé pour effecteur facilement des sections minces orientées dans les microfossiles: Rev. Micropaléontol., 7: 280–282.
Dupeuble, P. A., Mathieu, R., Momeni, I., Poignant, A., Rosset-Moulinier, M., Rouvillois, A. & Ubaldo, M. 1971. Recherces sur les Foraminifères actuels des côtes francaises de la Manche et de la Mer du Nord: *Ibid.*, 14, 5: 83–95.
Duplessy, J. C., Lalou, C. & Vinot, A. C. 1970. Differential isotopic fractionation in benthic foraminifera and paleotemperatures reassessed: Science, 168: 250.
Duplessy, J. C. & Vinot, A. C. 1972. L'interprétation des variations de la composition isotopique des tests de Foraminifères durant le Pléistocène, *in*, Coll. Meth. Tend. Strat., Bur. Rech. Géol. Minières, Mém., 77, 2: 759–765.
Dzhanelidze, O. L. 1957. O morfologicheskikh anomalijakh u foraminifer Tarkhanskogo gorizonta Gruzii: Akad. Nauk SSSR, Dokl., 113.

Eade, J. V. 1967. A checklist of Recent New Zealand Foraminifera: New Zeal. Oceanogr. Inst., Mem. 44: 9–72.
——. 1970. Present day ecology of benthonic Foraminifera from an Upper Miocene coral thicket from Wairarapa, New Zealand: Palaeogeogr., Palaeoclimat., Palaeoecol., 7: 33–39.
Earland, A. 1933. Foraminifera, Part 2, South Georgia: Discovery Rep., 7: 27–128.
——. 1934. Foraminifera, Part 3, The Falklands sector of the Antarctic (excluding South Georgia): *Ibid.*, 10: 1–208.
——. 1936. Foraminifera, Part 4. Additional records from the Weddell Sea sector from material obtained by the S.Y. *Scotia: Ibid.*, 13: 1–59.
Ebner, V. 1887. Über den feineren Bau der Skelettheile der Kalkschwämme nebst Bemerkungen über Kalkskelette überhaupt: Akad. Wiss. Wien, S.B., 95, 3.
Echols, D. J. & Schaeffer, K. M. M. 1960. Microforaminifera of the Marianna limestone (Oligocene) from Little Stave Creek, Alabama: Micropaleontol., 6, 4: 399–415.
Echols, D. J. & Wegweiser, A. E. 1967. Naked Foraminifera from shallow water environments: Amer. Assoc. Petrol. Geol., Bull., 51: 462.
Echols, R. J. 1971. Distribution of Foraminifera in sediments of the Scotia Sea Area, Antarctic waters; *in* Antarctic Oceanology, I: Antarct. Res. Ser., 15: 93–168.
Egger, J. C. 1893. Foraminiferen aus Meeresgrundproben gelothet von 1874 bis 1876 von S.M. Sch. *Gazelle:* K. Bayer. Akad. Wiss., Abhandl., Math.-Phys. Cl., 18: 193–458.
Ehrenberg, C. G. 1839. Über noch jetzt zahlreich lebende Thierarten der Kreidebildung und der Organismus der Polythalamien: Akad. Wiss. Berlin, Phys. Abh.: 81–174.
——. 1861. Über die Tiefgrund-Verhältnisse des Oceans am Eingange des Davisstrasse und bei Island: K. Preuss. Akad. Wiss. Berlin, Monatsber., Jahrg.: 275–315
Eicher, D. L. & Frush, M. P. 1974. Cenomanian benthic foraminifera: faunal changes with depth: Geol. Soc. Amer., Abstr. Prog., 6, 7: 719.
Eichwald, C. E. 1829. Zoologia Specialis: Vilna, 1.
Eimer, G. H. T. & Fickert, C. 1899. Die Artbildung und Verwandtschaft bei den Foraminiferen. Entwurft einer natürlichen Eintheilung derselben: Z. Wiss. Zool., 65, 4: 527–636.
Ekman, S. 1953. Zoogeography of the Sea: Sedgwick & Jackson, London, 417 pp.
Ellison, R. L. 1972. ***Ammobaculites*, foraminiferal proprietor of Chesapeake Bay estuaries:** *in* B. Nelson, ed., Environmental framework of coastal plain estuaries: Geol. Soc. Amer., Mem. 133: 247–262.
Ellison, R. I. & Nichols, M. N. 1970. Estuarine foraminifera from the Rappahannock River, Virginia: Cushman Found. Foram. Res., Contr., 21, 1: 1–17.
Ellison, R., Nichols, M. N. & Hughes, J. 1965. Distribution of Recent Foraminifera in the Rapphannock River estuary: Virginia Inst. Mar. Sci., Spec. Sci. Rept. no. 47: 1–35.
Ellison, S. 1953. An abnormal *Uvigerina* from the Upper Eocene, Mississippi: Cushman Found. Foram. Res., Contr., 4, 2: 66–67.
Emerson, B. 1916. Polarization of *Globigerina:* Science, 43, 1105: 316.
Emery, K. O. & Hulsemann, J. 1961. The relationship of sediments, life and water in a marine basin: Deep-Sea Res., 8: 165–180.
Emery, K. O. & Niino, H. 1963. Sediments of the Gulf of Thailand and adjacent continental shelf: Geol. Soc. America, Bull., 74: 541–544.
Emery, K. O. & Stevenson, R. E. 1957. Estuaries and lagoons; *in*, Treatise Mar. Ecol. Paleoecol., 1: *Ibid.*, Mem. 67: 673–750.
Emiliani, C. 1952. Systematic zoology. Nomenclature and grammar: Wash. Acad. Sci., 42: 137–141.
——. 1954. Depth habitats of some species of pelagic foraminifera as indicated by oxygen isotope ratios: Amer. Jour. Science, 252: 149–158.
——. 1955a. Mineralogical and chemical composition of the test of certain pelagic Foraminifera: Micropaleontol., 1, 4: 377–380.
——. 1955b. Pleistocene temperatures: Jour. Geol., 63: 538–578.

Emiliani, C. 1956. Oligocene and Miocene temperatures of the equatorial and subtropical Atlantic Ocean: *Ibid.*, 64: 281–288.
——. 1958. Paleotemperature analysis of core 280 and Pleistocene correlations: *Ibid.*, 66: 264–275.
——. 1961. The temperature decrease of surface sea-water in high latitudes and of abyssal-hadal water in open oceanic basins during the past 75 million years: Deep-Sea Res., 8: 144–147.
——. 1964. Paleotemperature analysis of Caribbean cores A-254-BR-C and CP-28: Geol. Soc. Amer., Bull., 75: 129–144.
——. 1966a. Isotopic paleotemperatures: Science, 154: 851–857.
——. 1966b. Paleotemperature analysis of Caribbean cores P6304-8 and P6304-9 and a generalized temperature curve for the last 425 000 years: Jour. Geol., 74, 2: 109–126.
——. 1969. A new paleontology: Micropaleontol., 15, 3: 265–300.
——. 1970. Pleistocene paleotemperatures: Science, 168: 822–825.
——. 1971a. Depth habitats of growth stages of pelagic Foraminifera: *Ibid.*, 173: 1122–1124.
——. 1971b. Paleotemperature variations across the Plio-Pleistocene boundary: *Ibid.*, 171: 60–62.
——. 1974. Isotopic paleotemperatures and shell morphology of *Globigerinoides rubra* in the Mediterranean deep-sea core 189: Micropaleontol., 20, 1: 106–109.
Emiliani, C. & Shackleton, N. J. 1974. The Brunhes Epoch: isotopic paleotemperatures and geochronology: Science, 183: 511–514.
Engelmann, T. W. 1879. Flimmer und Protoplasmabewegung: Hermann Handbuch Physiologie, 1: 343–408.
Ericson, D. B. 1953. Sediments of the Atlantic Ocean: Lamont Geol. Observ., Tech. Rept., 1: 1–34.
——. 1959. Coiling direction of *Globigerina pachyderma* as a climatic index: Science, 130: 219–220.
Ericson, D. B., Ewing, M. & Heezen, B. C. 1951. Deep-sea sands and submarine canyons: Geol. Soc. Amer., Bull., 62: 961–965.
——. 1952. Turbidity currents and sediments in North Atlantic: Amer. Assoc. Petrol. Geol., Bull., 36: 489–511.
Ericson, D. B., Ewing, M., Heezen, B. C. & Wollin, G. 1955. Sediment deposition in deep Atlantic; *in*, Crust of the Earth: Geol. Soc. Amer., Spec. Pap. 62: 205–219.
Ericson, D. B., Ewing, M. & Wollin, G. 1964. The Pleistocene Epoch in deep-sea sediments: Science, 146: 723–732.
Ericson, D. B., Ewing, M., Wollin, G., Heezen, B. C. 1961. Atlantic deep-sea sediment cores: Geol. Soc. Amer., Bull., 72: 193–286.
Ericson, D. B. & Wollin, G. 1956a. Micropaleontological and isotopic determinations of Pleistocene climates: Micropaleontol., 2, 3: 257–270.
——. 1956b. Correlation of six cores from the equatorial Atlantic and the Caribbean: Deep-Sea Res., 3: 104–125.
——. 1959. Micropaleontology and lithology of Arctic sediment cores: Cambridge Res. Contr., Geophys. Res. Pap. 63: 51–58.
——. 1968. Pleistocene climates and chronology in deep-sea sediments: Science, 162: 1227–1234.
Ericson, D. B., Wollin, G. & Wollin, J. 1954. Coiling direction of *Globorotalia truncatulinoides* in deep-sea cores: Deep-Sea Res., 2: 152–158.
Ewing, M., Ericson, D. B. & Heezen, B. C. 1958. Sediments and topography of the Gulf of Mexico; *in* Habitat of Oil: Amer. Assoc. Petrol. Geol., Tulsa: 995–1053.
Fager, E. W. & McGowan, J. A. 1963. Zooplankton species groups in the North Pacific: Science, 140: 453–460.

Fauré-Fremiet, E. 1914. Les Foraminifères de la seconde Mission Antarctique Francaise IIe Campagne de *Porquoi Pas* (note préliminaire): Soc. Géol. France, Bull., 38: 260–271.
Febvre-Chevalier, C. 1971. Constitution ultrastructurale de *Globigerina bulloides* d'Orbigny 1826 (Rhizopoda - Foraminifera): Protist., 3, 3: 311–324.
Feder, N. & Wolf, M. K. 1965. Studies on nucleic acid metachromasy. II. Metachromatic and orthochromatic staining by toluidine blue of nucleic acids in tissue sections: Jour. Cell Biol., 27: 327.
Ferreira, J. M. 1958. Foraminíferos actuais de Cabo Verde: Junta Invest. Ultramar, Mem., 4: 1–61.
Feyling-Hansen, R. W. 1972. The foraminifer *Elphidium excavatum* (Terquem) and its variant forms: Micropaleontol., 18, 3: 337–354.
Fichtel, L. & Moll, J. P. C. 1803. Testacea microscopica alliaque minuta ex generibus Argonauta et Nautilus ad naturam delineata et descripta: Pichler, Vienna, 2d ed., 123 pp.
Fillon, R. H. 1974. Late Cenozoic foraminiferal paleoecology of the Ross Sea, Antarctica: Micropaleontol., 20, 2: 129–151.
Fischer, P. H. 1966. Ecologie de certain foraminifères du haut niveau sur les côtes du Pacifique: Bull. Zool. France, 91, 3: 295–300.
Fisher, R. A., Corbett, A. S. & Williams, C. B. 1943. The relationship between the number of species and the number of individuals in a random sample of an animal population: Jour. Animal Ecol., 12: 42–58.
Forti, I. R. S. & Roettger, E. 1967. Further observations on the seasonal variation of mixohaline foraminifera from the Patos Lagoon, Southern Brazil: Arch. Oceanogr. Limnol., 15, 1: 55–61.
Fossatti Dutra Pereira, C. A. 1969. Recent foraminifera of Southern Brazil collected by hydrographic vessel 'Baependi': Iheringia, Zool., 37: 37–95.
Fournier, G. 1956. New methods and techniques in the photography of microfossils: Micropaleontol., 2, 1: 37–56.
——. 1957. Construction of pin-hole diaphragms for use in photomicrography: *Ibid.*, 3, 1: 85–87.
Fowler, G. A. & Kulm, L. D. 1966. A multiple corer: Limnol. Oceanogr., 11: 630–633.
Føyn, B. 1936a. Uber die Kernverhältnisse der Foraminifere *Myxotheca arenilaga* Schaudinn: Arch. Protist., 87: 272–295.
——. 1936b. Foraminferenstudien I. Der Lebenszyklus von *Discorbina vilardeboana* d'Orbigny: Bergens Mus. Arbok, Naturv. Rek., 2: 1–22.
——. 1937. Foraminferenstudien II. Zur Kenntnis der asexuellen Fortpflanzung und Entwicklung der Gamonten von *Discorbina vilardeboana*: *Ibid.*, 5: 1–16.
Frankel, J. J. 1964. Recent Foraminifera filled and encrusted with pyrite from Durban Bay, South Africa: South African Jour. Sci., 60, 10: 299–301.
Frankel, L. 1970. A technique for investigating microorganism associations: Jour. Paleontol., 44, 3: 575–577.
——. 1972. Surface reproduction in Foraminifera: *Ibid.*, 46, 1: 62–65.
——. 1974. Observations and speculations on the habitat and habits of *Trochammina ochracea* (Williamson) in subsurface sediments: *Ibid.*, 48, 1: 143–148.
Frerichs, W. E. 1968. Pleistocene-Recent boundary and Wisconsin glacial biostratigraphy in the Northern Indian Ocean: Science, 159: 1456–1468.
——. 1970. Distribution and ecology of benthonic foraminifera in the sediments of the Andaman Sea: Cushman Found. Foram. Res., Contr., 21, 4: 123–147.
Frerichs, W. E., Heiman, M. E., Borgman, L. E. & Bé, A. W. H. 1972. Latitudinal variations in planktonic foraminiferal test porosity: Part 1. Optical studies: Jour. Foram. Res., 2, 1: 6–13.
Freudenthal, H. D., Lee, J. J. & Pierce, S. 1963. Growth and physiology of foraminifera in the laboratory. 2., A tidal system for laboratory studies on eulittoral Foraminifera: Micropaleontol., 9, 4: 443–448.

Freydanck, H. 1955. Die Abhängigkeit einer rezenten Foraminiferen Vergesellschaftung von Sediment und Strömungsgeschwindigkeit des Wassers: N. Jahrb. Geol. Paläont., 100: 332–349.
Friese, H. 1951. Zur Foraminiferen-Fauna der Meeresmolasse des unteren Inngebietes: Dienst Berlin, Abh. N. F., 227: 1–52.
Frizzell, D. L. 1933. Terminology of types: Amer. Mid. Nat., 14: 637–668.
——. 1954. Handbook of Cretaceous Foraminifera of Texas: Univ. Texas, Bur. Econ. Geol., Rep. Inves., 22: 1–232.
Frizzell, D. L. & Schwartz, E. 1950. A new lituolid foraminiferal genus from the Cretaceous with an emendation of *Cribrostomoides* Cushman: Univ. Missouri, School Min. Metal., Bull., Tech. Ser., 76: 1–12.
Furssenko, A. V. 1959. Subclasis Foraminifera, obshchaja chast'; *in* Osnovy Paleontologii, 1: Akad. Nauk SSSR, Moscow: 115–168.
Furssenko, A. V. & Furssenko, K. B. 1970. O foraminiferakh laguny Busse i uslovijakh ikh sushchestvovanija: Akad. Nauk SSSR, Sibirskoe Otdel., Trudy Instit. Geol. Geofiz., 71: 114–135.
Gabel, B. 1971. Die Foraminiferen der Nordsee: Helgoländer Wiss. Meeresunters., 22, 1: 1–65.
Galhano, M. H. 1963. Foraminíferos da costa de Portugal (Algarve): Inst. Zool., Fac. Ciencias Porto, 89: 1–110.
Galloway, J. J. 1926. A revision of the classification and nomenclature of the Foraminifera [abstr.]: Geol. Soc. Amer., Bull., 37: 235–236.
——. 1933. A Manual of Foraminifera: Principia Press, Bloomington, 483 pp.
Gauthier-Lièvre, L. 1935. Sur une des singularités de L'Oued Rhir: des Foraminifères thalassoides vivants dans les eaux sahariennes: Soc. Hist. Nat. Afrique du Nord, Bull., 26: 142–147.
Gervais, P. 1847. Sur un point de la physiologie des Foraminifères: Acad. Sci. Paris, C. R., 25: 467–468.
Gevirtz, J. L., Park, R. A. & Friedman, G. M. 1971. Paraecology of benthonic Foraminifera and associated micro-organisms of the continental shelf off Long Island, New York: Jour. Paleontol., 45, 2: 153–177.
Ghose, B. K. 1974. Statistical analysis of the variation of *Pellatispira glabra* (foraminifer) and its bearing on the life cycle of the species: Jour. Geol., 82: 623–635.
Gibson, L. B. 1966. Some unifying characteristics of species diversity: Cushman Found. Foram. Res., Contr., 17, 4: 117–124.
Gibson, T. G. & Buzas, M. A. 1973. Species diversity: Patterns in modern and Miocene Foraminifera of the eastern margin of North America: Geol. Soc. Amer., Bull., 84: 217–238.
Gibson, T. G. & Walker, W. M. 1967. Flotation methods for obtaining Foraminifera from sediment samples: Jour. Paleontol., 41, 5: 1294–1297.
Giunta, M. 1955. Studio delle microfaune contenute in cinque saggi de fondo prelavati presso S. Margherita Ligura e Chiavari (Genova): Arch. Oceanogr. Limmol., 10.
Glaçon, G. 1964. Radiographie de Foraminifères: Géol. Soc. France, Bull., (7) 6, 1: 10–12.
Glaessner, M. F. 1945. Principles of Micropaleontology: Melbourne Univ. Press, 296 pp.
Goës, A. T. 1894. A synopsis of the Arctic and Scandinavian Recent marine Foraminifera hitherto discovered: K. Sven. Vet, Akad. Handl., 25, 9: 1–127.
Göke, G. 1963. Meeresprotozoen (Foraminiferen, Radíolarien, Tintinninen); Kosmos Verlag, Franckh-Stuttgart, 75 pp.
Golick, A. 1968. History of Holocene transgression in the Gulf of Panama: Jour. Geol., 76, 5: 497–507.
Gorycki, M. A. 1971. A compound embedding technique for the microsopical study of living and fossil foraminifera: Micropaleontol., 17, 4: 492–500.
Grabert, B. 1971. Zur Eignung von Foraminiferen als Indikatoren für Sandwanderung: Deutsch. Hydrogr. Zeit., 24, 1: 1–14.

Graham, J. & Militante, P. J. 1959. Recent Foraminifera from the Puerto Galera area, northern Mindoro, Philipinnes: Stanford Univ. Publ., Geol., 6: 1–132.
Grayson, J.F. 1956. The conversion of calcite to fluorite: Micropaleontol., 2, 1: 71–78.
Green, J. 1968. The Biology of Estuarine Animals: Sedgwick & Jackson, London, 401 pp.
Green, K. E. 1960. Ecology of some Arctic foraminifera: Micropaleontol., 6, 1: 57–78.
Gregory, M. R. 1973. Benthonic Foraminifera from a mangrove swamp, Whangaparapara, Great Barrier Island: Tane, 19: 193–204.
Greiner, G. O. G. 1969. Recent benthonic foraminifera; environmental factors controlling their distribution: Nature, 223: 168–170.
——. 1970. Distribution of major benthonic foraminiferal groups on the Gulf of Mexico continental shelf: Micropaleontol., 16, 1: 83–101.
——. 1974. Environmental factors controlling the distribution of Recent benthonic foraminifera: Brevoria, Mus. Comp. Zool., 420: 1–35.
Grell, K. G. 1952. Der Stand unserer Kenntnise über den Bau der Protistenkerne: Deutsch. Zool. Ges: Freiburg, Verhandl., 24: 212–251.
——. 1954a. Zur Sexualität der Foraminiferen: Naturwiss., 41: 44–45.
——. 1954b. Kerndualismus bei einer Foraminifere: Z. Naturforsch., 9b: 241.
——. 1954c. Der Generationswechsel der polythalamen Foraminifere *Rotaliella heterocaryotica:* Arch. Protist., 100, 2: 268–286.
——. 1956a. Protozoologie: Springer, Heidelberg, 284 pp.
——. 1956b. Der Kerndualismus der Foraminifere *Glabratella sulcata:* Z. Naturforsch., 11b, 6: 336–368.
——. 1956c. Protozoa and Algae: Ann. Rev. Microbiol., 10: 307–328.
——. 1956d. Über die Elimination somatischer Kerne bei heterokaryotischen Foraminiferen: Z. Naturforsch., 11b: 759–761.
——. 1957a. Sexuelle Differenzierung bei den Gamonten der Foraminifere *Patellina corrugata:* Z. Naturforsch., 12b: 414–415.
——. 1957b. Untersuchungen über die Fortpflanzung und Sexualität der Foraminiferen; I - *Rotaliella roscoffensis:* Arch. Protist., 102: 147–164.
——. 1958a. *Idem,* II - *Rubratella intermedia: Ibid.,* 103: 219–308.
——. 1958b. *Idem,* III - *Glabratella sulcata: Ibid.,* 103: 449–472.
——. 1958c. Studien zum Differenzierungsproblem an Foraminiferen: Naturwiss., 45: 3–32.
——. 1959. Untersuchungen über die Fortpflanzung und Sexualität der Foraminiferen; IV - *Patellina corrugata:* Arch. Protist., 104: 211–234.
——. 1962. Entwicklung und Geschlechts-Differenzierung einer neuen Foraminifere: Naturwiss., 49, 9: 214.
——. 1973. Protozoology: Springer, Berlin, 554 pp.
Griggs, G. B., Kulm, L. D., Duncan, J. R. & Fowler, G. A. 1970. Holocene faunal stratigraphy and paleoclimatic implications of deep-sea sediments in Cascadia Basin: Palaeogeogr., Palaeoclimat., Palaeoecol., 7: 5–12.
Grimsdale, T. R. & Morkhoven, F. P. C. M. 1955. The ratio between pelagic and benthonic Foraminifera as a means of estimating depth of deposition of sedimentary rocks: IV World Petrol. Congr., Proc., sect. I/D. Rept. 4: 473–491.
Grossman, S. 1967. Ecology of Rhizopoda and Ostracoda of southern Pamlico Sound Region, North Carolina–Part 1., Living and subfossil rhizopod and ostracode populations: Univ. Kansas Paleontol. Contr. 44, Ecology. 1, 82 pp.
Gruber, A. 1884. Über Kern und Kernteilung bei den Protozoen: Z. Wiss. Zool., 40: 121–153.
Guilcher, A., Berthois, L., Le Calvez, Y., Battistini, R. & Crosnier, A. 1965. Les récifs coralliens et le lagon de l'ile Mayotte: Off. Rech. Sci. Techn. Outre-Mer, Paris, 210 pp.
Gunter, G. 1957. Temperature; *in* Treatise Mar. Ecol. Paleoecol., 1; Geol. Soc. Amer., Mem. 67: 159–184.

Haake, F. W. 1962. Untersuchungen an der Foraminiferen-Fauna im Wattgebiet zwischen Langeoog und dem Festland: Meyniana, 12: 25–64.
——. 1964. Eine neue Methode zur Anfertigung von Dünnschnitten rezenter Foraminiferen: *Ibid.*, 14: 1–2.
——. 1967. Zum Jahresgang von Populationen einer Foraminiferen-Art in der westlichen Ostee: *Ibid.*, 17: 13–27.
——. 1970. Zur Tiefenverteilung von Miliolinen (Foram.) im Persischen Golf: Paläontol. Zeit., 44, 3–4: 196–200.
——. 1971. Ultrastructure of miliolid walls: Jour. Foram. Res., 1, 4: 187–189.
Häcker, V. 1908. Tiefseeradiolarien: Wiss. Ergebn. Deutsch. Tiefsee Exp. *Valdivia:* 417–706.
Hada, Y. 1936. Studies on the Foraminifera of brackish waters. I. Hijirippu and Mochirippu Lakes: Zool. Mag., 48: 847–860.
——. 1937. *Idem*, II. Hachiro-Gata. III. Koyama-Ike: *Ibid.*, 49: 341–347.
——. 1939. *Idem*, IV. The Foraminifera of Nakanoumi: *Ibid.*, 51: 135–140.
——. 1955. *Idem*, V. The Foraminifera from some lagoons in the region of Hokuriku: Biogeogr. Soc. Japan, Bull., 16–19: 1–6.
——. 1957. Biology of the arenaceous Foraminifera: Jour. Sci. Suzugamie Coll. 3, pt. B: 31–50.
Haga, H. 1963. Distribution of Foraminifera in the Gulf of Thailand sediments [abstract]; Journ. Paleontol., 37, 4: 977.
Haller, G., Ehret, C. F. & Naef, R. 1961. Technique d'inclusion et ultramicrotomie destinée a l'étude du developpment des organelles dans une cellule isolé: Experientia, 17: 524–529.
Haman, D. 1969. Seasonal occurrence of *Elphidium excavatum* (Terquem) in Llandanwg lagoon (North Wales, U.K.): Cushman Found. Foram. Res., Contr., 20, 4: 139–142.
——. 1971a. Foraminiferal assemblages in Tremadoc Bay, North Wales, U.K.: Jour. Foram. Res., 1, 3: 126–143.
——. 1971b. Morphologic variability of the genus *Technitella* Norman, 1878: Micropaleontol., 17, 4: 471–474.
Hamilton, E. L. 1953. Upper Cretaceous, Tertiary and Recent planktonic Foraminifera from Mid-Pacific flat-topped sediments: Jour. Paleontol., 27: 204–237.
——. 1957. Planktonic Foraminifera from an Equatorial Pacific core: Micropaleontol., 3, 1: 69–73.
Hamlin, W. H. & Bé, A. W. H. 1963. A plankton picker: Deep-Sea Res., 10: 459–461.
Hansen, H. J. 1965. On the sedimentology and the quantitative distribution of living foraminifera in the northern part of the Oresund: Ophelia, 2, 2: 323–331.
——. 1967. A technique for depiction of grind sections of foraminifera by aid of compiled electronmicrographs: Dan. Geol. Foren., Medd., 17, 1: 128–129.
——. 1968a. A technique for removing gold from plated calcareous microfossils: Micropaleontol., 14, 4: 499–500.
——. 1968b. X-ray diffractometer investigations of a radiate and a granulate foraminifera: Dan. Geol. Foren., Medd., 18, 3–4: 346–348.
——. 1969. Combined light and electron microscope studies of thin sections: Micropaleontol., 15, 4: 426.
——. 1970. Electronmicroscopical studies on the ultrastructure of some perforate calcitic radiate and granulate foraminifera: Dan Vidensk. Selsk., Biol. Skr., 17, 2: 1–16.
——. 1972a. Two species of Foraminifera of the genus *Turrilina* with different wall structure: Lethaia, 5: 39–45.
——. 1972b. Pore pseudopodia and sieve plates of *Amphistegina:* Micropaleontol., 18, 2: 223–230.
——. On feeding and supposed buoyancy mechanism in four Recent globigerinid foraminifera from the Gulf of Elat, Israel: Rev. Españ. Micropaleontol., 7, 2: 325–339.
Hansen, H. J. & Reiss, Z. 1971. Electron microscopy of rotaliacean wall structures: Geol. Soc. Den., Bull., 20, 4: 329–346.

Hansen, H. J. & Reiss, Z. 1972. Scanning electron microscopy of wall structures in some benthonic and planktonic Foraminiferida: Rev. Españ. Micropaleontal., 4, 2: 169–180.
Hansen, H. J., Reiss, Z. & Schneidermann, N. 1969. Ultramicrostructure of bilamellar walls in Foraminiferida: *Ibid.*, 1, 3: 293–316.
Hardy, A. 1970. The Open Sea: its natural history; part 1, The World of Plankton: Collins, London, 2d ed., 335 pp.
Harland, R. & Sutherland, G. 1972. A technique for the water mounting of microfossils for study and photography: Micropaleontol., 18, 1: 119–121.
Harman, R. A. 1964. Distribution of foraminifera in the Santa Barbara Basin, California: *Ibid.*, 10, 1: 81–96.
Harrington, G. 1956. *Ammobaculites*, migrant or Recent introduction to California?: Cushman Found. Foram. Res., Contr., 7 1: 29–30.
Harting, P. 1851. Die Macht des Kleinen: Englemann, Leipzig, 171 pp.
Hay, W. W. 1971. Scanning electron microscopy and information transfer in systematic micropaleontology; *in* Scanning electron microscopy, systematic and evolutionary applications: The Systematics Assoc., Spec. Vol. 4: 123–143.
Hay, W. W. & Sandberg, P. A. 1967. The scanning electron microscope, a major breakthrough for micropaleontology: Micropaleontol., 13, 4: 407–418.
Hay, W. W., Towe, K. M. & Wright, R. C. 1963. Ultramicrostructure of some selected foraminiferal tests: *Ibid.*, 9, 2: 171–195.
Haynes, J. 1964. Live and dead foraminifera between the Sarns, Cardigan Bay: Nature, 203: 774.
——. 1965. Symbiosis, wall structure and habitat in Foraminifera: Cushman Found. Foram. Res., Contr., 16, 1: 40–43.
——. 1973. Cardigan Bay Recent foraminifera (Cruises of the R. V. *Antur*, 1962–1964): Brit. Mus. (Nat. Hist.), Bull. Zool., Suppl. 4: 1–245.
Haynes, J. & Dobson, M. 1969. Physiography, foraminifera and sedimentation in the Dovey estuary (Wales): Geol. Jour., 6, 2: 217–256.
Hecht, A. D. 1973. A model for determining Pleistocene paleotemperatures from planktonic foraminiferal assemblages: Micropaleontol 19, 1: 68–77.
Hecht, A. D., Savin, S. M. 1970. Oxygen-18 studies of recent planktonic foraminifera; comparisons of phenotypes and of test parts: Science, 170: 69–71.
——. 1971. O^{18} isotopes of Recent foraminifera: *Ibid.*, 173: 168–169.
——. 1972. Phenotypic variation and oxygen isotope ratios in Recent planktonic foraminifera: Jour. Foram. Res., 2, 2: 55–67.
——. 1974. Intraspecific variation in Recent populations of *Globigerinoides ruber* and *Globigerinoides trilobus* and their application to paleoenvironmental analysis: Jour. Paleontol., 48, 6: 1217–1234.
Hedberg, H. D. 1934. Some Recent and fossil brackish to fresh-water Foraminifera: *Ibid.*, 8: 469–476.
Hedgpeth, J. W. 1957. Marine biogeography; *in* Treatise Mar. Ecol. Paleoecol., 1; Geol. Soc. Amer., Mem. 67: 359–382.
Hedley, R. H. 1957. Microradiography applied to the study of Foraminifera: Micropaleontol., 3, 1: 19–28.
——. 1958. A contribution to the biology and cytology of *Haliphysema* (Foraminifera): Zool. Soc. London, Proc., 130, 4: 569–576.
——. 1962a. *Gromia oviformis* (Rhizopoda) from New Zealand with comments on the fossil Chitinozoa: New Zeal. Jour. Sci., 5, 2: 427–434.
——. 1962b. The significance of an 'inner chitinous lining' in saccamminid organization with special reference to a new species of *Saccammina* (Foraminifera) from New Zealand: *Ibid.*, 5, 3: 375–389.
——. 1963. Cement and iron in the arenaceous foraminifera: Micropaleontol., 9, 4: 433–441.

Hedley, R. H. 1964. The biology of Foraminifera: Inter. Rev. Gen. Exper. Zool. 1: 1–45.
Hedley, R. H. & Bertaud, W. 1962. Electron-microscopic observations of *Gromia oviformis* (Sarcodina): Jour. Protozool., 9, 1: 79–87.
Hedley, R. H. Hurdle, C. M. & Burdett, I. D. J. 1965. A Foraminiferal fauna from the western continental shelf, North Island, New Zealand: New Zeal. Dept. Sci. Indus. Res. Bull. 163: 9–48.
——. 1967. The marine fauna of New Zealand: Intertidal Foraminifera of the *Corallina officinalis* Zone: *Ibid.*, 180: 9–86.
Hedley, R. H., Parry, D. M. & Wakefield, J. St. J. 1967. Fine structure of *Shepheardella taeniformis* (Foraminifera: Protozoa): Jour. Roy. Micro. Soc., 87: 445–456.
——. 1968. Reproduction in *Boderia turneri* (Foraminifera): Jour. Nat. Hist., 2: 147–151.
Hedley, R. H. & Wakefield, J. St. J. 1967a. A collagen-like sheath in the arenaceous foraminifer *Haliphysema* (Protozoa): Jour. Roy. Micro. Soc., 87, 3–4: 475–481.
——. 1967b. Clone culture studies of a new Rosalinid foraminifer from Plymouth, England and Wellington, New Zealand: Jour. Mar. Biol. Assoc. U.K., 47, 1: 121–128.
Hedley, R. H. & Underwood, C. M. 1957. Living Foraminifera and empty shells from the North Kent Coast: Ann. Mag. Nat. Hist., 10, 118: 721–732.
Helland-Hansen, B. 1916. Nogen hydrografishe metoder: Skand. Naturforsker, Kristiania, Oslo.
Hemleben, C. 1969a. Ultramicroscopic shell and spine structure of some spinose planktonic foraminifera: Int. Conf. Plank. Microfossils, 1st., Proc., Geneva, 2: 254–256.
——. 1969b. Zur Morphogenese planktonischer Foraminiferen: Zitteliana, 1: 91–106.
——. 1969c. Ultrastrukturen bei kalkschaligen Foraminiferen: Naturwiss., 56, 11: 534–538.
Hendrix, W. E. 1958. Foraminiferal shell form, a key to sedimentary environment: Jour. Paleontol., 32, 4: 649–659.
Henson, F. 1934. Portable sedimentary laboratory: Amer. Assoc. Petrol. Geol., Bull., 18: 1705–1709.
Hentschel, E. 1933. Allgemeine Biologie des Südatlantischen Ozeans: Wiss. Ergebn. Deutsch. Exp. *Meteor*, 11: 1–343.
Herb, R. 1968. Recent planktonic foraminifera from sediments of the Drake Passage, Southern Ocean: Eclogae Geol. Helv., 61, 2: 467–480.
——. 1971. Distribution of Recent benthonic foraminifera in the Drake Passage; *in* Biology of the Antarctic Seas IV: Antarct. Res. Ser., 17: 251–300.
Herman, Y. 1964. Temperate water planktonic Foraminifera in Quaternary sediments of the Arctic Ocean: Nature, 201: 386–387.
——. 1965. Étude des sédiments quaternaires de la Mer Rouge: Ann Inst. Océanogr. Monaco, 2: 1–341.
——. 1968. Evidence of climatic changes in Red Sea cores; *in* Means of Correlation of Quaternary Successions: Inter. Congr. Quater. Res., 8, proc. VII: 325–348.
——. 1969. Arctic Ocean Quaternary microfauna and its relation to paleoclimatology: Palaeogeogr., Palaeoclimat., Palaeoecol., 6: 251–276.
——. 1970. Arctic paleo-oceanography in Late Cenozoic time: Science, 169: 474–477.
——. 1972a. Quaternary eastern Mediterranean sediments: micropaleontology and climatic record; *in*, The Mediterranean Sea: A natural sedimentation laboratory: Dowen, Hutchinson and Ross, Stroudsburg, Pennsylvania: 129–147.
——. 1972b. *Globorotalia truncatulinoides:* a paleo-oceanographic indicator: Nature, 238: 394–396.
Herman, Y., O'Neil, J. R. & Drake, C. L. 1972. Micropaleontology and paleotemperatures of postglacial SW Greenland fjord cores; *in* Climatic Changes in Artic Areas during the last Ten Thousand Years: Acta Univ. Ouluensis, (A) Sci. Rerum Nat., 3, Geol. 1: 357–381.
Heron-Allen, E. 1915. Contributions to the study of the bionomics and reproductive processes of the Foraminifera: Roy. Soc. London, Phil. Trans., (B) 206: 227–279.

Heron-Allen, E. & Earland, A. 1912. On some foraminifera from the North Sea, etc., dredged by the Fisheries Cruiser *Goldseeker;* Part I - On some new Astrorhizidae and their shell structure: Jour. Roy. Micro. Soc.: 782–389.
——. 1914. The Foraminifera of the Kerimba Archipelago: Zool. Soc. London, Trans., 20, 1: 363–390.
——. 1915, The Foraminifera of the Kerimba Archipelago: *Ibid.*, 20, 2: 543–797.
——. 1920. An experimental study of the foraminiferal species, *Verneuilina polystropha* (Reuss), and some others, being a contribution to a discussion 'on the origin evolution and transmission of biological characters': Roy. Irish Acad., Proc., 35, sect. B: 153–175.
——. 1922. Protozoa. Part III - Foraminiferà: Brit. Antarct. (Terra Nova) Exp. 1910. Nat. Hist. Rep., Zool., 6, 2: 599–647.
——. 1929a. Some Foraminifera from the South Atlantic - I: Jour. Roy. Micr. Soc., 49: 101–108.
——. 1929b. *Idem* - II: *Ibid.*, 49: 324–334.
——. 1929c. *Idem - III: Ibid.*, 50: 38–45.
——. 1930. The Foraminifera of the Plymouth District: *Ibid.*, 50: 161–199.
——. 1932. Foraminifera, Part 1, The ice-free area of the Falkland Islands and adjacent seas: Discovery Rept., 4: 291–460.
Hertwig, R. 1876. Bemerkungen zur Organization und systematischen Stellug der Foraminiferen: Jenaische Z. Naturwiss., 10: 41–55.
——. 1877. Studien über Rhizopoden: *Ibid.*, 11: 324–348.
Hida, T. S. & King, J. E. 1955. Vertical distribution of zooplankton in the central equatorial Pacific, July to August. 1952: U.S. Fish. Wildl. Serv., Spec. Sci. Rept., 114: 1–22.
Hiltermann, H. 1954. Zur Artfassung in der Paläontologie: Roemeriana, 1: 385–392.
——. 1958. Anwendung der Mikropaläontologie in der Geologie entwickelt durch die Erdölgeologie; *in* Handbuch der Mikroskopie in der Technik, 2, 3: 15–40.
Hiltermann, H. & Koch, W. 1950. Taxonomie und Vertikalverbreitung von *Bolivinoides* Arten im Senon Nordwestdeutschland: Geol. Jahrb., 64: 595–632.
Himes, J. & Moriber, L. 1956. A triple stain for desoxyribonucleic acid, polysaccharides and proteins: Stain. Techn., 31: 67–70.
Hirsch, E. 1912. Die Entwicklungsgeschichte von *Saccammina:* Arch. Portist., 27: 219–253.
Hoffmeister, W. S. 1955. Microfossils provide new technique in exploration: World Oil, 140, 15.
Hofker, J. 1922. De Protozoen: Flora en Fauna der Zuiderzee, Monogr. van een Brakwatergebiet: 127–183.
——. 1925. On heterogamy in Foraminifera: Tijdschr. Ned. Dier. Ver., (2), 19: 68–70.
——. 1927. The Foraminifera of the Siboga Expedition: Siboga Exp., Mon. 4, 1: 1–78.
——. 1930a. Die Fortpflanzung der Foraminiferen: Ann. Protist., 3: 25–34.
——. 1930b. Der Generationswechsel von *Rotalia beccarii* var. *flevensis*, nov. var.: Z. Zellenf. Mikrosk. Anat., 10, 4: 756–768.
——. 1930c. Notizen über die Foraminiferen des Golfes von Neapel. I, II: Staz. Zool. Napoli, Publ., 10, 12: 365–406.
——. 1932. *Idem.* III. Die Foraminiferen Fauna der Ammontatura: *Ibid.*, 12, 1: 61–144.
——. 1933. Foraminifera of the Malay Archipelago: Dansk. Naturh. Foren., Vidensk., 93: 1–71.
——. 1951a. The Foraminifera of the Siboga Expedition, Part III: E. J. Brill, Leiden, 513 pp.
——. 1951b. The toothplate Foraminifera: Arch. Néerl. Zool. 8: 353–373.
——. 1951c. Pores of Foraminifera: Micropaleontologist, 5: 38.
——. 1956. Foraminifera dentata: Univ. Zool. Mus. København. Skrift., 15: 1–237.
——. 1962. Studien an planktonischen Foraminiferen: N. Jahrb. Geol. Paläont., 114: 81–134.
——. 1967. Hat die feinere Wandstruktur der Foraminiferen supragenerische Bedeutung?: Paläontol. Zeit., 41, 3–4: 194–198.

Hofker, J. 1968. Studies on foraminifera. Part 1 – General problems: Overdruk Publ. van het Natuurhist. Genootschap in Limburg, 18: 1–135.
——. 1971. Wall structure of globigerine and globorotalid foraminifera: Rev. Españ. Micropaleontol., 3, 1: 35–60.
——. 1972. Is the direction of coiling in the early stages of an evolution of planktonic Foraminifera at random? (50% right and 50% left): *Ibid.*, 4, 1: 11–17.
Höglund, H. 1947. Foraminifera in the Gullmar fjord and the Skagerak: Zool. Bidrag. Fran Uppsala, 26: 1–328.
Honjo, S. & Berggren, W. A. 1967. Scanning electron microscope studies of planktonic foraminifera: Micropaleontol. 13, 4: 393–406.
Honjo, S. & Okada, H. 1968. Scanning electron microscopy of planktonic foraminifera, a preparation technique: Hokkaido Univ., Fac. Sci., Jour., 4, 14, 1: 71–76.
Hooper, K. 1964. Electron Probe X-ray microanalysis of Foraminifera – An exploratory study: Jour. Paleontol., 38, 6: 1082–1092.
——. 1968. Benthonic foraminiferal depth-assemblages of the continental shelf off Eastern Canada: Maritime Seds., 4, 3: 96–99.
——. 1969. A re-evaluation of eastern Mediterranean foraminifera using factor-vector analysis: Cushman Found. Foram. Res., Contr., 20, 4: 147–151.
Hornibrook, N. B. 1968. Distribution of some warm benthic Foraminifera in the New Zealand Tertiary: Tuatara, 16, 1: 11–15.
Howarth, R. J. & Murray, J. W. 1969. The Foraminiferida of Christchurch Harbour, England: A reappraisal using multivariate techniques: Jour. Paleontol., 43, 3: 660–675.
Howe, H. V. 1959. Fifty years of micropaleontology: *Ibid.*, 33: 511–517.
Huang, T. 1971a. Foraminiferal trends in the surface sediments of Taiwan strait: U.N. Econ. Comm. Asia Far East, Comm. Coord. Joint Pros. Miner. Resour. Asian Offshore Areas, Tech. Bull., 4: 23–61.
——. 1971b. Species diversity of benthonic foraminifers in the Taiwan Strait, Taiwan, China: Geol. Soc. China. Proc., 15: 99–110.
Hulburt, E. M. 1963. The diversity of phytoplankton populations in oceanic, coastal, and estuarine regions: Jour. Mar. Res., 21, 2: 81–93.
Hulme, S. G. 1964. Recent foraminifera from Manakau Harbour, Aukland, New Zealand: New Zeal. Jour. Sci., 7, 3: 305–340.
Humsaon, G. 1962. Animal Tissue Techniques: W. H. Freeman, San Francisco, 569 pp.
Hunkins, K., Be, A. W. H., Opdyke, N. D. & Mathieu, G. 1971. The late Cenozoic history of the Arctic Oceans; *in* The Late Cenozoic Glacial Ages: Yale Univ. Press: 215–237.
Hurlbert, S. H. 1971. The nonconcept of species diversity: a critique and alternative parameters: Ecology, 52, 4: 577–586.
Hussey, K. M. & McNulty, C. L. 1950. *Planularia planotrochiformis*, a new species showing variation in the genus: Jour. Paleontol., 24: 472–473.
Hyde, P. & Krinsley, D. 1964. An improved technique for electron microscopic examination of foraminifera: Micropaleontol., 10, 4: 491–493.
Iaccarino, S. 1967. Ricerche sui foraminiferi dell'alto Adriatico: Arch. Oceanogr. Limnol., 15, 1: 11–54.
——. 1969. I foraminiferi de campioni de fondo prelevati nel Golfo di Taranto (M. Jonio): L'Ateneo Parmense, 5, 1: 1–27.
Ihering, H. 1927. Die Geschichte des Atlantischen Ozeans: Fischer, Jena, 237 pp.
Ikeya, N. 1970. Populational ecology of benthonic foraminifera in Ishikari Bay, Hokkaido, Japan: Rec. Oceanogr. Works Jap., 10, 2: 173–191.
——. 1971. Species diversity of benthonic foraminifera, off the Shimokita Peninsula, Pacific coast of north Japan: *Ibid.*, 11, 1: 27–37.
Illing, M. 1950. The mechanical distribution of Recent Foraminifera in Bahama Banks sediments: Ann. Mag. Nat. Hist., (12) 3: 757–761.

Imbrie, J. 1956. Biometrical methods in the study of invertebrate fossils: Amer. Mus. Nat. Hist., Bull., 108, 2: 211–252.
Imbrie, J., van Donk, J. & Kipp, N. G. 1973. Paleoclimatic investigation of a Late Pleistocene Caribbean Deep-sea core: comparison of isotopic and faunal methods; Quater. Res., 3, 1: 10–38.
Imbrie, J. & Kipp, N. E. 1971. A new micropaleontological method for quantitative paleoclimatology: Application to a Late Pleistocene Caribbean core; *in* The Late Cenozoic Glacial Ages: Yale Univ. Press.: 71–182.
Imbrie, J. & Phleger, F. B. 1963. Análisis por vectores de los foraminíferos bentónicos del área de San Diego, California: Soc. Geol. Méx., Bol., 26: 103–105.
Ingmanson, D. & Ross, A. 1969. Seasonal changes in Foraminifera at Seahorse Key: Quart. Jour. Fla. Acad. Sci., 32, 2: 108–118.
Iselin, C. 1936. A study of the circulation of the western North Atlantic: Mass. Inst. Tech. and Woods Hole Oceanogr. Inst., Pap. Phys. Oceanogr. Meteorol., 4, 4: 1–101.
Ishiwada, Y. 1958. Studies on the brackish water. III. Recent Foraminifera from the brackish lake Haman-Ko: Geol. Surv. Japan, Rept. 180: 1–19.
Jahn, B. 1953. Elektronenmikroskopische Untersuchungen an Foraminiferenschalen: Z. Wiss. Mikroskop., 61: 294–297.
Jahn, T. L. & Rinaldi, R. A. 1959. Protoplasmic movement in the foraminiferan, *Allogromia laticollaris*, and a theory of its mechanism: Biol. Bull. 117: 100–118.
Jarke, J. 1960. Beitrag zur Kenntnis der Foraminiferenfâuna der mittleren und westlichen Barents-See: Int. Rev. Ges. Hydrobiol., 45, 4: 581–654.
Jell, J. S., Maxwell, W. H. G. & McKellar, R. G. 1965. The significance of the larger Foraminifera in the Heron Island reef sediments: Jour. Paleontol., 39, 2: 273–279.
Jenkins, D. G. 1967. Recent distribution, origin, and coiling ratio changes in *Globorotalia pachyderma* (Ehrenberg): Micropaleontol., 13, 2: 195–203.
Jepps, M. W. 1926. Contributions to the study of *Gromia oviformis* Dujardin: Micr. Soc., Quart. Jour., 70, n.s.: 701–719.
——. 1942. Studies on *Polystomella* Lamarck: Jour. Mar. Biol. Assoc. U.K., 25: 607–666.
——. 1953. Nuclei of *Cycloclypeus carpenteri* Brady: Nature, 171: 1114–1115.
——. 1956. The Protozoa Sarcodina: Oliver and Boyd, Edinburgh, 183 pp.
Johnson, K. R. & Albani, A. D. 1973. Biotopes of Recent benthonic foraminifera in Pitt Water, Broken Bay, N.S.W. (Australia): Palaeogeogr., Palaeoclim., Palaeoecol., 14, 4: 265–276.
Johnson, R. G. 1965. Pelecypod death assemblages in Tomales Bay, California: Jour. Paleontol., 39: 80–85.
Jones, J. 1966. MS, Plantonic foraminifera as indicator organisms in the eastern North Atlantic Equatorial Current system: Univ. Miami Inst. Mar. Sci.
——. 1967. Significance of distribution of planktonic Foraminifera in the Equatorial Atlantic Undercurrent: Micropaleontol., 13, 4: 489–501.
——. 1971. The ecology and distribution of living planktonic Foraminifera in the Straits of Florida: Miami Geol. Soc., Mem. 1: 175–190.
Jousé, A. P. & Semina, G. I. 1955. Obshchie zakonomernosti v raspredelenii diatomovykh v planktone Beringova morja i v poverkhnostnykh donnykh osadkakh: Akad. Nauk SSSR, Dokl.,: 100.
Kabanova, J. G. 1961. O kultivirovanii v laboratornykh uslovijakh morskikh planktonnykh diatomovykh i peridinievykh vodoroslej: Inst. Okean., Trudy, 47: 203–216.
Kaesler, R. L. 1966. Quantitative re-evaluation of ecology and distribution of Recent Foraminifera and Ostracoda of Todos Santos Bay, Baja California, Mexico: Univ. Kansas Paleontol. Contr., Pap. 10, 50 pp.
Kaever, M. 1960. Biostratigraphische Auswertung der Windungsrichtung von Foraminiferen: Geol. Jahrb.,: 77.

Kafescioglu, I. A. 1971. Specific diversity of planktonic foraminifera on the continental shelves as a paleobathymetric tool: Micropaleontol., 17, 4: 455–470.
Kameswara„R. K. 1971. On some Foraminifera from the northeastern part of the Arabian Sea: Indian Acad. Sci., Proc., 73, sect. B, 4: 155–178.
Kane, H. E. 1967. Recent microfaunal biofacies in Sabine Lake and environs, Texas and Louisiana: Jour. Paleontol., 41, 4: 947–964.
Kane, J. 1956. North Atlantic planktonic Foraminifera as Pleistocene temperature indicators: Micropaleontol., 2, 3: 287–294.
Keany, J. & Kennett, J. P. 1972. Pliocene – early Pleistocene paleoclimatic history recorded in Antarctic-Subantarctic deep-sea cores: Deep-Sea Res., 19, 8: 529–548.
Keijzer, C. J. 1936. On variability in East Indian Foraminifera: Temminckia, 1: 75–153.
Kennard, M. C. & Smith, A. J. 1961. A simple micro-sample splitter: Jour. Paleontol., 25, 2: 396–397.
Kennett, J. P. 1966. Foraminiferal evidence of a shallow calcium carbonate solution boundary, Ross Sea, Antarctica: Science, 153: 191–193.
——. 1967. New foraminifera from the Ross Sea, Antarctica: Cushman Found. Foram Res., Contr., 18, 3: 133–135.
——. 1968a. *Globorotalia truncatulinoides* as a paleo-oceanographic index: Science, 159: 1461–1463.
——. 1968b. Latitudinal variation in *Globigerina pachyderma* (Ehrenberg) in surface sediments of the southwest Pacific Ocean: Micropaleontol., 14, 3: 305–318.
——. 1968c. The fauna of the Ross Sea. VI. Ecology and distribution of foraminifera: New Zeal. Dept. Sci. Industr. Res. Bull. 186: 7–48.
——. 1970a. Comparison of *Globigerina pachyderma* (Ehrenberg) in Arctic and Antarctic areas: Cushman Found. Foram. Res., Contr., 21, pt. 2: 47–49.
——. 1970b. Pleistocene paleoclimates and foraminiferal biostratigraphy in subantarctic deep-sea cores: Deep Sea Res., 17, no. 1: 125–140.
Kennett, J. P. & Huddlestun, P. 1972a. Late Pleistocene paleoclimatology, foraminiferal biostratigraphy and tephrochronology, western Gulf of Mexico: Quat. Res., 2, 1: 38–69.
——. 1972b. Abrupt climatic change at 90,000 years B.P.: faunal evidence from Gulf of Mexico cores: *Ibid.*, 3: 384–395.
Kennett, J. P. & Vella, P. 1971. Paleomagnetic chronology of Pliocene-early Pleistocene climates and the Plio-Pleistocene boundary in New Zealand: Science, 171: 276–279.
Kheradpir, A. 1970. Foraminiferal trends in the Quaternary of Tanner Basin, California: Micropaleontol., 16, 1: 102–116.
Khusid, T. A. 1971. Raspredelenie taksotsenozov foraminifer na materikovoj stupeni Juzhnoj Ameriki v Tikhom okeane: Okeanol., 11, 2: 266–269.
Kiaer, H. 1899. Thalamophora: Norwegian North-Atlantic Exp. 1876–1878. Zool., 7: 1–13.
——. 1908. Fortgnelse over Tromsosundets foraminifer, Tromso, Norway: Mus. Arshefter, 25.
——. 1909. On the bottom deposits from the second Norwegian Arctic Expedition in the *Fram:* 2d Norwegian Arctic Exped., *Fram*, Rept., 3, 17.
Kihle, R. 1971. Foraminifera in five sediment cores in a profile across the Norwegian Channel south of Mandal: Norsk Geol. Tidssk., 51: 261–286.
Kim, B. K., Kim, S. W. & Kim, J. J. 1970. Foraminifera in the bottom sediments off the southwestern coast of Korea: U.N. Econ. Comm. Asia and Far East, Comm. Coord. Joint Pros. Miner. Resour., Asian Offshore Areas, Tech. Bull., 3: 147–163.
King, J. E. & Demond, J. 1953. Zooplankton abundance in the Central Pacific: U.S. Fish Wildl. Serv., Fish. Bull., 82: 111–144.
King, K. & Hare, P. E. 1972. Amino acid composition of the test as a taxonomic character for living and fossil planktonic foraminifera: Micropaleontol., 18, 3: 285–293.
Kinne, O. 1963. The effects of temperature and salinity on marine and brackish water animals,

I. Temperature; *in*, Oceanogr. Mar. Biol. Ann. Rev., 1: George Allen & Unwin Ltd., London: 301–340.
Klinckowstroem, C. 1952. Chronik der Bäume: Kosmos, 48, 3: 114–117.
Knauff, W. 1963. Über Zuchtversuche an *Patellina corrugata* (Foram.) in unterschiedlichen Meerwasser-lösungen: Fortschritte in der Geol. von Rhein. Westfalen, Krefeld, 10: 363–366.
Kornicker, L. S. 1957. A spirally designed picking tray: Micropaleontol., 3, 2: 189.
Kraft, J. C. & Margules, G. 1971. Sediment patterns, physical characters of the water mass and Foraminiferida distribution in Indian River Bay, Coastal Delaware: Southeast. Geol., 12, 4: 223–252.
Krapin, V. O. 1929. O Foraminiferakh rek vpadajushchikh v ozero Ehlton: Russ. Gidrobiol. Zhurn., 8, 6–7.
Krasheninnikov, V. A. 1956. Mikrostruktura stenki nekotorykh kainozoiskikh foraminifer i metodika ee izuchenija v poljarizovannom svete: Vopr. Micropal., 1: 37–48.
——. 1958a. Stenofathyal'nye i evryfathyal'nye vidy foraminifer: VNIGRI, Trudy, 9, Paleontol. Sb., 2: 285–305.
——. 1958b. O stroenii ust'ja u nekotorykh predstavitelej nonionid i el'fidiid: Vopr. Micropal., 2: 105–120.
——. 1959. El'fidiidy v miothenovykh otlozhenijakh Podolii: Akad. Nauk SSSR, Geol. Inst., Trudy, 21: 3–136.
——. 1960. Izmenenie kompleksov foraminifer v ritmakh osadkonakoplenija miothenovykh otlozhenij jugo-zapada Russkoj platformy: Vopr. Micropal., 4: 33–70.
Krinsley, D. 1960. Trace elements in the tests of panktonic foraminifera: Micropaleontol., 6, 3: 297–300.
Krinsley, D. & Bé, A. W. H. 1965. Electron microscopy of internal structures of foraminifera; *in* Handbook of Paleontological Techniques: Freeman and Co., San Francisco: 335–434.
Krivoborskij, V. V. 1960. Dostizhenija v oblasti obwemnogo fotografirovanija: 1st Semin. Microfaun., Trudy: 138–169.
Kruit, C. 1955. Sediments of the Rhone Delta: Mouton and Co., Gravenhage, 141 pp.
Krümmel, O. 1907. Handbuch der Ozeanographie, Vol. 1: Engelhorn, 526 pp.
Kudo, R. R. 1969. Protozoología: Contiental, México, 905 pp.
Kudrass, H. R. & Thiede, J. 1970. Stratigraphische Untersuchungen an Sedimentkernen des ibero-marokkanischen Kontinentalrandes: Geol. Rundsch., 60, 1: 294–301.
Kugler, H. 1957. Preparation of microfossils for photography: Micropaleontol., 3, 2: 190.
Kummel, B. & Raup, D., eds. 1965. Handbook of Paleontological Techniques: Freeman and Co., San Francisco, 852 pp.
Kurc, G. 1961. Foraminifères et Ostracodes de l'étang de Thau: Inst. Pêches Marit., Rev. Trav., 25: 134–248.
Kureshy, A. A. 1969. Ecological studies of Foraminifera of Wash (England) and relationship between their distribution and sedimentation: Rev. Micropaléontol., 11, 4: 222–232.
Kurnick, N. B. 1955. Pyronin Y in the methyl green pyronin histological stain: Stain. Tech., 30: 213.
Kustanowich, S. 1964. Foraminifera of Milford Sound: New Zeal. Dept. Sci. Industr. Res. Bull., 157: 49–63.
Kuznethova, K. I. 1960a. Taksonomicheskoe znachenie nekotorykh morfologicheskikh priznakov jurskikh lenticulin: Vopr. Micropal., 4: 89–101.
Ladd, H. S. Hedgpeth, J. W. & Post, R. 1957. Environments and facies of existing bays of the Central Texas coast; *in* Treatise Mar. Ecol. Paleoecol., 2: Geol. Soc. Amer., Mem. 67: 599–640.
Lalicker, C. G. 1948. Dwarfed protozoan faunas: Jour. Sed. Petrol., 18: 51–55.
Lamarck, J. B. 1801. Système des Amimaux sams Vertèbres au Tableau Général des Classes, des Orders et des Genres de ces Animaux: Paris, 432 pp.
——. 1823. Recueil de Planches des Coquilles Fossiles des Environs de Paris: Paris.

Lamb, G. M. 1972. Distribution of Holocene Foraminiferida in Mobile Bay and the effect of salinity changes: Geol. Surv. Alab., Circ. 82: 1–12.
Lamotte, M. 1957. Introduction aux Méthodes Statistiques en Biologie: Masson & Cie., Paris.
Lamy, A. 1969. Hypothèse sur le cycle de vie de *Nonion dollfusi* Cushman dans le Miocène aquitain: Soc. Géol. France, C.R., 4: 123–124.
Lankford, R. R. 1959. Distribution and ecology of Foraminifera from east Mississippi delta margin: Amer. Assoc. Petrol. Geol., Bull., 43, 9: 2068–2099.
——. 1967. The depauperate fauna concept relative to Foraminifera: Lect. 8; *in*, Paleoecology, AGI Short Course, AGI, Washington,: RL1B–RL12B.
Lankford, R. R. & Phleger, F. B. 1973. Foraminifera from the nearshore turbulent zone, western North America: Jour. Foram. Res., 3, 3: 101–132.
Le Calvez, J. 1935. Les gamètes de quelques Foraminifères: Acad. Sci. Paris, C.R., 201: 1505–1507.
——. 1936a. Modifications de test des Foraminifères pélagiques en rapport avec la reproduction: *Orbulina universa* d'Orb. et *Tretomphalus bulloides* d'Orb.: Ann. Protist., 5: 125–133.
——. 1936b. Observation sur le genre *Iridia:* Arch. Zool. Exp. Gén., 78: 115–131.
——. 1937a. Processus schizogonique chez le Foraminifère *Planorbulina mediterranensis* d'Orb.: Acad. Sci. Paris, C.R., 204: 14–17.
——. 1937b. Les chromosomes spiraux de la première mitose schizogonique de Foraminifère *Patellina corrugata* Will.: *Ibid.*, 205: 1–3.
——. 1938a. Un Foraminifère géant *Bathysiphon filiformis* G.O. Sars: Arch Zool. Exp. Gén., 79: 82–88.
——. 1938b. Recherches sur les Foraminifères. I. Développement et reproduction: *Ibid.*, 80: 163–333.
——. 1939. *Trochosphaera planorbulina* n. gen, n sp., protiste parasite du Foraminifère *Planorbulina mediterranensis: Ibid.*: 425–443.
——. 1940. Une amibe *Wahlkampfia discorbini* n. sp., parasite de Foraminifère *Discorbis mediterranensis* (d'Orb.): *Ibid.*, 81: 123–129.
——. 1946. Place de la réduction chromatique et alternance de phases nucleaire dans le cycle des Foraminifères: Acad. Sci. Paris, C.R., 222: 612–614.
——. 1947a. *Entosolenia marginata*, Foraminifère apogamique ectoparasite d'un autre Foraminifère *Discorbis vilardeboanus: Ibid.*, 224: 1448–1450.
——. 1947b. Les perforations du test de *Discorbis erecta* (Foraminifère): Lab. Marit. Dinard, Bull., 29: 1–4.
——. 1950. Recherches sur les Foraminifères. II. Place de meiose et sexualité: Arch. Zool. Exp. Gén., 87: 211–243.
——. 1951. Dérèglement du metabolisme ferrugineuse chez des Foraminifères: Vie et Milieu, 2: 335–337.
——. 1953. Foraminifères; *in* Traité de Zoologie, v. 1, pt. 2: Masson, Paris.
Le Calvez, J. & Le Calvez, Y. 1951. Contribution a l'étude des Foraminifères des eaux saumatres. I. Etange de Canet et des Salses: Vie et Milieu, 2: 237–254.
——. 1958. Répartition des Foraminifères dans la baie de Villefranche. I. Miliolidae: Inst. Océanogr. Monaco, Ann., n.s., 35: 160–230.
Le Calvez, Y. 1958. Les Foraminifères de la mer Celtique: Inst. Pêches Marit., Rev. Trav., 22: 147–298.
——. 1963. Contribution a l'étude des Foraminifères de la région d'Abidjan (Côte d'Ivoire): Rev. Micropaléontol., 6, 1: 41–50.
——. 1965. Les Foraminifères; *in* Les Récifs Coralliens et le Lagon de l'ile Mayotte (Archipel des Comores, Océan Indien): Off. Rech. Sci. Tech. Outre-Mer, Paris: 181–201.
——. 1969. Remarques sur la conception et la taxinomie de quelques genres de Foraminifères: Cah. Micropaléontol., (1) 13: 1–13.

Le Calvez, Y. 1972. A propos de *Planorbulinopsis parasitica* Banner considere comme endoparasite d'un autre foraminifère: *Ibid.*, (3), 3, Arch. C.N.R.S., 516: 1–4.
Le Calvez, Y. & Cesana, D. 1972. Détection de l'état de vie chez les Foraminifères: Ann. Paléontol. (Invertebr.), 58, 2: 129–134.
Le Campion, J. 1968. Foraminifères des principaux biotopes du bassin d'Arcachon et du proche océan (Inventaire faunistique): Bull. Cent. Etud. Rech. Sci., 7, 2: 207–391.
——. 1970. Contribution a l'étude des Foraminifères du bassin d'Arcachon et du proche océan: Bull. Inst. Geol. Bassin Aquit., 8: 1–98.
Leclaire, L. & Vergnaud-Grazzini, C. 1972. Variations des associations de Foraminifères planctoniques et paléotemperatures isotopiques depuis le début de dernier Glaciare dans une carotte du bassin algéro-baléares: Acad. Sci. Paris, C.R., 275, (D), 19: 2111–2114.
Lee, J. J. 1974. Towards understanding the niche of foraminifera; *in* R. H. Hedley & C. G. Adams (eds.), Foraminifera, 1: Academic Press, London: 207–260.
Lee, J. J., Freudenthal, H. H., Kossoy, V. & Bé, A. W. H. 1965. Cytological observations on two planktonic foraminifera, *Globigerina bulloides* d'Orbigny. 1826. and *Globigerinoides ruber* (d'Orbigny. 1839) Cushman. 1927: Jour. Paleontol., 12, 4: 531–542.
Lee, J. J., Freudenthal, H. D., Muller, W. A., Kossoy, V., Pierce, S. & Grossman, R. 1963. Growth and physiology of foraminifera in the laboratory. 3. Initial studies of *Rosalina floridana* (Cushman): Micropaleontol., 9, 4: 449–466.
Lee, J. J. & McEnery, M. E. 1970. Autogamy in *Allogromia laticollaris* (Foraminifera): Jour. Protozool., 17, 2: 184–195.
Lee, J. J., McEnery, M. E., Pierce, S., Freudenthal, H. D. & Muller, W. A. 1966. Tracer experiments in feeding littoral foraminifera: *Ibid.*, 13, 4: 659–670.
Lee, J. J., McEnery, M. E. & Rubin, H. 1969. Quantitative studies on the growth of *Allogromia laticollaris: Ibid.*, 16, 2: 377–395.
Lee. J. J. & Muller, W. A. 1973. Trophic dynamics and niches of salt marsh Foraminifera: Amer. Zool, 13, 1: 215–223.
Lee. J. J., Muller, W. A., Stone, R. J., McEnery, M. E. & Zucker, W. 1969. Standing crop of foraminifera in sublittoral epiphytic communities of a Long Island salt marsh: Mar. Biol. 4, 1: 44–61.
Lee. J. J. & Pierce, S. 1963. Growth and physiology of Foraminifera in the laboratory; part 4 – Monoxenic culture of an allogromiid with notes on its morphology: Jour. Protozool., 10, 4: 401–411.
Lee. J. J., Pierce, S, Tentchoff, M. & McLaughlin, J. J. A. 1961. Growth and physiology of foraminifera in the laboratory; Part 1 – Collection and maintenance: Micropaleontol., 7, 4: 461–466.
Lee. J. J. & Zucker, W. 1969. Algal flagellate symbiosis in the foraminifer *Archaias:* Jour. Protozool., 16, 1: 71–81.
Legarde, E. & Arnold, Z. M. 1967a. Utilisation des antibiotiques dans la realisation des cultures de foraminifères sous faible volume, I. Etude de l'action des antibiotique sur les microflores hétérotrophes marines: Vie et Milieu, (A), Biol. Mar. 18, 1: 27–35.
——. 1967b. Utilisation des antibiotiques dans la réalisation des cultures de foraminifères sous faible volume. II. Application a la téchnique des cultures des foraminifères: *Ibid.*: 36–45.
Lena, H. 1966. Foraminíferos recientes de Ushuaia (Tierra del Fuego, Argentina): Ameghiniana, 4, 9: 311–331.
——. 1972. Cytological studies in *Allogromia flexibilis* (Wiesner): Int. Rev. ges. Hydrobiol., 57, 4: 637–644.
Lena, H. & Freire, F. 1974. Estudios citológicos en *Allogromia laticollaris* Arnold: Physis, 33, 86: 123–133.
Lengsfeld, A. M. 1969a. Nahrungsaufnahme und Verdauung bei der Foraminifere *Allogromia laticollaris:* Helgoländer Wiss. Meeresunters., 19, 3: 385–400.
——. 1969b. Zum Feinbau der Foraminifere *Allogromia laticollaris*. I. Mitteilung: Zellen mit

ausgestreckten und eingezogenen Rhizopodien: Helgoländer Wiss. Meeresunters, 19: 230–261.
LeRoy, D. O. & Crane, M. J. 1964. The Ultropak, an aid in photomicrography: Micropaleontol., 10, 1: 105–110.
LeRoy, D. O. & Levinson, S. A. 1974. A deep-water Pleistocene microfossil assemblage from a well in the northern Gulf of Mexico: Micropaleontol., 20, 1: 1–37.
Leslie, R. J. 1965. Ecology and paleoecology of Hudson Bay Foraminifera: Bedford Inst. Oceanogr., Rept. 65–6: 1–192.
Levine, N. D. 1962. Protozoology today: Jour. Protozool., 9, 1: 1–6.
Levy, A. 1967. Contribution a l'étude des Foraminifères des Rechs du Roussillon et du plateau continental de bordure: Vie et Milieu, (B), Océanogr., 18, 1-B: 63–102.
Lewis, K. B. 1966. A giant forâminifer; a new species of *Syringammina* from the New Zealand region: New Zeal. Jour. Sci., 9, 1: 114–123.
——. 1968. Size of fossil animals as an indicator of paleotemperatures: Tuatara, 16, 1: 62–68.
——. 1970. A key to the recent genera of the Foraminiferida: New Zeal. Dept. Sci. Industr., Res., Bull. 196: 1–88.
Lewis, K. B. & Jenkins, C. 1969. Geographical variation of *Nonionellina flemingi:* Micropaleontol., 15, 1: 1–12.
Lidz, B. 1972. *Globorotalia crassaformis* morphotype variations in Atlantic and Caribbean deep-sea cores: *Ibid.*, 18, 2: 194–211.
Lidz, B., Kehm, A. & Miller, H. 1968. Depth habitats of pelagic foraminifera during the Pleistocene: Nature, 217, 5125: 245–247.
Lidz, L. 1965. Sedimentary environment and foraminiferal parameters - Nantucket Bay, Massachusetts: Limnol. Oceanogr., 10, 3: 392–402.
——. 1966a. Deep-sea Pleistocene biostratigraphy: Science, 154: 1448–1452.
——. 1966b. Planktonic Foraminifera in the water column of the mainland shelf off Newport Beach, California: Limnol. Oceanogr., 11, 2: 257–263.
Lidz, L. & Lidz, B. 1966. Foraminiferal biofacies of Veracruz reefs: Amer. Assoc. Petrol. Geol. Bull., 50, 7: 1514–1517.
Linné, C. 1758. Systema Naturae: Stockholm, 10th ed., 824 pp.
Lipina, O. A. 1950. Foraminifery verkhnego devona Russkoj Platformy: Akad. Nauk SSSR, Inst. Geol., Trudy, 119, Geol., 43: 110–133.
——. 1961. Zavisimost foraminifer ot fathij v otlozhenijakh famenskogo jarusa verkhnego devona i turnejskogo jarusa karbona zapadnogo sklona Urala: Vopr. Micropal., 5: 147–161.
Lipps, J. H. 1973. Test structure in foraminifers: Ann. Rev. Microbiol., 27: 471–488.
Lipps, J. H. & Erskian, M. G. 1969. Plastogamy in foraminifera: *Glabratella ornatissima* (Cushman): Jour. Protozool., 16, 3: 422–425.
Lipps, J. H. & Ribbe, P. H. 1967. Electron-probe microanalysis of planktonic Foraminifera: *Ibid.*, 41, 2: 492–496.
Lipps, J. H. & Valentine, J. W. 1970. The role of Foraminifera in the trophic structure of marine communities: Lethaia, 3, 3: 279–286.
Lipps, J. H. & Warme, J. E. 1966. Planktonic foraminiferal biofacies in the Okhotsk Sea: Cushman Found. Foram. Res., Contr., 17, 4: 125–134.
Lister, J. J. 1894. Contributions to the life-history of the Foraminifera: Roy. Soc. London, Proc., 56: 155–160.
——. 1895. *Idem:* Roy. Soc. London, Phil Trans., 186, ser. B: 401–453.
——. 1903. The Foraminifera; *in* R. Lankaster, ed., A Treatise of Zoology, Part 1, 2: 47–149.
——. 1906. The Life History of the Foraminifera: Brit. Assoc., York.
Livan, M. 1937. Über Bohr-Löcher an rezenten und fossilen Invertebraten: Senckenberg., 19, 3–4: 138–149.
Livental. V. Eh. 1953. Materialy k paleontologicheskoj kharakteristike Buliminidae miothenovykh otlozhenij Prikarpat'ja: L'vovsk. Geol. O-vo, Trudy, Paleontol., 2.

Lizitzin, A. P. 1971. Skorost' sovremennogo osadkonaloplenija v okeanakh: Okeanol., 11, 6: 957–968.
Loeblich, A. R. & Tappan, H. 1953. Studies of Arctic Foraminifera: Smithson. Misc. Coll., 121 7: 1–24.
——. 1961. Suprageneric classification of the Rhizopoda: Jour. Paleontol., 35, 2: 245–330.
——. 1964. Sarcodina, chiefly "Thecamoebians" and Foraminiferida: Treatise on Invertebrate Paleontology, Part C, Protista 2, v. 1–2, Geol. Soc. Amer. and Univ. of Kansas Press, 900 pp.
——. 1974. Recent advances in the classification of the Foraminiferida; *in* R. H. Hedley & C. G. Adams, eds., Foraminifera, 1: Academic Press, London: 1–53.
Loep, K. J. 1965. A study of ecology and distribution of Recent Foraminifera in the northwest Gulf of Mexico: Gulf Coast Assoc. Geol. Socs., Trans., 15: 167–177.
Lohmann, H. 1903. Untersuchungen über die Tier und Planzenwelt sowie über die Bodensedimente des Nordatlantischen Ozeans zwischen den 38 und 50 Grade nord: Breite: Kon. Preuss. Akad. Wiss. S.B. Phys. Math. Kl., 26.
——. 1920. Die Bevölkerung des Ozeans mit Plankton nach den Ergebnissen der Zentrifugenfänge während der Ausreise der *Deutschland*, 1911. zugleich ein Beitrag zur Biologie des Atlantischen Ozeans: Ges. Naturf. Freund. Berlin, Arch. f. Biont., 4, 3: 1–617.
Longinelli, A. & Tongiorgi, E. 1960. Frequenza degli individui destrogiri in diverse popolazione di *Rotalia beccarii* Linné: Bol. Soc. Paleontol. Ital., 1, 1: 5–16.
——. 1964. Oxygen isotopic composition of some right and left coiled Foraminifera: Science, 144: 1004–1005.
Loose, T. L. 1970. Turbulent transport of benthonic Foraminifera: Cushman Found. Foram. Res., Contr., 21, 4: 164–166.
Lowman, S. 1949. Sedimentary facies in Gulf Coast: Amer. Assoc. Petrol. Geol., Bull., 33: 1939–1997.
Lubny-Gerthyk, E. A. 1953. Vesovaja kharakteristika osnovnykh predstavitelej zooplanktona Okhotskogo i Beringova morej: Akad. Nauk SSSR, Dokl., 91, 4: 949–952.
Lücke, F. 1910. *Saccammina sphaerica* M. Sars: Thesis, Univ. Kiel, Jensen.
Lukina, T. G. 1967a. O raspredelenii fauny izvestkovykh foraminifer v thentral'noj chasti Tikhogo okeana: Akad. Nauk SSSR, Dokl., 177, 5: 1205–1207.
——. 1967b. O raspredelenii glubokovodnoj fauny agglutinirujushchikh foraminifer v thentral'noj chasti Tikhogo okeana: *Ibid.*, 6: 1431–1433.
——. 1969. Raspredelenie foraminifer v thentral'noi chasti Tikhogo okeana: Zool. Zh., 48, 10: 1445–1450.
Lutze, G. F. 1962. Variationsstatistik und Ökologie bei rezenten Foraminiferen: Paläontol. Zeit., 36: 252–264.
——. 1964a. Statistical investigations on the variability of *Bolivina argentea* Cushman: Cushman Found. Foram. Res., Contr., 15, 3: 105–116.
——. 1964b. Zum Farben rezenten Foraminiferen: Meyniana, 14: 43–47.
——. 1965. Zur Foraminiferen-Fauna der Ostsee: *Ibid.*, 15: 75–142.
——. 1968a. Jahresgang der Foraminiferen-Fauna in der Bottsand-Laguna (westliche Ostsee): *Ibid.*, 18: 13–30.
——. 1968b. Siedlungs-Strukturen rezenter Foraminiferen: *Ibid.*: 31–34.
Lutze, G. F. Grabert, B. & Seibold, E. 1971. Lebend-beobachtungen an Gross-Foraminiferen (*Heterostegina*) aus dem Persischen Golf: *Meteor* Forsch. Ergebnisse, Reihe C, 6: 21–40.
Lwoff, A. 1925. La sporogenèse chez une Gromie marine: *Gromia dujardini:* Trav. Stat. Zool. Wimereux, 9: 140–145.
Lyman, J. & Fleming, R. H. 1940. Composition of sea water: Jour. Mar. Res., 3.
Lynts, G. W. 1962. Distribution of Recent Foraminifera in upper Florida Bay and associated Sounds: Cushman Found. Foram. Res., Contr., 13, 4: 127–144.
——. 1966a. Relationship of sediment-size distribution to ecologic factors in Buttonwood Sound, Florida Bay: Jour. Sed. Petrol., 36, 1: 66–74.

Lynts, G. W. 1966b. Variation of foraminiferal standing crop over short distances in Buttonwood Sound, Florida Bay: Limnol. Oceanogr., 11, 4: 562–566.
——. 1972. Factor-vector analysis models in ecology and paleoecology: Inter Geol. Congr., 24th, Montreal, 7: 227–237.
Lynts, G. W. & Judd, J. B. 1971. Late Pleistocene paleotemperatures at Tongue of the Ocean, Bahamas: Science, 171: 1143–1144.
Lynts. G. W., Judd, J. B. & Stehman, C. F. 1973. Late Pleistocene history of Tongue of the Ocean, Bahamas: Geol. Soc. Amer., Bull., 84: 2665–2684.
Lynts, G. W. & Pfister, R. M. 1967. Surface ultrastructure of some tests of Recent Foraminiferida from the Dry Tortugas, Florida: Jour. Protozool., 14, 3: 387–399.
MacDougall, I. & Tarling, D. H. 1963. Dating of polarity zones in the Hawaiian Islands: Nature, 200: 54–56.
Macfadyen. W. A. & Kenny, E. J. A. 1934. On the correct writing in form and gender of the names of the Foraminifera: Jour. Roy. Micr. Soc., (3), 54.
Madeira, M. L. 1969. Foraminifera from Sao Francisco do Sul, State of Santa Catarina, Brazil: Iheringia Zool., 37: 1–29.
Magné, J. & Dufaure, G. 1964. Une méthode nouvelle pour l'extraction rapide des microfossiles: Rev. Micropaléontol., 7, 1: 77–79.
Majer, E. M. 1962. Novye vidy Foraminifer melkovodij Kandalakshskogo zaliva: M.G.U., Belom. Biol. Stan., Trudy, 1: 70–87.
Majzon, L. 1936. Rendellanas Foraminiferehejak (Abnormalitäten an Foraminiferen-Schalen): Fold. Kozl, 64, 10–12.
Malapris, M. & Jannin, F. 1967. Utilisation du microscope électronique a balayage dans l'étude des foraminifères: Acad. Sci. Paris, C.R., ser. D, 264: 247–249.
Malmgren, B. & Kennett, J. P. 1972. Biometric analysis of phenotypic variation: *Globigerina pachyderma* (Ehrenberg) in the South Pacific Ocean: Micropaleontol., 18, 2: 241–248.
Mangin, J. P. Desbrosses, J. & Lecolle, J. 1969. Analyse de quelques populations de foraminifères du plateau de Côte d'Ivoire: Afr. Micropal. Coll., Proc. 3: 561–571.
Mangin, J. P., Mangin, M. 1969. Note préliminaire sur le renouvellement des foraminifères planctoniques vivant au débouche de la rade de Villefranche (Alpes-Maritimes). Méthodes et premiers résultats en surface: Protist., 5, 1: 121–125.
Margalef, R. 1961. Caracteristique et significance des zooxantelles du 'phytoplankton prisonier' des Acanthaire: Com. Inter. Explor. Scient. Mer. Medit., 16, 2: 141–142.
Margerel, J. P. 1966. Les petits foraminifères des sables du Bois-Gouet (Loire Atlantique): Soc. Scienc. Natur. l'Ouest France. Bull., 63: 3–11.
Marie, P. 1938. Sur la faune des Foraminifères de l'estuarie de la Rance: Lab. Dinard, Bull., 20: 68–86.
Marr, J. 1964. Some notes on the preservation of marine animals: Jour. Conseil, 28, 1: 121–125.
Marszalek, D. S. 1969. Observations on *Iridia diaphana*, a marine foraminifer: Jour. Protozool., 16, 4: 599–611.
Marszalek, D. S. & Hay, W. W. 1968. Marine aquaria for paleontology: Jour. Geol. Educ., 16: 159–163.
Marszalek, D. S., Wright, R. C. & Hay. W. W. 1969. Function of test in foraminifera: Gulf. Coast Assoc. Geol. Soc., Trans., 19: 341–352.
Martin, R. E. & Steinker, D. C. 1973. Evaluation of techniques for recognition of living Foraminifera: Compass, 50, 4: 26–30.
Matera, N. J. & Lee, J. J. 1972. Environmental factors affecting the standing crop of foraminifera in sublittoral and psammolittoral communities of a Long Island salt marsh: Mar. Biol., 14, 2: 89–103.
Mateu, G. 1965. Datos para el estudio de los Foraminíferos alguícolas del litoral de Blanes (Gerona): Inst. Biol. Aplic., Publ., 34: 129–135.

Mateu, G. 1968a. Contribución al conocimiento de los foraminíferos que sirven de alimentos a las holoturios: Soc. Hist. Nat. Baleares, 14: 5–17.
——. 1968b. Contribución al conocimiento de los foraminíferos que sirven de alimiento a los moluscos bivalvos: Inst. Biol. Aplic., 44: 81–102.
——. 1969a. Foraminíferos del contenido gástrico del *Spatangus purpureus* O. F. Muller y su degradación protoplasmática a través del aparato digestivo de este equínido: Soc. Hist. Nat. Baleares, Bol., 15: 75–84.
——. 1969b. El polimorfismo ontogenético de los foraminíferos Peneroplidae en relación con la sistemática y filogenia le los mismos: 1-st Symp. Inter. Zoofilogenia, Univ. Salamanca: 179–189.
——. 1970. Estudio sistemático y bioecológico de los foraminíferos vivientes de los litorales de Cataluña y Baleares: Inst. Españ. Oceanogr., Trab., 38: 1–84.
——. 1971. Foraminifères de la mer d'Alboran. Leur importance comme indicateur des différent masses d'eau qui confluent dans cette mer et l'influence des courants marins dans la distribution de leurs biocénoses planctoniques et benthoniques: Rapp. Comm. Int. Mer. Médit., 20, 3: 211–213.
Mathieu, R. 1971. Les associations de foraminifères du plateau continental Atlantique du Maroc au large de Casablanca: Rev. Micropaléontol., 14, 1: 55–61.
Matoba, Y. 1970. Distribution of Recent shallow water Foraminifera of Matsushima Bay, Miyagi Prefecture, northeast Japan: Sci. Rept., Tohoku Univ., (2) Geol., 42, 1: 1–85.
Matoba, Y. & Nakagawa, H. 1972. Recent foraminiferal assemblages from the continental shelf and slope off Akita, Japan Sea coast of Northeast Japan; *in* Prof. Jun-ichi Iwai Mem. Vol., Tohoku Univ., Inst. Geol. Paleontol., Sendai: 657–671.
Mattiat, B. & Zobel, B. 1968. *Edentostomina durrandii* (Millett) mit ingestierten Zirkonen: Geol. Jahrb., 86: 239–246.
Mayr, E. 1969. Principles of Systematic Zoology: McGraw-Hill, New York, 428 pp.
McCrone, A. W., Schafer, C. 1966. Geochemical and sedimentary environments of foraminifera in the Hudson River estuary, New York: Micropaleontol., 12, 4: 505–509.
McEnery, M. & Lee. J. J. 1970. Tracer studies on calcium and strontium mineralization and mineral cycling in two species of Foraminifera, *Rosalina leei* and *Spiroloculina hyalina:* Limnol. Oceanogr., 15, 2: 173–182.
McGlasson, R. H. 1959. Foraminiferal biofacies around Santa Catalina Island, California: Micropaleontol., 5, 2: 217–240.
McGowan, J. A. & Fraundorf, V. J. 1966. The relationship between size of net used and estimates of zooplankton diversity: Limnol. Oceanogr., 11, 4: 456–469.
McGowran, B. 1966. Bilamellar walls and septal flaps in the Robertinacea: Micropaleontol., 12, 4: 477–488.
——. 1968. Reclassification of early Tertiary *Globorotalia: Ibid.*, 14, 2: 179–198.
McIntyre, A., Ruddiman, W. F. & Jantzen, R. 1972. Southward penetration of the North Atlantic polar front: faunal and floral evidence of large scale surface water mass movements over the last 250,000 years: Deep Sea Res., 19, 1: 61–77.
McKnight, W. M. 1962. The distribution of Foraminifera off parts of the Antarctic Coast: Bull. Amer. Paleontol., 44, 201: 60–158.
McLean, J. D. 1959. Photomicrography of opaque specimens: Man. Micropaleontol. Techn., McLean, Alexandria, Va, USA.
Mello, J. F. & Buzas, M. A. 1968. An application of cluster analysis as a method of determining biofacies: Jour. Paleontol. 42, 3: 747–758.
Menard, H. W. 1964. Marine Geology of the Pacific: McGraw-Hill, New York, 271 pp.
Mereschkowsky, C. 1878. Ehtjudy nad prosteishimi zhivotnymi severa Rossii: Sankt Petersb. O-vo. Estestv.
——. 1879. Studien über Protozoen des nordlichen Russland: Arch Microsk. Anat., 16: 153–248.

Mesjathev, I. N. 1923. Materialy k zoogeografii russkikh severnykh morej: Plov. Morsk. Nauchn. Inst., Trudy, 1.
——. 1926a. 1-a ehkspedithija Plovuchego Morskogo Nauchnogo Instituta: *Ibid.*, 1, 1.
——. 1926b. 3-a ehkspedithija Plovuchego Morskogo Nauchnogo Instituta: *Ibid.*, 1, 1.
——. 1929. 11-a ehkspedithija Morskogo Nauchnogo Instituta: *Ibid.*, 4, 1.
Miklukho-Maclay, A. D. 1955. Nekotorye voprosy paleoehkologicheskikh issledovanij morskikh faun paleozoja: Leningrad Univ., Vestn. 1.
——. 1956. K sistematike paleozojskikh foraminifer: *Ibid.*, 6.
Miklukho-Maclay, A. D., Rauser-Chesnoussova, D. M. & Rozovskaja, S. E. 1958. Sistematika i filogenija fuzulinid: Vopr. Micropal., 2: 5–21.
Miklukho-Maclay, K. V. 1960. Nekotorye priemy shlifovanija vydelennykh iz porody rakovin foraminifer: 1 Semin. Microfaun., Trudy: 120–128.
Miller, D. N. 1953. Ecological study of the Foraminifera of Mason Inlet, North Carolina: Cushman Found. Foram. Res., Contr., 4, 2: 41–63.
Miller, R. L. & Kahn, J. S. 1963. Statistical Analysis in the Geological Sciences: John Wiley & Sons, New York, 483 pp.
Millett, F. W. 1898a. Report on the recent Foraminifera of the Malay Archipelago collected by Mr. A. Durrand, F.R.M.S.: Roy. Micr. Soc., Jour.: 285–269.
——. 1898. *Idem*: *Ibid.*, 3: 607–614.
——. 1899. *Idem*: *Ibid.*, 4: 249–255.
——. 1900a. *Idem*: *Ibid.*, 8: 273–281.
——. 1900b. *Idem*: *Ibid.*, 9: 539–549.
——. 1904. *Idem*: *Ibid.*, 17: 597–609.
Miró, M. D. 1965. Comparación de la fauna de foraminíferos de los sedimentos de la fosa de Cariaco con la del área oceánica adyacente: Soc. Cien. La Salle, Mem., 25, 70–72: 225–236.
——. 1971. MS, Los Foraminíferos planctónicos vivos y sedimentos del margen continental de Venezuela: Tésis, Univ. Barcelona, Fac. Cienc.
Miró, M. D. & Marval, J. A. 1967. Foraminíferos planctónicos vivos de la Fosa de Cariaco y del talud continental de Venezuela: Soc. Cienc. Nat. La Salle, Mem., 27, 76: 11–34.
Mjatljuk, E. V. 1974. K paleoehkologii foraminifer rànnemiothenovogo bassejna prikaspijskoj vpadiny: VNIGRI. Trudy, 349: 62–88.
Moebius, K. 1888. Bruchstücke einer Rhizopodenfauna d. Kieler Bucht: König. Preuss. Akad. Wiss., Abh., Phys. Abt., 2: 1–31.
Moncharmont-Zei, M. 1956. Foraminiferi di un campione de fondo prelevato all'Annontatura nel Golfo di Napoli: Soc. Natur. Napoli, Boll., 65: 1–11.
——. 1962. I Foraminiferi del Banco della Vedove (Golfo di Napoli): Publ. Staz. Zool. Napoli, 32 (suppl.): 1–12.
——. 1964. Studio ecologico sui Foraminiferi del Golfo di Pozzuoli (Napoli): *Ibid.*, 34: 160–184.
——. 1968. I foraminiferi di alcuni campioni di fondo prelevati lungo la costa di Beirut (Libano): Inst. Paleontol. Univ. Napoli, Publ. 24: 1–33.
Montagu, G. 1803–1808. Testacea Britannica, or Natural History of British Shells, Marine, Land, and Freshwater: London, 3 vols., 606 pp.
Montfort, P. D. 1808. Conchyologie Sistématique et Classification Méthodique des Coquilles, Paris, 409 pp.
Moore, T. C. 1973. Method of randomly distributing grains for microscope examination: Jour. Sed. Petrol., 43, 3: 904–906.
Moore, W. E. 1957. Ecology of Recent Foraminifera in Northern Florida Keys: Amer. Assoc. Petrol. Geol., Bull., 41: 727–741.
Morin, R. W. 1971. Late Quaternary biostratigraphy of cores from beneath the California Current: Micropaleontol., 17, 4: 475–491.

Morin, R. W. Theyer, F. & Vincent, E. 1970. Pleistocene climates in the Atlantic and Pacific Oceans: a reevaluated comparison based on deep-sea sediments: Science, 196: 365–366.
Morishima, M. 1948a. The accumulation of foraminiferal tests in inlets of Wakasa Bay on the inland sea of Japan: Rep. Comm. Treat. Mar. Ecol. Paleoecol., 7: 89–91.
——. 1948b. Foraminiferal thanatocoenosis of Ago Bay, Kii Peninsula, Japan: *Ibid.*, 8: 111–117.
——. 1955. Deposits of foraminiferal tests in the Tokyo Bay, Japan: Univ. Kyoto, Coll. Sci., Mem. B, 22, 2: 213–222.
Morkhoven, F. P. C. M. 1958. A simplified method of grinding Foraminifera: Micropaleontol., 4, 2: 209–210.
Morozova, V. G. 1956. Plastinki dlja otbora i razbora mikrofauny: Vopr. Micropal., 1: 79–87.
——. 1958. K sistematike i morfologii paleogenovykh predstavitelej nadsemejstva Globigerinidea: *Ibid.*, 2: 22–52.
Motomura, I. 1932. Statistical method in animal association: Zool Mag., Japan, 44, 528: 379–398.
Moulinier, M. 1966. Variabilité d'une population d'*Elphidium* de la rade de Brest (N. Finistère) apparentés a *Elphidium crispum* (Linné): Rev. Micropaléontol., 9, 3: 194–200.
Moura, A. R. 1965. Foraminíferos da Ilha da Inhaca (Moçambique): Estud. Gerais Univ. Moçambique, (2), 2: 1–74.
Muller, W. A. & Lee, J. J. 1969. Apparent indispensability of bacteria in foraminiferan nutrition: Jour. Protozool., 16, 3: 471–478.
Murray, J. 1876. Preliminary reports on work done on board the 'Challenger': Proc. Roy. Soc., 24: 471–544.
——. 1895. Summary of the scientific results obtained at the sounding, dredging and trawling stations of H.M.S. 'Challenger': Challenger Repts., Summary: 1431–1462.
——. 1897. On the distribution of the pelagic Foraminifera on the surface and on the floor of the ocean: Nat. Sci., 65: 17–27.
Murray, J. & Renard, A. F. 1891. Report of the Deep-Sea Deposits Based on the Specimens Collected During the Voyage of H.M.S. 'Challenger' in the Years 1873–1876: Longmans, London, 525 pp.
Murray, J. W. 1963. Ecological experiments on Foraminiferida: Jour. Mar. Biol. Assoc. U.K., 43, 3: 621–642.
——. 1965a. Significance of benthic foraminiferids in planktonic samples: Jour. Paleontol., 39, 1: 156–157.
——. 1965b. On the Foraminiferida of the Plymouth region: Jour Mar. Biol. Assoc. U.K., 45, 2: 481–505.
——. 1966a. The Foraminiferida of the Persian Gulf; 4, Khor al Bazam: Palaeogeogr., Palaeoclimat., Palaeoecol., 2, 2: 153–169.
——. 1966b. *Idem;* 5. The shelf off the Trucial Coast: *Ibid.*, 3: 267–278.
——. 1967a. Production in benthic foraminiferids: Jour. Nat. Hist., 1, 1: 61–68.
——. 1967b. Transparent and opaque foraminiferid tests: Jour. Paleontol., 41, 3: 791.
——. 1968a. The living Foraminiferida of Christchurch Harbour, England: Micropaleontol., 14, 1: 83–96.
——. 1968b. Living foraminifera of lagoons and estuaries: *Ibid.*, 14, 4: 435–455.
——. 1969. Recent foraminifers from the Atlantic continental shelf of the United States: *Ibid.*, 15, 4: 401–419.
——. 1970a. The foraminifera of the hypersaline Abu Dhabi lagoon, Persian Gulf: Lethaia, 3, 1: 51–68.
——. 1970b. Foraminifers of the western approaches to the English Channel: Micropaleontol., 16, 4: 471–485.
——. 1970c. The Foraminiferida of the Persian Gulf; 6, Living forms in the Abu Dhabi region: Jour. Nat. Hist., 4, 1: 55–67.
——. 1971. Living foraminiferids of tidal marshes: a review: Jour. Foram. Res., 1, 4: 153–161.

Murray, J. W. 1973a. Wall structure of some agglutinated Foraminiferida: Palaeontol., 16, 4: 777–786.
——. 1973b. Distribution and Ecology of Living Benthic Foraminiferids: Crane, Russak. & Co., New York, 274 pp.
Murray, J. W. & Wright, C. A. 1970. Surface textures of calcareous foraminiferids: Palaeontol., 13, 2: 184–187.
Myers, E. H. 1933. Multiple tests in the Foraminifera: Nat. Acad. Sci., Proc., 19: 893–899.
——. 1934. The life history of *Patellina corrugata*, a Foraminifer: Science, 79, 2054: 436–437.
——. 1935a. Culture methods for the marine Foraminifera of the littoral zone: Amer. Micr. Soc., Trans., 54: 264–267.
——. 1935b. The life history of *Patellina corrugata* Williamson, a Foraminifer: Scripps Inst. Oceanogr., Bull. Techn. Ser., 3, 15: 355–392.
——. 1935c. Morphogensis of the test and biological significance of dimorphism in the Foraminifer *Patellina corrugata* Williamson: *Ibid.*, 3: 393–404.
——. 1936. The life cycle of *Spirillina vivipara* Ehrenberg, with notes on morphogenesis, systematics and distribution of the Foraminifera: Jour. Roy. Micr. Soc., 56: 120–146.
——. 1937. Culture methods for marine Foraminifera of the littoral zone; *in* Culture Methods for Invertebrate Animals: Comstock, Ithaca, New York: 93–96.
——. 1938. The present state of our knowledge concerning the life cycle of the Foraminifera: Nat. Acad. Sci., Proc., 24: 10–12.
——. 1940. Observations of the origin and fate of flagellated gametes in multiple tests of *Discorbis* (Foraminifera): Jour. Mar. Biol. Assoc. U.K., 24: 201–226.
——. 1941. Ecological studies of Foraminifera: Subcomm. Ecol. Mar. Organ., Rept., 1: 43–46.
——. 1942a. A quantitative study of the productivity of the Foraminifera in the sea: Amer. Phil. Soc., Proc., 85, 4: 325–342.
——. 1942b. Rate at which Foraminifera are contributed to marine sediments: Jour. Sed. Petrol., 12, 2: 92–95.
——. 1942c. Ecologic relationship of some recent and fossil Foraminifera: Nat. Res. Counc., Div. Geol. Geogr., Ann. Rept., Appen. 2.
——. 1942d. Biological evidence as to the rate at which tests of Foraminifera are contributed to marine sediments: Jour. Paleontol., 16: 397–398.
——. 1943a. Ecologic relationship of larger Foraminifera: Comm. Mar. Ecol., Rept., 3: 26–36.
——. 1943b. Biology, ecology and morphogenesis of a pelagic Foraminifer: Stanford Univ., Biol. Sci., 9, 1: 5–30.
——. 1943c. Life activities of Foraminifera in relation to marine ecology: Amer. Phil. Soc., Proc., 86, 3: 439–458.
——. 1945. Collecting Recent Foraminifera: Com. Mar. Ecol. Paleoecol., Rept., 5: 24–28.
——. 1948. Bibliography on the ecology of the Foraminifera: *Ibid.*, 7: 28–37.
Myers, E. H. & Cole, W. S. 1957. Foraminifera; *in* Treatise Mar. Ecol. Paleoecol., 1: Geol. Soc. Amer., Mem. 67: 1075–1081.
Nagahama, M. 1951. Studies on the foraminiferal fauna of small or almost enclosed bays in Japan: Inst. Nat. Resour., Rept., 19–21: 142–148.
Nagappa, Y. 1957. Direction of coiling in *Globorotalia* as an aid in correlation: Micropaleontol., 3, 4: 393–398.
Nagy, J. 1965. Foraminifera in some bottom samples from shallow waters in Vestspitsbergen: Norsk Polarinst., Årbok: 109–127.
Napoli Alliata, E. di & Ornesi, M. R. 1970. Studio de carote prelevate nel Golfo dell'Asinara, Sardegna settentrionale: Soc. Geol. Ital., Boll., 89, 1: 181–208.
Napoli Alliata, E. di, Stefanini, S. & Fioramont, G. 1970. Étude de quelques carottes du plateau et du talus continental de la Sardaigne mériodionale: *Ibid.*: 209–232.
Narchi, W. 1956. Foraminíferos recentes do Brasil, Famílias Miliolidae, Peneroplidae, e Alveolinellidae: Inst. Oceanogr. São Paulo, Bol. 7, 1–2: 161–192.

Narchi, W. 1962a. Sôbre Nonionidae, Globorotalidae e Orbulindae recentes do Brasil: *Ibid.*, 12, 3: 23–47.
——. 1962b. Sôbre Lagenidae e Nodosariidae recentes do Brasil (Foraminífera): Univ. São Paulo, Fac. Fil. Letr., Bol. Zool., 24: 97–166.
——. 1963a. Locomoção em foraminíferos: Acad. Brasil Cienc., Ann., 35, 3: 463–464.
——. 1963b. Sôbre alguns foraminíferos arenáceous recentes do Brasil: *Ibid.*: 373–391.
Natland, M. L. 1933. The temperature and depth distribution of some Recent and fossil Foraminifera in the southern California region: Scripps Inst. Oceanogr., Techn. Ser., 3, 10: 225–230.
——. 1957. Paleoecology of west coast Tertiary sediments; *in* Treatise Mar. Ecol. Paleoecol., 2: Geol. Soc. Amer., Mem. 67: 543–572.
Natland, M. L. & Kuenen, P. H. 1951. Sedimentary history of the Ventura Basin, California, and the action of turbidity currents: Soc. Econ. Miner. Paleontol., Spec. Publ. 2: 76–107.
Nesteroff, W. D., Duplaix, S., Sauvage, J., Lancelot, Y., Milieres, F. & Vincent, E. 1968. Les dépôts recents du canyon de Cap-Breton: Soc. Géol. France, Bull., (7) 10 218–252.
Neumayr, M. 1887. Die natürlichen Verwandschaftsverhältnisse der schalentragenden Foraminiferen: Akad. Wiss. Wien., S.-B, 95: 156–186.
Newell, G. E. & Newell, R. C. 1963. Marine Plankton; a practical guide: Hutchinson Educ., London, 207 pp.
Nichols, M. M. & Ellison, R. L. 1967. Sedimentary patterns of microfauna in a coastal plain estuary; *in* Estuaries: Amer. Assoc. Adv. Sci., Publ. 83: 283–288.
Nichols, M. M. & Norton, W. 1969. Foraminiferal populations in a coastal plain estuary: Palaeogeogr., Palaeoclimat., Palaeoecol., 6, 3: 197–213.
Nicol, D. 1944. New west American species of the foraminiferal genus *Elphidium:* Jour. Paleontol., 18: 172–185.
Niitsuma, N. 1968. Analysis of the benthonic foraminiferal community: Fossils, 16: 25–33.
——. 1971. Detailed study of the sediments recording the Matuyama-Brunhes geomagnetic reversal: Tohoku Univ., Sci. Repts., Sendai, ser. 2, Geol., 43, 1: 1–39.
Niitsuma, N., Sakai, T. & Hattori, M. 1971. Statistical analysis of populations by electronic computer: *Ibid.*: 41–58.
Nikoljuk, V. F. 1947. Paleofauna vodonosnogo gorizonta pustyni Kara-Kum: Akad. Nauk Uzbek. SSR, Bjull., 8.
——. 1948. Reliktovye Foraminifera pustyni Kara-Kum: Akad. Nauk Uzbek. SSR, Izvest., 1: 79–88.
Niskin, S. & Jones, J. I. 1963. New collecting and recording devices for limnological and oceanographic research: Univ. Mich., Great Lakes Res. Div., Inst. Sci. Tech., Conf. Great Lakes Res., 6th, Proc.: 266.
Norman, A. M. 1876. Crustacea, Tunicata, Polyzoa, Echinodermata, Actinozoa, Foraminifera, Polycystina and Spongida; *in* Preliminary report of the biological results of a cruise in H.M.S. 'Valorous' to Davis strait in 1875: Roy. Soc. London, Proc., 25: 212–215.
Norton, R. D. 1930. Ecologic relations of some Foraminifera: Scripps Inst. Oceanogr., Bull., Tech. Ser., 2, 9: 331–388.
Nørvang, A. 1941. Notes on some Foraminifera from off Bergen: Bergens Mus. Årbok Naturvit, 11: 1–19.
——. 1945. The Zoology of Iceland; 2, 2, Munksgaard, Copenhagen & Reykjavik: 1–79.
Nyholm, K. G. 1951. A monothalamous Foraminifer, *Marenda nematoides* n. gen, n. sp.: Cushman Found. Foram. Res., Contr., 2, 3: 91–95.
——. 1952. Studies on Recent Allogromiidae. I. *Micrometula hyalostriata* n. gen., n. sp., from the Gullmar Fjord, Sweden: *Ibid.*, 3, 1: 14–17.
——. 1953. *Idem*, II. *Nemogullmia longevariabilis* n. gen., n. sp. from the Gullmar Fjord: *Ibid.*, 4, 3: 105–106.
——. 1954. *Idem*, III – *Tinogullmia hyalina* n. gen., n. sp., from the Gullmar Fjord: *Ibid.*, 5, 1: 7.

Nyholm, K. G. 1955a. *Idem*, IV – *Phainogullmia aurata* n. gen., n. sp.: Zool. Bidr. Uppsala, 30: 465–474.
——. 1955b. Observations on the monothalamous *Hippocrepinella alba* Heron-Allen & Earland: *Ibid.*: 475–484.
——. 1956. On the life cycle and cytology of the Foraminiferan *Nemogullmia longevariabilis*: *Ibid.*, 31: 483–495.
——. 1957. Orientation and binding power of Recent monothalamous Foraminifera in soft sediments: Micropaleontol., 3, 1: 75–76.
——. 1961. Morphogenesis and biology of the Foraminifer *Cibicides lobatulus*: Zool. Bidr. Uppsala, 33: 157–192.
——. 1962. A study of the Foraminifer *Gypsina*: *Ibid.*: 201–206.
——. 1973. To the study of vertical and horizontal taxonomy in Foraminifera: *Ibid.*, 1, 1: 3–9.
Nyholm, K. G. & Gertz, I. 1973. To the biology of the monothalamous foraminifer *Allogromia marina* n. sp.: 1: 89–93.
Oba, T. 1967. Planktonic Foraminifera from the deep-sea cores of the Indian Ocean: Tohoku Univ., Sci. Rept., ser. 2, Geol., 38, 2: 193–219.
——. 1969. Biostratigraphy and isotopic paleotemperatures of some deep-sea cores from the Indian Ocean: *Ibid.*, 41, 2: 129–195.
Okropidze, O. V. 1957. Nekotorye dannye v oblasti biologii iskopaemykh foraminifer: Akad. Nauk Gruz. SSR, Soobshch, 19, 2.
Olausson, E. 1961. Remarks on some Cenozoic core sequences from the central Pacific, with a discussion of the role of Coccolithophorids and Foraminifera in carbonate deposition: Oceanogr. Inst. Göteborg, Medd., ser. B, 8, 10: 1–35.
——. 1965. Evidence of climatic changes in north Atlantic deep-sea cores, with remarks on isotopic paleotemperature analysis: Progress in Oceanogr., 3: 221–252.
Olausson, E., Bilal ul Haq, U. Z., Karlsson, G. B. & Olsson, I. U. 1971. Evidence in Indian Ocean cores of Late Pleistocene oceanic and atmospheric circulation: Geol. Foren i Stockholm Forhand, 93: 51–84.
Olivier, R., Very, J. M. & Vuagnat, P. 1972. Analyse statistique de micrographies digitalisées, appliquées aux parois de foraminifères porcelanes actuels: Inter. Symp. Prob, Biomineral., Mainz, 6: 101–106.
Olsson, I., On the distribution and ecology of the meiofauna, with special reference to Foraminifera, in Byfjorden on the west coast of Sweden (in press).
Olsson, I. & Eriksson, B. 1974. Horizontal distribution of meiofauna within a small area, with special reference fo Foraminifera: Zoon, 2: 67–84.
Olsson, R. K. 1971a. The logarithmic spire in planktonic foraminifera: its use in taxonomy, evolution, and paleoecology: Gulf Coast Assoc. Geol. Socs., Trans., 21: 419–432.
——. 1971b. Pliocene-Pleistocene planktonic foraminiferal biostratigraphy of the northeastern Pacific: Inter. Conf. Plank. Microfossils, 2d, Rome, Proc.: 921–928.
——. 1973. What is a kummerform planktonic foraminifer?: Jour. Paleontol., 47, 2: 327–329.
——. 1974. Pleistocene paleooceanography and *Globigerina pachyderma* (Ehrenberg) in Site 36, DSDP, northeastern Pacific: Jour. Foram. Res., 4, 2: 47–60.
Omaly, N. 1966. Moyens de prélèvement du zooplancton. Essai historique et critique: Pelagos, Bull. Inst. Océanogr. d'Alger, 5: 1–169.
Orbigny, A. D., de. 1826. Tableau méthodique de la classe des Céphalopodes. 3me ordre-Foraminifères: Ann. Sci. Nat., 7: 245–314.
——. 1839a. Foraminifères; *in* Barker-Webb and Berthelot, Histoire naturelle des îles Canaries, 2, 2: 119–146.
——. 1839b. Foraminifères: Voyage dans l'Amérique Méridionale, 5, 5, 86 pp.
——. 1839c. Foraminifères; *in* De la Sagra, Histoire physique, politique et naturelle de l'île de Cuba, Paris, 224 pp.
——. 1846. Foraminifères Fossiles du Bassin Tertiaire de Vienne, Paris, 312 pp.

Orbigny, A. D. de. 1849–1852. Prodrome de Paléontologie Stratigraphique Universelle des Animaux Rayonnés et Mollusques, 2 vols., Paris, 819 pp.
Orr, W. N. 1969. Variation and distribution of *Globigerinoides ruber* in the Gulf of Mexico: Micropaleontol., 15, 3: 373–379.
Ottestad, P. 1942. On periodical variations in the yield of the great sea-fisheries and the possibility of establishing yield prognoses: Rept. Nor. Fish. Mar. Inv., 7, 5.
Otto, G. H. 1933. Comparative tests of several methods of sampling heavy mineral concentrates: Jour. Sed. Petrol., 3: 30–39.
Ovey, C D. 1938. Difficulties in establishing relationships in the Foraminifera: Geol. Assoc., Proc., 49.
——. 1950. On the interpretation of climatic variations as revealed by a study of samples from an equatorial Atlantic deep-sea core; *in* Centenary Symposium on Climatic Change: Roy. Meteor. Soc., Cent. Proc.
Owen, S. R. I. 1868. On the surface-fauna of mid-ocean. No. 2 – Foraminifera: Jour. Linn. Soc., Zool., 9: 147–157.
Patin, C. F. A. 1960. Notes on Microscopical Technique for Zoologists: Cambridge Univ. Press, 76 pp.
Parducz, B. 1966. Ciliary movement and coordination in ciliates: Int. Rev. Cytol., 21: 91–128.
Parker, F. L. 1948. Foraminifera of the continental shelf from the Gulf of Maine to Maryland: Harvard Col., Mus. Comp. Zool., Bull., 100, 2: 213–241.
——. 1952a. Foraminifera species off Portsmouth, New Hampshire: *Ibid.*, 106, 9: 391–423.
——. 1952b. Foraminiferal distribution in the Long Island sound, Buzzards Bay area: *Ibid.*, 106, 10: 214–241.
——. 1954. Distribution of the Foraminifera in the northeastern Gulf of Mexico: *Ibid.*, 111, 10: 453–588.
——. 1958. Eastern Mediterranean Foraminifera: Swedish Deep-Sea Exp., Rept., 8, 4: 219–283.
——. 1960. Living planktonic Foraminifera from the equatorial and south east Pacific: Tohoku Univ., Sci. Rept., Geol., Spec., 4: 71–82.
——. 1962. Planktonic foraminiferal species in Pacific sediments: Micropaleontol., 8, 2: 219–254.
——. 1971. Distribution of planktonic foraminifera in Recent deep-sea sediments; *in* The Micropaleontology of the Oceans: Cambridge Univ. Press: 289–307.
——. 1973. Living planktonic foraminifera from the Gulf of California: Jour. Foram Res., 3, 2: 70–77.
Parker, F. L. & Athearn, W. D. 1959. Ecology of marsh Foraminifera in Poponesset Bay, Massachusetts: Jour. Paleontol., 33, 2: 333–343.
Parker, F. L. & Berger, W. H. 1971. Faunal and solution patterns of planktonic foraminifera in surface sediments of the South Pacific: Deep-Sea Res., Oceanogr. Abstr., 18, 1: 73–107.
Parker, F. L. Phleger, F. B. & Peirson, J. F. 1953. Ecology of Foraminifera from San Antonio bay and environs, Southwest Texas: Cushman Found. Foram. Res., Spec. Publ., 2: 1–19.
Parker, W. K. & Jones, T. R. 1857. Description of some Foraminifera from the coast of Norway: Ann. Mag. Nat. Hist., 2, 112: 10–11.
——. 1865. On some Foraminifera of the North Atlantic and Arctic ocean, including Davis straits and Baffin bay: Roy. Soc. London, Phil Trans., 155: 325–441.
Parr, W. J. 1950. Foraminifera: Rept. B.A.N.Z. Antarct. Res. Exp. 1929–1931 (Zool. Bot.), 5, 6: 237–392.
Pazdrowa, O. 1958. *Ophtalmidium* wzulu i batonu oklic czenstochowy: Inst. Geol. Micropaleontol., 3: 91–162.
Pearcey, F. G. 1914. Foraminifera on the Scottish National Antarctic Expedition: Roy. Soc. Edinburgh, 49, 4: 991–1044.
Pearse, A. G. R. 1960. Histochemistry, Theoretical and Applied: Brown and Co., Boston.

Pease, D. C. 1964. Histological Techniques for Electron Microscopy: Academic Press, New York.
Perath, I. 1966. Living Foraminifera in the littoral zone of Achziv (Northern Israel): Israel Jour. Earth Sci., 15, 2: 64–70.
Perconig, E. 1966. The "Adaro" apparatus for the rapid disintegration of hard rock samples: Micropaleontol., 12, 1: 104–108.
Perkins, E. J. 1958. The food relationships of the microbenthos, with particular reference to that found at Whitstable, Kent: Ann. Mag. Nat. Hist., 13: 64–77.
Pessagno. E. A. 1960. Thin sectioning and photographing smaller Foraminifera: Micropaleontol., 6, 4: 419–423.
Petelin, V. P. & Fotiev, A. V. 1952. O novoj tjazheloj zhidkosti dlja sporovopyl'thevogo i diatomovogo analizov: Avtoref Dokl. "Bjull. MOIP" Otd. Geol, 27, 3.
Peterson, M. N. A. 1966. Calcite: rates of dissolution in a vertical profile in the central Pacific: Science, 140: 1542–1544.
Pflaumann, Uwe. 1971. Porositäten von Plankton-Foraminiferen als Klimaanzeiger?: *Meteor* Forsch., Reihe C, 7: 4–14.
Pflum, C. E. 1963. The distribution of Foraminifera in the eastern Ross sea, Amundsen sea, and Bellingshausen sea, Antarctica: Florida State Univ., Dept. Geol., Sed. Res. Lab., Čontr., 6, 106 pp.
Philippi, E. 1910. Die Grundproben der Deutschen Südpolar-Expedition 1901–1903: Deutsch. Südpol. Exped., 2, 6: 415–616.
Phleger, F. B. 1939. Foraminifera of submarine cores from the continental slope: Geol. Soc. Amer., Bull., 50: 1395–1422.
——. 1942. Note in current activities: Rep. Comm. Mar. Ecol., 2.
——. 1945. Vertical distribution of pelagic Foraminifera: Amer. Jour. Sci., 243: 377–383.
——. 1947. Foraminifera of three submarine cores from the Thyrrenian Sea: Göteborgs Kungl. Vetensk. Vitterh.-Sam. Handl., Sjat. Fol., (B), 5: 3–19.
——. 1948. Foraminifera of a submarine core from the Caribbean sea: *Ibid.:* 1–9.
——. 1951a. Displaced Foraminifera faunas: Soc. Econ. Paleontol. Mineral., Spec. Publ. 2: 66–75.
——. 1951b. Foraminifera distribution; *in* Ecology of Foraminifera, northwest Gulf of Mexico: Geol. Soc. Amer., Mem. 46: 1–88.
——. 1952a. Foraminifera ecology of Portsmouth, New Hampshire: Harvard Col., Mus. Comp. Zool., Bull., 106: 315–390.
——. 1952b. Foraminifera distribution in some sediment samples from the Canadian and Greenland Arctic: Cushman Found. Foram. Res., Contr., 3, 2: 80–89.
——. 1954a. Foraminifera and deep-sea research: Deep-Sea Res., 2: 1–23.
——. 1954b. Ecology of Foraminifera and associated micro-organisms from Mississippi Sound and environs: Amer. Assoc. Petrol. Geol., Bull., 38, 4: 584–647.
——. 1955. Ecology of Foraminifera in South Eastern Mississippi delta area: *Ibid.*, 39, 5: 712–752.
——. 1956. Significance of living foraminiferal populations along the central Texas coast: Cushman Found. Foram. Res., Contr., 7, 4: 106–151.
——. 1960a. Foraminiferal populations in Laguna Madre, Texas: Tohoku Univ., Sci. Rept., Geol., Spec., 4: 83–91.
——. 1960b. Ecology and Distribution of Recent Foraminifera: Johns Hopkins Press, Baltimore, 297 pp.
——. 1962. Sedimentology and oceanography of coastal lagoons in Baja California, Mexico: Geol. Soc. Amer., Bull., 73: 145–182.
——. 1963. Oceanografía física y ecologia de los foraminíferos del Golfo de California: Soc. Geol. Mex., Bol., 26, 2: 75–83.

Phleger, F. B. 1964. Patterns of living benthoonic Foraminifera, Gulf of California: Amer. Assoc. Petrol, Geol., Mem. 3: 377–394.
——. 1965a. Patterns of marsh Foraminifera, Galveston Bay, Texas: Limnol. Oceanogr., 10 (suppl.): R 169–R 184.
——. 1965b. Depth patterns of benthonic Foraminifera in the eastern Pacific; *in*, Progress in Oceanography, 3: Pergamon Press, London: 273–287.
——. 1965c. Living foraminifera from coastal marsh, southwest Florida: Soc. Geol. Mex., Bol., 28, 1: 45–60.
——. 1966. Patterns of living marsh Foraminifera in south Texas coastal lagoons: *Ibid.*: 1–44.
——. 1967. Marsh foraminiferal patterns, Pacific coast of North America: Ann. Inst. Biol. Univ. Nac. Méx., Ser. Cienc. Mar. Limnol., 38, 1: 11–39.
——. 1970. Foraminiferal populations and marine marsh processes: Limnol. Oceanogr., 15, 4: 522–534.
Phleger, F. B. & Bradshaw, J. S. 1966. Sedimentary environments in a marine marsh: Science, 154: 1551–1553.
Phleger, F. B. & Ewing, G. C. 1962. Sedimentology and oceanography of coastal lagoons in Baja California, Mexico: Geol. Soc. Amer., Bull., 73: 145–181.
Phleger, F. B. & Hamilton, W. A. 1946. Foraminifera of two submarine cores from the North Atlantic Basin: *Ibid*, 57: 951–966.
Phleger, F. B. & Lankford, R. R. 1957. Seasonal occurrences of living benthonic Foraminifera in some Texas bays: Cushman Found Foram. Res., Contr., 8, 3: 93–105.
Phleger, F. B. & Parker, F. L. 1951. Foraminifera species; *in* Ecology of Foraminifera northwest Gulf of Mexico: Geol. Soc. Amer., Mem 46: 1–69.
Phleger, F. B., Parker, F. L. & Peirson, J. F. 1953. North Atlantic Foraminifera: Swedish Deep-Sea Exp., Rept., 7, 122 pp.
Phleger, F. B. & Soutar, A. 1973. Production of benthic foraminifera in three east Pacific oxygen minima: Micropaleontol., 19, 1: 110–115.
Phleger, F. B. & Walton, W. R. 1950. Ecology of marsh and bay Foraminifera, Barnstable, Massachusetts: Amer. Jour. Sci., 248, 4: 274–294.
Pickford, G. E. 1946. *Vampyrotheutis infernalis* Chun, an arctic dibranchiate Cephalapod. I - Natural history and distribution: Dana Rept., 29.
Pierce, S., Kossoy, V., Valenti, R. & Smetana, D. G. 1968. Cytochemical studies on the test of *Allogromia laticollaris:* Micropaleontol., 14, 2: 242–246.
Pirie, J. H. H. 1913. Scottish National Antarctic Expedition, 1902–1904. Deep-sea deposits: Roy. Soc. Edinburgh, Trans., 49, 3, 10: 645–686.
Poag, C. W. 1972. Shelf-edge submarine banks in the Gulf of Mexico: paleoecology and biostratigraphy: Gulf Coast. Assoc. Geol. Soc., Trans., 22: 267–287.
Poag, C. W. & Sweet, W. E. 1972. Claypile Bank, Texas continental shelf; *in* Contributions on the Geological and Geophysical Oceanography of the Gulf of Mexico: Texas A & M Univ. Oceanogr. Stud., 3: 223–261.
Pohl, E. R. & Browne, R. G. 1973. A thin section method for ultrasmall paleontologic specimens: Jour. Paleontol., 47, 2: 221–225.
Pokorny, V. 1958. Grundzüge der Zoologischen Mikropaläontologie, v. 1: Deutsch. Verl. Wiss., Berlin, 429 pp.
Polski, W. 1959. Foraminiferal biofacies off the North Asiatic coast: Jour. Paleontol., 33, 4: 569–587.
Ponder, R. W. & Glendinning, I. G. 1974. The magnesian content of some miliolicean foraminifera in relation to their ecology and classification: Palaeogeogr., Palaeoclimat., Palaeoecol., 15, 1: 29–32.
Poore, R. Z. & Berggren, W. A. 1974. Pliocene biostratigraphy of the Labrador Sea: Calcareous plankton: Jour. Foram. Res., 4, 3: 91–108.

Popper, R. J. 1958. MS, Final report of the study of the Microforaminifera project: Amer. Mus. Nat. Hist., Dept. Micropaleontol.
Post, R. 1951. Foraminifera of the south Texas coast: Inst. Mar. Sci., 2, 1: 165–176.
Pozaryska, K. 1957. Lagenidae du Crétacé Supérieur de Pologne: Acad. Pol. Sci., Paléontol. Polon. 8: 1–190.
Prandtl, H. 1907. Der Entwicklungskreis von *Allogromia* sp.: Arch. Protist., 9: 209–213.
Pratje, O. 1930. Die Beziehungen der Foraminiferen der Deutschen Bucht (Nordsee) zu iher Umgebung: Paläont. Zeit., 12, 2: 1–126.
——. 1931. Die Sedimente der Deutschen Bucht. Eine regional-statistische Untersuchung: Wiss. Meersunt., N. F., Abt. Helgoland, 18, 2.
Premoli-Silva, I. 1966. La struttura della parete di alcuni Foraminiferi planctonici: Eclog. Geol. Helv., 59, 1: 219–233.
Pringsheim, E. H. 1946. Pure Cultures of Algae: Cambridge Univ. Press.
——. 1957. Methods for the cultivation of algae; *in* Manual of Physiology, Chronica Botanica, Mass.
Provasoli, L. 1958. Nutrition and ecology of protozoa and algae: Ann. Rev. Mikrobiol., 12.
Provasoli, L., McLaughlin, J. J. A. & Dropp. M. R. 1957. The development of artificial media for marine algae: Arch. Mikrobiol., 25: 292–428.
Pujos, M. 1972. Répartition des biocoenoses de Foraminifères benthiques sur le plateau continental du Golfe de Gascogne a l'ouest de l'embouchure de la Gironde: Rev. Españ. Micropaleontol., 4, 2: 141–156.
——. 1973a. Mise en évidence des biocoenoses faunes déplacées et paléothanatocoenoses de Foraminifères benthiques sur un plateau continental; applications à la zone ouest-Gironde (Golfe de Gascogne): Soc. Géol. France, Bull, 13, 3–4: 251–256.
——. 1973b. Les biocoenoses de Foraminifères benthiques et de thécamoebiens dans le complexe Garonne-Dordogne-Gironde: manifestations de l'influence des facteurs de l'environment sur les microfaunes: Bull. Inst. Géol. Bassin Aquit., 13: 3–19.
Pujos-Lamy, A. 1973. Répartition bathymétrique des Foraminifères benthiques profonds du Golfe de Gascogne. Comparaison avec d'autre aires océaniques: Rev. Españ. Micropaleontol., 5, 2: 213–234.
Pytkowicz, R. M. & Fowler, G. A. 1967. Solubility of foraminifera in sea water at high pressure: Geochem. Jour. (Geochem. Soc. Japan), 1, 4: 169–182.
Rae, K. M. 1960. Parameters of the marine environment; *in* Perspectives in Marine Biology: Univ. Calif. Press, Berkeley: 3–16.
Ramanathan, R. 1969. Seasonal variations in foraminiferan abundance and their relations to the substrate in Vellar Estuary, Madras: Jour. Geol. Soc. India, 10: 254–272.
——. 1970. Quantitative differences in the living benthic foraminifera of Vellar Estuary, Tamil Nadu: *Ibid.*, 11: 127–141.
Rao, K. K. 1969. Foraminifera of the Gulf of Cambay: Jour. Bombay Nat. Hist. Soc., 66, 3: 584–596.
——. 1970. *Idem: Ibid.*, 2: 259–273.
——. 1971. On some Foraminifera from the northeastern part of the Arabian Sea: Ind. Acad. Sci., Proc., sect. B, 73, 4: 155–178.
Ratnovskaja, K. I. 1967. Nekotorye pravila obrazovanija vidovykh nazvananij dlja foraminifer i drugikh mikroorganizmov: VNIGRI, Trudy, 252: 5–35.
Pau, W. W. 1948. Foraminifera from the Porter shale (Lincoln formation), Grays Harbour county, Washington: Jour. Paleontol., 22, 2: 152–174.
Rauser-Chernoussova, D. M. & Furssenko, A. V. 1937. Opredelitel' foraminifer neftenosnykh rajonov SSSR: ONTI, 1, 320 pp.
——. 1959. Vol. Obshchaja chast'. Prosteishie; *in* Osnovy Paleontologii: Akad. Nauk SSSR, Moscow: 114–367.

Rauser-Chernoussova, D. M. & Kulik, E. L. 1949. Ob otnoshenii fusilinid k fathijam i o periodichnosti ikh razvitija: Akad. Nauk SSSR, Izv., Geol., 6.
Reichel, M. 1946. Sur quelques Foraminifères nouveaux du Permien méditerranéen: Eclog. Geol. Helv., 38, 2: 524–560.
Reiss, Z. 1957. The Billamellidea nov. superfam., and remarks on Cretaceous Globorotaliids: Cushman Found. Foram. Res., Contr., 8, 4: 127–145.
——. 1958. Classification of lamellar foraminifera: Micropaleontol., 4, 1: 51–70.
——. 1959. The wall structure of *Cibicides*, *Planulina*, *Gyroidinoides* and *Globorotalites*: *Ibid.*, 5, 3: 355–357.
——. 1960. Structure of so-called *Eponides* and some other rotaliform foraminifera: Geol. Surv. Israel, Bull., 29: 1–28.
——. 1963a. Note sur la structure des foraminifères planctoniques: Rev. Micropaléontol., 6, 3: 127–129.
——. 1963b. Reclassification of perforate Foraminifera: Geol. Surv. Israel, Bull., 35: 1–111.
Reiss, Z. & Luz, B. 1970. Test formation pattern in planktonic foraminiferids: Rev. Españ. Micropaleontol., 2, 1: 85–96.
Reiss, Z. & Schneidermann, N. 1969. Ultramicrostructure of *Hoeglundina*: Micropaleontol., 15, 2: 135–144.
Reiter, M. 1959. Seasonal variation in intertidal Foraminifera of Santa Monica Bay, California: Jour. Paleontol., 33: 606–630.
Reitlinger, E. A. 1950. Foraminifery srednekamennougol'nykh otlozhenij zentral'noj chasti Russkoj platformy: Akad. Nauk SSSR, Inst. Geol. Nauk, Trudy, 126: 1–126.
——. 1958. K voprosu sistematiki i filogenii nadsemejstva Endothyridae: Vopr. Micropal., 2: 53–73.
Renz, H. H. 1948. Stratigraphy and fauna of the Agua Salada group, state of Falcon, Venezuela: Geol. Soc. Amer., Mem. 32: 1–129.
Resig, J. M. 1959. Ecology of Foraminifera of the Santa Cruz Basin, California: Micropaleontol., 4, 3: 287–308.
——. 1960. Foraminiferal ecology around ocean outfalls off southern California; *in* Waste Disposal in the Marine Environment: Pergamon Press, London: 104–121.
——. 1962. The morphological development of *Eponides repandus* (Fichtel & Moll), 1798: Cushman Found. Foram. Res., Contr., 13, 2: 55–57.
——. 1963. Size relationships of *Eggerella advena* to sediment and depth of substratum; *in* Essays in Marine Geology in Honor of K. O. Emery: Univ. South. Calif. Press: 121–126.
——. 1974. Recent foraminifera from a landlocked Hawaiian lake: Jour. Foram. Res., 4, 2: 69–76.
Reuss, A. 1862. Entwurf einer systematischen Zusammenstellung der Foraminiferen: Akad. Wiss. Wien, S.-B., Math. Nat. Kl., Abt. 1, 44: 355–396.
Revelle, R. R. 1944. Scientific results of Cruise VII of the *Carnegie* during 1928–1929. Marine bottom samples collected in the Pacific Ocean: Carnegie Inst. Wash., Publ. 556, Ocean., II, 1: 1–182.
Reyment, R. A. 1959. A note on the Foraminifera in a sediment core from the Mindanao Trough: Swedish Deep-Sea Exp., Rept., 6, 4: 139–156.
——. 1961. Quadrivariate principal component analysis of *Globigerina yeguaensis*: Stockholm Contr. Geol., 8, 2: 17–26.
——. 1966. Preliminary observations on gastropod predation in the western Niger delta: Palaeogeogr., Palaeoclimat., Palaeoecol., 2: 81–102.
——. 1969. *Textilina mexicana* (Cushman) from the western Niger delta: Univ. Uppsala, Geol. Inst., Bull., 1, 1–6: 75–81.
——. 1972. Application of multivariate morphometrics in paleontology: Inter. Geol. Congr., 24th, Montreal (7): 238–245.

Rezvoj, P. D. 1951. O nakhozhdenii foraminifer v bol'shom Gulikovskom ozere: Akad. Nauk Tadzh. SSSR, Dokl., 1.
Rhumbler, L. 1893. Beiträge zur Kenntnis der Rhizopoden. II. *Saccammina sphaerica* M. Sars: Z. Wiss. Zool., 57: 133–617.
——. 1895. Entwurf eines natürluchen Systems der Thalamophoren: Ges. Wiss. Göttingen, Math. Phys. Kl., Nachr., 1: 51–98.
——. 1905. Mitteilungen über Foraminiferen: Deutsch. Zool. Ges., Verh., 15: 97–106.
——. 1911. Die Foraminiferen (Thalamophoren) der Plankton-Expedition: Erg. Plank. Exp. Humboldt Stift., 3, L.C., 331 pp.
——. 1923. Die Foraminiferen; *in* Kukenthal and Krumbach, Handbuch der Zoologie, 1, 1, Berlin: 50–89.
——. 1935. Rhizopoden der Kieler Bucht, gesammelt durch A. Remane. I. Teil. Schrift. Naturwiss. Verein f. Schleswig-Holstein, 21, 2: 143–145.
——. 1938. Foraminiferen aus dem Meeressand von Helgoland, gesammelt von A. Remane (Kiel): Kieler Meersf., 2: 157–222.
Richter, G. 1961. Beobachtungen zur Ökologie einiger Foraminiferen des Jade-Gebietes: Natur u. Museum, 91: 163–170.
——. 1964a. Zur Ökologie der Foraminiferen. I. Die Foraminiferen-Gesellschaften des Jadegebietes: Natur u. Museum, 94, 9: 343–353.
——. 1964b. *Idem*, II. Lebensraum und Lebensweise von *Nonion depressulum*, *Elphidium excavatum* und *E. selseyense*: *Ibid.*, 11: 421–430.
——. 1965. *Idem*, III. Verdriftung und Transport in der Gezeitenzone: *Ibid.*, 95, 2: 51–62.
——. 1967. Faziebereiche rezenter und subrzenter Wattensedimente nach ihren Foraminiferen-Gemeinschaften: Senckenberg. Leth., 48, 3–4: 291–335.
Richter, R. 1948. Einführung in die zoologische Nomenclatur durch Erläuterung der Internationalen Regeln: Kramer, Frankfurt a. M., 2d ed., 252 pp.
Rieth, A. 1958. Vorkommen einer Foraminifere im Asterner Salzgebiet: Mikrokosmos, 47, 5.
Riveroll, D. D. & Jones, B. C. 1954. Varves and Foraminifera of a portion of the Upper Puente Formation (Upper Miocene): Jour. Paleontol., 28, 2: 122–131.
Rocha, A. T. & Mateu, G. 1971. Contribuição paro conhecimento dos Foraminíferos actuais da Ilha de Maio (Arquipélago de Cabo Verde): Inst. Invest. Cien. Angola, 108 pp.
Rocha, A. T. & Ubaldo, M. L. 1964a. Contribution for the study of Foraminifera from sands of Diu, Gogola and Simbot: Garcia de Orta, Lisboa, 12, 3: 407–420.
——. 1964b. Note of the Recent Foraminifera from the beach sands of Jampur (Damao) and Baga (Goa): *Ibid.*, 12, 4: 645–649.
Rodionov, D. A. 1968. Statisticheskie metody razgranichenija geologicheskikh obwektov po kompleksu priznakov: Izd. Nedra, 158 pp.
Rodrigues, M. A. 1968. Foraminíferos recentes da Barra de Itabapoana, Estado do Rio de Janeiro: Acad. Brasil, Cienc., Ann., 40, 4: 555–569.
——. 1972. Foraminíferos dos sedimentos recentes da Ilha de Fernando de Noronha: *Ibid.*, 44, 2: 281–316.
Roettger, E. U. 1970. Recent foraminifera from the continental shelf of Rio Grande do Sul collected by the hydrographic vessel *Canopus*: Iheringia, Zool., 38: 3–71.
Ronai, P. H. 1955. Brackish-water Foraminifera of the New York bay: Cushman Found. Foram. Res., Contr., 6, 4: 140–148.
Ross, C. A. 1972. Biology and ecology of *Marginopora vertebralis* (Foraminiferida), Great Barrier Reef: Jour. Protozool., 19, 1: 181–192.
Ross, G. C. 1963. Chromatographic technique for alcohol-preserved foraminiferal material: Micropaleontol, 19, 4: 442.
Rosset-Moulinier, M. 1971. Étude systemátique et écologie des Elphidiidae et des Nonionidae (Foraminifères) du littoral Breton 1; Les *Elphidium* du groupe *crispum* (Linné): Rev. Micropaléontol., 14, 1: 76–81.

Rosset-Moulinier, M. 1972. Étude des Foraminifères des côtes nord et ouest de Bretagne: Thèse de Doc. d'État, Univ. Paris, 225 pp.
Roth, E. 1962. Regernerations-Bildung an einem Dentalina-Gehäuse (Foraminifera): Senckenberg. Leth., 43, 3: 231–234.
Rotschy, R., Grazzini, C. V., Bellaiche, G. & Chamley, H. 1972. Étude paléoclimatologique d'une carotte prélevée sur un dome de la plaine abyssale ligure (structure alinat): Palaeogeogr., Palaeoclimat., Palaeoecol., 11: 125–145.
Röttgardt, D. 1952. Mikropaläontologisch wichtige Bestandteile rezenter brackische Sedimente an der Küsten Schleswig-Holsteins: Univ. Kiel, Geol. Inst., 1: 169–228.
Röttger, R. 1972a. Die Kultur von *Heterostegina depressa* (Foraminifera: Nummulitidae): Mar. Biol., 15: 150–159.
——. 1972b. Die Bedeutung der Symbiose von *Heterostegina depressa* (Foraminifera, Nummulitidae) für hohe Siedlungsdichte und Karbonatproduktion: Sonder. Verhand. Deutsch. Zool. Gesell., 65: 42–47.
——. 1972c. Analyse von Wachstumskurven von *Heterostegina depressa* (Foraminifera: Nummulitidae): Mar. Biol., 17: 228–242.
——. 1973. Die Ektoplasmahülle von *Heterostegina depressa* (Foraminifera: Nummulitidae): *Ibid.*, 21: 127–138.
Röttger, R. & Berger, W. H. 1972. Benthic Foraminifera: morphology and growth in clone cultures of *Heterostegina depressa: Ibid.*, 15: 89–94.
Rouvillois, A. 1966. Contribution a l'étude micropaléontologique de la baie du Roi, au Spitzberg: Rev. Micropaléontol., 9, 3: 169–176.
——. 1970. Variations de la biocoenose de la microfaune des vases entre un milieu marin artificiel, le bassin a Flot Vauban, et l'avant-port de Saint-Malo: Ouest Med., 23e ann., 20: 1695–1701.
——. 1971. Biocoenose et taphocoenose de Foraminifères sur le plateau continental Atlantique au large de l'Ile d'Yeu: Rev. Micropaléontol., 13, 3: 188–204.
——. 1972. Biocoenose de Foraminifères en relation avec les conditions physicochimiques du milieu dans les bassins et l'avant-port de Saint-Malo (Ille-et-Vilaine): Cah. Micropaléontol., (3), 1, Arch. orig. Centre Docum., C.N.R.S., 514: 1–10.
Rouvillois, A. & Rosset-Moulinier, M. 1969. Mise au point d'un petit carottier pour le prélèvement sans perturbation de la partie superficielle des sédiments marins: Cah. Océanogr., 21, 10: 1–9.
Ruddiman, W. F. 1971. Pleistocene sedimentation in the Equatorial Atlantic: stratigraphy and faunal paleoclimatology: Geol. Soc. Amer., Bull., 82: 283–302.
Ruddiman, W. F., Tolderlund, D. S. & Bé, A. W. H. 1970. Foraminiferal evidence of a modern warming of the north Atlantic Ocean: Deep-Sea Res., 17, 1: 141–155.
Ruthmann, A. 1966. Methoden der Zellforschung: Kosmos, Stuttgart.
Růžička, B., Hajkr, O. & Grmela, A. 1966. Study of the growth spiral of *Ammonia beccarii tepida* (Cushman): Inst. Mining Metal. Ostrava, Trans., Mining Geol. Ser., 12, 1: 85–93.
Sachs, K. N. 1965. Removal of ash from plankton samples concentrated by ignition: Deep-Sea Res., 12: 697.
Sachs, K. N., Cifelli, R. & Bowen, V. T. 1964. Ignition to concentrate shelled organisms in plankton samples: Deep-Sea Res., 11, 4: 621–622.
Saedeleer, H. 1932. Recherches sur les pseudopodes des Rhizopodes testacés: Arch. Zool. Expér. Gén., 74: 597–626.
Said, R. 1950. The distribution of Foraminifera in the northern Red Sea: Cushman Found. Foram. Res., Contr., 1, 1: 9–29.
——. 1951a. Preliminary note on the spectroscopic distribution of elements in the shells of some Recent calcareous Foraminifera: Cushman Found. Foram. Res., Contr., 2, 2: 11–13.
——. 1951b. Foraminifera of Narrangasett Bay: *Ibid.*, 3: 76–86.
——. 1953. Foraminifera of Great Pond, East Falmouth, Massachusetts: *Ibid.*, 4, 1: 7–14.

Said, R. & Kamel, T. 1957. The distribution of Foraminifera in the Egyptian Mediterranean coast: Egypt. Jour. Geol., 1, 2: 143–155.
Saidova, Kh. M. 1953. O znachenii bentosnykh foraminifer v tolshche osadkov Okhotskogo morja: Akad. Nauk SSSR, Dokl., 93, 1: 143–146.
——. 1956. Metodika izuchenija foraminifer iz donnykh otlozhenij: Akad. Nauk SSSR, Inst. Okeanol., Trudy, 19: 294–296.
——. 1957a. Kolichestvennoe raspredelenie foraminifer v Okhotskom more: Akad. Nauk SSSR, Dokl., 114, 6: 1302–1305.
——. 1957b. Raspredelenie foraminifer v tolsche osadkov Okhotskogo morja: *Ibid.*, 115, 6: 1213–1216.
——. 1958a. Foraminifera; *in* Ob Osadkakh i Donnoj Faune Severnoj Chasti Vostochno-Kitajskogo Morja: Ocean. Limnol. Sinica, 1, 3: 269–315.
——. 1958b. Novye dannye po ehkologii foraminifer: Priroda, 10: 107–110.
——. 1959. Raspredelenie foraminifer v donnykh otlozhenijakh i paleogeografia severozapadnoj chasti Tikhogo okeana: Akad. Nauk SSSR, Dokl., 129, 6: 1401–1404.
——. 1960a. Raspredelenie foraminifer v donnykh otlozhenijakh Okhotskogo morja: Akad. Nauk SSSR, Inst. Okeanol., Trudy, 32: 96–157.
——. 1960b. Stratigrafia osadkov i paleogeografia severo-zapadnoj chasti Tikhogo okeana i dal'nevostochykh morej SSSR po donnym foraminiferam: Mezhd. Geol. Congr., XXI Ses., Dokl. Sov. Geol., Mor. Geol.: 59–68.
——. 1961a. Ehkologia foraminfer i paleogeografia dal'nevostochnykh morej SSSR i severo-zapadnoj chasti Tikhogo okeana: Akad. Nauk SSSR, Inst. Okeanol., 182 pp.
——. 1961b. Kolichestvennoe raspredelenie donnykh foraminifer v Antarktike: Akad. Nauk SSSR, Dokl., 139, 4: 967–969.
——. 1962. Raspredelenie osnovnykh bentosnykh vidov sekrethionnykh (izvestkovykh) foraminifer v severo-zapadnoj chasti Tikhogo okeana: Vopr. Mikropal., 6: 31–63.
——. 1963. O zonal'nom kolichestvennom raspredelenii donnykh foraminifer v Tikhom okeane: *Ibid.*, 7: 196–208.
——. 1964. Raspredelenie donnykh foraminifer i stratigrafija osadkov v severo-vostochnoj chasti Tikhogo okeana: Inst. Okeanol., Trudy, 68: 84–119.
——. 1965. Raspredelenie donnykh foraminifer v Tikhom okeane: Okeanol., 5, 1: 99–110.
——. 1966a. Fauna donnykh foraminifer Tikhogo okeana: *Ibid.*, 6, 2: 276–284.
——. 1966b. Rasprostranenie vidov bentosnykh aggljutinirujushchikh foraminifer v Tikhom okeane: *Ibid.*, 1: 144–147.
——. 1967a. Depth changes in Bering Sea during the Upper Quaternary, as indicated by benthonic foraminifera; *in*, The Bering Land Bridge: Stanford Univ. Press: 364–368.
——. 1967b. Sediment stratigraphy and paleogeography of the Pacific Ocean by benthonic foraminifera during the Quaternary; *in* Progress in Oceanography, v. 4, The Quaternary history of the ocean basins: Pergamon Press, London: 143–151.
——. 1967c. Biomassa i kolichestvennoe raspredelenie zhivykh foraminifer v rajone Kurilo-Kamchatskoj vpadiny: Akad. Nauk SSSR, Dokl., 174, 1: 207–209.
——. 1968. Sokhrannost' foraminifer v vode, grunte i kishechnikakh gruntoedov: *Ibid.*, 182, 2: 453–455.
——. 1970a. Rajonirovanie dna Tikhogo okeana po bentosnym foraminiferam: *Ibid.*, 192, 5: 1145–1148.
——. 1970b. Planktonnye foraminifery iz rajona Kurilo-Kamchatskogo zheloba: Inst. Okeanol., Trudy, 86: 162–164.
——. 1970c. Bentosnye foraminifery rajona Kurilo-Kamchatskogo zheloba (po materialam 39-go reisa e/s 'Vitjaz'): *Ibid.*, 186: 134–161.
——. 1971. O rasprostranenii foraminifer u tikhookeanskogo poberezh'ja Juzhnoj Ameriki: Okeanol., 11, 2: 256–265.

Sancetta, C., Imbrie, J. & Kipp, N. G. 1972. Climatic record in North Atlantic Deep-Sea Core V23-82; comparison of the last and present interglacials based on quantitative time series: Quater. Res., 2, 3: 363–367.
Sanders, H. L. 1969. Benthic marine diversity and the stability-time hypothesis; *in* Diversity and stability in ecological systems: Brookhaven Symposium in Biology, 22: 71–81.
Sandon, H. 1932. The food of Protoza: Univ. Egypt, Fac. Sci., Publ., 1, 187 pp.
——. 1934. Pseudopodial movements of Foraminifera: Nature, 133: 761–762.
Saunders, J. B. 1957. Trochamminidae and certain Lituolidae (Foraminifera) from Recent brackish water sediments of Trinidad, British West Indies: Smithson. Misc. Coll., 134, 5: 1–16.
——. 1958. Recent Foraminifera of mangrove swamps and river estuaries and their fossil counterparts in Trinidad: Micropaleontol., 4, 1: 79–92.
Savin, S. M. & Douglas, R. G. 1973. Stable isotope and magnesium geochemistry of Recent planktonic Foraminifera from the South Pacific: Geol. Soc. Amer., Bull., 84: 2327–2342.
Schafer, C. T. 1968. Lateral and temporal variation of foraminifera populations living in near-shore water areas: Atlantic Oceanogr. Lab., Rpt. A.O.L. 68-4: 1–28.
——. 1970. Studies of benthonic foraminifera in the Restigouche estuary: 1, Faunal distribution patterns near pollution sources: Maritime Sediments, 6, 3: 121–134.
——. 1971. Sampling and spatial distribution of benthonic foraminifera: Limnol. Oceanogr., 16, 6: 944–951.
——. 1973. Distribution of foraminifera near pollution sources in Chaleur Bay: Water, Air, Soil Pollut., 2: 219–233.
Schafer, C. T. & Cole, F. E. 1974. Distributions of benthonic foraminifera: their use in delimiting local nearshore environments: Geol. Surv. Canada, Pap. 74-30, 1: 103–108.
Schafer, C. T. & Prakash, A. 1968. Current transport and deposition of foraminiferal tests, planktonic organisms and lithogenic particles in Bedford Basin, Nova Scotia; Maritime Sediments, 4, 3: 100–103.
Schafer, C. T. & Sen Gupta, B. K. 1969. Foraminiferal ecology in polluted estuaries of New Brunswick and Maine: Atlantic Oceanogr. Lab. Rpt., A.O.L. 69-1: 1–24.
Schaudinn, F. 1893. *Myxotheca arenilega* n.g., n. sp. ein neuer mariner Rhizopode: Z. Wiss. Zool., 57: 18–31.
——. 1894. Die Fortpflanzung der Foraminiferen und eine neue Art der Kernvermehrung: Biol. Zentrabl., 14: 161–166.
——. 1895a. Über den Dimorphismus der Foraminiferen: Ges. Naturf. Freunde, Berlin, S.-B., 5: 87–97.
——. 1895b. Über Plastogamie bei Foraminiferen: *Ibid.* 5: 179–190.
——. 1895c. Untersuchungen an Foraminiferen. I. *Calcituba polymorpha:* Z. Wiss. Zool., 59: 191–232.
——. 1896. Verzeichniss der während des Sommers 1894 aus dem Pudde - By und Hjeltefjord bei Bergen gesammelten Foraminiferen: Bergens Mus. Aarborg, 9.
——. 1911. Fritz Schaudinn's Arbeiten: Leopold Voss, Hamburg and Leipzig.
Schlumberger, C. 1894. Note sur les Foraminifères des mers arctiques russes: Soc. Zool. France, Mém., 7: 252–258.
Schmalhausen, O. I. 1950. Novyj vid foraminifer iz ozera Balpash-Sor: Akad. Nauk SSSR, Dokl., 75, 6: 869–872.
Schmid, K. 1934. Biometrische Untersuchungen an Foraminiferen *Globorotalia menardii* (d'Orbigny) - *Globorotalia tumìda* (Brady) und *Truncatulina margaritifera* Brady - *Truncatulina granulosa* Fischer aus dem Pliocaen von Ceram (Nieder. Indien): Eclogae Geol. Helv., 27: 1–91.
Schmidt, H. 1953. Ökologische Beobachtangen an den Foraminiferen des Golfes von Neapel: Paläont. Zeit., 27: 123–128.

Schmidt, R. A. M. 1952. Microradiography of microfossils with x-ray diffraction equipment: Science, 115: 94–95.
Schmidt, W. J. 1929. Rheoplasma und Stereoplasma. Nach Beobachtungen an einer neuen monothalamen Foraminifere *Rhumblerinella bacillifera* n.g., n.sp.: Protoplasma, 7: 353–394.
——. 1937. Über den Feinbau der Filopodien, insbesondere ihre Doppelbrechung bei *Miliola*: *Ibid.*, 27.
Schnitker, D. 1967. Variation in test morphology of *Triloculina linneiana* d'Orbigny in laboratory cultures: Cushman Found. Foram. Res., Contr., 18, 2: 83–87.
——. 1971a. Distribution of Foraminifera on the North Carolina continental shelf: Tulane Stud. Geol. Paleontol., 8, 4: 169–215.
——. 1971b. Das Magnet-Tonbandgerät als Hiflmittel in der Micropaläontologie: Paläont. Zeit., 45, 1-2: 92–93.
——. 1973. Abyssal benthonic foraminifera as indicators of present and past deep-sea circulation in north Atlantic Ocean [abstr.]: Amer. Assoc. petrol. Geol., Bull., 57, 4: 803.
——. 1974a. Ecotypic variation in *Ammonia beccarii* (Linné): Jour. Foram. Res., 4, 4: 216–223.
——. 1974b. West Atlantic abyssal circulation during the past 120,000 years: Nature, 248: 385–387.
Schott, W. 1935. Die Foraminiferen in dem äquatorialen Teil des Atlantischen Ozeans: Wiss. Ergebn. Deutsch. Atlant. Exp. "Meteor", 4: 43–134.
——. 1938. Über die Sedimentationsgeschwindigkeit rezenter Tiefseesedimente: Geol. Rund., 39: 322–329.
——. 1939. Rate of sedimentation of recent deep-sea sediments; *in* Recent Marine Sediments: Amer. Assoc. Petrol. Geol., Symposium: 409–415.
——. 1944. Die Bodenbedeckung des Atlantischen Ozeans; *in* Schott, G., Geographie des Atlantischen Ozeans, 3d ed., C. Boysen, Berlin, 435 pp.
——. 1952a. Zur Klimaschichtung der Tiefseesedimente im äquatorialen Atlantischen Ozean: Geol. Rund., 40: 20–31.
——. 1952b. On the sequence of deposits in the Equatorial Atlantic Ocean: Göteborgs Kungl. Vetensk. Vitterh. -Sam. Handl., Sjat. Folt., ser. B, 6, 2: 3–15.
——. 1954. Über stratigraphische Untersuchungsmethoden in rezenten Tiefseesedimenten: Heidelberger Beitr. Miner. Petrogr., 4: 192–197.
Schubert, R. 1907. Beiträge zu einer natürlichen Systematik der Foraminiferen: N. Jahrb. Miner. Geol. Paläont., 25: 232–260.
Schultz, E. 1915. Die Hyle des Lebens. I. Beobachtungen und Experimente an *Astrorhiza limicola*: Arch. Entw. Mech. Org., 41: 215–236.
Schultze, M. S. 1854. Über den Organismus der Polythalamien (Foraminiferen) nebst Bemerkungen über die Rhizopoden im Allgemeinen: Engelmann, Leipzig, 68 pp.
——. 1856. Beobachtungen über die Fortpflanzung der Polythalamien: Arch. Anat. Phys., vol. J. Muller: 165–173.
——. 1860. Die Gattung *Cornuspira* unter den Monothalamien, und Bemerkungen über die Organisation und Fortplanzung der Polythalamien: Arch. Naturgesch., 26: 287–310.
Schulze, F. E. 1875. Rhizopodenstudien III: Arch. Mikrosc. Anat., 11: 94–139.
——. 1877. *Idem*, IV: *Ibid.*, 13: 329–353.
Schümmelfeder, N. 1957. Fluoreszenzmikroskopische und cytochemische Untersuchungen über Frühschaden am Kleinhirm der Maus nach Röntgenbestrahlung: 3d Congr. Inter. Neuropathologie, 'Acta Medica Belgica': 295.
Schuster, D. 1954. Verzwergung durch Eisen? Untersuchungen an einer heute noch lebenden Zwergfauna in Mittelamerika: Umschau, 54, 21.
Schwab, D. 1969. Elektronenmikroskopische Untersuchung an der Foraminifera *Myxotheca arenilega* Schaudinn: Z. Zellforsch., 96: 295–324.

Schwab, D. & Schlobach, H. 1973. Carotene wax pigments in the foraminiferan *Allogromia laticollaris* Arnold: Jour. Protozool., 20, 2: 339–341.
Schwager, C. 1877. Quadro del proposito sistema de classificatione del Foraminiferi con guscio: R. Com. Geol. Ital., Boll., 8: 18–27.
Scorziello, R. 1966. Osservazioni sulla distribuzione batimetrica nel Golfo di Napoli di *Amphicoryna scalaris* (Batsch), Foraminifera: Estratto Publ. Staz. Zool. Napoli, 35: 115–129.
Scott, G. H. 1974. Biometry of the foraminiferal shell; *in* R. H. Hedley & C. G. Adams (eds.), Foraminifera, 1: Academic Press, London: 55–151.
Seal, H. 1964. Multivariate statistical analysis for biologists: Methuen, London, 207 pp.
Segura, L. R. 1963. Sistematica y distribución de los foraminíferos litorales de la "Playa Washington", al sureste de Matamores, Tamaulipas, México: Univ. Nac. Autónoma Méx., Inst. Geol., Bol. 68, 92 pp.
Seibold, E. 1967. La mer Baltique prise comme modèle de géologie marine: Rev. Géogr. Phys. Géol. Dyn., (2), 9, 5: 371–384.
Seiglie, G. A. 1964. Significación de los foraminíferos anormales de la Laguna de Unare: Lagena, 1: 6.
——. 1965. Some observations on Recent foraminifers from Venezuela, part 1: Cushman Found. Foram. Res., Contr., 16, 2: 70–73.
——. 1966. Distribution of foraminifers in the sediments of Araya–Los Testigos (Caribbean) shelf and upper slope: Caribbean Jour. Science, 6, 3-4: 93–117.
——. 1968a. Relationships between the distribution of *Amphistegina* and the submerged Pleistocene reefs off western Puerto Rico: Tulane Stud. Geol., 6, 4: 139–147.
——. 1968b. Foraminiferal assemblages as indicators of high organic carbon content in sediments and of polluted waters: Amer. Assoc. Petrol. Geol., 52, 11: 2231–2241.
——. 1970a. Additional observations on the foraminiferal genus *Buliminoides* Cushman: Cushman Found. Foram. Res., Contr., 21, 3: 112–115.
——. 1970b. The distribution of the foraminifers in the Yabucoa Bay southeastern Puerto Rico and its paleoecological significance: Rev. Españ. Micropaleontol., 2, 2: 183–208.
——. 1971a. Foraminíferos de las bahias de Mayagüez y Añasco, y sus alrededores, oeste de Puerto Rico, Parte II. Foraminíferos aglutinados: *Ibid.*, 3, 3: 255–276.
——. 1971b. A preliminary note on the relationships between foraminifers and pollution in two Puerto Rican bays: Caribbean Jour. Science, 11, 1-2: 93–98.
——. 1973. Pyritization in living foraminifers: Jour. Foram. Res., 3, 1: 1–6.
Seiglie, G. A. & Bermúdez, P. J. 1963. Distribución de los foraminíferos del Golfo de Cariaco: Univ. Oriente, Inst. Oceanogr., Bol., 2, 1: 7–87.
Sellier de Civrieux, J. M. 1968. Un método de observaciones y representaciones gráficas del metabolismo (foraminíferos) con aplicación al estudio de la ecología en biotopes de salinidad altamente variable: Univ. Oriente, Inst. Oceanogr., Bol., 7, 2: 99–109.
——. 1970a. Mutaciones recientes del género *Peneroplis* y relaciones filogénicas con otros Soritidae: Rev. Españ. Micropaleontol., 2, 1: 5–12.
——. 1970b. Biofacies bentónicas de foraminíferos en la plataforma continental de Cumaná, Venezuela: Univ. Oriente, Inst. Oceanogr., Bol., 9, 1-2: 21–70.
Sellier de Civrieux, J. M. & Bonilla, J. 1971. La influencia de los parámetros físico-químicos del fondo en las facies de foraminíferos bentónicos: *Ibid.*, 10, 2: 15–34.
Sellier de Civrieux, J. M. & Bermúdez, P. J. 1973. Ecología y distribución de foraminíferos bentónicos del Golfo de Santa Fé (Venezuela): Rev. Españ. Micropaleontol., 5, 1: 33–80.
Sen Gupta, B. K. 1967. Distribution of Foraminifera in the sediments of the Grand Bank - A preliminary report: Maritime Sediments, 3, 2-3: 61–63.
——. 1971. The benthonic foraminifera of the tail of the Grand Banks: Micropaleontol., 17, 1: 69–98.

Sen Gupta, B. K. 1972. Distribution of Holocene benthonic foraminifera on the Atlantic continental shelf of North America: Inter. Geol. Congr., 24th, Montreal, (8): 123–134.
Sen Gupta, B. K. & Kilbourne, R. T. 1974. Diversity of benthic Foraminifera on the Georgia continental shelf: Geol. Soc. Amer., Bull., 85: 969–972.
Sen Gupta, B. K. & McMullen, R. M. 1969. Foraminiferal distribution and sedimentary facies on the Grand Banks of Newfoundland: Canadian Jour. Earth Sci., 6, 3: 475–487.
Sen Gupta, B. K. & Schafer, C. T. 1973. Holocene benthonic foraminifera in leeward bays of St. Lucia, West Indies: Micropaleontol., 19, 3: 341–365.
Serova, M. Ja. 1961a. Vid *Trochammina vitrea* Serova sp.n., ego paleoehkologia; stratigrafischeskoe znachenie: Vopr. Mikropaleontol., 5: 69–82.
——. 1961b. Taksonomicheskoe znachenie nekotorykh osobennostej mikrostruktury stenki i stroenija rakovin Miliolid: *Ibid.*: 129–134.
Servicio Hidrografia Naval. 1958. Manual de instrucciones para observaciones oceanográficas: Serv. Hidrogr. Naval, Argentina, H 601, 451 pp.
Shackleton, N. J. 1968. Depth of pelagic foraminifera and isotope changes in Pleistocene oceans: Nature, 218: 79–80.
Shackleton, N. J., Wiseman, J. D. H. & Buckley, H. A. 1973. Non-equilibrium isotopic fractionation between seawater and planktonic foraminiferal tests: *Ibid.*, 242: 177–179.
Shannon, C. E. & Weaver, W. 1963. The mathematical theory of communication: Univ. Illinois Press, Urbana, 117 pp.
Shaw, A. B. 1969. Adam and Eve, paleontology, and the non-objective arts: Jour. Paleontol., 43, 5: 1085–1098.
——. 1971. The butterfingered handmaiden: *Ibid.*, 45, 1: 1–5.
Sheehan, R. & Banner, F. T. 1972. The pseudopodia of *Elphidium incertum* (Will.): Rev. Españ. Micropaleontol., 4, 1: 31–63.
Shenton, E. H. 1957. A study of the Foraminifera and sediments of Matagorda Bay, Texas: Gulf Coast Assoc. Geol. Soc., Trans., 7: 135–150.
Shifflett, E. 1961. Living, dead, and total foraminiferal faunas, Heald Bank, Gulf of Mexico: Micropaleontol., 7: 45–54.
Shishkevish, L. J. 1964 Ms. The distribution of foraminifera in cores from the northern Scotia Arc, Scotia Sea: unpubl. Ph.D. thesis, New York Univ.
Shupak, B. 1934. Some foraminifera from Western Long Island and New York harbor: Amer. Mus. Novit., 737: 1–12.
Sidner, B. R. & Poag, C. W. 1972. Late Quaternary climates indicated by foraminifers from the southwestern Gulf of Mexico: Gulf Coast Assoc. Geol. Soc., Trans., 22: 305–313.
Sigal, J. 1952. Odre des Foraminifera; *in* Piveteau, Traité de Paléontologie: Masson, Paris: 133–301.
——. 1959. Trimorphisme morphologique et trimorphisme structural – un exemple: *Choffatella dicipiens* Schlumberger 1905 (Foraminifères): Soc. Géol. France, Bull., (7), 1: 662–668.
——. 1966. Le concept taxinomique de spectre: Soc. Géol. France, Mém. Hors-Ser., 3: 1–126.
Simpson, E. H. 1949. Measurement of diversity: Nature, 163: 688.
Sissingh, W. 1973. Foraminifera of the Breid Bay area (Antarctica): Neth. Jour. Sea Res., 6, 3: 355–364.
Slama, D. 1954. Arenaceous tests in Foraminifera: an experiment: The Micropaleontologist, 8, 1: 33–34.
Sliter, W. V. 1965. Laboratory experiments on the life cycle and ecologic controls of *Rosalina globularis* d'Orbigny: Jour. Paleontol., 12, 2: 210–215.
——. 1968. Shell material variation in the agglutinated foraminifer *Trochammina pacifica* Cushman: Tulane Stud. Geol., 6, 2: 80–84.
——. 1970a. Inner-neritic Bolivinitidae from the eastern Pacific margin: Micropaleontol., 16, 2: 155–174.

Sliter, W. V. 1970b. *Bolivina doniezi* Cushman and Wickenden in clone culture: Cushman Found. Foram. Res., Contr., 21, 3: 87–99.
——. 1971. Predation on benthic foraminifers: Jour. Foram. Res., 1, 1: 20–29.
——. 1972a. Cretaceous foraminifers – depth habitats and their origin: Nature, 239: 514–515.
——. 1972b. Upper Cretaceous planktonic foraminiferal zoogeography and ecology – eastern Pacific margin: Palaeogeogr., Palaeoclimat., Palaeoecol., 12: 15–31.
——. 1974. Test ultrastructure of some living benthic foraminifers: Lethaia, 7: 5–16.
Sliter, W. V. & Baker, R. A. 1972. Cretaceous bathymetric distribution of benthic foraminifers: Jour. Foram. Res., 2, 4: 167–183.
Small, E. B. & Marszalek, D. S. 1969. Scanning electron microscopy of fixed, frozen and dried protozoa: Science, 163: 1064–1065.
Smayda, J. T. 1970. The suspension and sinking of phytoplankton in the sea: Oceanogr. Mar. Biol. Ann. Rev., 8: 353–414.
Smith, A. B. 1963. Distribution of living planktonic Foraminifera in the northeastern Pacific: Cushman Found. Foram. Res., Contr., 14, 1: 1–15.
Smith, M. A. & Kaesler, R. L. 1970. Selection of adventitious test material by *Reophax curtus* (Foraminiferida): Jour. Paleontol., 44, 5: 953–957.
Smith, P. B. 1963. Quantitative and qualitative analysis of the family Bolivinidae: U.S. Geol. Surv. Prof. Pap. 429-A: A1–A39.
——. 1964. Ecology of benthonic species: *Ibid.*, Pap. 429-B: B1–B55.
Smith, P. B. & Emiliani, C. 1968. Oxygen-isotope analysis of Recent tropical Pacific benthonic Foraminifera: Science, 160: 1335–1336.
Smith, R. K. 1967. Ignition and filter methods of concentration of shelled organisms: Jour. Paleontol., 41, 5: 1288–1291.
——. 1968. An intertidal *Marginopora* colony in Suva Harbor, Fiji: Cushman Found. Foram. Res., Contr., 19, 1: 12–17.
Smout, A. H. 1954. Lower Tertiary Foraminifera of the Qatar Peninsula: Brit. Mus. (Nat. Hist.), Mon., 96 pp.
Sneath, P. H. A. & Sokal, R. R. 1973. Numerical Taxonomy: Freeman, San Francisco, 573 pp.
Snedecor, C. W. & Cochran, W. G. 1967. Statistical Methods: Iowa State Univ. Press, Ames, 596 pp.
Sokal, R. R. & Rohlf, F. J. 1969. Biometry: Freeman, San Francisco, 776 pp.
Sokal, R. R. & Sneath, P. H. A. 1963. Principles of Numerical Taxonomy: *Ibid.*, 359 pp.
Soldani, A. 1789. Testaceografia et Zoophytografia parva et microscopica, 1, Siena, 1–80.
——. 1798. *Idem*, 2, *Ibid.*: 201–289.
Sollas, W. J. 1921. On *Saccammina carteri* Brady and the minute structure of the foraminiferal shell: Geol. Soc. London, Quart. Jour., 77: 193–212.
Sorby, H. C. 1879. Anniversary address of the President: Geol. Soc. London, Proc., 1878–1879: Geol. Soc. London, Quart. Jour., 35, appendix: 56–93.
Sosnina, M. I. 1960. Izuchenie Lagenid metodom posledovatel'nykh prishlifovok:1st Semin. Mikrofaun., Trudy: 88–119.
Sozeri, B. 1966. The actual Foraminifera and their variations on the beach sands of Cesmi Illicasi, Izmir: Geol. Soc. Turkey, Bull., 10, 1-2: 148–154.
Spärk, R. 1933. Contributions to the animal ecology of Franz Joseph fiord and adjacent east Greenland waters: Med. om Grønland, 100, 1.
Spindler, M. & Röttger, R. 1973. Der Kammer-Bauvorgang der Grossforaminifere *Heterostegina depressa* (Nummulitidae): Mar. Geol., 18: 146–159.
Srinivasan, M. S. & Kennett, J. P. 1974. Secondary calcification of the planktonic foraminifer *Neogloboquadrina pachyderma* as a climatic index: Science, 186: 630–632.
Stainforth, R. M. 1952a. Classification of uniserial calcareous Foraminifera: Cushman Found. Foram. Res., Contr., 3, 1: 6–14.
——. 1952b. Ecology of arenaceous Foraminifera: The Micropaleontologist, 6, 1: 42–44.

Stapleton, R. P. 1973. Ultrastructure of tests of some Recent benthic hyaline Foraminifera: Palaeontolographica, (A), 142, 1–3: 16–49.
Stehli, F. G. 1965. Paleontologic techniques for defining ancient ocean currents: Science, 148: 943–946.
——. 1966. Some applications of foraminiferal ecology: West African Micropaleontol. Colloq., 2d, Proc.: 223–240.
Stehli, F. G. & Creath, W. B. 1964. Foraminiferal ratios and regional environments: Amer. Assoc. Petrol. Geol., Bull., 48, 11: 1810–1827.
Stehman, C. F. 1972. Planktonic foraminifera in Baffin Bay, Davis Strait and the Labrador Sea: Maritime Sediments, 8, 1: 13–19.
Steinker, D. C. 1973. Test dimorphism in *Glabratella ornatissima* (Cushman): Compass, 50, 3: 10–21.
Stephenson, W. & Rees, M. 1965a. Ecological and life history studies upon a large foraminiferan (*Discobotellina biperforata* Collins, 1958) from Moreton Bay, Queensland; I. The life cycle and nature of the test: Univ. Queensland Pap. (Zool.), 2: 207–223.
——. 1965b. *Idem;* Aquarium observations: *Ibid.*: 239–255.
Sterba, G. 1963. Ankündigung eines neuen histochemischen DNS-Nachweis mit hoher, die Feulgen-Reaktion übertreffender Empfindlichkeit: Acta Biol. Med. Ger., 10: 694.
Stetson, H. C. 1953. The sediments of the western Gulf of Mexico. Part 1, The continental terrace of the western Gulf of Mexico; its surface sediments, origin and development: Papers in Phys. Oceanogr. Meteor., Mass. Inst. Tech. and Woods Hole Oceanogr. Inst., 8, 4: 1–45.
Steuerwald, B. A. & Clark, D. L. 1972. *Globigerina pachyderma* in Pleistocene and Recent Arctic Ocean sediment: Jour. Paleontol., 46, 4: 573–580.
Stinemeyr, E. H. 1965. Microfossil vacuum-needle segregating pick; *in* B. Kummel & D. Raup (eds.), Handbook of Paleontological Techniques: Freeman, San Francisco: 276–283.
Stockton, W. L. 1973. Distribution of benthic foraminifera at Arthur Harbor, Anvers Island: Antarctic Jour., 8, 6: 348–350.
Stone, S. W. 1956. Some ecologic data relating to pelagic foraminifera: Micropaleontol., 2, 4: 361–370.
Streeter, S. 1964. Foraminiferal distribution on the Orinoco-Trinidad-Paria shelf: a restudy using vector analysis [abstr.]: Geol. Soc. Amer., Prog. Abstr., 1964: 200.
——. 1972. Living benthonic Foraminifera of the Gulf of California, a factor analysis of Phleger's (1964) data: Micropaleontol., 18, 1: 64–73.
——. 1973. Bottom water and benthonic Foraminifera in the North Atlantic - Glacial-Interglacial contrasts: Jour. Quater. Res., 3, 1.
Stschedrina, Z. 1936a. *Alveolophragmium orbiculatum*, n.g., n. sp.; Zool. Anz., 114: 312–319.
——. 1936b. K faune kornenozhek poljarnykh morej SSSR: Arktich. Inst., Trudy, 33: 51–64.
——. 1938. K raspredeleniju foraminifer v Karskom more: Akad. Nauk SSSR, Dokl., 19: 321–324.
——. 1939. Novyj rod peshchanistykh foraminifer iz arkticheskikh morej: *Ibid.*, 24: 94–96.
——. 1946. Novye formy foraminifer iz Severnogo Ledovitogo okeana: Dreif. Ehksp. Glavsevmorputi led. 'G. Sedov' 1937–1940, Trudy, 3: 138–148.
——. 1947. K raspredeleniju foraminifer v Grenlandskom more: Akad. Nauk SSSR, Dokl., 55: 871–874.
——. 1950a. K faune foraminifer Okhotskogo morja: Issled. Dal'nevost. Morej SSSR, 2: 248–280.
——. 1950b. K raspredeleniju morskikh kornenozhek v svjazi s uslovijami ikh obitanija: Akad. Nauk SSSR, Dokl., 70: 711–713.
——. 1952a. O razlichnykh formakh u foraminifer (*Rhabdammina abyssorum* Carpenter): Akad. Nauk SSSR, Zool. Inst., Trudy, 12: 7–24.
——. 1952b. Novye vidy foraminifer roda *Rhabdammina* M. Sars: Akad. Nauk SSSR, Zool. Inst., Trudy, 12: 25–33.

Stschedrina, Z. 1952c. Foraminifery Chukotskogo morja i Beringova proliva: Simp. Krajnij Severo-Vostok SSSR. 3: 98–100.
——. 1953a. Novye dannye po faune foraminifer Okhotskogo morja i ee raspredeleniju: Akad. Nauk SSSR, Zool. Inst., Trudy, 13: 12–32.
——. 1953b. K izusheniju foraminifer glubokovodhnykh donnykh otlozhenij Okhotskogo morja: Akad. Nauk SSSR, Dokl., 90: 287–289.
——. 1955. Otrjad foraminifery-Foraminifera: Akad. Nauk SSSR, Atlas Bespozv. Dal'nevost. Morej SSSR.
——. 1956a. Itogi izuchenija fauny foraminifer morej SSSR: Vopr. Micropaleontol., 1: 23–36.
——. 1956b. Fauna foraminifer dal'nevostochnykh morej Sovetskogo Sojuza: Trudy Prob. Temat. Soveshch. ZIN, 6: 66–71.
——. 1957. Izuchenie zakonomernostej raspredelenija sovremennykh foraminifer (Foraminifera): Leningr. O-vo Estestv., Trudy, 73: 99–105.
——. 1958a. Fauna foraminifer (Foraminifera) morskikh vod juzhnogo Sakhalina i juzhnykh Kuril'skikh ostrovov: Issled. Dal'nevost. Morej SSSR, 5: 6–41.
——. 1958b. Foraminifery vod vostochnogo Murmana: Murmansk Biol. Stan., Trudy, 4: 118–129.
——. 1958c. O faune foraminifer (Foraminifera) Kurilo-Kamchatskoj vpadiny: Akad. Nauk SSSR, Inst. Okeanol., Trudy, 27: 161–179.
——. 1958d. Ob iskopaemykh foraminiferakh v donnykh otlozhenijakh Karskogo morja: Nauchn. Issled. Inst. Geol. Arktiki, Sborn. Stat. Paleont. Biostrat., 2: 66–72.
——. 1958e. Fauna foraminifer vostochnogo sektora Antarktiki: Sovet. Antarkt. Ehksp., Bull., 3: 51–54.
——. 1962. Foraminifery zalivov Belogo morja: MGU, Belomor. Biol. Stan., Trudy, 1: 51–69.
——. 1964a. Foraminifery (Foraminifera) severnoj chasti Grenlandskogo morja: Arkt. Antarkt. Nauchn. Issled. Inst., Trudy, 259: 120–142.
——. 1964b. Foraminifery vysokikh shirot Arkticheskogo bassejna: *Ibid.*: 79–119.
——. 1972. Sopostavlenie fauny agglutinirujushchikh foramihifer vysokikh shirot severnogo i juzhnogo polusharij: Antarktika, 11: 174–186.
Stuart, A. 1866. Über *Coscinosphaera ciliosa*, eine neue Radiolarie: Z. Wiss. Zool., 16: 328–345.
Stubbings, H. G. 1939. Stratification of biological remains in marine deposits: John Murray Exp. 1933–1934, Sci. Repts., 3: 159–192.
Subbotina, N. N. 1953. Globigerinidy, Khantkenidy i Globorotalidy: VNIGRI, Iskop. Foram. SSSR, 76, 296 pp.
Sverdrup, H. U., Johnson, M. W. & Fleming, R. H. 1942. The Oceans; their physics, chemistry and general biology: Prentice-Hall, Englewood Cliffs, N.J., 1087 pp.
Swarczewsky, B. 1909. Zur Kenntnis der *Allogromia ovoidea* (Rhumbler): Arch. Protist., 14: 396–416.
Swinnerton, H. H. 1940. The study of variations in fossils: Geol. Soc. London, Quart. Jour., Proc., 3, 96: LXXVII–CXVIII.
Switzer, G. & Boucot, A. J. 1955. The mineral composition of some microfossils: Jour. Paleontol., 29: 525–533.
Sylvester-Brady, P. C. 1969. Aluminum coating in scanning electron microscopy: Micropaleontol., 15, 3: 366.
Szidat, L. 1955. Beiträge zur Kenntnis der Reliktfauna des La Plata-Strom-System: Arch. Hidrobiol., 51, 2: 209–260.
Takayanagi, Y. 1953. Distribution of the Recent Foraminifera from the adjacent seas of Japan, I.: Rec. Oceanogr. Works Japan, 1, 2, n. ser.: 78–85.
——. 1955. Studies on the ecology and sedimentation of Matsukawa-Ura, Soma City, Fukushima Prefecture: Tohoku Univ., Inst. Geol. Paleontol., Contr., 45: 18–52.

Takayanagi, Y., Niitsuma, N. & Saki, T. 1968. Wall microstructure of *Globorotalia truncatulinoides* (d'Orbigny): Tohoku Univ., Sci. Rept., (2), Geol., 40, 2: 141–170.
Tamanova, S. V. & Stschedrina, Z. G. 1965. Fauna foraminifer v donnykh otlozhenijakh Norvezhskogo morja: VNIRO, Trudy, 57: 285–296.
Tan, Sin Hok. 1932. On the genus *Cycloclypeus* Carpenter: Wetensch. Med., 19: 1–194.
Tapley, S. 1969. Foraminiferal analysis of the Miramichi estuary: Maritime Sediments, 5, 1: 30–38.
Tappan, H. 1951. Foraminifera from the Arctic slope of Alaska, general introduction and part 1, Triassic Foraminifera: U.S. Geol. Surv., Prof. Pap. 236-A: 1–20.
Tasch, P. 1955. The concept of psephonecronosis applied to 'dwarf' fossil faunas: Amer. Mid. Nat., 53, 1: 205–212.
Taugourdeau-Lantz, J. & Poignant, A. 1964. La membrane chitinoide de quelques Foraminifères: Rev. Micropaléontol., 7, 1: 68–71.
Terquem, O. & Terquem, E. 1886. Foraminifères et Ostracodes de l'Islande et du sud de la Norvège: Soc. Zool. France, Bull., 10: 328–339.
Thalmann, H. E. 1952. Twenty years of 'Foraminiferal statistics': 1931 to 1950: Cushman Found. Foram. Res., Contr., 3, 4: 145–146.
Theyer, F. 1966. Variationsstatistiche Untersuchungen zur Verbreitung der Gattung *Buccella* Anderson im Südlichen Teil Sudamerikas (Protozoa, Foraminifera): Zool. Jahrb. Abt. Syst. Ökol. Geogr. Tiere, 93, 2: 203–222.
——. 1970. Benthic foraminiferal studies in the Pacific-Antarctic Basin: Antarctic Jour., 5, 5: 180–181.
——. 1971a. Benthic foraminifera; Bathymetric patterns of calcareous and arenaceous assemblages: Nature, Phys. Sci., 229, 7: 207–209.
——. 1971b. Benthic foraminiferal trends, Pacific-Antarctic Basin: Deep-Sea Res., 18, 7: 723–738.
——. 1971c. Size-depth variation in *Cyclammina cancellata* Brady, Peru-Chile Trench area: Antarct. Res. Ser., 15: 309–313.
Thiede, J. 1971a. Planktonische Foraminiferen in sedimenten von ibero-marokkanischen Kontinentalrand: 'Meteor' Forsch. Ergebn., Reihe C, 7: 15–102.
——. 1971b. Planktonische Foraminiferen in Sedimentkernen vom portugiesischen Kontinentalhang: Inter. Conf. Plank. Microfossils, 2d, Proc., Rome, 2: 1201–1212.
——. 1971c. Variations in coiling ratios of Holocene planktonic foraminifera: Deep-Sea Res., 18: 823–831.
——. 1973. Planktonic foraminifera in hemipelagic sediments: Shell preservation off Portugal and Morocco: Geol. Soc. Amer., Bull., 84: 2749–2754.
Thompson, P. R. & Saito, T. 1974. Pacific Pleistocene sediments: planktonic foraminifera dissolution cycles and geochronology: Geology, 2, 7: 333–335.
Tinoco, I. de Medeiros. 1955. Foraminíferos recentes de Cabo Frío, estado do Rio de Janeiro: Dep. Nac. Prod. Min., Div. Geol. Miner., Bol., 159: 7–43.
——. 1958. Observações sobre a microfauna de Forminíferos de lagoa de Araruama, estado do Rio de Janeiro: Acad. Brazil. Cienc., An., 30: 576–583.
——. 1959. Classificação sistemática dos foraminíferos dos testemunhos de sondagens submarinas efetuadas das pelo navio escola "Almirante Saldanha" na embocadura do rio Amazonas: Univ. Recife, Inst. Biol. Mar. Oceanogr., Trab., 1: 107-112.
——. 1967a. Foraminíferos do Atol das Rocas: Univ. Fed. Pernambuco, Inst. Oceanogr., Trab., 7–8: 91–114.
——. 1967b. Contribuiçâo a sedimentologia e microfauna da Bahia de Sepetiba (Estado do Rio de Janeiro): *Ibid.*: 123–136.
——. 1971. Distribuição dos Foraminíferos na plataforma continental do norte-nordeste do Brasil: Arquivos do Museu Nac., 54: 93–97.

Todd, R. 1957. Foraminifera from Carter Creek, northeastern Alaska: U.S. Geol. Surv., Prof. Pap. 294-F: 223–235.
——. 1958. Foraminifera from western Mediterranean deep-sea cores: Swedish Deep-Sea Exp., 8, 3: 169–215.
——. 1963. Nomenclature of Foraminifera: Cushman Found. Foram. Res., Contr., 14, 3: 109–111.
——. 1964. Planktonic Foraminifera from deep-sea cores off Eniwetok Atoll: U.S. Geol. Surv., Prof. Pap. 260-CC: 1067–1100.
——. 1965. A new *Rosalina* (Foraminifera) parasite on a bivalve: Deep-Sea Res., 12, 6: 831–837.
——. 1971. *Tretomphalus* (Foraminifera) from Midway: Jour. Foram. Res., 1, 4: 162–169.
Todd, R. & Blackmon, P. 1956. Calcite and aragonite in Foraminifera: Jour. Paleontol., 30: 217–219.
Todd, R. & Low, D. 1961. Near-shore Foraminifera of Martha's Island, Massachusetts: Cushman Found Foram. Res., Contr, 12: 5–21.
——. 1966. Foraminifera from the Arctic Ocean off the eastern Siberian Coast: U.S. Geol. Surv., Prof. Pap. 550-C: 79–85.
——. 1967. Recent Foraminifera from the Gulf of Alaska and southeastern Alaska: *Ibid.*, 573-A: 1–46.
——. 1971. Foraminifera from the Bahama Bank west of Andros Island: *Ibid.*, 683-C: 1–22.
Tolderlund, D. S. & Bé, A. W. H. 1971. Seasonal distribution of planktonic foraminifera in the western North Atlantic: Micropaleontol., 17, 3: 297–329.
Towe, K. M. 1967. Wall structure and cementation in *Haplophragmoides canariensis:* Cushman Found. Foram. Res., Contr., 18, 4: 147–151.
——. 1971. Lamellar wall construction in planktonic Foraminifera: Inter. Conf. Plank. Microfossils, 2d, Proc., Rome, 2: 1213–1224.
Towe, K. M. & Cifelli, R. 1967. Wall ultrastructure in the calcareous Foraminifera, crystallographic aspects and a model for calcification: Jour. Paleontol., 41, 3: 742–762.
Triebel, E. 1947. Methodische und technische Fragen der Mikropaläontologie: Senckenberg-Buch, 19: 1–47.
——. 1958. Die Photographie im Dienste der Mikropaläontologie; *in* Freund; Handbuch der Mikroskopie in der Technik, 2, 3: 85–144.
Troithkaja, T. S. 1970. Uslovija obitanija i raspredelenie foraminifer v Japonskom more: Akad. Nauk SSSR, Sibirsk. Otd., Inst. Geol. Geofis., Trudy: 136–172.
Tsi-chung, C. & Sau-yee, C. 1962. On the ecology of the planktonic foraminifera of the Yellow Sea and the East China Sea: Ocean. Limnol. Sinica, 4, 1–2: 60–85.
——. 1963. An attempt at a preliminary delineation of the geographic regions of planktonic foraminifera of the Yellow Sea, the East China Sea and adjacent waters: *Ibid.*, 5, 3: 207–214.
——. 1964. The planktonic Foraminifera of the northern South China Sea: *Ibid.*, 6, 1: 38–78.
Tufescu, M. 1967. Schiţă de dinamică sezonieră a genului *Ammonia* Brünnich în zona de spargere a valurilor litoralului românesc: Univ. Bucureşti, Ser. Stiinţe Nat., Biol., Anal., 16: 145–150.
——. 1968. *Ammonia tepida* (Cushman) (ord. Foraminifera), some features of its variability in the Black Sea basin: Rev. Roum. Biol. Sér. Zool., 13, 3: 169–177.
——. 1973. Les associations des Foraminifères du Nord-Ouest de la mer Noire: Rev. Españ. Micropaleontol., 5, 1: 15–32.
——. 1974. Populaţiile de foraminifer din apele litorale româneşti: Acad. Rep. Soc. România, Bucureşti, 176 pp.
Ubaldo, M. L. 1967. Contribuição para o estudo dos foraminíferos recentes do Arquipélago dos Açores: Rev. Fac. Cienc. Lisboa, (C), 15, 1: 35–63.
——. 1973. Étude des Foraminifères de sondages du Canal de Mozambique (Océan Indien): Mem. Junta Invest. do Ultramar, 2d ser., 62, 278 pp.

Uchio, T. 1959. Ecology of shallow-water Foraminifera off the coast of Noborobetsu, Southwestern Hokkaido, Japan: Seto Marin. Biol. Lab., Publ., 7, 3: 295–302.
——. 1960a. Ecology of living benthonic Foraminifera from the San Diego area: Cushman Found. Foram. Res., Spec. Pap. 5, 72 pp.
——. 1960b. Planktonic Foraminifera of the Antarctic Ocean: Japan. Antarct. Res. Exp., Biol. Res., 11: 3–9.
——. 1960c. Benthonic Foraminifera of the Antarctic Ocean: Japan. Antarct. Res. Exp., Biol. Res., 12: 1–20.
——. 1964. Influence of the river Shinano on foraminifera and sediment grain-size distributions; *in* Papers in Marine Geology: Univ. of South. Calif. Press: 411–428.
——. 1967. Foraminiferal assemblages in the vicinity of the Seto Marine Biological Laboratory, Shirahama-Cho, Wakayama-Ken, Japan: Publ. Seto. Mar. Biol. Lab., 5, 5: 399–417.
Ujiié, H. 1956. The internal structures of some Elphidiidae: Tokyo Kyoiku Daigaku, Sci. Repts., sect. C., 38: 267–282.
——. 1962. Introduction to statistical foraminiferal zonation: Jour. Geol. Soc. Japan, 68, 803: 431–450.
——. 1965. Shell structure of Japanese smaller foraminifera. Part 1. *Ammonia tochigiensis* (Uchio, 1951): Paleontol. Soc. Japan, Trans. Proc., 5, 60: 156–165.
——. 1968. Distribution of living planktonic foraminifera in the Southeast Indian Ocean: Nat. Sci. Mus. Tokyo, Bull., 11, 1: 97–125.
——. 1973. Sedimentation of planktonic foraminiferal shells in the Tsushima and Korea Straits between Japan and Korea: Micropaleontol., 19, 4: 444–460.
Ujiié, H. & Kusukawa, T. 1969. Analysis of foraminiferal assemblages from Miyako and Yamada Bays, northeastern Japan: Nat. Sci. Mus. Tokyo, Bull., 12, 3: 735–772.
Ujiié, H. & Nagase, K. 1971. Cluster analysis of living planktonic Foraminifera from the southeastern Indian Ocean: Inter. Conf. Plank. Microfossils, 2d., Proc., Rome, 2: 1251–1258.
Umbgrove, J. H. F. 1928. Het genus *Pellatispira* in het indo-pacifische gebied: Nederland-Indie, Dienst. Mijnb., 10: 43–71.
UNESCO. 1968. Zooplankton sampling: UNESCO, Geneva, 174 pp.
Urey, H. C. 1947. The thermodynamic properties of isotopic substances: Jour. Chem. Soc., 1: 562–581.
Vašíček, M. 1953. Change in the ratio of sinistral and dextral individuals of the Foraminifer *Globororalia scitula* (Brady) and their use in stratigraphy: Sborn. Ústř. Úst. Geol., Paleontol., 20: 1–76.
Vaughan, T. W. 1923. On the relative value of species of smaller Foraminifera for the recognition of stratigraphic zones: Amer. Assoc. Petrol. Geol., Bull., 7: 517–531.
Vedantam, D. & Rao, M. S. 1970. Recent foraminifera from off Pentakaṭa, east coast of India: Micropaleontol., 16, 3: 325–344.
Vella, P. 1962. Determining depths of New Zealand Tertiary seas, and introduction to depth paleoecology: Tuatara, 10, 1: 19–40.
——. 1964. Foraminifera and other fossils from Late Tertiary deep-water coral thickets, Wairarapa, New Zealand: Jour. Paleontol., 38: 916–928.
——. 1974. Coiling ratios of *Neogloboquadrina pachyderma* (Ehrenberg): variations in different size fractions: Geol. Soc. Amer., Bull., 85, 9: 1421–1424.
Verworn, M. 1888a. Biological studies of Protista: Ann. Mag. Nat. Hist., 6, 2: 155–169.
——. 1888b. Biologische Protisten-Studien: Z. Wiss. Zool., 46: 455–470.
——. 1889. Psycho-physiologische Protistenstudien: Jena, 218 pp.
Very, J. M. 1970. Présence de diatomées dans la membrane organique des tests de foraminifères porcelanés actuels: Soc. Phy. Hist. Nat. Genève, C.R. Séances, n.s., 5, 1: 25–36.

Vialli, V. 1963. Il problema della specie e la paleontologia (con cenni sulla variabilita intraspecifica): Mus. Geol. Bologna, Ann., (2), 31: 1–88.
Vilks, G. 1964. Foraminiferal study of East Bay, Mackenzie King Island, District of Franklin (Polar Continental Shelf Project): Canada Geol. Surv. Pap. 64-53: 1–25.
——. 1967. Quantitative analysis of Foraminifera in Bras d'Or Lakes: Bedford Inst. Oceanogr. Rep. 67-1: 1–83.
——. 1968. Foraminiferal study of the Magdalen Shallows, Gulf of St. Lawrence: Maritime Sediments, 4, 1: 14–21.
——. 1969. Recent Foraminifera in the Canadian Arctic: Micropaleontol., 15, 1: 35–60.
——. 1970a. New method for separation of skeletal material from a plankton sample: Maritime Sediments, 6, 2: 72–73.
——. 1970b. Circulation of surface waters in parts of the Canadian Arctic Archipelago based on foraminiferal evidence: Arctic, 23, 2: 100–111.
——. 1973 Ms. A study of *Globorotalia pachyderma* (Ehrenberg) – *Globigerina pachyderma* (Ehrenberg) in the Canadian Arctic: Ph.D. thesis, Dalhousie University.
——. 1974. The distribution of planktonic foraminifera in the sediments and water of the Northwest Passage and northern Baffin Bay: a tool for paleooceanographic analysis: Geol. Surv. Canada, Pap. 74-30, 1: 109–121.
Vilks, G. & Walker, D. A. 1974. Morphology of *Orbulina universa* d'Orbigny in relation to other spinose planktonic foraminifera: Jour. Foram. Res., 4, 1: 1–8.
Vilks, G., Anthony, E. H. & Williams, W. T. 1970. Application of association-analysis to distribution studies of Recent foraminifera: Canadian Jour. Earth Sci., 7, 6: 1462–1469.
Vincent, E. 1972. Climatic change at the Pleistocene-Holocene boundary in the southwestern Indian Ocean: Palaeoecol. Africa, 6: 45–54.
Vinogradov, A. P. 1953. The elementary chemical composition of marine organisms: Sears Found. Marine Res., Mem. 2, Yale Univ., 647 pp.
Vinogradov, L. G. 1948. O zoogeografisheskom rajonirovanii dal'nevostochnykh morej: Izv. TINRO, 28: 162–164.
Vinogradov, M. E. 1961. Istochniki pitanija glubokovodnoj fauny (o skorosti razlozhenija otmershikh Pteropoda): Akad. Nauk SSSR, Dokl., 138, 6: 1439–1442.
Vinot-Bertouille, A. C. & Duplessy, J. C. 1973. Individual isotopic fractionation of carbon and oxygen in benthic foraminifera: Earth Planet. Sci. Lett., 18, 2: 247–252.
Volozhanina, P. P. 1960. Vzaimootnoshenija fuzilinid i fathij v razreze srednego karbona juzhnogo Timana: Vopr. Mikropal., 4: 71–82.
Voorthuysen, J. H. 1960. Die Foraminiferen des Dollart-Ems-Estuarium: Verh. Kon. Ned. Geol. Minjb. k. Gen., Geol. (19): 237–269.
Vyalov, O. S. & Kantolinskaja I. I. 1968. Sledy sverlenij khishchnykh gastropod v rakovinakh miothenovykh foraminifer: Paleontol. Sbornik, 2, 5: 88–94.
Waldron, R. P. 1963. A seasonal ecological study of Foraminifera from Timbalier Bay, Louisiana: Office Naval Res., Geogr. Br., Tech. Rept. 16, pt. B, La. State Univ., Univ. Coastal Stud. Inst. Contr. 63-3: 132–188.
Walker, G. 1784. Testacea Minuta Rariora Nuperrime Detecta in Arena Litoris Sandvicensis a G. Boys: London, 25 pp.
Walker, G. & Jacob, E. 1798. *in* Essays On The Microscope: F. Kanmacher, 2d ed., London: 629–645.
Walker, D. A. 1971 Etching of the test surface of benthonic foraminifers due to ingestion by the gastropod *Littorina littorea* Linné: Canadian Jour. Earth Sci., 8, 11: 1487–1491.
Walker, D. A., Linton, A. E. & Schafer, C. T. 1974. Sudan Black B: a superior stain to Rose Bengal for distinguishing living from non-living foraminifera: Jour. Foram. Res., 4, 4: 205–215.
Walker, D. A. & Vilks, G. 1973. Spinal ultrastructure of the planktonic foraminifers *Hastigerina* Thompson and *Globigerinella* Cushman: *Ibid.*, 3, 4: 196–198.

Waller, H. O. 1960. Foraminiferal biofacies of the south China coast: Jour. Paleontol., 34: 1164–1182.
Waller, H. O. & Polski, W. 1959. Planktonic Foraminifera of the Asiatic shelf: Cushman Found. Foram. Res., Contr., 10, 4: 123–126.
Wallis, T. E. 1955. Drawing from the microscope: Roy. Microsc. Soc., 75, 2.
Walther, J. 1893–1894. Einleitung in die Geologie als historische Wissenschaft. I. Bionomie des Meeres; II. Lebenweise der Meerestiere, Fischer, Jena, 1055 pp.
Walton, W. R. 1952. Techniques for recognition of living Foraminifera: Cushman Found. Foram. Res., Contr., 3, 2: 56–60.
——. 1955. Ecology of living benthonic Foraminifera, Todos Santos Bay, Baja, California: Jour. Paleontol., 29: 952–1018.
——. 1960. Diagnostic faunal characteristics on and near a barrier island; Horn Island, Mississippi: Gulf Coast Assoc. Geol. Socs., Trans., 10: 7–24.
Wantland, K. F. 1969a. Foraminiferal assemblages of the coastal lagoons of British Honduras; *in*, A. Ayala-Castañares & F. B. Phleger, eds., Coastal Lagoons: Univ. Nac. Auto. México: 621–643.
Warren, A. D. 1957. Foraminifera of the Buras-Scofield Bayou region, southeast Louisiana: Cushman Found. Foram. Res., Contr., 8, 1: 29–40.
Warthin, A. S. 1934. Foraminifera from the Ross Sea: Amer. Mus. Novit, 721: 1–4.
Watkins, J. G. 1961. Foraminiferal ecology around the Orange County, California ocean sewer outfall: Micropaleontol., 7, 2: 199–206.
Wedekind, R. 1913. Rhizopoda; *in* Handwörterbuch der Naturwissenschaften.
Weiss, D. 1974. Late Pleistocene stratigraphy and paleoecology of the lower Hudson River estuary: Geol. Soc. Amer., Bull., 85: 1561–1570.
Wetzel, O. 1957. Fossil 'microforaminifera' in various sediments and their reaction to acid treatment: Micropaleontol., 3, 1: 61–64.
Wicher, E. A. 1942. Praktikum der Angewandten Mikropaläontologie: 143 pp., Bornträger, Berlin.
Wiebe, P. H. & Holland, W. R. 1968. Plankton patchiness: effects on repeated net tows: Limnol. Oceanogr., 13, 2: 315–321.
Wiesner, H. 1931. Die Foraminiferen der Deutschen Südpolar-Expedition, 1901–1903: Deutsche Südpol. Exp., 20 (Zool. 12): 53–165.
Wilcoxon, J. A. 1964. Distribution of Foraminifera off the southern Atlantic Coast of the United States: Cushman Found. Foram. Res., Contr., 15, 1: 1–24.
Wiles, W. W. 1967. Pleistocene changes in the pore concentration of a planktonic foraminiferal species from the Pacific Ocean; *in* M. Sears, ed.; Progress in Oceanography, 4: 153–160.
Williams, D. F. 1973. Late Pleistocene water mass fluctuations in the southern Indian Ocean: Antarct. Jour., 8, 5: 284–285.
Williamson, W. C. 1848. On the structure of the shell and soft animal of *Polystomella crispa*, with some remarks on the zoological position of the Foraminifera: Microsc. Soc. London, Trans., 2: 159–178.
——. 1851. On the minute structure of the calcareous shells of some recent species of Foraminifera: *Ibid.*, 3: 105–128.
Wilson, D. P. 1951. A biological difference between natural sea waters: Jour. Biol. Assoc., 30: 1–26.
Wilson, D. P. & Armstrong, F. A. J. 1952. Further experiments on biological differences between natural sea waters: Jour. Biol. Assoc., 31, 2: 335–349.
——. 1954. Biological differences between sea waters; experiments in 1953: *Ibid.*, 33: 347–360.
——. 1958. *Idem*; experiments in 1954 and 1955: *Ibid.*, 37: 331–348.
——. 1961. *Idem*; experiments in 1960: *Ibid.*, 41: 663–681.
Wilson, L. R. & Hoffmeister, W. S. 1952. Small Foraminifera: The Micropaleontologist, 6, 2: 26–28.

Wingstrand, K. C. & Melander, Y. 1953. Gomori's hematoxylin as a chromosome stain: Stain Techn., 28: 217.

Winter, F. W. 1907. Zur Kenntnis der Thalamophoren. I. Untersuchungen über *Peneroplis pertusus* (Forskål): Arch. Protist., 10: 1–113.

Wohlfarth-Bottermann, K. E. 1957. Die Kontrastierung tierischer Zeller und Gewebe im Rahmen ihren elektronenmikroskopischen Untersuchung an ultradünnen Schnitten; Naturwiss., 44: 287–288.

——. 1961. Cytologische Studien VIII. Zum Mechanismus der Cytoplaströmung in dünnen Fäden: Protoplasma, 54: 1–25.

——. 1964. Cell structures and their significance for ameboid movement: Inter. Rev. Cytology, 16: 61–131.

Wollin, G., Ericson, D. B. & Ewing, M. 1971. Late Pleistocene climates recorded in Atlantic and Pacific deep-sea sediments; *in* K. K. Turekian, ed., The Late Cenozoic Glacial Ages: Yale Univ. Press, New Haven: 199–214.

Wood, A. 1949. The structure or wall of the test in the Foraminifera, its value in classification: Geol. Soc. London, Quart. Jour., 104: 229–255.

——. 1963. Wall structure of foraminifera in polarized light: Micropaleontol., 9, 4: 432.

Wood, A. & Haynes, J. 1957. Certain smaller British Paleocene Foraminifera, Part 2. *Cibicides* and its allies: Cushman Found. Foram. Res., Contr., 8, 2: 45–53.

Wright, C. A. & Murray, J. W. 1972. Comparisons of modern and Paleogene foraminiferid distributions and their environmental implications: Extrait du Mém. du B.R.G.M., 79: 87–96.

Wright, R. C. 1968. Miliolidae (Foraminíferos) recientes del estuario del Río Quequén Grande (Prov. de Bs. As.): Rev. Mus. Argentino Cienc. Nat., Hidrobiol., 2, 7: 225–256.

Wright, R. C. & Hay, W. W. 1971. The abundance and distribution of foraminifers in a back-reef environment, Molasses Reef, Florida: Miami Geol. Soc., Mem. 1: 121–174.

Yule, G. U. 1944. The Statistical Study of Literary Vocabulary: Cambridge Univ. Press, London, 306 pp.

Zalesny, E. R. 1959. Foraminiferal ecology of Santa Monica Bay, California: Micropaleontol., 5, 1: 101–126.

Zenkevich, L. A. 1951. Fauna i Biologischeskaja Produktivnost' Morja, Mirovoj Okean: 1, Sov. Nauka, Moscow.

——. 1967. Tridthat' devjatyj rejs: Priroda, 2: 12–26.

Zhadin, V. I. 1960. Metody Gidrobiologicheskogo Issledovanija: Vysshaja Shkola, Moscow.

Zittel, K. A. 1876. Handbuch der Paläontologie, v. 1: Oldenbourg, Munich and Leipzig, 765 pp.

Zobel, B. 1968. Phänotypische Varianten von *Globigerina dutertrei* d'Orbigny (Foram.); ihre Bedeutung für die Stratigraphie im quartären Tiefsee-Sedimenten: Geol. Jahrb., 85: 97–121.

——. 1971. Foraminifera from plankton tows, Arabian Sea; areal distribution as influenced by ocean water masses: Inter. Conf. Plank. Microfossils, 2d, Proc., Rome, 1: 1323–1333.

——. 1973. Biostratigraphische Untersuchungen an Sedimenten des indischpakistanichen Kontinentalrande (Arabisches Meer): 'Meteor' Forsch. Ergebnis., Reihe C., Geol. Geophys., 12: 9–73.

Zobell, C. E. 1946. Redox potentials of marine sediments: Amer. Assoc. Petrol. Geol., Bull., 30, 4: 477–513.

Zoppi, E. 1961. Distribución vertical del zooplancton en el golfo y extremo este de la fosa de Cariaco: Inst. Oceanogr., Cumaná, Bol., 1, 1: 219–247.

SPECIES INDEX

AUTHOR INDEX

Every author cited in the text appears in the index with the page number on which he/she is cited. Should the examination of a listed page fail to reveal the desired name, it is because the person is coauthor of a publication with more than 3 authors, a publication which is cited by the first author's name and *et al.* For example, after the name Vilks, G. in the index, page 49 is listed, but the name Vilks does not appear on page 49 in the text because he is coauthor of a multiple author paper cited as Buckley *et al.*, 1974.

SUBJECT INDEX

MIX
Papier aus verantwortungsvollen Quellen
Paper from responsible sources
FSC® C105338

FSC
www.fsc.org

If you have any concerns about our products,
you can contact us on
ProductSafety@springernature.com

In case Publisher is established outside the EU,
the EU authorized representative is:
Springer Nature Customer Service Center GmbH
Europaplatz 3, 69115 Heidelberg, Germany

Printed by Libri Plureos GmbH
in Hamburg, Germany